Notes on symbols and nomenclature

Symbols

The symbols used in this book agree generally with those recommended by the International Organization for Standardization. The only important changes made are the use of τ for volume (instead of V, used for potential) and **N** for the Poynting vector (instead of **S**, used for area). Symbols for the principal quantities and their units and for certain physical constants are given in the tables inside the back cover. Note the following possible sources of confusion:

ρ is used both for resistivity and volume charge density.
σ is used both for conductivity and surface charge density.
p is used both for linear momentum and electric dipole moment.
l is used for lengths, L for displacement in a general direction.
d**l** is used for vector lengths of elements of wire, etc., d**L** for vector elements of path.
A is used for areas of plates, cross-sections, etc., **S** for general surface areas.
N is used for a total number of entities, n for a number per unit volume, length, etc., and n also for the nth object in a set.
U_E, U_M and U_{EM} are electric, magnetic and electromagnetic energies.
u_E, u_M and u_{EM} are the corresponding energies per unit volume.
Vector quantities are indicated by bold type thus: **A**; *complex numbers* by bold sans-serif type thus: **A**.

Diagrammatic convention

The following convention is used to indicate the direction of quantities perpendicular to the page (representing the tip or tail of an arrow viewed end-on):

$\odot$ current, field, etc., *out of* diagram

$\otimes$ current, field, etc., *into* diagram

Nomenclature

The recommended names for the various electric and magnetic field quantities are those given in the table of SI units. Because these names are confusing and incongruous, it is strongly recommended that both **B** and **H** be referred to as magnetic fields, being distinguished merely by the terms **B**-field and **H**-field as appropriate. The same should apply to the electric **E**-field and **D**-field.

Prefixes for multiples or sub-multiples

E	exa	10^{18}	k	kilo	10^{3}	n	nano	10^{-9}
P	peta	10^{15}	c	centi	10^{-2}	p	pico	10^{-12}
T	tera	10^{12}	m	milli	10^{-3}	f	femto	10^{-15}
G	giga	10^{9}	μ	micro	10^{-6}	a	atto	10^{-18}
M	mega	10^{6}						

Electricity and Magnetism

Fourth Edition

Electricity and Magnetism

Fourth Edition

W.J. DUFFIN
Fellow of the University of Hull

McGRAW-HILL BOOK COMPANY

London · New York · St Louis · San Francisco · Auckland
Bogotá · Guatemala · Hamburg · Lisbon · Madrid · Mexico
Montreal · New Delhi · Panama · Paris · San Juan · São Paulo
Singapore · Sydney · Tokyo · Toronto

Published by
McGRAW-HILL Book Company (UK) Limited
SHOPPENHANGERS ROAD · MAIDENHEAD · BERKSHIRE · ENGLAND
TEL: 0628-23432; FAX: 0628-35895

British Library Cataloguing in Publication Data

Duffin, W.J. (William John)
Electricity and magnetism – 4th ed.
1. Electricity 2. Magnetism
I. Title
537

ISBN 0-07-707209-X

Library of Congress Cataloging-in-Publication Data

Duffin, W.J.
Electricity and magnetism / W.J. Duffin. – 4th ed.
p. cm.
ISBN 0-07-707209-X
1. Electricity. 2. Magnetism. I. Title.
QC522.D83 1990
537–dc20

12345 RC 93210

Typeset by Vision Typesetting, Manchester
and printed and bound in Great Britain by Richard Clay Ltd, Bungay, Suffolk

Contents

List of Commentaries

Preface

Although a substantial proportion of the text has been rewritten for this new edition, the basic structure adopted in previous editions has been retained, with an even greater emphasis on assistance for students with limited backgrounds in physics and mathematics. Electromagnetism is essentially a three-dimensional subject, and this causes difficulties in the early stages, both in the physical concepts themselves and in their mathematical expression. It is therefore vital that those embarking on a course of study should not be swamped by too great a pace in the initial stages, and for that reason the early chapters introduce the new concepts gradually and with adequate explanations. Students starting with a more detailed knowledge of the subject can profit from the revision summaries at the ends of the chapters to check that they have fully understood the material in the early part of the book. In any case, as the chapters pass, fewer concessions are made and the pace quickens in line with a student's increasing mastery, so that the final standard is one which should be sufficient for the first two years of an honours degree in physics, for the final year of combined honours degrees and for the early stages of electronic and electrical engineering courses.

Students have traditionally found electromagnetism a difficult and demanding subject, so that an important factor during their study of it is their motivation. It is not good enough simply to present the material as a body of knowledge that must be assimilated because teachers and lecturers have decided that it is important, with the implication that the student can like it or lump it. The students themselves must be convinced that the study is worth while, and this is not an easy task to undertake, given the variability in outlook among them. For that reason, I have always adopted a whole set of different methods to appeal to a wide variety of readers:

1. The text particularly emphasizes the experimental bases of the subject without neglecting its theoretical importance.
2. The development of new concepts is always accompanied by reasons for introducing them, instead of baldly defining them as if it were obvious why we do so.
3. General laws are carefully formulated and limitations in their field of application discussed, so that the subject appears less 'cut and dried' than might appear at first sight.

4. Specific problems are worked out in the text to illustrate how the laws are applied and bring out their significance in tackling practical situations.
5. There is continual reference to the general direction in which the study of the subject is proceeding. The 'Introduction for the Student' at the beginning of the book outlines the structure of the whole book and the reasons for studying the subject. Each chapter starts and finishes with a discussion of how the material fits into the general structure.
6. For the more practically minded, there is continual reference throughout the text to applications of the material in present-day science and technology.
7. For those who are inclined to question accepted versions of any topic, there is no shirking of difficult points, some of which are discussed in the text, some in the Commentaries.
8. Self-study is assisted by providing problems at the ends of chapters divided into groups, each related to the various numbered sections, and the answers at the end are in many cases given at some length.
9. Examinations are likely to loom at some stage in any course, and it is hoped that the revision summaries will help to give a quick overview of the important results from each chapter.
10. Every opportunity is taken to emphasize that the subject is alive and highly relevant. The Commentaries in particular are intended to show that this is so and to indicate the sort of topics that are still developing, with references to encourage students in the very desirable activity of consulting original papers.

In the present edition, I have benefited from both favourable and adverse criticism of the previous edition, for which I am extremely grateful. I have rewritten those parts of the text where, on reflection, the material was less than clearly expressed and more help has been provided where difficult new mathematical topics are introduced. The major changes from the third edition are

1. The introduction of many more worked examples and practical applications throughout.
2. An updating of material dealing with recent developments.
3. The insertion of informal treatments dealing with a special case before many of the formal and general proofs. It is often more important that students appreciate what is happening than that they should learn a formal proof.
4. The labelling of many equations so that it is clear at a glance what situation they refer to. Many of those expressing general laws are also given in words immediately before or after the equation.

A relatively trivial change, but one that needs justification, is the choice of $\mathrm{d}L$ as a symbol for a line element of a path in space instead of $\mathrm{d}s$ as in previous editions. As Burge (1987) points out, the lower case s is difficult to distinguish from S (for area) in students' handwritten work and I felt that this was an important consideration. Although $\mathrm{d}r$ is often used in mathematical treatments for the same quantity, it seems to me confusing when r is also used as a polar coordinate, so I have adopted Burge's recommendation.

Introduction for the student

Anybody starting on the long journey through the subject of electromagnetism has a right to know where it is leading and why it is being studied. Indeed, without some sense of direction and purpose it is easy to lose the way. This introduction attempts to locate the destination, to give some idea of the reasons for choosing a particular path to reach it, and to indicate why it is worth following.

The goal: Maxwell's equations

The whole of electromagnetism at the level of this book can be summarized by a set of equations, finally completed in Chapter 13, known as Maxwell's equations. They form the core of the subject, not only being a powerful summary of all that has led up to them, but also providing a base for deducing consequences as far-reaching as the existence and properties of electromagnetic waves (Chapter 14).

The following is a summary of the equations as they apply in free space (vacuum) together with a non-rigorous description of their meaning. The picturesque device of using lines of force is adopted here because the reader is likely to be familiar with it, but the lines are intended only as an aid to understanding and not as a means of solving problems.

1: $\mathrm{div}\,\mathbf{E} = \dfrac{\rho}{\varepsilon_0}$

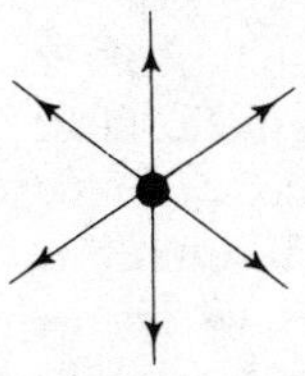

$\mathbf{E}$ is the symbol for electric field strength, printed in heavy type because it is a vector (i.e. it has direction as well as magnitude). The prefix 'div' is an operator which tells us how much divergence the field has at a point, i.e. how many lines of $\mathbf{E}$ start at the point (as in the small sketch). ρ is the density of electric charge and the equation tells us that the divergence of $\mathbf{E}$ is proportional to the charge density at the point and therefore that electric field lines begin and end on charges. This equation stems from experiments on the law of force between charges.

2: $\mathrm{div}\,\mathbf{B} = 0$

$\mathbf{B}$ is the symbol for magnetic field and the equation tells us that the lines of $\mathbf{B}$ neither start nor end at any point. They are therefore either continuous and close on themselves or continuous and go to infinity.

3: $\text{curl } \mathbf{E} = -\dfrac{\partial \mathbf{B}}{\partial t}$

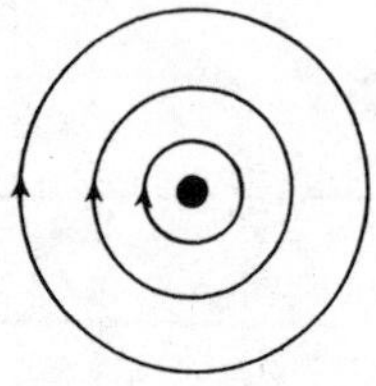

This relation stems from the experimental observation that single magnetic poles do not exist. 'Curl' is an operator telling us how much **E**-field is generated at a point with lines which can close on themselves as in the sketch. The equation tells us that such an **E**-field arises from time-varying magnetic fields. The relation stems from the experimental laws of electromagnetic induction.

4: $\text{curl } \mathbf{B} = \mu_0 \mathbf{J} + \dfrac{1}{c^2}\dfrac{\partial \mathbf{E}}{\partial t}$

J is the electric current density at a point, and this equation tells us that closed lines of **B** can arise both from currents (first term on the right) and from **E**-fields changing in time. The first term comes from the experimental laws of force between electric currents and the second from the existence of electromagnetic waves.

When various media are present instead of the vacuum, these four equations take the form on page 341 in which two new vectors **D** and **H** are introduced for convenience, but the relations remain substantially as above.

The path to Maxwell's equations

Naturally, the reader will not appreciate all that has been said above until well into the study of the subject, but the following points should be clear. Firstly, that we cannot fully understand Maxwell's equations until we can appreciate the meaning of the **fields** **E** and **B** and the **sources** of the fields, the **charges** and the **currents**. Secondly, that in spite of their abstract nature, all the equations are based on experimental observations. Thirdly, that the mathematical symbolism is not elementary and its connection with the physical laws summarized by it is not obvious. For all these reasons, it is not desirable to start the study of electromagnetism by *postulating* Maxwell's equations and deducing everything from them rather as mechanics is developed from Newton's laws of motion. That approach is more suitable for an advanced treatment when the concepts are already clear.

We therefore adopt an approach which builds a gradual understanding of the concepts of charge and current (Chapter 1), of electric and magnetic fields (Chapters 3 and 7) and of the fundamental laws on which the subject is based (Chapters 1, 2, 7, 9, and 13). Instead of starting with the most general situations, we begin in fact with the simplest, dealing in turn with charges which are stationary, then with charges moving with constant velocity (forming steady currents) and finally with accelerating charges or varying currents. In a similar way, we first look at the laws *in vacuo* and only introduce the effects of various media at a later stage (Chapters 11 and 12).

The path to Maxwell's equations in this book is not the only right one and the student will undoubtedly encounter alternatives. The objective should be to obtain sufficient mastery of the subject to choose that path which best suits one's own style of thinking. Occasional references are made to other treatments in case the course followed by the reader does not coincide with the order in this book.

A second goal: network analysis

The reader will be aware of a more practical side to electromagnetism: one which concerns networks of various electrical and electronic *components* connected together by wires or cables or placed in intimate contact. Much of the behaviour of such networks can be analysed in terms of idealized components, exhibiting the properties of resistance, capacitance and inductance, together with sources of energy which generate currents or voltages. Like the rest of electromagnetism, network theory is also based on Maxwell's equations, but it does not demand details about what occurs inside the components. Instead, it only concerns itself with the currents and voltages in the connecting leads entering and leaving each component. For that reason it is possible to simplify the solutions of network problems to matters of algebra. These aspects are explored in Chapters 6 and 10.

Why is electromagnetism worth studying?

For an engineer, the great technical importance of the subject in the generation and distribution of power, in the construction and development of electrical and electronic equipment and in the fields of telecommunications and computing are too obvious to need stressing. The physical scientist, too, needs the bases of the subject for intelligent use of the engineers' design in the above fields. However, there are reasons beyond the purely technical why physicists cannot afford to be unaware of electromagnetism.

In the first place, only four types of force or interaction are known to occur between pieces of matter. One is the **gravitational** force, the weakest of all, and usually only significant on an astronomical scale. Two others are **nuclear** forces, one weak, one strong, but both with an extremely short effective range, making them important sub-atomically but not otherwise. The fourth is **electromagnetic**, and this is effective over the whole range of distances we investigate: from those between sub-atomic particles to those between astronomical bodies. It is the interaction responsible for most of the forces we experience, including those holding atoms together to form large pieces of matter. Electromagnetism must clearly therefore hold a central place in the physical sciences.

Secondly, because light and other forms of radiation consist of electromagnetic waves, the whole science of optics can only be fully understood through electromagnetic theory.

Finally, electromagnetism is an excellent example of the development of a physical theory from basic experiments and concepts to a general mathematical formulation. Because it took its present form largely in the nineteenth century, it is possible to see it as a whole and as a significant part of classical physics. It is also, however, an integral part of modern physics as well: related to atomic and nuclear forces through the interaction between charged particles; related to the development and fundamental ideas of relativity; and intimately related to quantum physics through the photon.

Structure of the book

Each chapter consists of

1. A list of section headings showing how they are related, followed by a brief outline of the aim and subject-matter of the chapter.
2. Numbered sections developing the concepts, laws, and applications, in a systematic way. Equations which are mere definitions of a quantity are so labelled for emphasis.
3. A final numbered section in most chapters devoted to a summary of the important results of the chapter in a form suitable both for revision and for a quick overview of the subject to that point.
4. A set of commentaries on the material in the chapter to show that the subject is still alive and well, and to counteract the feeling, often engendered by a textbook, that elementary classical physics is closed to further argument and discussion. References for further reading are included here in the form *author* (*date*) and details of all such references are collected at the end of the book.
5. A collection of problems related to the appropriate sections, the more difficult being starred. The answers are collected together at the end of the book.

The interrelationship of the various chapters is most easily brought out by arranging them in the following pattern:

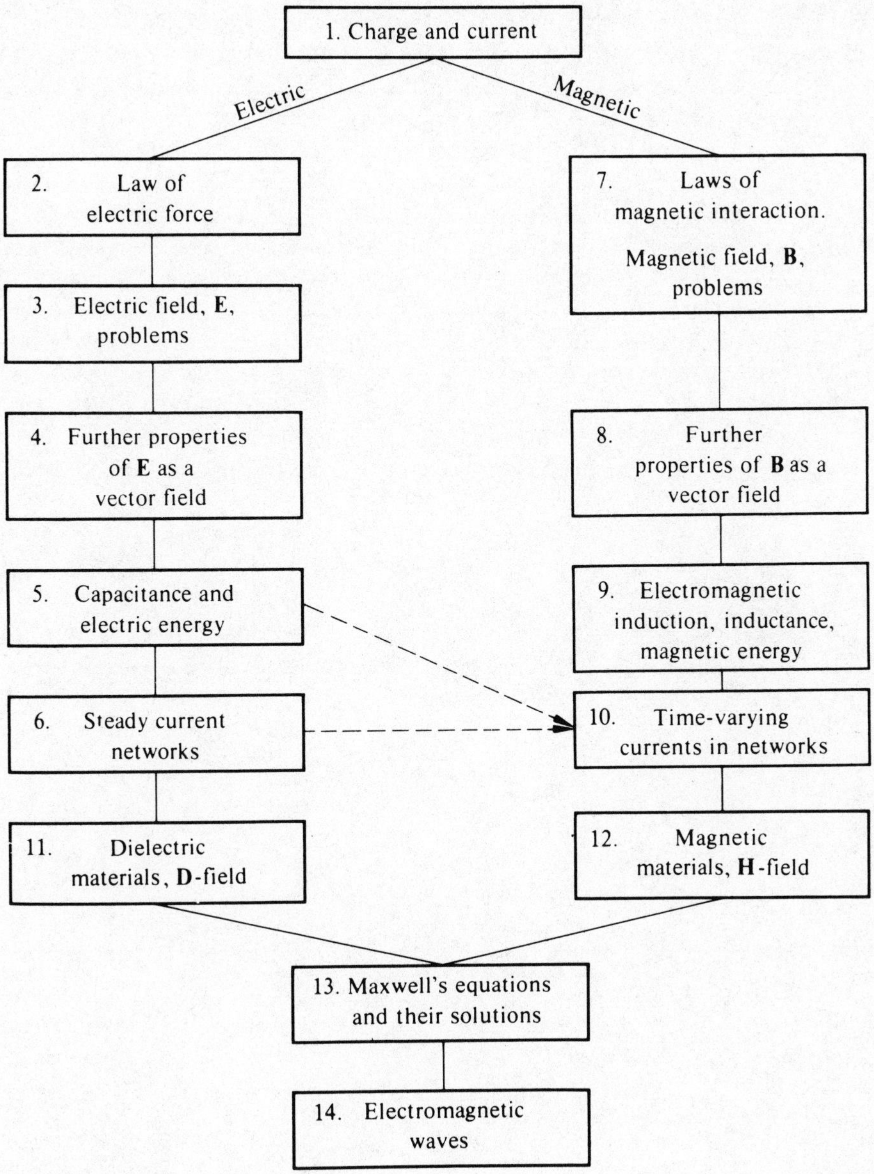

The three appendices at the end of the book deal with various mathematical points and with electromagnetic standards and units.

1

Electric charge and current

Electricity and magnetism, as branches of physics, are principally concerned with the production of, and with interactions between, **electric charges** and **electric currents**. Our first aim in Secs 1.1 and 1.2 must therefore be to establish what is meant by these terms so that we know what we are talking about, and we shall do this by recognizing charges and currents by the effects they produce. Mere *recognition* of them, however, does not allow us to go very far in discovering physical laws describing their behaviour. So the first two sections of the chapter also explain how some of the effects can be used immediately to make *measurements*. Even though such measurements are made initially in arbitrary units, we show that they can still lead to the discovery of experimental laws.

Some of the properties of charge and current suggest that there is a relationship between the two and Sec. 1.3 demonstrates what this is. We can then, in Sec. 1.4, agree to define a unit for one in terms of the other: in particular, the **ampere** is established as the unit of current, so linking all measurements to SI units.

At that point, questions about the nature of charge and current from an atomic aspect are raised and discussed in an elementary way (Sec. 1.5), and the view of current as a flow of charge is seen to be fully justified. We introduce in Secs 1.6 and 1.7 some extensions of the concepts—the **point charge**, the **current element** and various **charge densities** and **current densities**—which are used extensively in later chapters to formulate and solve problems. The chapter concludes with a brief historical note.

The first two chapters present a very thorough introduction to some basic ideas in the subject, and are intended for those whose previous acquaintance with it is minimal. Readers with a good background in elementary physics, however, might well find it sufficient to scan the revision summaries to ensure that they have understood the fundamental concepts and results covered in them.

1.1 Electric charge

Recognition of charge We recognize a body as **charged** when it attracts nearby light objects, such as small pieces of paper or cork, without touching them. A dry plastic rod, for instance, can be charged by rubbing it on a coat sleeve, or a glass rod by rubbing it with silk, and careful investigation shows that the coat sleeve and the silk are also charged. Friction between two surfaces invariably leaves both of them charged, and even mere contact and separation, as with a vehicle tyre and the road, will usually have the same effect.

Comment C1.3

The charged condition of one body can be transferred to another, which is initially uncharged, either by contact between the two or by connecting them together with a metallic wire. The idea thus arises of some entity which is transferred, distinct from the material of the body itself and giving it the property of attracting other bodies. This we call **electric charge**. Materials through which charge moves freely, like the above metallic wire, are known as **electric conductors** and are usually distinguished from non-conductors or **insulators**, although the difference is often one of degree rather than of kind.

A similar notion of 'something material' being transferred also arose in eighteenth century theories of heat, in which a fluid known as caloric was assumed to pass from hot to cold bodies in contact. However, this theory did not survive nineteenth century experiments which led to the concept of heat as a form of energy in transit. In contrast, the original idea of charge as an entity in its own right which passes from one body to another was reinforced by later work and survives to the present day.

When the forces between two charged bodies are examined we find that repulsion as well as attraction occurs and that two types of charge can be distinguished. If the type of charge found on the plastic rod above is denoted by R (for resinous, its original name) and that on the glass rod by V (for vitreous), it is found that

V repels V but attracts R
R repels R but attracts V

Any other charged body is found either to repel V and attract R, and is thus indistinguishable from V, or to repel R and attract V, and is thus indistinguishable from R. Hence all charges are either resinous or vitreous, and obey the general law that

Like charges repel each other, unlike charges attract each other.

Numerous practical applications of this simple law exist today, ranging from the use of attraction in xerographic printing to the use of repulsion in ion drives for artificial satellites (see page 63).

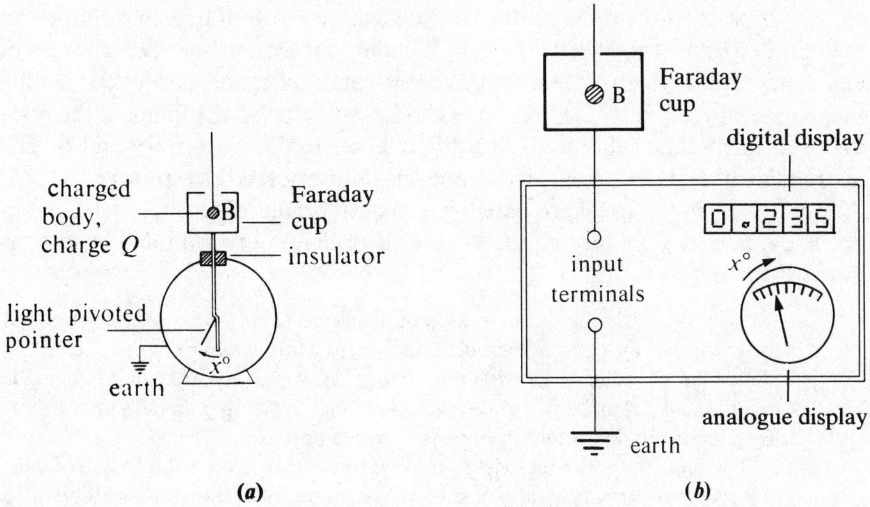

Figure 1.1 Electrometers. (a) A simple electroscope; (b) representation of a modern electrometer.

Charge measurement We now require a method of detecting charge which is more systematic and sensitive than the above test of attraction for light objects. If possible, the method should also *measure* charge since only then can quantitative laws be discovered. In principle it would be possible to use forces between charged bodies directly, but such a method is impracticable (see Problem 1.1 and solution). Instead, we use instruments known as **electrometers**, as shown in Fig. 1.1.

The simple version of Fig. 1.1a is often called an **electroscope** because it is more useful for detecting charge than for measuring it. The light pivoted conducting pointer is connected to a rigid conducting rod all mounted in, but insulated from, a case connected to the earth. The rod is surmounted by a metallic container with a small opening in the top known as a **Faraday cup**. The instrument represented in Fig. 1.1b is far more precise, reliable, and versatile than the simple electroscope, but a description of its detailed inner construction is too complex to be given here. However, when fitted with a Faraday cup it responds to charge in exactly the same way as the electroscope and the principles of its operation are irrelevant for the moment.

A charged body B, completely within the cup, produces a deflection of the pointer which does not depend on the exact position of B and which remains even when B touches the inside of the cup. If B is of conducting material, touching the inside of the cup in this way leaves a permanent charge on the electrometer and discharges B. (The deflection of the pointer in Fig. 1.1a can be explained in elementary terms as a repulsion between like charges on the pointer and rod arising from electrostatic induction, which is explained later in this section. For a fuller explanation see Sec. 5.7.)

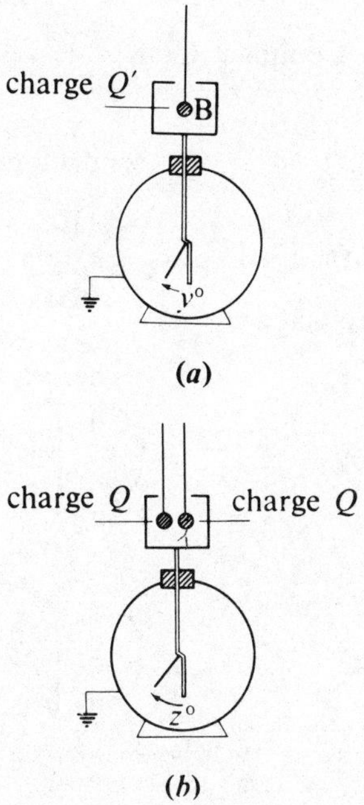

Figure 1.2 Measurement of charge. (a) Q' equals Q of Fig. 1.1 if $y = x$; (b) the deflection z corresponds to a charge $2Q$, although z is not necessarily $2x$.

Comment C1.2

Measurement of charge proceeds in the following way. Two charges are said to be equal in magnitude if they give equal deflections when placed separately in the cup of the electrometer (Fig. 1.2a). If two such equal charges are now placed in the cup at the same time, we find no deflection at all if they are unlike charges, but an increased deflection if they are like charges. We define the magnitude of a single charge to be $2Q$ if it produces the same deflection by itself that two charges of magnitude Q give together (Fig. 1.2b). Similarly, if we can find two charges of equal magnitude which together produce the same deflection as does Q, then the magnitude of each is $\frac{1}{2}Q$. This process can be extended to the limits of the scale and an electroscope calibrated accordingly in terms of an arbitrary unit Q. The instrument of Fig. 1.1a can thus be converted into a true electrometer.

In practice, modern electrometers are pre-calibrated against electrical standards (Appendix C) although they could in principle be calibrated exactly as described above.

Note particularly that the deflection or displayed value for a charge $2Q$ need not be twice that for Q. In other words, the scale is not necessarily *linear*, i.e. with a deflection or value proportional to charge. However, linearity of an analogue scale (Fig. 1.2a) makes accurate reading so much easier that modern instruments for measuring any quantity possess it where possible.

The fact that a digital display may not be linear is only too easy to forget. Although makers generally ensure the accuracy of readings to within stated limits, a digital instrument can still become non-linear outside certain ranges.

Comment C1.4

Positive and negative charges Experiments with a calibrated electrometer show that when two unlike charges are placed in the Faraday cup without contact between them, the resultant deflection corresponds to the *difference* between their magnitudes. Extending the experiment further, we can add any number of charges to the cup. When we do this, we find that the resultant magnitude can be obtained by treating one type of charge as a **positive** quantity and the other as **negative**. It is a universal convention that the vitreous charge shall be the positive one so that we can abandon the original names from now on and treat amounts of charge as algebraic quantities.

Conservation of charge An experiment superficially similar to that just described, but in fact quite distinct from it, *brings into contact* two charged bodies placed in the cage of an electrometer. Even if they are conductors, when some redistribution can be expected, there is no change in the total charge indicated by the electrometer. Repetition of this experiment with a number of charges leads us to think that

In any isolated system, the algebraic sum of the charges is constant.

This law, that of the **conservation of charge**, is confirmed by other facts such as the equal but opposite sign of the charges produced on a glass rod and on a piece of silk when the two are rubbed together: the total charge is zero before and after the rubbing. Although these experiments are relatively crude there are good reasons for thinking that the law holds exactly (see Sec. 1.5).

Electrostatic induction An important phenomenon can be demonstrated by bringing a charged body B near to an uncharged piece of metal in two detachable parts as in Fig. 1.3. If the contact between the two parts is broken while B is kept in position, they are found to possess equal and opposite charges as shown. This

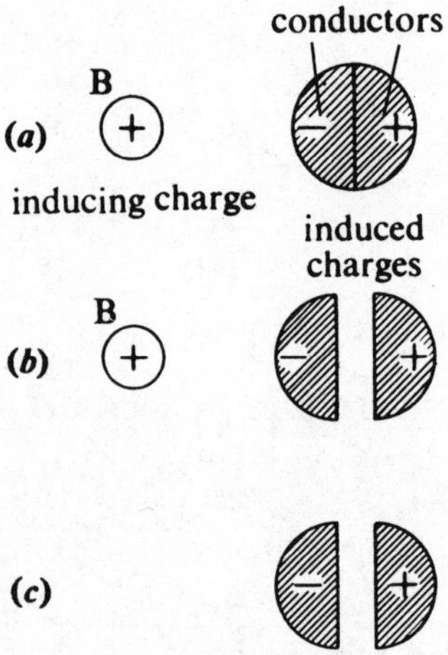

Figure 1.3 Induced charges in conductors. (a) Electrostatic induction; (b) separation of conductors while B is still present; (c) removal of B leaves two charged conductors. If the conductors in (a) were temporarily earthed with B present, the positive charge would run away to earth, leaving a negative charge even when B is removed.

separation of charge taking place under the influence of a nearby charge is called **electrostatic induction.**

All conductors behave in the same way and we can infer that they must contain charges free to move under the attractive or repulsive force of the inducing charge, though it is not possible at this stage to tell whether these free charges are positive or negative or both.

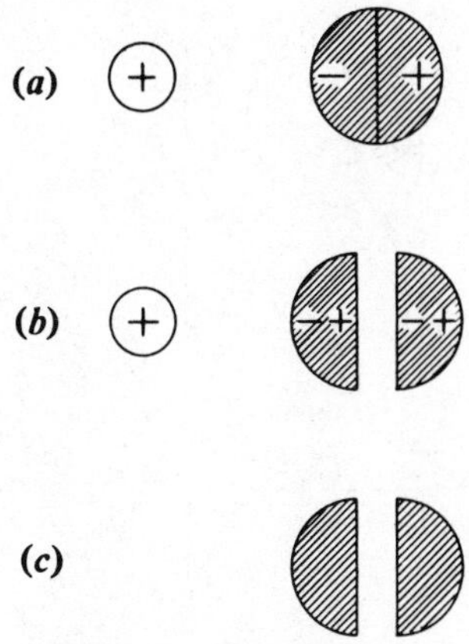

Figure 1.4 Induced charges in insulators: the same operations as in Fig. 1.3 leave two uncharged insulators.

The attraction between a charged and an uncharged conductor can be explained by induction provided it is assumed that the forces between charges diminish with increasing distance apart—an assumption to be tested in Chapter 2. Although good insulators are also attracted by a charged body, the experiment of Fig. 1.3 repeated with a split insulator does not show resultant charges on the two pieces afterwards. The inference to be drawn is that insulators contain charges which can move small distances so that attraction still occurs, but that they are bound in equal and opposite amounts so that no splitting of the body can separate two kinds of charge (Fig. 1.4). This is discussed further in Sec. 3.8.

Suppose the two-part conductor in Fig. 1.4a is replaced by a single conducting body which is connected to an earthing point while the positive inducing charge is still in place. The induced positive charge will then be conducted away to earth, leaving the negative charge behind. *If the earth connection is broken before the inducing charge is removed*, the body is left with a net negative charge: it has been **charged by induction**. This effect has a number of practical consequences and applications also discussed in Sec. 3.8.

Production of charge by batteries The disadvantages of the production of charge by friction are that the dampness of the apparatus usually allows charge to leak away (water being a conductor) and that it is difficult to produce a charge of some preassigned magnitude: in short, that steady charges of a required size are not easily available for quantitative experiments. Mechanically driven machines which will replenish charge as it leaks away can be constructed (Sec. 5.8) but these are inconvenient for our present purpose.

It is found, however, that a battery of voltaic cells connected between the upper terminal and the earthed case of an electrometer produces a charge proportional to the number of cells (the Faraday cup is found to be no longer necessary). This charge, moreover, is steady as long as the battery is connected and remains if the connection is broken, although it will now generally leak slowly away because of imperfect insulation. The terminal of the battery giving positive charge to the upper terminal of the electrometer when connected to it is known as the **positive terminal** and the other as the **negative** (Fig. 1.5).

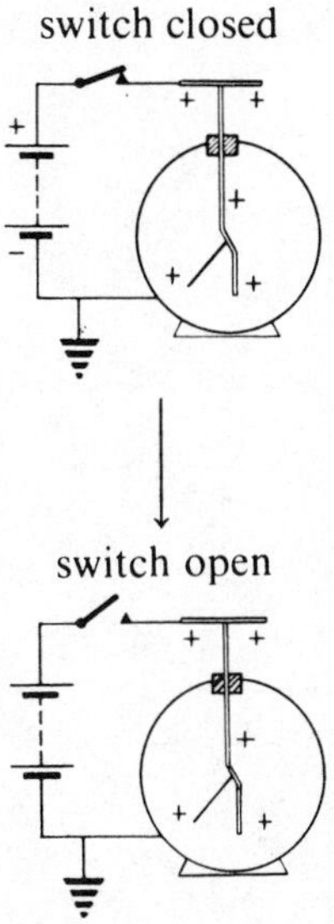

Figure 1.5 A battery of voltaic cells charges an electrometer. The upper terminal need not now be a Faraday cup.

We are not at the moment considering the mechanism whereby charges are separated either by a battery or by friction: our immediate concern has been to find a satisfactory way of producing and measuring charge, and this we have achieved. It would now be possible to establish properly the laws of force between charges, laws which would allow a universally agreed *unit charge* to be adopted (Sec. 2.4). In practice, however, the unit charge is defined in a more roundabout way to be specified in Sec. 1.4, and for this we need first to examine electric currents. Whatever unit is chosen, **we shall use the symbol Q to denote a quantity of charge measuring by the method of this section or by any equivalent method.**

1.2 Electric current

Recognition of current A piece of wire made from a conducting material rises in temperature if connected between the terminals of a battery of voltaic cells: a **heating effect** is observed. Furthermore, two such wires arranged approximately parallel to each other as in Fig. 1.6 are found to attract or repel according to the way the connections are made to the battery terminals: this is known as the **magnetic effect** and the forces are sometimes called **Lorentz forces**.

The occurrence of these two effects indicates the presence of an **electric current** in the wire. The magnetic effect shows that there is a general law applying to magnetic forces which can be expressed in the form:

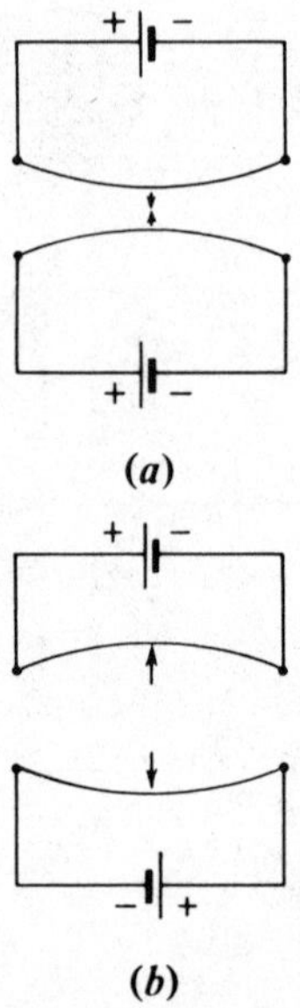

Figure 1.6 Magnetic forces: (a) like currents attract; (b) unlike currents repel.

> Like currents attract each other, unlike currents repel each other.

This is the opposite of the similar law applying to charges on page 2, and is the first of many examples illustrating the contrasting nature of electric and magnetic effects.

It is important to realize that we have no evidence at this stage of a *flow* of anything. The word 'current' is merely a name for whatever it is in the wire that produces the heating and magnetic effects.

The Lorentz force is the basis of many practical applications: almost all electric motors depend on the magnetic forces between the currents in two sets of coils, the armature and field windings, with suitable switching arrangements to ensure the continuous rotation of the driving shaft.

We also find that, if the wire between the battery terminals is broken and the ends dipped into the solution of a salt in water, chemical effects occur at the ends of the wires: substances are deposited from the solution (or evolved if gaseous) or the wires themselves dissolve. This is **electrolysis**, a **chemical** effect, and it is assumed that the reader is familiar with its terminology and with simple examples of it. The terms **electrode**, **anode** (connected to the positive terminal) and **cathode** (connected to the negative), although originating in connection with electrolysis, are now used in cases where substances other than salt solutions separate the wires.

Current measurement Any of the three effects can be used to detect a current and in each instance there is a quantity which can be used as a measure of the strength of the current. The heating effect can be used by measuring the heat generated per unit time in a given piece of wire; the chemical effect by measuring the mass of a certain substance deposited per unit time from a certain salt solution; and the magnetic effect by measuring the force on a standard current, although the setting up of this standard current will clearly cause difficulty.

Whichever of the three methods is used, it becomes possible to recognize the existence of a **steady** current—one which does not decay in time—and to establish the important quantitative result that

> The strength of a steady current in a given arrangement is the same at all points along the wire.

For example, if the wire is broken at any point whatsoever and the ends dipped into an electrolyte, the same deposition of mass per unit time occurs. This result is still obtained if several pieces of wire of different materials and cross-sections are connected in series across the terminals of a battery.

Using the magnetic effect as a measure of current Before we go on to compare the three methods we must look a little more closely at the magnetic effect. If a steady current were easy to maintain, then an arrangement as in Fig. 1.7a would be sufficient as a current-measuring device. The mutual force of attraction between coils A and B in standard positions carrying the two currents is balanced by weights and, if unit current in A is defined as that which attracts B with a force W, then a current in A exerting nW on the *same* current in B is of n units. This arrangement is one form of **current balance**.

Such a steady current for B is, however, impossible to produce over indefinite periods of time and to avoid this difficulty the arrangement is modified slightly. In Fig. 1.7a, if the current in A were first increased until the weight was $2W$ we should say that the current in A had doubled: if then the current in B were increased until the weight was $4W$, keeping the current in A constant, we should say that the current in B had doubled. Thus when both currents are doubled, the weight increases fourfold: more generally, an m-fold increase in one current and an n-fold increase in the other is accompanied by an mn-fold increase in the force.

If, then, the current to be measured is passed through the two coils in series (Fig. 1.7b) and a balancing weight W represents unit current, an n^2-fold increase in this weight means an n-fold increase in current. The current, in other words, is proportional to the square root of the force. It cannot be too strongly emphasized that this is true only because we are using the magnetic force as a measure of current—the result is not a law of nature but a consequence of our method of measurement.

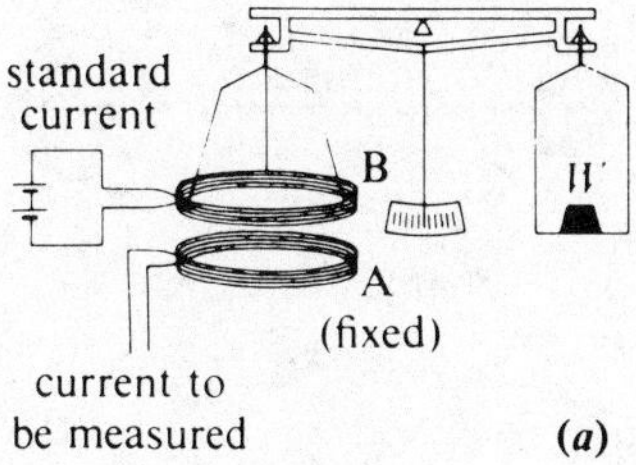

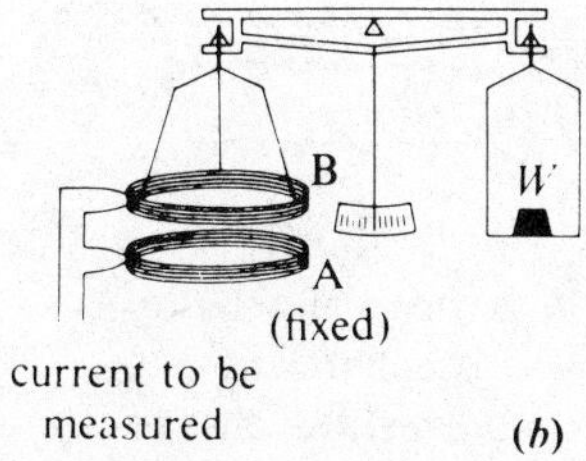

Figure 1.7 Current balances: (a) use of force between current to be measured and a standard steady current, the latter not easy to produce; (b) a practical current balance using the force between two coils, both carrying the current to be measured.

Comment C1.2

Comparison of the three possible measures of current Methods used in practice for measuring currents in modern laboratories are discussed in Chapter 6. Our concern now is to see whether the values of a given current will agree if measured using each of the three effects in turn. It is found *experimentally* that current measured electrolytically and magnetically are proportional, but that both are proportional to the **square root** of the current measured using the heating effect in metals.

It therefore follows that we choose either

1. Current measured by the magnetic or chemical effect, denoted by I: the heating effect is then proportional to I^2, or
2. Current measured by the heating effect, denoted by I_H: then the magnetic and chemical effects are proportional to $I_H^{1/2}$.

It is universally agreed that the first choice be made and that the magnetic effect in particular be used to define a measure for current. Provided this is recognized, we can restate the results of measurements on the chemical and heating effects in the form of two experimental laws named after their discoverers:

Faraday's law of electrolysis: mass deposited per unit time $= k'I$ (1.1)

Joule's law: rate of generation of heat in a metallic wire $= k''I^2$ (1.2)

where k' and k'' are constants discussed in Sec. 1.4.

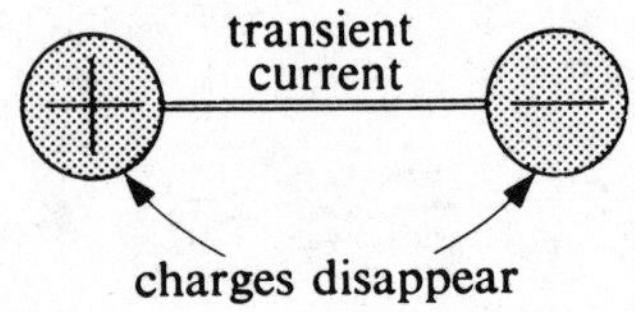

(a)

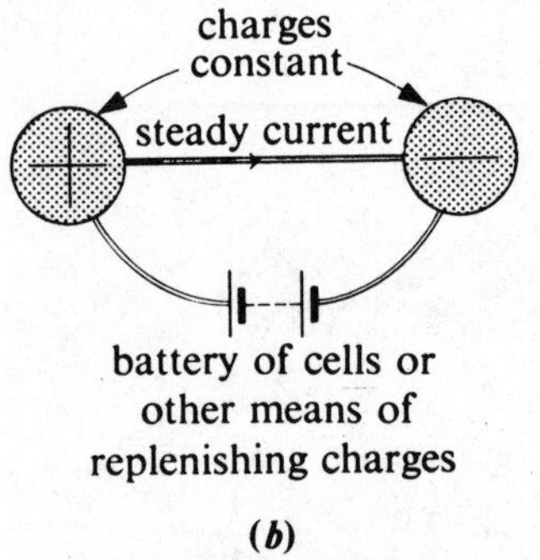

(b)

Figure 1.8 (a) Discharge of conductors is accompanied by a transient current; (b) a steady current is maintained by an external source that completes a circuit. The conductors now remain charged.

Note that, while heating and chemical effects do not always accompany a current (e.g. from cathode to screen of a cathode ray tube), there is *always* a magnetic effect.

1.3 The relationship between charge and current

Materials which act as conductors of charge are also carriers of electric current. This fact suggests an intimate connection between the two quantities, a suggestion reinforced by the following experiment. If two bodies carrying equal and opposite charges are connected by a conducting wire as in Fig. 1.8a, the charges disappear in a very short time and while they are doing so a transient current (detected by its magnetic effect) occurs in the wire. Since charge is a conserved quantity, we conclude that the charges combine by moving along the wire and that **this movement constitutes the current**.

If this conclusion is correct it should apply to steady currents as well, and these are far more convenient for investigations. We assume that a battery of cells, or any other source of such a steady current, will continually replenish the charges on the two bodies of Fig. 1.8a so as to keep them constant. This produces the **circuit** of Fig. 1.8b in which the two bodies become the **terminals** of the source.

With such a steady current available, it should be possible to establish the relationship that exists between the value of the current as measured by the magnetic effect (I) and the rate of flow of charge in the wire ($\mathrm{d}Q/\mathrm{d}t$ as discussed below). Rowland's experiment to investigate this is described below.

> Some readers at this stage may well ask: why should we bother to establish the connection experimentally when we could *define* current as a rate of flow of charge? They should realize that this would be a different primary definition of current from that in Sec. 1.2. It is a legitimate alternative, but its connection with measurement of current using the magnetic effect would still have to be established. Rowland's experiment is therefore still relevant.

The expression $\mathrm{d}Q/\mathrm{d}t$ If Q represents the charge which flows in a time t across any area of a conducting path (such as the cross-section of a wire), the **rate of flow** is given by $\mathrm{d}Q/\mathrm{d}t$. There is a sign convention incorporated in this expression in that we must first agree on a positive sense for the direction of flow. $\mathrm{d}Q/\mathrm{d}t$ then gives the **resultant** rate of flow of charge in this positive sense, even though in practice there may be contributions to the flow from negative charge moving in the opposite direction. We establish the conventional direction in a diagram by drawing an arrow as in Fig. 1.8b.

Another meaning can be attached to $\mathrm{d}Q/\mathrm{d}t$ when a situation like that of Fig. 1.8a arises. If Q is now the charge at any time t **on one of the conducting bodies**, then $\mathrm{d}Q/\mathrm{d}t$ here gives the rate at which the bodies are gaining or losing charge. The law of conservation of charge ensures that the magnitude of $\mathrm{d}Q/\mathrm{d}t$ for both bodies is the same but that their sign is different. Conservation also ensures that the rate of flow of charge along the path between the conductors is given by the same value of $\mathrm{d}Q/\mathrm{d}t$. We can summarize the relationship by

$$\left(\frac{\mathrm{d}Q}{\mathrm{d}t}\right)_{\text{current flow}} = \pm\left(\frac{\mathrm{d}Q}{\mathrm{d}t}\right)_{\text{conductors}} \tag{1.3}$$

where the negative sign is taken if the conventional current is flowing *away* from the conducting body.

Rowland's experiment We now turn to the relation between the rate of passage of charge along a path (dQ/dt) and the current in the path (I) measured by its magnetic effect. (Note that the mere association of I and dQ/dt does not imply their proportionality.)

Although the idea of current as essentially something moving rather than as a peculiar condition of matter seems convincing, we have so far provided no **direct** evidence of any motion accompanying a current. Hence, even a qualitative experiment showing that a charged body in motion produces a magnetic effect would be valuable in reinforcing this picture. Rowland's experiment, named after its originator, shows the equivalence of currents and moving charges both qualitatively and quantitatively.

The apparatus used by Eichenwald (1903) is shown in Fig. 1.9. An annular strip of tinfoil with a thin sector removed was fixed on to an insulating disc which could rotate at a known number of revolutions n per second about an axis through its centre. The ring of foil could either be charged and rotated (Fig. 1.9b) or be made to carry a current I while stationary (Fig. 1.9c). In both instances the magnetic effect was measured by a magnetometer, a current-measuring device equivalent to that illustrated in Fig. 1.7a. The charge Q was measured by an electrometer calibrated in terms of voltaic cells, a method equivalent to that in Sec. 1.1. Figure 1.9a shows the method of making the connections and of screening the disc from external effects (Sec. 4.1).

Eichenwald found, by varying both n and Q, that the magnetic effect produced by a positive charge rotated in a clockwise sense was the same as that produced by a negative charge of the same magnitude rotated counterclockwise at the same speed, showing that we cannot yet distinguish between these cases. His other results may be summarized as follows: a charge moving with speeds up to 150 m/s produces magnetic effects which in every respect (direction, magnitude, distribution in space) are equivalent to those of an electric current.

To summarize, if I is the current in any conductor measured by its magnetic effect and if dQ is the charge crossing a section of the conductor in an element of time dt and giving rise to I, then

$$I = k\,dQ/dt \tag{1.4}$$

where k is a constant depending only on the particular units chosen.

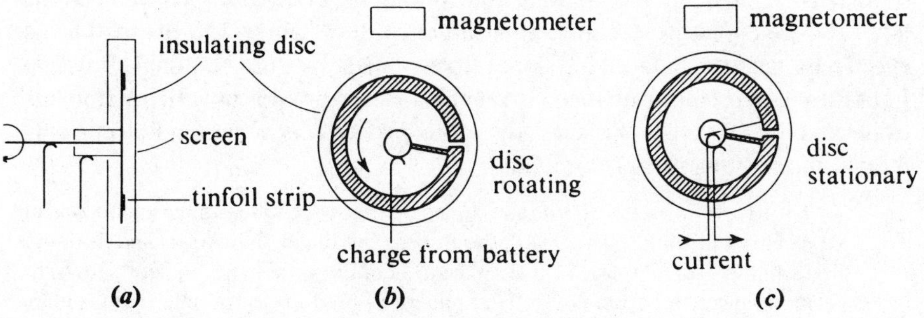

Figure 1.9 Eichenwald's version of Rowland's experiment: (a) method of making connections; (b) connections for moving charge; (c) connections for current.

1.4 Units of current and charge

As we have seen, any arbitrary units may be used to establish laws and as long as everybody agrees to use the same phenomena to measure the quantities involved, the same laws will be discovered. Thus, if all laboratories use a current balance to measure current they will all find that Faraday's and Joule's laws are obeyed: using current balances which differ in their dimensions will only affect the values of the constants k' and k'' in (1.1) and (1.2) or k in Eq. (1.4).

However, the values of these constants are of considerable interest as well as the form of the laws. Thus, k' is found to have values characteristic of the substance deposited in electrolysis and is therefore a **property of the substance**; the constant k'' in Joule's law is found to take values characteristic not only of the material through which the current passes but also of the dimensions and the temperature and is thus a **property of the system** at a given temperature; while the constant k in (1.4) does not appear to take different values and is a **universal constant**. Unless we all agree about the units to use we shall not be able to use other people's values of these properties or of the universal constants. The need thus arises for universally agreed units for every measurable quantity.

Our problem concerns charge and current. If we continue to use completely independent units for these, then (1.4) will remain with k as a constant having a value to be determined by experiment. There is, however, almost universal agreement to choose units for I and Q so that $k=1$ and

$$\boxed{I=\mathrm{d}Q/\mathrm{d}t} \qquad (1.5)$$

so that we need only agree on a unit **either** of I **or** of Q.

The ampere, A The following is the definition of the unit of current now established by international agreement and based on the magnetic effect:

> The ampere is that constant current which, when maintained in two parallel infinitely long rectilinear conductors of negligible circular section placed at a distance of 1 metre apart *in vacuo*, produces between these conductors a force of 2×10^{-7} newtons per metre length.

This definition is not intended as an exact prescription for the measurement of a current, but it fixes the value of a constant in the general law of force between currents (Sec. 7.4), just as the choice made before Eq. (1.5) fixed the value of k at unity. This general law enables a measurement made with one current balance to be related to that made with another and, although we have not yet developed the law here, we could, if we chose, go straight on to Chapter 7 to do so. This is effectively what standardizing laboratories such as the National Physical Laboratory do in setting up their current balances to provide us with instruments calibrated in amperes. We may therefore proceed as if we had a current-measuring instrument so calibrated.

The SI system used in this book is based on the metre, kilogramme and second as mechanical units, together with the ampere, the kelvin, the candela and the mole. Definitions of all these, with the exception of the ampere (given here), are to be found inside the front cover. That part of SI used in electromagnetism is often known as the MKSA system.

It was for a long time conventional to use systems of units in physics based on the cm–g–s (CGS) mechanical units and they are still occasionally to be found in texts. For that reason, and because readers may wish to consult older texts or original papers, a full discussion of the conversion between SI and CGS formulae will be found at the end of Appendix C.

Charge and the coulomb, C Integration of both sides of Eq. (1.5) with respect to time yields

$$Q = \int_0^{t_0} I \, dt \qquad (1.6)$$

giving the total charge flowing in time t_0 due to a current I.

If I is constant in time, (1.6) becomes simply $Q = It_0$ and enables a unit of charge to be defined. The **coulomb** is the charge crossing the section of a conductor in which a steady current of 1 A flows for 1 s. Measurement of charge can now be related to that of current through (1.6).

If I is not steady but has a known variation with time, then the integration of (1.6) is needed to calculate the total charge that has passed.

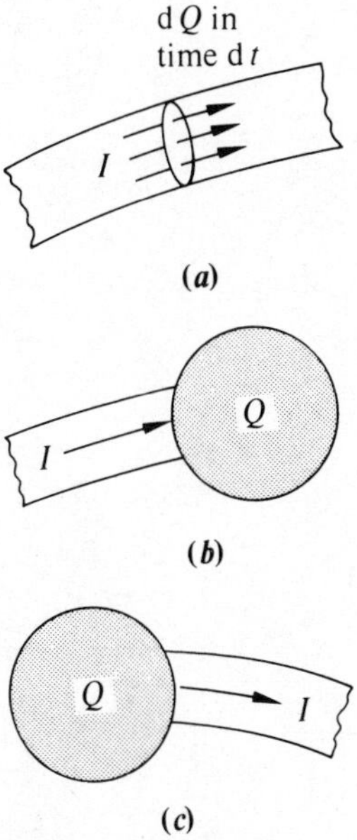

Figure 1.10 Interpretation of I and Q. (a) I is the rate of flow of Q across an area; (b) I charges the region containing charge Q at any instant and $I = dQ/dt$; (c) I discharges the region and $I = -dQ/dt$.

Worked example Suppose it is known that the current I in a conducting wire varies with time t according to the equation $I = a + bt$, where a and b are constants. The total charge passing across any section of the wire between the times $t = 0$ and $t = t_0$ is

$$\int_0^{t_0} (a + bt) dt = at_0 + bt_0^2/2$$

Interpretation of I and Q in $I = dQ/dt$ Equations (1.5) and (1.6) above refer to the current I across any section as in Fig. 1.10a, and to the corresponding time rate at which charge Q is crossing the same section. However, the equations also hold if I denotes the current flowing out of or into a conductor whose total charge is Q, as in Fig. 1.10b and c. This follows from our discussion of the expression dQ/dt in Sec. 1.3. Two additional points now need to be made. First, the question of sign arises as in Eq. (1.3), so that for a current **discharging** a conductor we should have $I = -dQ/dt$. Thus

$$I_{\text{charging}} = +dQ/dt \qquad I_{\text{discharging}} = -dQ/dt \qquad (1.7)$$

The second point is that the body or region being charged or discharged need not even be conducting. The total current I flowing into *any* volume containing Q increases the charge at a rate dQ/dt equal to I.

1.5 Atomic aspects of charge and current

The nature of charge Is electric charge a continuous fluid or are there 'atoms' of electricity as there are of matter? Is charge weightless or is it always associated with mass? Are there distinct positive and negative charges, or is one merely matter with a deficiency of the other? These questions relate to the nature of charge and cannot be answered as a result of any phenomena we have so far considered: either alternative is possible in every case. For instance, while the measurements of charge using an electrometer seem to indicate that any magnitude is possible, there could exist elementary charges so small that the addition of one of them produces no detectable change in deflection. Much of this book is concerned with situations in which a few of such charges produce no noticeable effects: in those cases we shall not worry too much about the existence of discrete charges. However, we shall also frequently ask how large-scale observations are explained by events on an atomic scale, so that the answers to the above questions must be sought.

It was realized early in the nineteenth century that the laws of electrolysis suggest very strongly what the answers are likely to be. As we saw in Eq. (1.1), Faraday found that the mass of any substance deposited at an electrode was proportional to the product of current and time, a quantity which we now know is the total charge passed, Q. In addition, he also found that the mass was proportional to the chemical equivalent, i.e. the atomic mass of the substance divided by a small integer, its valency n. Thus, for a substance of relative atomic mass A (in atomic mass units u)† and of valency n, the mass deposited by a charge Q would be

$$M = KQAm_u/n \tag{1.8}$$

where m_u is the mass of 1 u in kg and K is a universal constant. Now Am_u is the mass of one atom of the substance, so that M/Am_u would be the number of atoms deposited. Using (1.8) we then have

$$\text{Number of atoms of any substance deposited} = KQ/n \tag{1.9}$$

Hence the passage of a given charge Q deposits the same number of atoms of *any* monovalent substance ($n=1$), half the number of any divalent atoms and so on. The inference is that all monovalent atoms are associated with the same elementary charge, and n-valent atoms with n times this amount. We denote this elementary charge by e. From (1.9), the charge carried by KQ monovalent atoms is KQe and this is equal to Q itself, so that $Ke=1$ or $e=1/K$.

Unfortunately, measurements of electrolytic deposition will only give the values of M, Q and A in (1.8), and K cannot be found explicitly. Only Km_u or its reciprocal e/m_u can be determined by this method. It is found experimentally that

$$\boxed{e/m_u = 9.65 \times 10^7\ \mathrm{C\,kg^{-1}}}$$

Electrolysis thus suggests that electric charge *is* discrete; that it is associated in electrolytes with atoms to form what are called **ions**; and that both positive and negative charges are so associated since deposition may occur on both electrodes.

The electron and the atom Experiments on conduction in gases confirm the existence of ions there as well, also carrying small multiples of an elementary

---

† The atomic mass unit, 1 u, is defined as *exactly* one-twelfth of the mass of the ^{12}C isotope. A hydrogen atom or ion has a mass of *approximately* 1 u.

charge. These experiments also show that a negative particle, known as the **electron**, with Q/m 1823 times larger than e/m_u, is a universal constituent of matter and has a charge-to-mass ratio of

$$e/m_e = -1.76 \times 10^{11}\,\mathrm{C\,kg^{-1}}$$

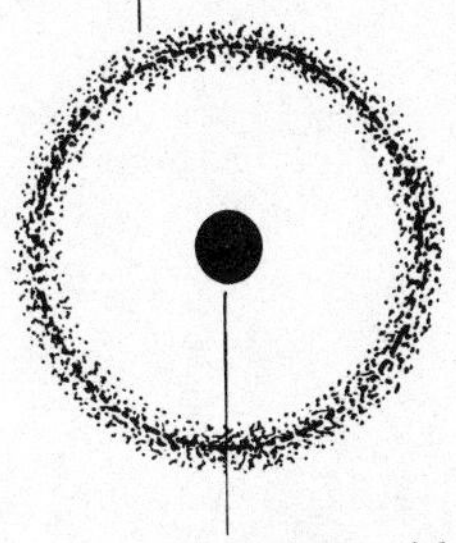

Figure 1.11 Representation of the structure of the nuclear atom.

Subsequently, the elementary or electronic charge e was determined independently and found to be approximately 1.602×10^{-19} C for both monovalent ions and electrons (Sec. 3.7). The electronic mass is thus 1823 times smaller than that of the atomic mass unit, m_u, and it is also found to be 1836 times smaller than that of the lightest positive ion, the hydrogen ion or **proton** (see the table inside the back cover for precise values).

The neutral atoms of normal matter have long been known to consist of a nucleus with charge $+Ze$ together with Z electrons outside it, Z being the **atomic number** of the element concerned. The nucleus contains Z protons with charge $+e$ and $A-Z$ **neutrons** with no charge (Fig. 1.11). Since the neutron and proton have approximately the same mass of about 1 u, A gives the total number of **nucleons** (protons + neutrons) in the nucleus and is called its **mass number**. The extra-nuclear electrons contribute so little to the mass of the whole atom that A also gives the atomic mass in u to the nearest integer. Ions are single atoms with $Z \pm n$ electrons, or groups containing such atoms, where n is a small integer determining the ionic charge. The following table summarizes the various cases:

	Neutral atom		n-valent + ion		n-valent − ion	
	Number	Charge	Number	Charge	Number	Charge
Nuclear protons	Z	$+Ze$	Z	$+Ze$	Z	$+Ze$
Nuclear neutrons	A-Z	0	A-Z	0	A-Z	0
Extra-nuclear electrons	Z	$-Ze$	$Z-n$	$-(Z-n)e$	$Z+n$	$-(Z+n)e$
Total charge		0		$+ne$		$-ne$

Identifying the carriers of current in conductors The charges carrying current are particles that can conveniently be characterized by their **specific charge** Q/m since many experiments yield this quantity rather than Q and m separately (see, for instance, Secs 3.7 and 7.5). When we turn from electrolytes and gases discussed above to metallic conductors, we find abundant evidence that electrons ($e/m = -1.76 \times 10^{11}\,\mathrm{C\,kg^{-1}}$) are not tightly bound. They are emitted from the surface of many metals when at high temperatures (**thermionic emission**), when irradiated with visible or ultraviolet light (**photoelectric emission**) or when bombarded with other electrons (**secondary emission**). There is thus a strong presumption that electrons are able to move freely within a metal and are thus responsible for electric conduction within them.

There is, however, more direct evidence for this in the form of **electromechanical experiments**. The idea behind these is that the *mass* of any free charges in a metal should give rise to inertial effects at the same time as the *charge* gives rise to currents. In one type of experiment, a conductor with no current is given a sudden acceleration or deceleration resulting in a measurable pulse of current. Tolman and Stewart (1917) and others conducted numerous investigations on these lines and showed that the carriers in copper have e/m about $-1.93 \times 10^{11}\,\mathrm{C\,kg^{-1}}$, with similar values for silver and aluminium. These experiments are difficult to perform with great accuracy, and a converse type has yielded more convincing results. In this, a circular coil is suspended in a horizontal plane so that it can rotate about its axis. The current in it is suddenly changed, and this produces a measurable change in its angular momentum. Kettering and Scott (1944) found that the carriers in copper and aluminium had $e/m = -1.77 \times 10^{11}\,\mathrm{C\,kg^{-1}}$ and could thus be identified as electrons, while Brown and Barnett (1952) and Scott (1951) found the same for molybdenum, zinc and cadmium.

These experiments suggest that a neutral metallic conductor consists of a set of massive positive ions held together in a three-dimensional array which is on the average stationary. Through these positive ions a loosely bound cloud of electrons moves freely and randomly, so that there is no net charge over any large volume. The electrons are, however, easily moved to produce currents and are easily detached from the surface. We pursue this model in Sec. 6.12 in an attempt to find a theory of conduction.

By contrast with metals, electromechanical experiments with electrolytes by Tolman, Osgerby and Stewart (1914) confirmed that the positive and negative **ions** in solution are responsible for the carriage of current.

Finally, we mention the important class of **semiconducting materials**. In these, electrons moving *en masse* through a lattice sometimes behave as if they were positively charged carriers called **holes** (rather as the motion of a liquid in a spirit-level is described as that of the bubble). In a *p*-type semiconductor, the predominant or majority carriers are holes, while in an *n*-type semiconductor the electrons behaving as negative carriers have the higher concentration. Evidence for this behaviour comes from such phenomena as the Hall effect (Sec. 7.6).

Atomic interpretation of the conservation of charge From all the above evidence, we see that charge in normal matter is carried ultimately by electrons and protons. Conservation of charge on a laboratory scale thus seems to depend on (a) the stability of electrons and protons so that their number remains constant and (b) the invariance of their charge. Point (b) is covered in Comments C3.2 and C3.3, which also show that the charges on electron and proton are exactly equal in magnitude. As for point (a), the physics of fundamental particles shows that electrons are stable as far as we know and that proton decay would take more than 10^{31} years according to the latest data, and this is stable enough for our purposes. Experiments have also revealed a great variety of particles, both stable and unstable, possessing a wide range of masses but all carrying charges of $\pm e$ or small multiples of it. Searches for particles carrying less than e have so far (1989) failed to reveal any convincing evidence for their independent existence. Moreover, the algebraic sum of elementary charges in any isolated system of fundamental particles is found to be constant, no matter how the number and type of particle may change through interactions. The law of conservation of charge clearly rests on very firm foundations.

The quantization of electric charge into units which are multiples of $\pm e$ is an important and fundamental feature of today's physics, but it would be presumptuous to say that it is fully understood. The question is further complicated by experimental evidence for the existence of quarks carrying charges of $e/3$ and $2e/3$ within nucleons, but not yet observed in the free state. These points are also discussed in Comments C1.4, C3.3 and C7.2 and in Sec. 4.2.

Macroscopic charge in terms of elementary charges If we have a finite volume τ containing n elementary charges e per unit volume, the total charge in the volume is $Q=ne\tau$ (we use τ for volume rather than V since the latter is needed later for electric potential).

More generally, there will be n_1 particles per unit volume with charge e_1, n_2 with e_2, etc., in which case:

for each set: $Q=ne\tau$ (1.10)

for all sets: $Q=(n_1e_1+n_2e_2+\cdots)\tau=(\sum_i n_ie_i)\tau$ (1.11)

where we use a notation for summation which should be familiar to the reader. Some of the e_i may be negative.

Electric current in terms of the velocity of elementary charges Suppose that in a region there are n charges per unit volume each of magnitude e and with velocity v. Figure 1.12 shows that the charge crossing an area S in unit time is that contained in a cylinder of base area S and vertical height $v\cos\alpha$, i.e. $nevS\cos\alpha$. This is therefore the current I, so that

$$I=nevS\cos\alpha=nevS_{\mathrm{n}} \tag{1.12}$$

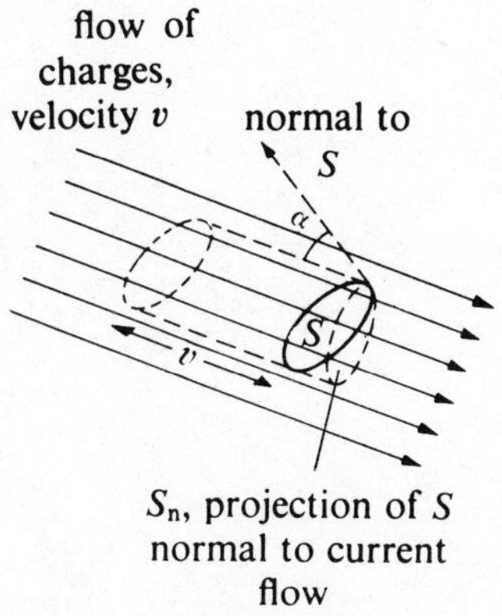

Figure 1.12 The velocity of charges and its relationship to current.

where S_{n} is the area projected normal to the current flow. This would be appropriate for metallic conductors in which only one type of carrier exists.

More generally, as in electrolytes or semiconductors, there will be several types of charged particles with densities n_1, n_2, etc., and velocities v_1, v_2, etc. In this case the total current would be

$$I=n_1e_1v_1S_{\mathrm{n}}+n_2e_2v_2S_{\mathrm{n}}+\cdots=S_{\mathrm{n}}\sum_i n_ie_iv_i \tag{1.13}$$

Here both e_i and v_i may be negative and the summations are algebraic. Negative charges with negative velocities thus give positive currents, in line with the results of Eichenwald's experiments.

More generally still, it is possible that each set of charged particles has a range of velocities which are a combination of (a) **random** velocities whose mean is zero and (b) **drift** velocities whose mean is the v in (1.12) or whose means are the v's of (1.13). Currents encountered in normal laboratory experiments have charges whose drift velocities are of the order of $0.01\ \mathrm{cm\,s^{-1}}$ only (Problem 1.9).

Convection currents Since currents consist of moving charges, an expression like *convection current* would seem unnecessary. However, we shall find it useful occasionally to distinguish those currents in which the elementary particles pass through a vacuum. Examples would be the electron beam in a cathode-ray tube or the currents of positive and negative ions in a gaseous discharge. Currents in metals, electrolytes, etc., are, on the other hand, called **conduction currents**.

Comment C1.5

1.6 The point charge and charge densities

The need for extension of fundamental concepts Although previous sections of this chapter have established the concepts of charge and current quite firmly, we have not yet considered how to tackle concrete problems about interactions. For instance, how are we to begin the investigation of forces between charged bodies? Or, if we know the forces in a simple case how can we use that result to calculate a more complicated one? To make headway with such problems we need to extend the concept of charge as discussed below. These extensions allow us to formulate laws simply and to form theoretical models upon which we can operate with the laws. They also give us freedom to choose simple models at first, yielding sufficient agreement with experiment and enough insight into what is happening, before proceeding to greater complexity, better agreement, further insight. All this is important whether we are trying to solve problems on a large scale or exploring matters at the atomic level.

The point charge Except for the electron and muon, all charge is ultimately spread over a finite volume whose absolute size may range from that of a proton to that of a star or of a large region of interstellar space. If, however, the linear dimensions of two or more such finite volumes should become very small compared with the distances between them, all the elementary charges in each volume can be considered together. In the limit, the resultant charge Q in each volume is effectively concentrated at a point and we arrive at the concept of the **point charge**.

The importance of the idea is twofold. First, we can expect laws of interaction between point charges to be especially simple because complex factors like the shape, size and non-uniformity of the charge distribution have been eliminated. Second, it yields a method of dealing with complex distributions by splitting them into a number of small elements, each equivalent to a point charge, and then summing the results in an appropriate way.

Volume charge density When a finite volume containing charge is sufficiently large compared with other dimensions, it cannot be treated as a single point charge. It is often still possible, however, to ignore the fact that many individual elementary charges are present and to treat the charge distribution as if it were continuous. An analogous situation is that of a solid or fluid whose mass is concentrated largely in the nuclei of their atoms but which is treated as continuous distributions of mass for many problems. The continuous distributions are **macroscopic models** of the systems of atoms.

We can define a volume density of charge as follows. If a volume $\Delta\tau$ contains a charge ΔQ, the mean charge per unit volume $\langle\rho\rangle$ is $\Delta Q/\Delta\tau$ so that

$$\Delta Q = \langle\rho\rangle\Delta\tau \tag{1.14}$$

If $\Delta\tau$ is shrunk to an infinitesimal volume $d\tau$ around a point (Fig. 1.13a) while the charge within it becomes dQ, then in this limit the charge per unit volume defines the **volume charge density at the point**, ρ. This definition is clearly equivalent to

$$dQ = \rho\, d\tau \qquad \text{(definition of } \rho\text{)} \tag{1.15}$$

where we adopt the practice, followed throughout the book, of labelling an equation that follows directly from, or embodies, a definition. The **SI unit** of ρ is the $\mathrm{C\,m^{-3}}$.

Note that ρ is not necessarily constant over a charge distribution any more than a mass density is in a solid or fluid. In general, ρ will vary with position within the volume. If so, the total charge Q in a region such as that of Fig. 1.13a must be obtained by summing all the charges in the elementary volumes. Thus

volume charge: $$Q=\int_\tau \rho\,\mathrm{d}\tau \qquad (1.16)$$

This can only be integrated if ρ is a known function of position. Integration over surfaces and volumes is discussed in Appendix A.5.

Surface charge density If the charge is spread thinly over a surface or a sheet of material, a useful quantity is the amount of charge per unit area or the **surface density of charge**, σ. It is defined by

$$\mathrm{d}Q=\sigma\,\mathrm{d}S \qquad \text{(definition of } \sigma) \qquad (1.17)$$

where $\mathrm{d}Q$ is the charge occupying an area $\mathrm{d}S$ as in Fig. 1.13b. The total charge on a whole surface S is then given by the integral

surface charge: $$Q=\int_S \sigma\,\mathrm{d}S \qquad (1.18)$$

Note that we use the symbol S to denote a general area, especially when it can occur anywhere in a given region of space. However, where we need a second symbol or where the area is that of a real object, we shall often use A (e.g. the area of the plates in a capacitor, Chapter 5).

Further, if we wish to regard the area as a vector quantity, we need to give it a direction as well as a magnitude. The only unique direction associated with a plane is its normal, and so our only hope is to see whether this is a suitable direction. It turns out that if we give a small area $\mathrm{d}S$ a direction along its normal and denote the resultant quantity in bold type as $\mathbf{dS}$, then several such quantities do add like vectors (i.e. obey the parallelogram rule for addition).

Line charge density Similarly, if a charge is distributed in such a way that one linear dimension is much larger than any other, as in Fig. 1.13c, it is useful to define a charge per unit length of a **line charge density**, λ, by

$$\mathrm{d}Q=\lambda\,\mathrm{d}L \qquad \text{(definition of } \lambda) \qquad (1.19)$$

where $\mathrm{d}Q$ occupies a length $\mathrm{d}L$. The total charge in a total length L is then

line charge: $$Q=\int_L \lambda\,\mathrm{d}L \qquad (1.20)$$

Note that we use the symbol L to denote a general length or displacement in space, and when its direction is included it becomes the vector $\mathbf{L}$. However, when

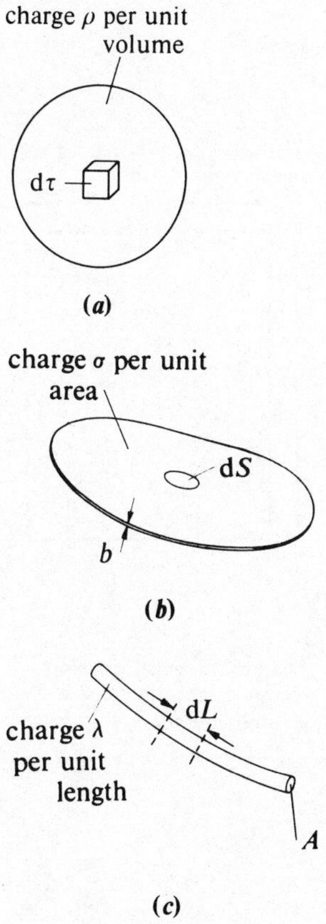

Figure 1.13 Continuous distributions of charge. (a) Volume density: charge in element is $\rho\,\mathrm{d}\tau$; (b) surface density: charge in element is $\sigma\,\mathrm{d}S$; (c) line density: charge in element is $\lambda\,\mathrm{d}L$.

the length is that of a real object, we shall often use l (e.g. the current element in Sec. 1.6, which often corresponds to a section of current-carrying wire, has a length denoted as dl, or vectorially as d**l**).

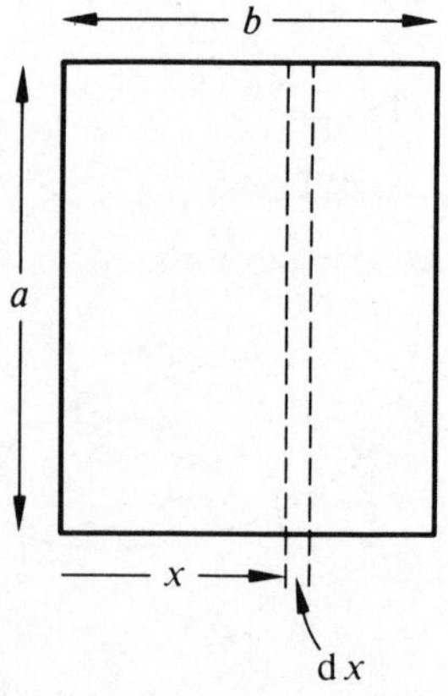

Figure 1.14 Sheet carrying charge of surface density given by $\sigma = kx$.

Worked example As a simple example to illustrate the type of integration involved above, consider a rectangular sheet as in Fig. 1.14, having sides a and b and possessing a charge whose surface density varies only with x along the side of dimension b. The variation of σ with x is given as $\sigma = kx$. To find the total charge on the sheet, take an elementary strip as shown of width dx and area $\mathrm{d}S = a\,\mathrm{d}x$. Along this strip, the surface density will be sufficiently close to the value kx at all points, so that the charge on the strip will be

$$\mathrm{d}Q = \sigma\,\mathrm{d}S = kxa\,\mathrm{d}x \tag{1.21}$$

The total charge is thus the integral of this from $x=0$ to $x=b$, which the reader should be able to show is equal to $kab^2/2$.

Elementary charges The total charge Q and the densities ρ, σ, λ will usually refer to macroscopic (i.e. large-scale) systems. It is sometimes necessary to take into account the elementary charges of which matter is composed, and we must then be clear what meaning is to be attached to a quantity like ρ. If we intend it to refer to a microscopic (i.e. atomic-scale) model then its value will fluctuate quite violently between a point within a nucleus, say, and another point just outside it. More often in this book we shall intend ρ to refer to a smoothed-out macroscopic model, when it will be the average over many atoms of the microscopic ρ.

Even the macroscopic ρ can still be expressed in terms of elementary charges, for if there are n of them per unit volume each with charge e, then

$$\boxed{\rho = ne} \tag{1.22}$$

If there are several different types of particle with densities n_1, n_2, etc., and charges e_1, e_2, etc., then

$$\rho = n_1e_1 + n_2e_2 + \cdots = \sum_i n_ie_i \tag{1.23}$$

Equations (1.22) and (1.23) follow from (1.10) and (1.11).

1.7 The current element and current densities

Although the results of this section will not be used until Chapter 6, it is convenient to consider them here because problems occur with currents similar to those with charges in the last section. Since charge is known to be discrete, its flow (i.e. an electric current) will strictly be discontinuous. Nevertheless, when there is a large density of very small charges the current can often be treated as if it were the flow of a continuous fluid.

What we must remember is that currents are not necessarily confined to wires of well-defined cross-sections but can flow in conductors of large volume. They might also flow as convection currents (end of Sec. 1.5). In all these more general examples the total current is given by the rate of passage of charge **across a specified area** even though the area may be understood, as in the case of a wire, rather than explicitly defined.

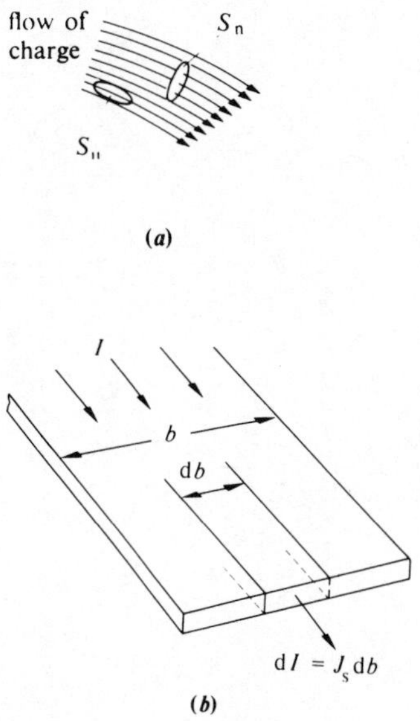

Figure 1.15 Current densities. (a) Current is associated with an area: across S_{11} the current is zero, across S_n it is a maximum; (b) surface current density J_s is the current per unit width perpendicular to the flow.

Volume current density Figure 1.15a shows that for a given flow of charge the current across a fixed area depends on the orientation of the area: across $S_{\parallel}$ it is zero, across S_n it is a maximum. The **volume current density** J at a point in the flow is defined as the current per unit area normal to the flow. Thus in Fig. 1.15a if the area S_n is shrunk to dS_n about a point and the current across it is dI, then

$$dI = J\,dS_n \qquad \text{(definition of } J) \tag{1.24}$$

The **SI unit** of J is the A m^{-2}, although current densities are more realistically expressed in A mm^{-2}. The total current across any area S in general given by

volume current:
$$I = \int_S J dS_n \tag{1.25}$$

We shall see later that J is a vector quantity in the direction of flow and that an area dS has a 'direction' along its normal. If we use **J** and **dS** to represent these as vectors, then we shall find that (1.25) can be written as $I = \int \mathbf{J} \cdot \mathbf{dS}$, where the dot indicates a scalar product. This indicates that I is a scalar quantity, unlike **J**. See Appendix B.3.

Surface current density Current sometimes flows in thin surface layers or in sheets as in Fig. 1.15b, and here the **surface current density** J_s is a useful quantity. It is defined as the current per unit width perpendicular to the flow. Thus, if b is the dimension of the conductor across the flow, the current dI in an elementary strip of width db is $J_s\,db$, so that the total current is

$$dI = J_s\,db \qquad \text{(definition of } J_s) \tag{1.26}$$

The current element Sometimes we need to split a current into small elements, calculate the effect of each element, and sum the contributions. If the current is confined to a filament, as in a wire, then the appropriate element is a small section of length dl carrying the current I as in Fig. 1.16. The product $I\,dl$ occurs frequently in formulae involving such elements.

If the current is not filamentary but flows in large volumes or sheets, then the element is better expressed in terms of J or J_s. The current across the area dA in Fig. 1.16 is now part of a large flow and is properly denoted by dI. The product $I\,dl$ becomes $dI\,dl$. By using Eqs (1.24) and (1.26) this expression can be converted into $J\,d\tau$ or $J_s\,dS$, as the reader can verify. We thus have the equivalent products

$$I\,dl \equiv J\,d\tau \equiv J_s\,dS \tag{1.27}$$

provided the proper integrations are subsequently carried out. This proviso is necessary, since (1.27) will otherwise be equating a first-order infinitesimal (dl) to a second-order one (dS: for example $dx\,dy$) or a third-order one ($d\tau$: for example $dx\,dy\,dz$).

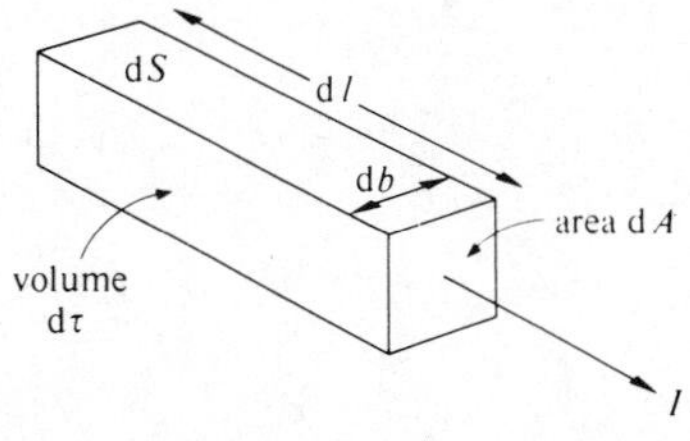

Figure 1.16 A current element.

If the current element is to be regarded as a vector quantity in the direction of the length $\mathbf{dl}$, then (1.27) should be written in the form

$$\boxed{I\,\mathbf{dl} \equiv \mathbf{J}\,d\tau} \tag{1.28}$$

Current densities and elements in terms of moving charges The current densities J and J_s will usually refer to macroscopic systems. If we wish to express these densities in terms of the elementary charges we must decide how detailed a model we need. For some purposes the discrete nature of charge will have to be recognized, and current density, like charge density, will fluctuate violently on a microscopic scale from one point to another or from one instant of time to another. Such fluctuations are responsible for what is called 'shot noise' in electronic networks.

More often we shall find that a smoothed-out model is quite sufficient so that J and J_s will be average values over very many charges. Thus if we have charges of volume density $\rho = ne$ moving with velocity v, Eqs (1.12) and (1.24) give

$$J = nev = \rho v \tag{1.29}$$

or if several types of charge are present

$$J = \sum_i n_i e_i v_i = \sum_i \rho_i v_i \tag{1.30}$$

Similarly, it is easy to show that the surface current density is related to the surface density σ of moving charge by

$$J_s = \sum_i \sigma_i v_i \tag{1.31}$$

Finally, the current element product $I\,dl$ can be expressed as $nevS\,dl$, assuming only one type of moving charge. Since $neS\,dl$ gives the total moving charge in the element, Q, we can replace $I\,dl$ by Qv if we wish:

$$I\,dl = Q_{\text{moving}} v \tag{1.32}$$

although this is likely to be an oversimplified model of an actual current of elementary particles.

Once again, there are vector forms of these equations. Equation (1.29) can be written as

$$\boxed{\mathbf{J} = \rho \mathbf{v}} \tag{1.33}$$

and Eq. (1.32) as

$$\boxed{I\,\mathbf{dl} = Q_{\text{moving}} \mathbf{v}} \tag{1.34}$$

1.8 Historical note

Although the ancient Greeks were aware that amber ($\eta\lambda\epsilon\kappa\tau\rho o\nu$) could be electrified by rubbing, it was not until A.D. 1600 that William Gilbert (1540–1603) showed that many other substances behaved in the same way, and only in 1731 did Stephen Gray (d. 1736) announce the discovery of conductors of electricity. Charles-François du Fay (1698–1739) distinguished vitreous and resinous electricity in 1733, while 14 years later Benjamin Franklin (1706–1790) suggested the law of conservation of charge and the use of + and − to distinguish two different charged conditions. From Gilbert onwards there was much speculation about the nature of electricity, but little quantitative work of any description was carried out until late in the eighteenth century: this will be discussed in the next chapters.

Luigi Galvani (1737–1798) announced in 1791 the discovery of what was called animal electricity—the convulsion of a frog's leg when a connection was made across its ends through two dissimilar metals. However, Alessandro Volta (1745–1827) showed with his pile (of copper, zinc and moistened pasteboard discs repeated in that order) that the animal was not essential and that the pile produced effects similar to those of frictional electricity, viz. the charging of an electroscope and the production of electric shocks and sparks. These last two phenomena associated with discharge were then much used for identifying electricity even though they do not lend themselves to exact measurement.

Comment C1.1

Current from the voltaic pile enabled the chemical effect to be studied closely, first by Nicholson and Carlisle in 1820 and most fruitfully by Michael Faraday (1791–1867) who stated his laws of electrolysis and conjectured the atomicity of electricity in 1833. The magnetic effect as we have studied it was discovered by André-Marie Ampère (1775–1836) in 1820, while the heating effect was first measured by James Prescott Joule (1818–1889) in 1841. The first experiment to determine whether moving electrified bodies were equivalent to currents was carried out by H.A. Rowland (1848–1901) in 1876 and was subsequently repeated several times by him and others between then and 1903.

The discovery of the electron was largely the result of the work of J.J. Thomson (1856–1940) who measured the specific charge (e/m) for a number of charged particles and in the period 1897–1899 established the basis of atomic physics summarized in Sec. 1.5.

1.9 Conclusion

The next steps We cannot proceed usefully any further until we have looked at the quantitative laws governing the electric force between charges and the magnetic force between currents. Which is taken first is a matter of taste, but we expect the interaction between stationary charges to be less complex than one between moving charges, where velocities are an additional variable. We shall therefore consider first (Chapters 2–5) the electric force and then, before tackling the magnetic force, look at the production and properties of steady currents in more detail (Chapter 6).

References For more explanation and description of very elementary electricity and magnetism, textbooks of pre-degree level should be consulted, e.g. Bennett (1974) and Nelkon and Parker (1987). The historical bare bones of Sec. 1.8 can be covered by selective reading of the standard work by Whittaker (1951). More directly, Magie (1964) gives extracts from the original papers of many of the pioneers, while the *Experimental Researches in Electricity* of Faraday (1951 edition) gives a fascinating account of investigations relevant to this chapter.

1.10 Revision summary

Electric charge is recognized by forces of attraction exerted on *any* other uncharged body. All charges fall into one of two classes, each repelling any of its own class and attracting any of the other.

Quantities of charge of either class can be measured by an *electrometer*, and when so measured are denoted by the symbol Q. Electrometer measurements show that

- The two classes of charge can be labelled *positive* and *negative* because they add like positive and negative numbers.
- In any isolated system the algebraic sum of the charges is constant (*the law of conservation of charge*).

Electric current is recognized by heating, chemical or magnetic effects. It is universally agreed that magnetic forces shall be used to measure the magnitude of a current, denoted by I. With this measure, chemical deposition in electrolysis is proportional to I while the heating rate in a conductor is proportional to I^2.

Experiments show that currents consist of moving charges and that positive charge moving in one direction is equivalent as a current to the same magnitude of negative charge moving in the opposite direction with the same speed. A sign convention is adopted that a *positive* current has the same direction as the flow of any positive charge it contains. Experiments further show that

- An electric current across any area is the rate of flow of resultant positive charge across the area, a relationship expressed as

$$I = \frac{dQ}{dt} \tag{1.5}$$

or its inverse

$$Q = \int I \, dt \tag{1.6}$$

It then follows that

- If a current I flows *into* a region of space or into a conductor, producing in it a resultant charge Q at any time t, then

$$I = +dQ/dt$$

If the current flows *out*, then

$$I = -dQ/dt \tag{1.7}$$

- A steady current flowing along a path has the same value at all cross-sections of the path. In particular, a steady current in a thin wire has the same value at all points along it.

The **SI unit** of current is the *ampere*, symbol A, defined in Sec. 1.4. The **SI unit** of charge is the ampere-second or *coulomb*, symbol C.

Atomic aspects Experiment shows that charge is not a continuous quantity but is discrete. All charges in matter are multiples of the *elementary charge*

$$e = 1.602 \times 10^{-19}\,\text{C}$$

Uncharged matter consists of exactly equal numbers of positive and negative elementary charges. The resultant charge in any system is the algebraic sum of the elementary charges:

$$Q=\sum N_i e_i=\left(\sum n_i e_i\right)\tau \tag{1.11}$$

The moving charges that constitute currents may be electrons (as in a metal or a cathode-ray tube), ions (as in electrolytes) or both (as in discharge tubes). In all cases, the current I consists of n elementary charges per unit volume moving with a mean drift velocity v across a perpendicular area S, and it is given by

$$I=nevS \tag{1.12}$$

or by $S\sum n_i e_i v_i$ if more than one type of elementary charge is present.

Extension of fundamental concepts The following quantities are necessary to deal with problems involving charges and currents:

Point charge, Q When the volume containing Q has linear dimensions negligible in comparison with distances from other charges or from observers, it can be treated as a single point charge Q.

Charge densities, ρ, σ, λ These are appropriate when the linear dimensions of the volume containing the charge cannot be neglected, but the charge can be assumed to be spread continuously throughout a volume, over a surface or along a line. The charge per unit volume at a point defines the *volume charge density* ρ with corresponding definitions for σ (surface) and λ (line). The total charge in a region is given by one of (1.16), (1.18), or (1.20):

$$Q=\int_\tau \rho\,d\tau; \qquad Q=\int_S \sigma\,dS; \qquad Q=\int_L \lambda\,dL$$

In terms of *elementary charges* e_i, whose total number in a region is N_i and whose number per unit volume is n_i:

for each set: $\quad Q=Ne; \quad \rho=ne \quad$ (1.22)

for all sets: $\quad Q=\sum_i N_i e_i; \quad \rho=\sum_i n_i e_i \quad$ (1.23)

Current densities, J, J_s The current is assumed to be a flow of charge across an area. The *volume current density* J at a point is the current per unit area normal to the flow, i.e. $dI=J dS_n$. The total current across an area S is

$$I=\int_S J\,dS_n \tag{1.25}$$

For current flowing in a sheet, J_s is the current per unit length perpendicular to the flow, i.e.

$$dJ=J_s\,db \tag{1.26}$$

In terms of *moving elementary charges* with velocites v_i and either volume densities ρ_i or surface densities σ_i:

for each set: $\quad I=nevS_n=\rho v S_n$

$\qquad J=nev=\rho v \quad$ or $\quad J_s=\sigma v \quad$ (1.29)

for all sets: $\quad I=S_n\sum n_i e_i v_i$, etc.

If $\mathbf{J}$ is treated as a vector in the direction of flow and $\mathbf{dS}$ is a vector area normal to its plane, we can write

$$\mathbf{I}=\int_S \mathbf{J}\cdot\mathbf{dS} \quad \text{and} \quad \mathbf{J}=\rho\mathbf{v} \tag{1.25)(1.33}$$

The current element, $I\,\mathrm{d}l$, $J\,\mathrm{d}\tau$, $J_s\,\mathrm{d}S$ An element of current used for calculation is characterized by the equivalent products

$$I\,\mathrm{d}l \equiv J\,\mathrm{d}\tau \equiv J_s\,\mathrm{d}S \tag{1.27}$$

If there is a total moving charge Q with velocity v in the element, then

$$I\,\mathrm{d}l \equiv Qv \tag{1.32}$$

When $\mathbf{dl}$ is a vector quantity, we can write

$$I\,\mathbf{dl} \equiv \mathbf{J}\,\mathrm{d}\tau \equiv Q\mathbf{v} \qquad (1.28)(1.34)$$

Commentary

C1.1 On Historical Treatments If this chapter has appeared straightforward we should realize from Sec. 1.8 that the work of centuries has been compressed into a few pages. The concepts of charge and current presented here were only developed through the efforts of many outstanding people, and some of the possible lines of investigation which they had to eliminate have been omitted. For instance, the laws of electrolysis have no mention of the concentration of the electrolytic solution or of the size of the electrodes: it is not obvious *a priori* that these would have no effect on the mass deposited and Faraday had to eliminate them by experiment. But to include every detail of this sort would be both wearisome and a disservice to the work of people like Faraday which was done so that we could avoid their sidetracks. We are concerned, moreover, with the subject as it stands today and it is right that our account should be more direct than a strictly historical one.

C1.2 On Measurement Many readers, in spite of the last comment, will undoubtedly feel that the account is not at all direct and that a lot of fuss has been made about what are essentially simple concepts. They may feel that they already understand charge and current intuitively, a feeling which is certainly desirable, but not by itself sufficient. Most concepts in physics are the combination of a whole cluster of ideas (e.g. current is recognized by many effects other than those mentioned, or can be thought of as a fluid or a flow of charged particles) but precision of *definition* and agreement about *measurement* must enter at some stage if progress is to be made through the cooperation of many people.

There is a general tendency towards agreement about what are fundamental and what are derived quantities and units. By a **fundamental** quantity we mean that the only precise way of defining it lies in the prescription for its measurement. Such a prescription invariably involves two stages: the recognition of the **equality** of two similar quantities and the setting up of a standard scale by some form of **addition**. The methods described for current and charge illustrate both stages—for charges, addition is 'putting both charges into the cup'; for currents, addition is recognized by the n^2 increase in force. For reasons given in Sec. 1.4 it is then desirable to agree on a unit. On the other hand, **derived** quantities can be defined in terms of fundamental ones. Experimental laws not only enable us to reduce the number of basic units needed to establish a system like SI, but give us a larger number of methods available for measurement of a given derived quantity.

One purpose of this chapter has been to draw some attention to the nature of physical concepts and measurement as well as to make clear the meaning of some fundamental electrical quantities. Except in quantum mechanics, where the role of the observer as a disturbing factor is in question, the subject of measurement has received little attention since the classic works of Campbell (1920, 1928). The interested reader is referred to Ellis (1966) for a greater depth of discussion. See also Cook (1977).

C1.3 On Contact and Frictional Electrification 'The contact electrification of solids can claim to be, by many centuries, the oldest branch of electrical studies, yet it is certainly now the most backward' (Henry, 1967). Since that was written, recent studies have clarified the various mechanisms responsible for the transfer of charge from one solid to another, but it remains a complex process to understand. We shall comment further on this work at the end of Chapter 5 when new concepts have been developed.

C1.4 On Positive and Negative Charge It seems a remarkable fact that the two types of charge should have properties that permit them to be labelled with the signs + and −. An obvious possibility is that in some fundamental way one type is merely a deficiency of the other, and this idea has been exploited in theories as far apart as Franklin's one-fluid theory of the eighteenth century and Dirac's theory of the electron and positron of 1936. Recent ideas put electric charge on the same footing as other quantum numbers describing the state of fundamental particles, such as spin, strangeness, lepton number, etc. Neutral particles are described by charge quantum number 0, charged particles ± 1, etc. This idea in no way offers any deeper explanation of the nature of charge, but it does put the question on much the same level as other fundamental problems in modern physics.

It is even more remarkable that the labels + and − should work not only for the **addition** of charges but, as we see in the next chapter, for their interactions as well, using now the **product** of the quantities. The law *like charges repel, unlike charges attract* is translated into

$$\left.\begin{array}{c} + \times + \\ - \times - \end{array}\right\} \rightarrow \begin{array}{c}\text{positive numbers,} \\ \text{repulsion}\end{array} \qquad \left.\begin{array}{c} - \times + \\ + \times - \end{array}\right\} \rightarrow \begin{array}{c}\text{negative numbers,} \\ \text{attraction}\end{array}$$

This is an excellent example of the way physical properties and laws can be expressed using mathematical structures with an independent existence. Once the physical attributes have been shown to bear a one-to-one correspondence with the fundamental axioms of the mathematical structure, all the subsequent *mathematical* operations can be carried out with confidence that they are valid for the physical situation as well.

C1.5 On Convection and Conduction Current It is not easy to give exact meanings to these terms and authors differ in their usage of them. Even with the explanation given in Sec. 1.5 doubtful cases can occur. For instance, how would the current produced by rotating the charged annulus in Eichenwald's experiment (Sec. 1.3) be classified? A more fruitful distinction can be made between **neutral** currents (in which the resultant charge density is zero as in a wire) and **charged** currents. (This is the subject of Problem 1.16.)

For many purposes these distinctions are unnecessary since all that matters is thatboth types of current are produced by actual moving charges. When, later in the book, we meet other 'currents' that are the result of theoretical models (amperian currents in Chapter 12, displacement current in Chapter 13) we shall distinguish between them all by the use of subscripts. Thus conduction and convection currents will be denoted by I_c, J_c or J_{sc}.

Problems

Section 1.1

1.1 Devise a method for measuring charge based on the force between charges and similar to that used for measuring current in the current balance of Sec. 1.2. Do you need to assume conservation?

1.2 You are shown two identical rods which repel each other when placed end to end, but which show attraction for both resinous and vitreous electricity. You are *told* that the rods are of Spurite and that they are charged with a third form of electric charge. What do you suspect to be the truth and how would you test your suspicion?

Section 1.2

1.3 Assess the relative merits of the three possible methods of measuring current, considering both theoretical and practical aspects.

Section 1.4

1.4 The charge on a conductor at time t is given by $Q = Q_0(1 - e^{-t/T})$ where Q_0 and T are constants. Find the current I at time t and sketch the variations of Q and I with t. If the charge were given instead by $Q = Q_0 e^{-t/T}$ how would the current differ from that in the first case?

1.5 The current flowing into a conductor varies in time according to the equation $I = I_0 e^{-\alpha t}$, where I_0 and α are constants. Find the charge Q which has accumulated on the conductor after a time t_0. Sketch the variation of I and Q with time.

Section 1.5

1.6 If it were possible to charge a conductor weighing 10 g with 1 C, what change in mass would occur on discharge to earth? How does this compare with that detectable by a chemical balance? (As to the feasibility of a charge of 1 C, see Problems 2.3 and 4.6.)

1.7 From the definition of the mole, show that the molar mass of any element in kg is 10^{-3} times its atomic mass in u, i.e. that the mass of 1 mol in g is *equal* to its atomic mass in u. The Avogadro constant N_A is the number of elementary units of any substance in 1 mol. Show that if m_u is the mass of 1 u in kg, then $N_A = 1/10^3 m_u$. (For convenience, we denote $10^3 N_A$ in this book by N'_A.) Show that a substance with relative atomic or molecular mass M and density ρ has a number of atoms or molecules per unit *volume* equal to $N'_A\rho/M$.

1.8 The faraday, F, is that quantity of charge which deposits 1 mol of any substance in electrolysis. Show that F is a universal constant equal to $N_A e$ and find its value from the data in Sec. 1.5.

1.9 Assuming that each copper atom contributes one free electronic charge to the current in a wire, *estimate* the mean drift velocity of charges when the wire has a diameter of 1 mm and carries a current of 1 A. (Relative atomic mass of copper = 63.6; density of copper = 8.9×10^3 kg m^{-3}.)

1.10 Stow (1969) gives the total transfer of positive charge from the ionosphere to earth due to currents in fine weather as about 90 C km^{-2} per year. What is the approximate fine weather current falling on 1 mm^2 of the earth? How many electronic charges per second does this represent?

Section 1.6

1.11 Show, with the symbols used in Fig. 1.13b and c, that the relations between σ and λ and the equivalent volume density ρ are $\rho b = \sigma$ and $\rho A = \lambda$.

1.12 The linear charge density along a thin rod of length L varies with distance l along the rod according to $\lambda = \lambda_0 l^2/L^2$ where λ_0 is a constant. What is the total charge along the rod? What is the mean line charge density?

1.13 A circular disc of radius a has on one side a surface density of charge σ which varies with distance r from the centre but is constant for a given r. What is the total charge on the disc of (a) $\sigma=\sigma_0 r/a$, (b) $\sigma=\sigma_0(1-r^2/a^2)$, (c) $\sigma=\sigma_0 e^{-r/a}$?
1.14 A spherical region of space of radius a is filled with positive charge of volume density ρ which varies with distance r from the centre of the sphere but is constant for a given r. Find the total charge within the sphere if (a) $\rho=\rho_0 r/a$, (b) $\rho=\rho_0(1-r^2/a^2)$, (c) $\rho=\rho_0 e^{-r/a}$. Interpret the constant ρ_0 in each case. (The last two have been suggested as approximate charge distributions in atomic nuclei.)

Section 1.7

1.15 The current of electrons in a certain cathode-ray tube forms a thin cylindrical beam of constant radius a in which the charges move parallel to the axis. The current density in a cross-section is found not to be constant but to vary with the distance r from the axis of the beam according to $J=J_0(1-r/a)$. What is the total current in the beam, assuming that it can be treated as the flow of a continuous distribution of charge? J_0 is clearly the current density on the axis at $r=0$. What is the mean current density in any cross-section in terms of J_0 only?
1.16 A **neutral** current is defined as one in which the densities of positive and negative carriers, n_p and n_n, are equal so that $n_p e_p + n_n e_n = 0$ because $e_n = -e_p$. Show that the magnitude of a neutral current (e.g. in a copper wire) depends only on the *relative* drift velocity of the two types of carrier and is independent of any motion of the observer. Show also that the magnitude of a **charged** current (for which $n_p e_p + n_n e_n \neq 0$) depends on the velocity v of an observer O relative to an origin P fixed in the laboratory.
1.17 An electron of charge e travels round a circular orbit with an angular velocity ω. To what current is this equivalent?

2

The law of force between charges (*in vacuo*)

We now go on to investigate more fully the force between charged bodies described in Sec. 1.1. Such a force should depend, among other factors, on the size, shape, and orientation of the bodies if their sizes are of the same order of magnitude as their distances apart. If we look instead for a law of force between what are effectively two **point** charges, the only factors involved ought to be the magnitude and sign of the charges, their distance apart, and the direction of the line joining them, and the law should be correspondingly simple. Because few problems concern only **two** point charges we shall want to know how these electric forces combine when, for instance, both of two point charges exert forces on a third: once this is established, the law can be used to find forces between large bodies carrying charge by the usual process of considering elements of volume, and integrating.

To establish the complete law, we shall examine first the direction of the force and the combination of forces; secondly, the dependence on magnitude and sign of the charges; and thirdly, the dependence on distance. The law will then be applied to particular problems, with an eye on any limits in its range of application. Finally, we examine the interaction between charges in terms of **energy** and relate this to the forces calculated earlier in the chapter.

2.1 Direction of the force and superposition

Direction of the electric force Two point charges situated at O and P, as in Fig. 2.1, will exert forces on each other which must, by Newton's third law, be equal and opposite. Since there is, in addition, symmetry about the line OP, the force F_{OP} (acting on P due to O) must act along the joining line and so must F_{PO}. The force system will therefore be as indicated in Fig. 2.1 if the charges are of the same sign. The result means that the electric force falls into the very important class of **central forces**, i.e. forces whose lines of action always pass through their point of origin. Thus, wherever P was situated, the force on it due to O would always pass through O.

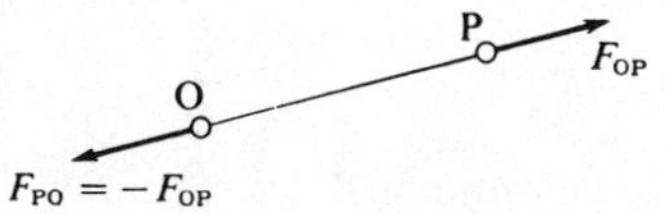

Figure 2.1 Central forces. The line of F_{OP} passes through O wherever P may be situated.

Crude experimental verification can be obtained by suspending from insulating threads two light conductors, small compared with their distance apart, and charging them. The mutual force causes a displacement of each which is, as far as can be detected, in the same direction in space as the line joining the centres, whatever the sign of the charges.

This part of the general law is as important as the others to be considered immediately but is often neglected, possibly because it appears self-evident. It should be emphasized, however, that the symmetry about OP which allows us to deduce the result assumes that charge is a *scalar* quantity (see Problem 2.1 and Appendix B.1). The experiments of Sec. 1.1 show that charge can be specified by a single number and thus justify the assumption.

Superposition Connected with the direction of the force is the question of superposition. This can best be explained with reference to Fig. 2.2 in which O, P, and Q are three charges, F_{OP} and F_{PO} are the forces between O and P *in the absence of Q*, and similarly for the other forces. If the resultant force on Q when all three charges are present is the vector sum (see Appendix B.1) of F_{PQ} and F_{OQ}, then these forces are said to superpose or to obey the **principle of superposition.** In elementary mechanics we can show that elastic and gravitational forces obey the principle by verifying experimentally the parallelogram law of addition, but this does not mean that electric forces will necessarily behave in the same way. Direct experiments can be performed by finding the resultant displacement of a small charged body Q, suspended again by a light thread, due to two other charged bodies, O and P, as in Fig. 2.2. Such experiments are subject to errors due to leakage and the finite size of the bodies, but do show that superposition is probably valid. The assumption that the principle is strictly true is made very frequently in solving problems and no disagreement with experiment has been found that would need its rejection.

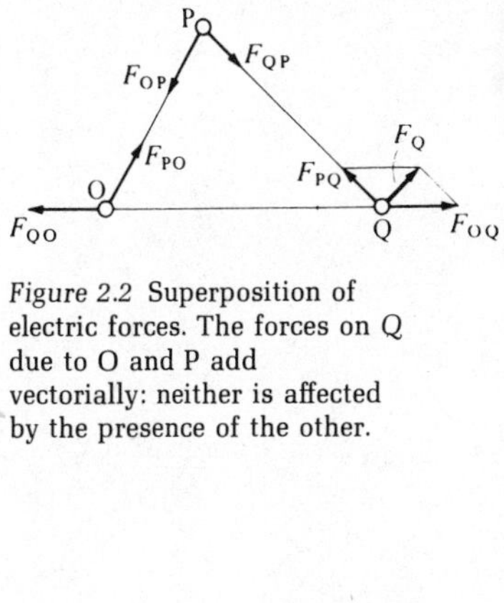

Figure 2.2 Superposition of electric forces. The forces on Q due to O and P add vectorially: neither is affected by the presence of the other.

2.2 Dependence on magnitude and sign of charge

Had we chosen to define a unit of charge in terms of the force exerted on a standard charge at a standard position, much as we did with current, then it would follow that the force between charges of magnitude Q_1 and Q_2 a fixed distance apart would be proportional to Q_1Q_2. We have chosen a more practical method of measuring charge (by means of an electrometer) and must therefore appeal to an experiment like that of Henry Cavendish (1731–1810) who appears to have been the first to show that the force between two small conductors with equal charges Q, as measured by an electrometer, varied as Q^2. [See Nicola (1972) for a comment.]

Comment C1.4

We shall assume that the force is directly proportional to the product of the magnitudes of the charges and shall justify the assumption by testing accurately a consequence of it later (Sec. 5.2). If the sign of the charges is included in the product, the sign of the force conforms to the normal convention—the positive repulsive force F_{OP} in Fig. 2.1, for instance, is in the direction of positive displacement from O. There is no evidence that the sign of a charge affects any other part of the law.

2.3 Dependence on distance: the inverse square law

History It is well known from the theory of gravitation that a mass inside a uniform spherical shell of matter experiences no resultant force due to the material of the shell itself. This can be shown to follow from the inverse square law of force between two point masses. Joseph Priestley (1773–1804) observed that the charges on a hollow conductor resided on the outer surface and apparently exerted no force on a small charged body placed inside. He conjectured from this that the electric force also varied inversely as the square of the distance between charges, although the parallel with gravitation is not obvious unless the conductor is known to be a uniformly charged spherical shell. (We shall see in Sec. 4.1 that Priestley's observations are in fact a consequence of the inverse square law.)

A direct test of the law by measurement of the force between small charges seems first to have been carried out by John Robison (1739–1805) in 1769, but the best-known demonstration is that of Charles-Augustin Coulomb (1736–1806) who in 1785 verified the law both for repulsion, using his torsion balance, and for attraction, with an oscillation method (see Problem 2.2). The confidence of theoretical workers in the law was based mainly on his results.

Our confidence in it lies rather in a series of experiments designed to check Priestley's observation of the absence of electric forces inside a hollow charge distribution on a conductor, using a uniform density of charge over a sphere. The first of these experiments was carried out by Cavendish in 1772, although the results were not published until 1879. It was repeated by Clerk Maxwell (1831–1879) and by several others in this century (see later in this section).

Basic aim of the Cavendish experiment To see the point of what is called the Cavendish experiment, we first assume from Coulomb's rough measurements that the law is an inverse nth power law. We then show that a small charge placed somewhere inside a charged conducting spherical shell will in general experience a force either towards or away from the centre exerted on it by the shell, **except that, when $n=2$ exactly, no force exists.** In other words, if an experiment can establish that such a force is zero, then $n=2$.

Theory of the Cavendish method Consider a small charge Q inside a sphere, centre O (Fig. 2.3), and divide the sphere into two segments by a plane through Q perpendicular to OQ. The charge on the sphere will, because of the symmetry, spread uniformly over the surface with a density σ provided that Q is sufficiently

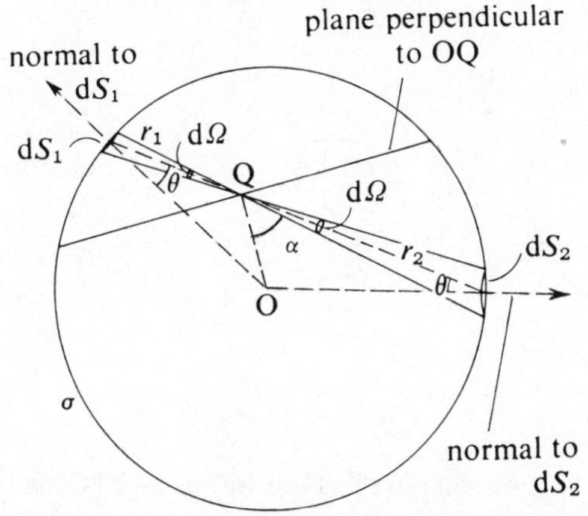

Figure 2.3 Deduction of the force on Q due to a spherical shell of charge with surface density σ. Areas dS_1 and dS_2 are exaggerated for clarity.

small to produce an insignificant change in the distribution. Consider first the force on Q due to two small surface elements dS_1 and dS_2, both cut off by a solid angle $d\Omega$† with apex at Q, and let dS_1 refer to the element on the smaller segment so that $r_1 < r_2$.

Because of our assumed inverse nth law, the force on Q due to the charge on the two elements will be

$$dF \propto \sigma\, dS_1 Q/r_1^n - \sigma\, dS_2 Q/r_2^n$$

which, since $d\Omega = (dS_1 \cos\theta)/r_1^2 = (dS_2 \cos\theta)/r_2^2$, is also

$$dF \propto \frac{\sigma Q d\Omega}{\cos\theta}\left(\frac{1}{r_1^{n-2}} - \frac{1}{r_2^{n-2}}\right)$$

the direction of this being towards dS_2 if both σ and Q are positive and $n > 2$. All pairs of surface elements whose joining line makes the same angle α with OQ produce similar forces directed into the larger segment, and the components along the plane perpendicular to OQ will cancel, leaving a resultant force towards the centre O. The whole sphere can be divided in this way and we arrive at the following conclusions:

$n > 2$, $\sigma +$, the force on a positive Q is towards the centre, on a negative Q away from the centre,

$n < 2$, $\sigma +$, the force on a positive Q is away from the centre, on a negative Q towards the centre,

$n = 2$, no force.

Now suppose we place a conductor inside the sphere and connect it to the sphere by a metallic wire. Because the wire and conductor are electrically neutral they do not affect the spherical distribution in any way, but if the electric force exists it would act on both the positive and the negative charges in the wire, although in opposite directions. The effect would be to cause a flow of charge away from or towards the inner conductor, and with the wire this would of course be entirely a flow of electrons. If the wire is then removed, the inner conductor is left charged unless, as we have seen, n is exactly equal to 2.

Comment C2.2

† For solid angle formulae, see Appendix A.1, where Eq. (A.3) shows that $d\Omega$ will be given by both $(dS_1 \cos\theta)/r_1^2$ and $(dS_2 \cos\theta)/r_2^2$.

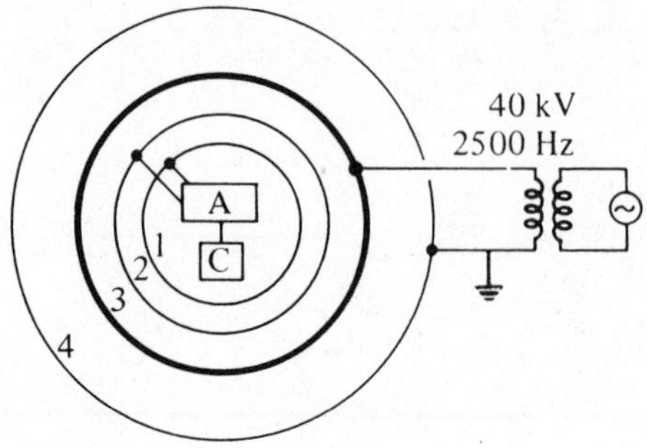

Figure 2.4 The apparatus of Bartlett, Goldhagen and Phillips used to carry out the Cavendish experiment (schematic). Sphere 4 is an earthed shield, sphere 3 the effective outer sphere. Sphere 1 contains an amplifier A, a chart recorder C and their power supplies.

Experimental results In all experiments of this type, any charge left on the inner conductor was less than the smallest detectable by the most sensitive methods available at the time. This lower limit to the detectable charge determines the amount by which n can be said to differ from 2. The calculation of this amount is easy if the inner conductor is also spherical and concentric with the outer. Cavendish was able to show that if $n=2\pm q$, then $q\leq 1/50$, while Maxwell used a more sensitive electrometer to show that $q\leq 1/21\,600$. This is about the best that can be achieved if contact is made and broken because contact potentials are set up. To avoid these, more recent workers have used permanent contact between the spheres, have charged and discharged the outer sphere periodically, and have mounted the sensitive detecting, amplifying, and recording apparatus inside the inner conductor. A typical arrangement is that of Bartlett, Goldhagen and Phillips (1970) shown in outline in Fig. 2.4, in which any current between the *two* inner concentric spheres would be amplified and recorded on a chart recorder for later examination. A null result was obtained and subsequent experiments were carried out to ensure that this was not due simply to faulty apparatus. They found that $q\leq 1.3\times 10^{-13}$ and since then Williams, Faller and Hill (1971) have pushed back the limit of q to 3×10^{-16}.

It might reasonably be asked why so much effort is still being made to establish the validity of Coulomb's law with the high precision described in this section. We discuss this in Comment C4.2.

Comments It should be emphasized that this experiment only demonstrates the absence of force due to a hollow **uniform sphere of charge** on a charge inside it. It does **not** show that a hollow charge distribution, even a spherical one, shields an internal charge from the effects of charges placed outside—no such shielding effect occurs, any more than it does in the corresponding gravitational case. It does **not** show that a hollow charged conductor of **any** shape exerts no force on an internal charge. Finally, it does **not** show that a hollow conductor shields internal charges from external influences, although this is true under certain circumstances discussed in Sec. 4.1. (Note that there is no gravitational analogue to an electric conductor.)

2.4 Forms of Coulomb's law

The laws established with varying degrees of reliability in the previous three sections show that the force $\mathbf{F}$ between two point charges Q_1 and Q_2 separated by

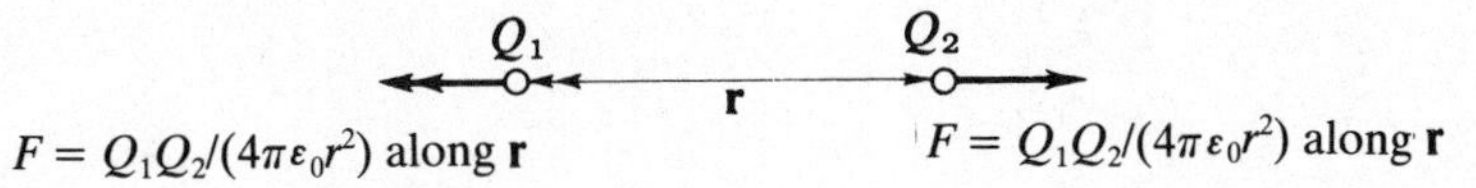

Figure 2.5 Coulomb's law. For Q_2, the origin is at Q_1 and $\mathbf{r}$ is directed from Q_1 to Q_2, and vice versa.

a distance r is proportional to Q_1Q_2/r^2 and is along the joining line. This we shall refer to as **Coulomb's law** and the force as a **Coulomb force**. The complete law is summarized by

$$F = kQ_1Q_2/r^2 \qquad \text{in magnitude, along } r \text{ in direction} \tag{2.1}$$

(see Fig. 2.5), where k depends only on the units chosen for the various quantities and is thus a universal constant. (Note that k is often used to denote a constant needed by a writer for a short time: this k is *not* the same as the universal constant of Sec. 1.3.)

We have already chosen for ourselves the coulomb as the unit of charge and the SI mechanical units: k is thus a quantity to be determined in SI units by experiment. It is usual to write k as $1/(4\pi\varepsilon_0)$, where ε_0 now becomes the universal constant to be determined by experiment. The reasons for choosing a more complicated expression are the following: the factor ε_0 is in the denominator rather than the numerator because we then find that a property of materials known as the relative permittivity always occurs as a multiplier of ε_0, while the factor 4π makes the system a *rationalized* one as recommended by successive international conferences. One advantage of this we note at this stage: formulae involving spherical or cylindrical symmetry usually contain a 4π or 2π, respectively, while those with planar symmetry do not contain a π.

Coulomb's law can thus be expressed in the form

$$F = Q_1Q_2/(4\pi\varepsilon_0 r^2) \qquad \text{in magnitude, along } r \text{ in direction} \tag{2.2}$$

In order to avoid the addition of the words denoting direction while retaining an indication that the force is a vector, the notation outlined in Appendix B.1 is used and the law written

Coulomb's law: $$\boxed{\mathbf{F} = Q_1Q_2\hat{\mathbf{r}}/(4\pi\varepsilon_0 r^2)} \tag{2.3}$$

where $\hat{\mathbf{r}}$ is a unit vector along r, or

Coulomb's law: $$\boxed{\mathbf{F} = Q_1Q_2\mathbf{r}/(4\pi\varepsilon_0 r^3)} \tag{2.4}$$

in which the numerator and denominator of (2.3) have been multiplied by r.

Superposition When more than two charges interact, each exerts Coulomb forces on all the others. The principle of superposition can then be applied to find the resultant of the forces on any one of the charges by vector summation, as in Fig. 2.2. In general, the force on any *one* charge Q of a collection as in Fig. 2.6 is the vector sum of all the individual forces due to all the others. Thus, in that case,

$$\begin{aligned}\mathbf{F} &= \mathbf{F}_1 + \mathbf{F}_2 + \mathbf{F}_3 + \cdots \\ &= Q[Q_1\hat{\mathbf{r}}_1/(4\pi\varepsilon_0 r_1^2) + Q_2\hat{\mathbf{r}}_2/(4\pi\varepsilon_0 r_2^2) + \cdots] \\ &= Q\sum_i Q_i\hat{\mathbf{r}}_i/(4\pi\varepsilon_0 r_i^2)\end{aligned} \tag{2.5}$$

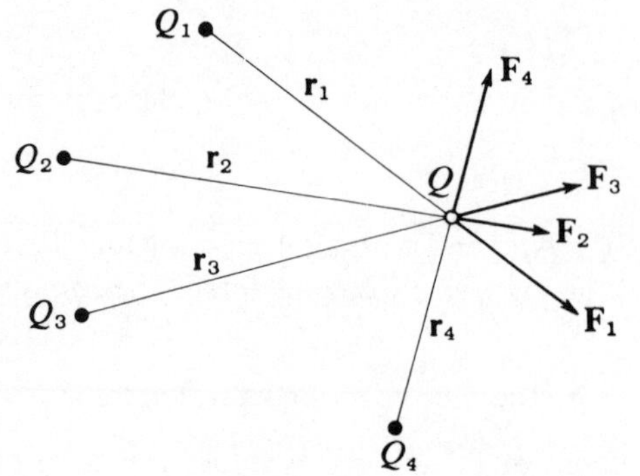

Figure 2.6 The resultant force on Q due to a collection of charges is the vector sum of the $\mathbf{F}_s$.

Note that this is purely a formal statement of the resultant. In many simple problems it is enough to use a diagram marked with the directions and magnitudes of the forces and to add by resolving into components (see Sec. 2.6 for an example). Note also that the charge Q exerts forces on Q_1, Q_2, etc., and Q_1, Q_2, etc., on each other, but we have simply chosen one charge Q as the object of interest.

Value of ε_0 The constant ε_0 is variously known as the **permittivity of free space**, the **electric constant**, or simply as **epsilon nought**. Its value in SI units can be determined in principle by measuring the force between two point charges of known magnitude and distance apart, but this is not a practicable method. Instead, we use formulae involving ε_0 that are ultimately deduced from Coulomb's law and which lend themselves more readily to precise measurement. Two such formulae are of overriding importance to the value of ε_0 and we shall meet them in later chapters of this book. They are anticipated here so that numerical calculations using Coulomb's law can be carried out.

The first of the equations is that for the capacitance of an air capacitor which, as we shall see in Chapter 5, is simply given by ε_0 multiplied by a geometrical factor with the dimensions of a length. This was originally used to obtain an experimental value for ε_0, but it now concerns us only because it simplifies the units we can use. Although the SI unit for ε_0, using Coulomb's law, is the $\mathrm{C\,m^{-2}\,N^{-1}}$, it can clearly be more concisely expressed as the farad per metre ($\mathrm{F\,m^{-1}}$).

The second important equation involving ε_0, which we meet in Chapter 13, relates it to the velocity of light, c, which is now specified to nine significant figures. To five figures only, this gives a value for ε_0 of $8.8542 \times 10^{-12}\,\mathrm{F\,m^{-1}}$. More useful, perhaps, is the information that since c is so close to $3 \times 10^8\,\mathrm{m\,s^{-1}}$, the value of $4\pi\varepsilon_0$ is very nearly $1/(9 \times 10^9)\,\mathrm{F\,m^{-1}}$. To summarize, therefore:

$$\begin{array}{ll} \text{to about 1 in } 10^5: & \varepsilon_0 = 8.8542 \times 10^{-12}\,\mathrm{F/m} \\ \text{to about 1 in 500:} & 4\pi\varepsilon_0 = 1/(9 \times 10^9)\,\mathrm{F/m} \end{array} \tag{2.6}$$

In an older system of units known as CGS electrostatic units (e.s.u.), Coulomb's law itself is used to choose a unit for Q. This e.s.u. of charge is such that $k = 1$ when F and r are in dyn and cm, respectively. All formulae may clearly be converted to this system by the substitution $4\pi\varepsilon_0 = 1$. See Appendix C.

2.5 Range of application of Coulomb's law

A law is usually established experimentally over a limited range of the variables included in it and is then applied with caution outside this range, with an eye open for disagreements with experiment which might indicate its breakdown. Coulomb's law is no exception and we consider in turn several limitations imposed by our method of establishing it.

First, it has not been verified, as far as we are concerned, for huge values of Q or, perhaps more important, for very small ones. In particular, we have by no means verified it for single elementary charges since our conductors must contain very many of them and our result may be the effect of averaging some different law over a vast number.

Secondly, the distances used have been in the centimetre–metre range and any extrapolations to atomic or astronomical distances are tentative. Mason and Weaver (1929), for instance, have pointed out that if the law were such that the force was proportional to $e^{-a^2/r^2}/r^2$, where a is a length comparable with the size of an atom, then for values of r of a centimetre or so the law would be indistinguishable by experiment from inverse square, but at atomic distances would be very different. A similar argument clearly applies to $e^{r^2/a^2}/r^2$, where a is now of astronomical size. Comment C4.2

Thirdly, the charges in our experiments were stationary relative to each other: hence the term **electrostatic** to describe the forces. However, what happens if the charges move? We know from Chapter 1 that the moving charges constitute currents between which a magnetic force operates, but this is in addition to the electrostatic force which must still be acting. What we do not know is whether this electrostatic force is unaltered by the motion. Moreover, the charges in the experiments are only stationary taken as a whole: there may be considerable random motion of the elementary charges present and once again Coulomb's law may be only an average over a period of time of some more fundamental law.

Finally, we have assumed that the presence of air between the charges can be ignored—a point we shall take up when considering the effect of a medium on electric forces (Chapter 11).

Although this may seem too gloomy a prospect for a law we have only just established, it is reassuring to note that it is still applied today with some confidence over wide ranges of the variables, and with success. For the moment, we shall be making two important assumptions. The first is that Coulomb's law is valid for any value of r, a subject discussed for atomic distances in Sec. 4.2 and for very large distances in Comment C4.2. The second is that the law is valid for moving charges, a matter discussed further in Secs. 7.6 and 13.10. We must just take care that consequences of these assumptions do not lead to disagreement with experiment or inconsistency with well-established results.

2.6 Application of Coulomb's law to specific problems

Point charges We deal first with the use of Coulomb's law in the calculation of forces between point charges. In practice the results will be sufficiently accurate in cases where the charges occupy small regions rather than points, provided these regions are separated by large enough distances.

The **magnitude** of the force between **two** charges is calculated from Eqs (2.2) and (2.6).

Worked example Problem 2.3 shows that net positive or negative charges of 1 C would be impossibly large under typical laboratory conditions: 1 μC is a more reasonable size. To demonstrate this, we calculate the force between two charges of 1 μC separated by 1 cm. In Eq. (2.2), $Q_1 = Q_2 = 10^{-6}$ C and $r = 10^{-2}$m. Using Eq. (2.6), we obtain $F = (9 \times 10^9 \times 10^{-6} \times 10^{-6})/(10^{-2})^2$ N or 90 N (a newton is a force equal to weight of a medium-sized apple so that even 90 N is not all that small).

Although electronic, ionic and nuclear charges are very small compared with 1 μC, the distances between them in matter are also much smaller than 1 cm, and Coulomb forces are large enough to play a significant and sometimes dominant role in the behaviour of matter at the nuclear and atomic levels. Two examples where this is so are (a) the hydrogen atom and (b) ionic crystals. In general, however, electric forces are much more complex than simple inverse square type Coulomb forces between point charges. For example, the chemical bonds responsible for the cohesion of solids and liquids can be broadly classified into covalent, metallic, Van der Waals, hydrogen and ionic. Only in the last-named are inverse square law forces dominant.

Application *ionic crystals* In an NaCl crystal, the Na^+ and Cl^- ions are monovalent and lie alternately at the nodes of cubic lattice a distance $a_0 = 0.28$ nm apart. The Coulomb force between two such ions is therefore about 3×10^{-9} N using the method of the above worked example. However, in attempting to break ('cleave') a crystal with a cross-section as small as, say, 1 mm² (10^{-6} m²), the number of bonds between nearest neighbours that will be broken is about $10^{-6}/a_0^2$ or approximately 10^{13}, so that the Coulomb force to be overcome is some 30 kN: the crystals are quite strong. Of course, this is a grossly oversimplified model of what actually happens since (a) it is a three-dimensional problem, (b) other forces, albeit smaller, are involved and (c) the crystal structure is not perfect.

When we turn to the calculation of forces where **more than two charges** are involved (as in the real ionic crystal), we shall concentrate more on the method than the numerical values. To avoid the elementary mistake of forgetting **directions**, it is usually best to make a sketch showing the forces whose sum is required.

Worked example Figure 2.7a shows a group of three equal charges placed at the corners of an equilateral triangle of side a. The resultant force on the top charge is required. Since both F_1 and F_2 are equal to $Q^2/(4\pi\varepsilon_0 a^2) = F$, say, their resolved parts along the broken line add to give $2F \cos 30°$ or $\sqrt{3}Q^2/(4\pi\varepsilon_0 a^2)$. The resolved components at right angles to the broken line are equal and opposite, and so cancel. Similar problems will be found in the examples at the end of the chapter.

The resultant of the forces $\mathbf{F}_1$ and $\mathbf{F}_2$ can also be found by vector addition (see Fig. B.2 in Appendix B). If they are drawn head to tail so as to form the two sides of a triangle, the resultant force is given by the third side as a vector drawn from the tail of $\mathbf{F}_1$ to the head of $\mathbf{F}_2$, i.e. along the broken line in Fig. 2.7a. The magnitude of this resultant is obtained by an elementary calculation of the length of this third side and leads to the same result as above.

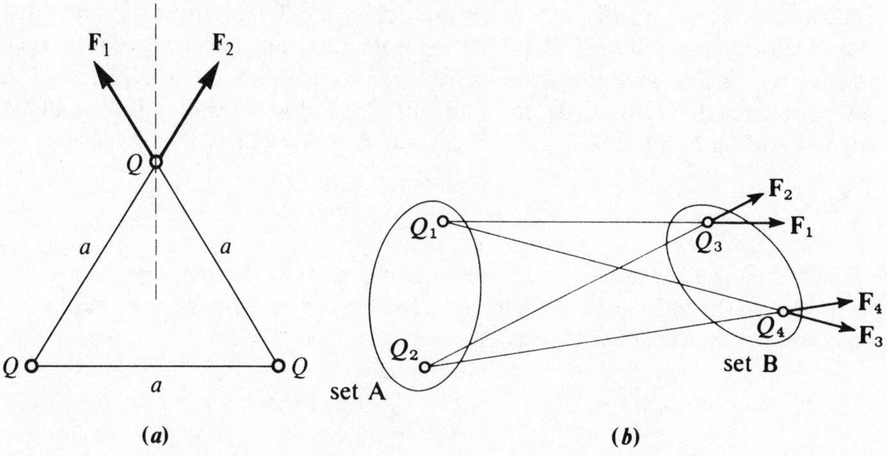

Figure 2.7 (a) A group of three charges: the resultant force on the top charge is required and is the vector sum of $\mathbf{F}_1$ and $\mathbf{F}_2$; (b) two sets of charges: the resultant force on set B is the vector sum of $\mathbf{F}_1$, $\mathbf{F}_2$, $\mathbf{F}_3$ and $\mathbf{F}_4$.

Occasionally the force exerted by one **group** of charges on another may be required but, as shown in Fig. 2.7b, the method is still the same.

Continuous distributions If we are given continuous distributions, then they must be split into small elements of charge each of which behaves like a point charge. The forces between the elements can then be calculated, resolved into components and added by **integrating** over each distribution. Problems of this kind are often more easily solved by methods to be developed in succeeding chapters, but an example is given here to illustrate the technique we have just described.

Worked example We take a problem involving a one-dimensional distribution: a charge with uniform line density λ along a thin rod of total length L. The force on a point charge Q is required, where Q is situated a distance a from the rod along its perpendicular bisector as in Fig. 2.8. We divide the rod into elements of length $\mathrm{d}l$, a typical one being a distance l from the midpoint. All the charge in the element can be considered as a point charge of magnitude $\lambda \mathrm{d}l$ which exerts a force $\mathrm{d}F$ on Q given by

$$\mathrm{d}F = Q\lambda \mathrm{d}l/(4\pi\varepsilon_0 r^2) \qquad \text{along } r$$

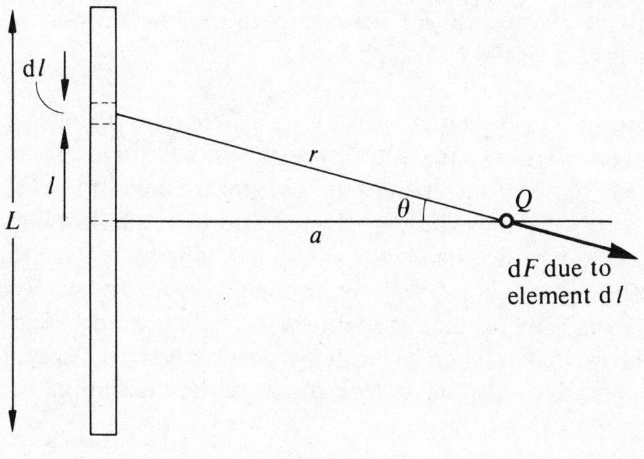

Figure 2.8 Calculation of the force on Q due to a uniformly charged rod.

The forces due to the different elements are thus in different directions and must be added as vectors. If each dF is resolved both along and perpendicular to a, we see by symmetry that the components perpendicular to a will cancel. The component of dF along a is $dF\cos\theta$, and all such components may now be added by integration because they are in the same direction. Thus

$$F=\int_{-L/2}^{+L/2}\frac{Q\lambda\,dl\cos\theta}{4\pi\varepsilon_0 r^2}$$

Two of the three variables l, r and θ must be eliminated before the integration can be carried out. Since $l=a\tan\theta$, we have $dl=-a\sec^2\theta\,d\theta$. In addition, $r=a\sec\theta$, and so after cancellation we are left with

$$F=\frac{Q\lambda}{4\pi\varepsilon_0 a}\int_{l=-L/2}^{l=+L/2}-\cos\theta\,d\theta$$

$$=\frac{Q\lambda}{4\pi\varepsilon_0 a}[\sin\theta]_{l=-L/2}^{l=+L/2}$$

or

$$F=Q\lambda L/4\pi\varepsilon_0 a(a^2+L^2/4)^{1/2}$$

Several checks on such a result are easily carried out. Firstly, it is dimensionally correct since the right-hand side has a $(\text{charge})^2$ in the numerator and a $4\pi\varepsilon_0(\text{distance})^2$ in the denominator. Secondly, we can look at extreme cases. If the rod is very short ($L\to 0$) it should behave like a point charge λL: this is confirmed because $F\to Q\lambda L/(4\pi\varepsilon_0 a^2)$. For a very long rod, $L\to\infty$, and in the limit

$$F=Q\lambda/2\pi\varepsilon_0 a \tag{2.7}$$

This is a result we shall be looking at again.

2.7 The mutual potential energy of charges

Electric interactions are involved in many scientific fields such as astrophysics, space science, biology and chemistry, as well as in other branches of physics and engineering. Although the expression of these interactions in terms of **forces** is sometimes useful, a far more fruitful approach is to consider them in terms of **energy**. This is because each of the various forms of energy can be converted into any of the other forms provided the overall conservation law is satisfied. It follows that the relationship between electricity and other areas of science depends very heavily on having available expressions for the energy of the interactions.

We now look at this aspect of the Coulomb force.

Potential energy in general The potential energy of a system is the energy associated with its configuration in space, i.e. with its position relative to other systems and with the relative positions of its parts. The precise definition of the difference in potential energy ΔU between two configurations is as follows: it is the work done by forces external to the system solely in changing it from one configuration to the other. The word 'solely' is inserted because the external forces could do work in increasing the kinetic energy of the system and we clearly want to exclude that. The value of ΔU will be positive if positive work is done and the final configuration then has a greater potential energy than the initial one.

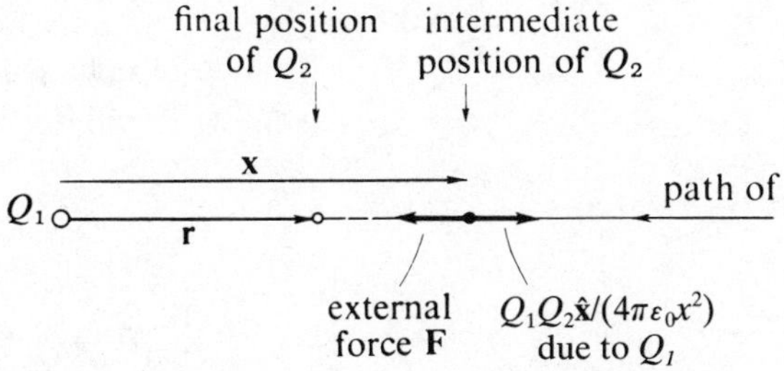

Figure 2.9 Calculation of the potential energy of two interacting point charges. Q_2 is moved from r' to r against the force from Q_1 by an external force **F** doing work.

When potential energy (U) rather than ΔU is quoted, the initial configuration is a standard one to be used as a zero. Strictly speaking, this standard should always be specified, but it is often omitted in well-known formulae.

Calculation of U for two charges Let us take two point charges Q_1 and Q_2 initially separated by a distance r' and move Q_2 directly towards Q_1 by an external force until the final separation is r (Fig. 2.9). A second force keeps Q_1 fixed in position. The potential energy difference ΔU between the initial and final configurations is then equal to the work done by the external forces.

We need not worry about the work done by the external force needed to keep Q_1 in position: its point of application is stationary and the work is zero. Referring to Fig. 2.9, the path taken by Q_2 is chosen to be always along the same direction as r and the external force needed to bring in Q_2 is always equal and opposite to the force repelling it from Q_1; Q_2 then moves with a constant velocity and no kinetic energy is created. At some intermediate stage when Q_2 is at a distance x from Q_1, the work done by the external force F in a small displacement dx is

$$dW = F\,dx = -Q_1Q_2\,dx/(4\pi\varepsilon_0 x^2) \tag{2.8}$$

Hence, the total work done over the whole path, W, or the potential energy difference, ΔU, is given by

$$W = \Delta U = \int_{r'}^{r} -Q_1Q_2\,dx/(4\pi\varepsilon_0 x^2) \tag{2.9}$$

or

$$\Delta U = \frac{Q_1Q_2}{4\pi\varepsilon_0}\left(\frac{1}{r} - \frac{1}{r'}\right) \tag{2.10}$$

This expression still applies (a) when $r' < r$ and (b) if the appropriate signs are inserted for the Q's when substituting numerical values. Should ΔU be negative, any external force is having work done **on** it by the internal ones. In general, it should be noted, **a reversal of the path reverses the sign of ΔU.**

A natural zero for the potential energy of a pair of charges would seem to be when they are so far apart that their interaction is negligible. Letting $r' \to \infty$ in (2.10) we have the following expression for the potential energy of point charges Q_1 and Q_2 separated by a distance r, the zero being taken at an infinite separation:

two point charges:

$$\boxed{U = Q_1Q_2/(4\pi\varepsilon_0 r)} \tag{2.11}$$

With any other zero, U would differ from this value only by a constant.

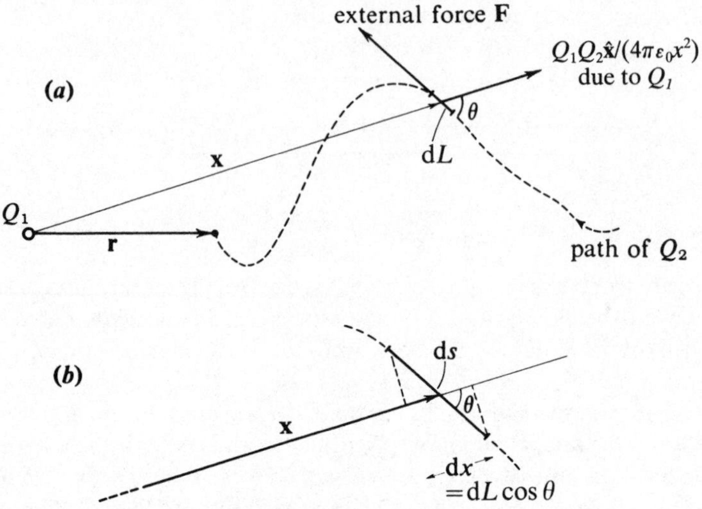

Figure 2.10 (a) Calculation of potential energy with a general path; (b) detail of an element of the path.

Path-independence The expressions (2.10) and (2.11) contain reference only to the initial and final configurations and none to the path taken. We chose a path which made the calculation very simple, but what happens if we choose a different one as in Fig. 2.10?

Suppose at some intermediate stage the shortest distance of Q_1 from Q_2 is x so that the internal force is $Q_1Q_2/(4\pi\varepsilon_0x^2)$ along **x**. The external force **F** needed is now only $-(Q_1Q_2\cos\theta)/(4\pi\varepsilon_0x^2)$ along d**L**. Hence the work done by **F** in the increment of path is $F\,\mathrm{d}L$ or $-(Q_1Q_2\,\mathrm{d}L\cos\theta)/(4\pi\varepsilon_0x^2)$. Out of this expression we pick $\mathrm{d}L\cos\theta$ and see that it equals dx, the increment in the radial distance from Q_1. The increment of work, dW, is thus given as before by Eq. (2.8), and (2.10) and (2.11) both follow: U and ΔU are independent of the path. (Forces for which this is true are described as *conservative*: see Sec. 3.4.)

Physically, this arises because the force is a central one (Sec. 2.1) and work is done only in the radial component of a displacement. Any tangential displacement (movement at a constant distance from Q_1) involves no work at all and this allows us to generalize even further. The two positions of Q_2 in (2.10) need not be on the same radial line from Q_1 and the expression is valid for *any* two points at the distances r and r'.

Finally, it is clear that the same expressions would arise if Q_1 were moved and Q_2 fixed. The potential energy is **mutual** and does not belong to one charge more than the other.

***U* for a collection of charges** Extension to a collection of more than two charges by the same arguments gives the potential energy of the system **as a whole**. Thus, because the forces superpose, the charges in Fig. 2.11 have potential energy

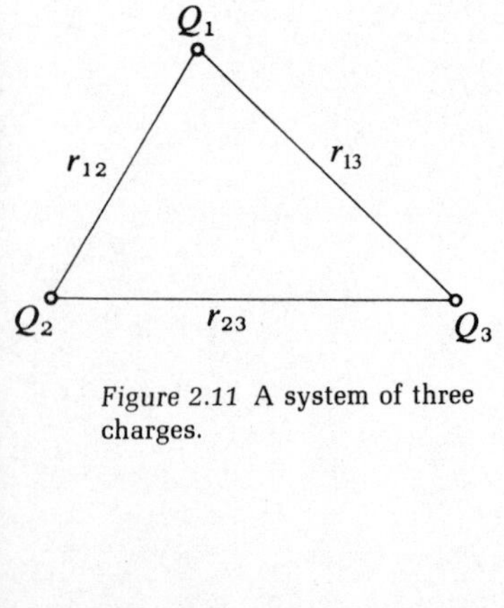

Figure 2.11 A system of three charges.

three point charges:

$$U=Q_1Q_2/(4\pi\varepsilon_0r_{12})+Q_2Q_3/(4\pi\varepsilon_0r_{13})+Q_3Q_1/(4\pi\varepsilon_0r_{23}) \qquad (2.12)$$

with respect to the arrangement with all three at infinite distances from each other. This expression can clearly be extended to any number of charges. The path-independence of U has the physical consequence that we cannot extract energy from a system of charges by moving one of them in a closed path: if we could, we should have an inexhaustible supply of energy.

Relationship between U and electric force The value of U, once the zero is chosen, is perfectly definite for any configuration and does not depend on the existence of the external forces which were introduced for the purposes of the definition. Indeed, an equivalent definition of U is 'the work done by the internal forces in taking a system **from** the given configuration **to** the standard one'.

The relation between the internal forces and U is a perfectly general one in mechanics and follows from the definition of U. Suppose the internal force on any one charge in a direction x is F_x and that this force is allowed to move the charge by an amount $\mathrm{d}x$. The work done, $F_x \mathrm{d}x$, is equal to the decrease in potential energy $-\mathrm{d}U$, and so

$$\boxed{F_x = -\frac{\mathrm{d}U}{\mathrm{d}x}} \qquad (2.13)$$

It follows that:

1. If no external forces exist to keep charges in their positions, they will move under the action of internal forces in such a way as to reduce U.
2. A given charge will be in equilibrium if $F_x = 0$, i.e. if $\mathrm{d}U/\mathrm{d}x = 0$, which means that U has a stationary value (maximum, minimum or point of inflection).
3. The equilibrium in 2 will be stable if U is a minimum, since a small displacement of the charge will bring forces into play acting towards its original position [$\mathrm{d}U$ is positive for a positive $\mathrm{d}x$, so that F_x is negative by Eq. (2.13)].

An important graphical representation of these statements is provided by the potential diagram (Appendix A.6). Plotting U against x reveals any 'potential wells' where U is a minimum and at the bottom of which a particle would be in stable equilibrium.

It should be noted that Eq. (2.13) assumes that U is a function of only one variable denoted by x. In general, U will depend on more than one position coordinate and (2.13) should then properly be a partial derivative. For instance, if U is given in terms of x, y and z, then $F_x = -\partial U/\partial x$, $F_y = -\partial U/\partial y$, $F_z = -\partial U/\partial z$ (see Appendix A.3), but these matters are taken up in the next chapter and for the moment the one-dimensional case illustrates the important point that internal forces act so as to minimize U.

Worked example Consider the problem of finding the ionization energy of the hydrogen atom using the Bohr theory, in which the average distance of the electron from the proton is $r_0 = 5.3 \times 10^{-11}$ m. Since ionization involves the complete removal of the electron, the potential energy of the atom in its ground state from Eq. (2.11) is $-e^2/(4\pi\varepsilon_0 r_0)$. The kinetic energy $mv^2/2$ is positive and is obtained from the law of motion $F = mv^2/r_0$, that is $mv^2/2 = e^2/(8\pi\varepsilon_0 r_0)$. It follows that the total energy is $-e^2/(8\pi\varepsilon_0 r_0)$, and substitution of numerical values gives this as 2.17×10^{-18} J or 2.17 aJ, which is equal to 13.6 eV in the normal units used for energy on the atomic scale (see page 63). This is the energy that must be given to the atom by bombardment in order to ionize it.

2.8 Conclusion

The next steps The whole of electrostatics is contained in Coulomb's law and the principle of superposition, and the next three chapters use no other basic laws than these. Coulombs law is often cumbersome to use, however, in many practical and theoretical situations. We now proceed to develop more powerful concepts which not only make many problems easier to deal with but form the basis for future developments outside electrostatics.

References Further details of experimental verification of Coulomb's law can be found in Magie (1964) for Coulomb's own work, Maxwell (1904) for Cavendish's and Maxwell's and in Plimpton and Lawton (1936). Goldhaber and Nieto (1971) give a complete list of references up to the date of their paper.

2.9 Revision summary

Electric force between charges Two laws describing the forces between stationary charged bodies have been established, laws which together form the complete basis of electrostatics:

- **Coulomb's law** The Coulomb force between two stationary charged bodies carrying Q_1 and Q_2 with dimensions negligible compared with their distance r apart is

$$\mathbf{F} = \frac{Q_1 Q_2}{4\pi\varepsilon_0 r^2}\hat{\mathbf{r}} \qquad (2.3)$$

  where ε_0 is a constant determined by the units. In SI units:

$$4\pi\varepsilon_0 = 1/(9 \times 10^9) \text{ approx.} \qquad (2.6)$$

  The law has been established very precisely for a limited range of distances and magnitudes of charge. Assumptions of its validity outside these ranges does not lead to any contradiction with experiment. Whether the law also applies if one or more of the charges is moving is a more difficult question. For the moment we assume that it will apply for sufficiently small velocities, reserving more detailed discussion until Chapters 7 and 13.
- **The principle of superposition** If several charges separately exert forces $\mathbf{F}_1$, $\mathbf{F}_2$, etc., on a single charge, then the total force when all the charges act is the vector sum $\mathbf{F}_1 + \mathbf{F}_2 + \cdots$. This enables the forces in collections of charge and in continuous distributions to be calculated.

Electric energy of charges The following can be deduced from Coulomb's law: when two charges Q_1 and Q_2 change their distance apart from r' to r, the change in potential energy ΔU is

$$\Delta U = \frac{Q_1 Q_2}{4\pi\varepsilon_0}\left(\frac{1}{r} - \frac{1}{r'}\right) \qquad (2.10)$$

Note that ΔU

- is independent of the path of either charge in changing the distance from r' to r, because the Coulomb force is a central force,
- reverses its sign if the path is reversed, i.e. if r and r' are interchanged,
- is a mutual potential energy of both charges.

If a zero of potential energy is chosen to be when Q_1 and Q_2 are an infinite distance apart ($r' \to \infty$), then

$$U = Q_1Q_2/(4\pi\varepsilon_0 r) \qquad \text{(zero at infinite separation)} \tag{2.11}$$

gives the potential energy U of two charges a distance r apart.

The principle of superposition allows this to be extended to any number of charges in the form

$$U = Q_1Q_2/(4\pi\varepsilon_0 r_{12}) + Q_2Q_3/(4\pi\varepsilon_0 r_{13}) + Q_3Q_1/(4\pi\varepsilon_0 r_{23}) + \cdots \tag{2.12}$$

The relationship between F and U is that $F_L = -\mathrm{d}U/\mathrm{d}L$, where F_L is the force on any charge in a general direction L and $\mathrm{d}U/\mathrm{d}L$ is the rate of change of U with displacement in the same direction.

Commentary

C2.1 On the Forces in the Universe The four basic forces at present known to exist are, in order of increasing strength, the gravitational, the weak nuclear, the electromagnetic and the strong nuclear (see Davies, 1986; Quigg, 1985). Since the weak and strong forces are short range, their effect is only dominant within the nucleus and between interacting elementary particles. Both gravitational and electromagnetic forces are long range, but the Coulomb force exceeds the gravitational by such a factor (Problem 2.5) as to make the latter negligible in comparison. So at the atomic level, electromagnetic forces are dominant. As soon as large amounts of matter are involved, however, the situation changes. The existence of both positive and negative charges, which may cancel each other's effects to a high degree, means that electric forces can remain very small or zero. Mass, on the other hand, is always positive and the gravitational forces always increase in proportion to the amount of matter. At the terrestrial and stellar level, therefore, gravitational effects become important.

In matter itself the electric forces are entirely responsible for all the atomic and molecular bonding that occurs. However, the distances are such that atoms, ions and molecules can no longer be regarded always as points and their complex electronic structure has to be taken into account. Coulomb's law in its simple form cannot therefore be used directly to calculate general interatomic forces, though it will give order-of-magnitude results for the attractive forces in ionically bonded solids. An elementary account of atomic and molecular bonding is given in Holden (1971), but brief accounts are given in many texts on materials.

Suggestions have recently been made that a fifth fundamental force (and even a sixth) might exist in nature to explain certain experimental observations made in mines and boreholes. This is likely to remain a contentious area for some time, although current thinking inclines to the view that the measurements are explicable in terms of a modification to the gravitational force rather than an entirely new type (Gribbin, 1988; Stacey and Tuck, 1988).

Ever since the four basic forces were first recognized, attempts have been made to produce theories relating one or more of them to each other, thus reducing their number. A unified theory of the weak and electromagnetic interactions, the electroweak theory, has proved successful enough to be generally accepted. Grand unified theories (GUTs) linking the electroweak and strong forces are, on the other hand, much more speculative, but are of great importance in describing the early history of the universe (Close, 1983). The gravitational interaction still seems to defy attempts to incorporate it in a super unified theory.

C2.2 On Symmetry Arguments Shaw (1965) points out that symmetry alone is not sufficient to guarantee that the charge on a conducting spherical shell spreads uniformly over it: it would not, after all, on a similar insulating shell. It is necessary, in addition, that there shall be *only one* possible distribution, i.e. there must be uniqueness as well as symmetry. In Chapter 13 we prove a uniqueness theorem, that there is only one equilibrium distribution of charge on conductors with a given total charge. The proof of the theorem, however, requires the inverse square law.

We thus have the following subtle point: the argument about electric force inside the spherical conducting shell is correct for the case $n=2$ exactly. However, the estimates of the limits of q in Sec. 2.3 are based on the assumption by the experimentalists that the charge still spreads uniformly if n differs slightly from 2. This is apparently not easy to prove and may be impossible. Fortunately, as long as any deviations from uniformity are small, the effect on the estimate of q is of the second order of smallness.

Shaw's paper is a valuable reminder of the correct way to present symmetry arguments. It also considers, with the freedom available to mathematicians, the consequences of 'wild' laws of force in which n differs greatly from 2.

Problems

Section 2.1

2.1* Examine the consequences of mutual forces between point charges not being directed along the joining line.

Section 2.3

2.2 A short insulating rod of negligible mass carries a small charged body at one end. The rod is pivoted about its centre so that it can rotate about a vertical axis. A fixed charge of opposite sign is placed in the same horizontal plane as the rod and at a large distance d from it. Show that if Coulomb's law holds, small oscillations of the rod would be simple harmonic with a period proportional to d.

Section 2.4

2.3 Estimate the force in tonnes weight which would be exerted between two point charges each of 1 C if they could be placed 1 m apart (1 tonne = 1000 kg). A thunder cloud may typically hold a negative charge of 100 C at a mean height of 5 km and a similar positive charge at a mean height of 9 km. Estimate the force between these two.

2.4 Estimate the charge occurring on a rubbed body capable of picking up small pieces of paper.

Section 2.5

2.5 Compare the electric and gravitational forces between two electrons or between two protons a certain distance apart. Why do gravitational forces predominate at astronomical distances and electric forces at atomic distances?

Section 2.6

2.6 Four equal charges each of magnitude Q are placed at the corners of a square of side a. Find the resultant force on any one charge.

2.7 To the four charges of Problem 2.6 is added a charge q at the centre of the square. If all the Q's at the corners are positive, can a value of q be so chosen that the resultant force on all charges is zero? If so, what is its value? If two diagonally opposite charges were negative instead of positive, could a value for q now be found to make the resultant forces zero? If so, what?

2.8 Point charges $+4Q$ and $-Q$ are separated by a distance a. Show that the only positions where a third charge $+Q$ could be in equilibrium are along the line joining the first two, and find any such positions. Investigate the stability of the equilibrium by finding the direction of the force for a small displacement of $+Q$.

2.9 A thin rod of length L possesses a charge λ per unit length distributed uniformly along it. Find the force on a charge Q lying in the same line as the rod a distance h from the nearest end. Comment on the behaviour of the force as $L \to 0$ and as $L \to \infty$, in each case keeping h constant.

2.10 A horizontal thin rod of length L possesses a charge λ per unit length distributed uniformly along it. A point charge Q is situated vertically above one end of the rod at a height h above it. What force does Q exert on the rod?

2.11 A plane circular sheet of radius b has charge distributed uniformly over its surface with a density σ per unit area. Find the force exerted on a charge Q situated perpendicular distance a from the centre of the sheet. What does the force become as $b \to 0$ and $b \to \infty$? (See Fig. 3.4 for method of dividing the sheet into elementary strips.)

Section 2.7

2.12 Two fixed point charges $-Q$ are a distance $2a$ apart and a third point charge $+Q$ is situated midway between them. Find the potential energy of the system when $+Q$ is displaced a distance x along the line of charges, taking the zero for U when $+Q$ is at the midpoint. Investigate the stability of $+Q$ when $x=0$.

2.13 Show that the solution of Problem 2.8 can also be obtained by finding the potential energy U of the system for a general position of $+Q$ along the joining line and determining the positions for which U is a minimum.

2.14 Ionic crystals consist of a three-dimensional array of positive and negative ions and the calculation of its potential energy due to Coulomb forces is tedious. As a simpler example, consider a one-dimensional array consisting of a straight row of N point charges, alternately $+e$ and $-e$ and each a distance a from its nearest neighbours. If N is very large, find the potential energy of a charge in the middle of the row and of one at the end in the form $\alpha e^2/(4\pi\varepsilon_0 a)$. This sets upper and lower bounds to the Coulomb energy of the whole array. α is known as the **Madelung constant** for such systems.

2.15 Estimate the cohesive energy in aJ per NaCl 'molecule' in a crystal, assuming that the cohesive forces are entirely Coulombic and that the interionic distance is 0.28 nm. (The cohesive energy is the difference between the energy of the crystal and that of the separated ions.)

3

Electric field and electric potential

Although this chapter and the next two are largely concerned with consequences of Coulomb's law, new concepts make many problems simpler and more vivid. Two of these concepts, **electric field E** and **potential** V, are introduced in Secs 3.1 and 3.4, respectively, and related to each other in Secs 3.5 and 3.6. Both are field quantities, i.e. they have values at every point in a region, but they differ in that **E** is a vector quantity and V a scalar. We shall see how to calculate them when they are produced by given distributions of charge (Secs 3.2 and 3.5). However, the great advantages of introducing these new concepts first become evident when examining the motion of charges *in* **E**-fields. Section 3.7 deals with motion in free space and Sec. 3.8 with effects in conductors and insulators.

Finally, we take a look in detail at some special arrangements of charge. The **electric dipole** occurs frequently in many aspects of atomic physics and this warrants the full treatment given in Sec. 3.9. Moreover, most of the results are equally applicable to magnetic dipoles and will be needed in similar forms in Chapter 7. The quadrupole and higher multipoles are not normally introduced at this stage but the brief outline of their meaning and properties in Sec. 3.10 will be helpful in understanding some aspects of nuclear properties and spectra.

3.1 Electric field strength (E-field)

Concept It frequently happens in physics that we have a given set of charges fixed in position, and we wish to examine the behaviour of another charge placed at *various* points in relation to the given set. As a practical example, we might wish to know what forces are exerted on an electron passing through a set of fixed electrodes in a cathode-ray tube. As a theoretical example, we may propose models of atoms or molecules with fixed charge distributions and wish to know the forces exerted on an electron in its path around and near them. It is of course possible to use Coulomb's law to find the force for every position of the wandering electron, but this is a cumbersome way of proceeding and we prefer to introduce the idea of an electric field. This tells us what force *would be* exerted on a charge *if* it were placed at *any* point in a region. We now proceed to make this concept more precise.

Definition Suppose that a stationary charge Q_t placed at any point in a region experiences a force **F**. We then say that there is an **electric field** in the region whose **strength E** at the point is given by the force per unit positive charge or by

$$\boxed{\mathbf{E}=\mathbf{F}/Q_t \qquad \text{(definition of } \mathbf{E}\text{)}} \tag{3.1}$$

Some comments on this definition need to be made. In the first place, the condition that Q_t shall be stationary is necessary in order to exclude magnetic fields. We have seen that currents are moving charges which exert magnetic forces on each other. If we allowed Q_t to move, (3.1) would include these as well.

The second comment refers to the size of Q_t. Clearly (3.1) is intended to give values for **E** that are independent of the size of Q_t. If **E** is itself produced by a set of charges *rigidly fixed in space* then we shall see in the next section that it is indeed independent of Q_t since $|\mathbf{F}| \propto Q_t$. On the other hand, **E** may in practice be produced by conductors carrying charges *free to move* within them. In that case, the introduction of Q_t could cause a redistribution of these charges and the value of **E** will depend on the size of Q_t. While we *might* be interested in the field **E** for large values of Q_t the most useful concept is that of the electric field which would exist at a point due to the original undisturbed charges. We define this field by making Q_t smaller until its effect becomes negligible, when Comment C3.1

$$\mathbf{E}=\lim_{Q_t \to 0} \mathbf{F}/Q_t \tag{3.2}$$

Under these circumstances, Q_t is called a **test charge** and *the use of the symbol Q_t will always be taken to imply that its presence does not disturb the sources of the field in any way*, i.e. it will imply the limiting process of (3.2) where necessary. In fact, the definitions (3.1) and (3.2) are not used in the direct measurement of **E** but are used to calculate its value. In such a theoretical situation the size of Q_t can be *assumed* not to affect the calculation. It should also be noted that the original definition prior to (3.1) was carefully framed as the force *per* unit charge rather than the force *on* unit charge. The latter statement would imply the introduction of a charge of 1 C which is unrealistically large.

A final comment is that the charge Q_t could be possessed by a particle or body subject to other forces that are completely independent of Q_t, such as their weight. Comment C4.3

We should normally wish to exclude these forces from the **F** of (3.1).

The electric field strength defined by (3.1) or (3.2) will often be referred to simply as the **electric field** or the **E-field.**

Units of E The **SI unit** for **E** is clearly the newton per coulomb ($N\,C^{-1}$), but it is not the one used conventionally. We shall see in Sec. 3.6 that an equivalent unit is the **volt per metre** ($V\,m^{-1}$), and this is more usual. The magnitudes of electric fields encountered in practice are also discussed in Sec. 3.6.

Sources of E The concept of an electric field and the definitions of (3.1) and (3.2) are certainly suggested by the form of Coulomb's law. Because of this law, we see that stationary charges themselves will be sources giving rise to **E**-fields known as **electrostatic fields**. Our study of how such fields are calculated begins in Sec. 3.2.

However, the definitions do not in fact specify anything about the origin of **E** and we must keep in mind the possibility that **E-fields other than electrostatic ones** may exist. Indeed, a discussion in Sec. 6.1 shows that they *must* exist if steady currents are to be maintained in conducting circuits. We leave this point for the moment, noting only that if we wish to distinguish electrostatic from other types of field we shall do so by using the symbol $\mathbf{E}_Q$. The distinction is useful because electrostatic fields are unique among other **E**'s in being central force fields (from Sec. 2.1). The rest of this chapter and the next two are concerned entirely with electrostatic fields and no such distinction is necessary.

Visualization of E: lines of force It is often helpful to visualize electric fields in terms of lines of force, i.e. lines in space drawn so that the tangent to them at any point gives the **direction** of **E** at the point. This is the kind of representation used for any vector field (Appendix B.7). Figure 3.1 gives some examples which can be calculated from Coulomb's law and Problem 3.10 shows how it is possible to find the equations for **E**-lines in simple cases. This is not, however, a particularly useful exercise at this stage, and it is more important to be able to sketch the form of such lines using more general ideas developed later.

It is also possible to use the density of lines (the number crossing unit area perpendicular to the field) as an indication of the **magnitude** of **E**. It is clear from Fig. 3.1 that the stronger fields near the charges are accompanied by a greater density of lines, and this idea can be developed more exactly by using the concept of **flux** developed in Sec. 4.1. It is not advisable, however, to use lines of force as anything more than a pictorial aid (see Comment C.8.1).

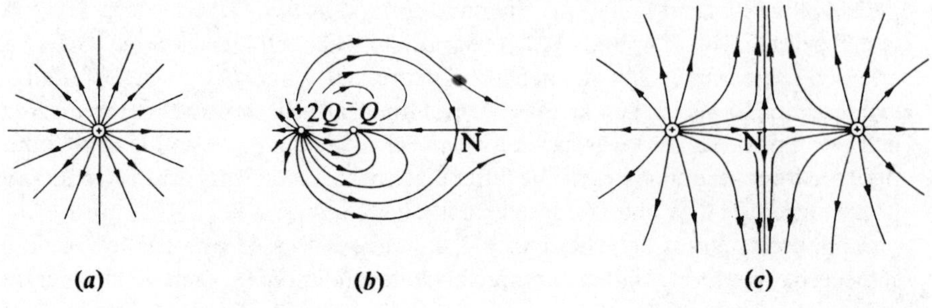

Figure 3.1 Lines of **E** due to simple systems of point charges: (a) a single positive charge; (b) point charges of $+2Q$ and $-Q$; (c) two equal positive charges. N denotes a neutral point where $\mathbf{E}=0$.

3.2 Finding E-fields due to charges

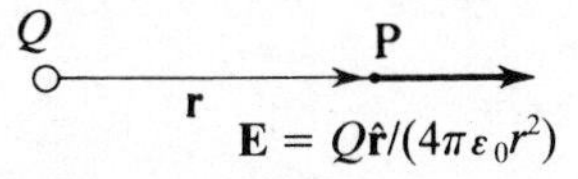

Figure 3.2 E-field at P due to a point charge.

Point charges When there is only a **single point charge** Q (Fig. 3.2), the E-field at a distance **r** from it is, from Coulomb's law,

point charge:
$$\mathbf{E}=\frac{Q}{4\pi\varepsilon_0 r^2}\hat{\mathbf{r}} \tag{3.3}$$

giving lines of force as in Fig. 3.1a. Thus the E-field at 1 cm from 1 μC is $(9\times10^9)\times10^{-6}/10^{-4}\,\mathrm{V\,m^{-1}}$ or $9\times10^7\,\mathrm{V\,m^{-1}}$. We comment on this value in Sec. 3.6.

When the **E**-field at a point is due to a **collection of point charges**, as in Fig. 3.3, then the principle of superposition gives

$$\mathbf{E}=\frac{Q_1}{4\pi\varepsilon_0 r_1^2}\hat{\mathbf{r}}_1+\frac{Q_2}{4\pi\varepsilon_0 r_2^2}\hat{\mathbf{r}}_2+\cdots=\sum_i\frac{Q_i}{4\pi\varepsilon_0 r_i^2}\hat{\mathbf{r}}_i \tag{3.4}$$

This is a formal statement meaning that the resultant is found by the addition of vectors just as with the electric forces in Sec. 2.4. In practice the addition may be carried out by repeated application of the parallelogram law as illustrated in Fig. 3.3a. This method can be shortened for more than two vectors by the polygon method of Fig. 3.3b. Alternatively, and most often, each separate field may be resolved into components along mutually perpendicular directions, for example, x, y and z. Each set of components can then be added, e.g. the x components to give a total E_x and similarly for the others. The resultant field may either be specified as a set of components like E_x, E_y and E_z or they may be added by vector addition to give **E** in magnitude and direction. Figure 3.3c shows how this method works in two dimensions.

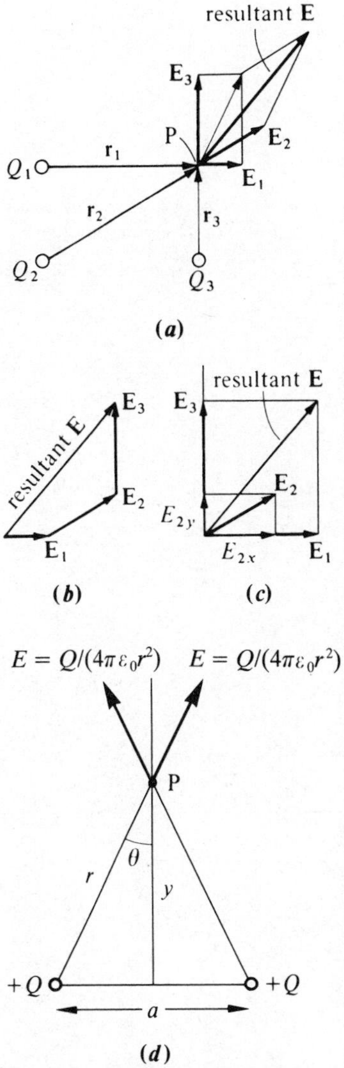

Figure 3.3 Superposition of E-fields. (a) Addition of $\mathbf{E}_1$, $\mathbf{E}_2$ and $\mathbf{E}_3$ by repeated application of parallelogram law; (b) the same **E**'s added by the polygon method; (c) the same **E**'s added by resolving and adding components; (d) an example discussed in the text.

Worked examples Occasionally the **E**-field at *one specific point* is required. For instance, in Fig. 2.7a of Chapter 2, suppose the field at the point occupied by the upper charge is required (this would be like Fig. 3.3d with $r=a$). There will be two fields in the directions shown, each equal to $Q/(4\pi\varepsilon_0 a^2)$. Resolving and adding as with the **F**'s in Sec. 2.6 shows that the resultant is a vertical field of magnitude $\sqrt{3}Q/(4\pi\varepsilon_0 a^2)$.

More often, however, it is useful to find the field at a *general point*, say along the perpendicular bisector of the line joining two charges as in Fig. 3.3d. The **E**-field is then obtained in terms of the variable y by resolving each separate one along y and adding, to give $2[Q/(4\pi\varepsilon_0 r^2)]\cos\theta$ or $2Qy/[4\pi\varepsilon_0(y^2+a^2/4)^{3/2}]$ along y. The components at right angles to y cancel. This is more useful than the first example since it gives the **E**-field for any y and not just for $y=\sqrt{3}a/2$.

Source and field points Before considering continuous distributions, we note a useful way of talking about field problems. The points P in Figs 3.2 and 3.3 are positions in space at which the field is required and are known as **field points**. The positions occupied by charges, Q in Fig. 3.2 and Q_1, Q_2, Q_3 in Fig. 3.3, are known as **source points** because the fields originate there. The reason why it is important to make the distinction at this stage is that we are about to embark on several

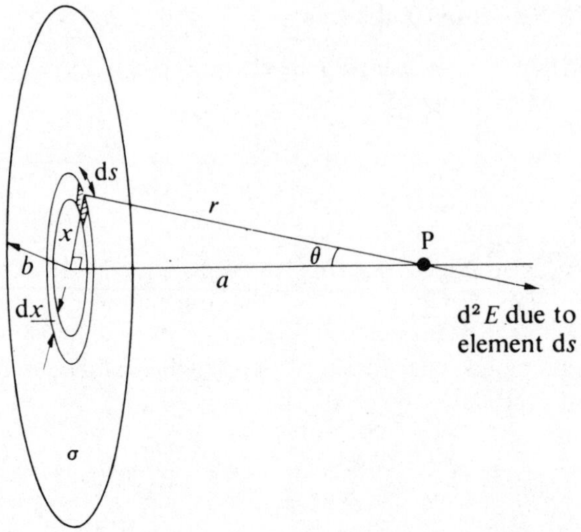

Figure 3.4 Finding the **E**-field at P due to a plane circular sheet of charge.

topics in which integration is involved. Sometimes, as in the immediately following paragraph, the integration will be over source points to find the total field at a single typical field point. At other times, as in Sec. 3.5, the integration will be through a set of field points with the sources fixed, and we should be clear about the difference.

Continuous distributions When a charge is distributed continuously, it must be divided into suitable elements each producing a field at the field point P. These fields must then be added vectorially as for point charges, but generally speaking the addition will involve integration. The only suitable method is then to resolve the field from the typical element into perpendicular components, adding each set of components by integration giving, say, E_x, E_y and E_z from which **E** can be found.

The method can be illustrated by the problem posed in Chapter 2 and Fig. 2.8. The charge Q is replaced by the field point P and the element dl produces an element of field dE at P in the direction shown for d**F**. The resolution and integration can be carried out exactly as for **F**. As we should expect, the field **E** is given by $\lambda L/[4\pi\varepsilon_0 a(a^2+L^2/4)^{1/2}]$ which is the value of $\mathbf{F}/Q$ in the example.

Worked example As a more challenging example, consider the plane circular sheet of radius b shown in Fig. 3.4. Charge is distributed uniformly over the sheet with a surface density σ and the **E**-field is required at P, a perpendicular distance a from the centre of the sheet. The plane is divided into annular strips as shown, a typical one having internal and external radii x and $x+\mathrm{d}x$, respectively. This choice of element is governed by the fact that all parts of it are at the same distance r from P. An element of the annulus of length ds carries a charge $\sigma\,\mathrm{d}s\,\mathrm{d}x$ and so produces a field

$$\mathrm{d}^2\mathbf{E}=\frac{\sigma\,\mathrm{d}s\,\mathrm{d}x}{4\pi\varepsilon_0 r^2} \qquad \text{along } r$$

The field due to the various elements of the ring forms a cone of semi-vertical angle θ at P, and their components perpendicular to a will cancel. The component parallel to a is $d^2E \cos\theta$ for every element of length ds, and hence for a total length of ring $2\pi x$ the field is

$$dE = \frac{\sigma 2\pi x dx \cos\theta}{4\pi\varepsilon_0 r^2} \quad \text{along } a$$

$$= \frac{\sigma a x dx}{2\varepsilon_0 r^3} \quad \text{along } a$$

The total field at P is thus perpendicular to the sheet in direction and is given in magnitude by integrating from zero to b:

$$E = \frac{\sigma a}{2\varepsilon_0}\int_0^b \frac{x\,dx}{(a^2+x^2)^{3/2}}$$

$$= \frac{\sigma a}{4\varepsilon_0}\int_{x=0}^{x=b} \frac{d(a^2+x^2)}{(a^2+x^2)^{3/2}}$$

$$= \frac{\sigma a}{2\varepsilon_0}\left[-\frac{1}{(a^2+x^2)^{1/2}}\right]_{x=0}^{x=b}$$

$$= \frac{\sigma a}{2\varepsilon_0}\left[\frac{1}{a} - \frac{1}{(a^2+b^2)^{1/2}}\right]$$

For a very large plane, $b \rightarrow \infty$ and in the limit

plane sheet of charge: $\mathbf{E} = \sigma/2\varepsilon_0$ along a

is the field due to a single plane sheet of charge of infinite extent. Note that it is independent of a.

This example shows how important it is to choose the appropriate set of elementary pieces to start with. For instance, it would be useless to attempt a solution by dividing the circular plane into parallel straight strips since different parts of each strip would be at different distances from P.

Uniform fields The example just completed shows that a very large plane sheet of charge with uniform surface density σ gives an electric field, **on both sides**, of $\sigma/(2\varepsilon_0)$ away from the sheet, the lines of force being as in Fig. 3.5a. This particular example illustrates a **uniform** electric field, which is one with the same magnitude

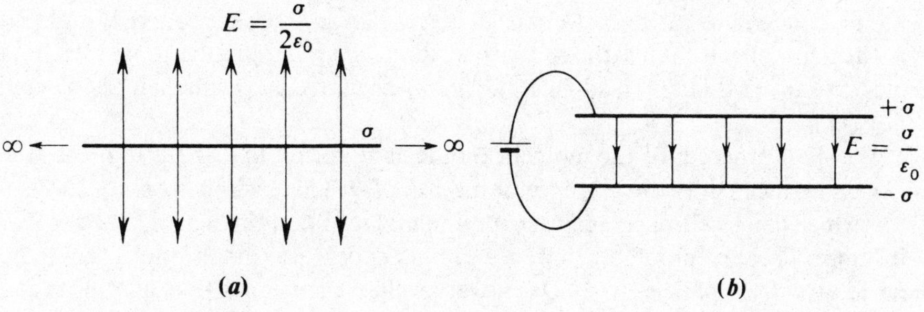

Figure 3.5 (a) **E**-field due to a large isolated sheet of charge with a uniform surface density σ; (b) **E**-field between charged parallel plates.

and direction at all points over a region of space. The charge distribution giving rise to it is, however, hardly a practical one because it consists of a sheet of charge isolated in space. Since uniform electric fields are frequently useful in practice, it is convenient to consider here how to produce one. A battery of cells is connected across *two* thin parallel metallic plates (Fig. 3.5b) having linear dimensions large compared with the distance apart. The battery transfers charge as usual, so that one plate is charged positively and the other negatively, the surface densities being approximately uniform. The fields produced by each plate are as in Fig. 3.5a with the directions of the lines reversed for the negative one. The fields will thus cancel in regions outside the plates but will reinforce between them, giving a total uniform field **E** of magnitude between parallel sheets of charge:

between parallel sheets of charge: $$E=\sigma/\varepsilon_0 \tag{3.5}$$

as the field between two very large plane parallel sheets of charge with surface densities $\pm\sigma$ (see Fig. 3.5b). This important result will be obtained by less roundabout methods in Chapter 4 when we deal with other methods for calculating fields.

3.3 Charges in E-fields – preliminary discussion

The whole point of introducing the concept of an electric field was to enable us to find what happens to a charge when placed at any point in a region of space where **E** is already given. So we now turn to this problem: assuming that **E** is known at a typical point, what effect does this have on a charge Q at the same point? The definitions (3.1) and (3.2) lead us to believe that Q would experience a force given by

force on Q in **E**: $$\mathbf{F}=Q\mathbf{E} \tag{3.6}$$

If the sources of **E** are charges rigidly fixed in their positions, then this expression is nothing more than a consequence of Coulomb's law. It would be a mistake, however, to assume that (3.6) is always intended to be used in that way and is nothing more than a rearrangement of (3.1). In the definitions of **E**, Q_t is a test charge which is stationary and, if necessary, small enough to have no effect on the sources of **E** whatever they may be. Equation (3.6), on the other hand, is intended to apply to an actual Q which in practice will be of given magnitude (e.g. on electrons, charged bodies, etc.) and may be moving. For example, suppose a charged conductor produces a field $\mathbf{E}_1$ at a nearby point, where the value of $\mathbf{E}_1$ has been calculated using an infinitesimal test charge. If a large charge Q is placed at the point, it will change the charge distribution on the conductor in such a way as to produce a different field $\mathbf{E}_2$ at the point. The force on Q will then be $Q\mathbf{E}_2$ and not $Q\mathbf{E}_1$.

We shall proceed for the moment on the assumption that the **E** to be used in (3.6) is unaffected by the presence or motion of Q. There will be consequences of this which can be checked against experimental results, as indicated in Sec. 3.7 for instance. These results show that, as we might expect, our assumptions lead to no detectable discrepancies when Q is no larger than a small multiple of e and is at a macroscopic distance from the sources of **E**. (It is entirely reasonable that where Q is as large as the source charges, or is small but very close to the sources, we might have to recalculate an effective **E** with Q in position. Section 4.8 will provide an example.)

We now have, through (3.6), a means of calculating the force on a charge if we know the **E**-field in which it is placed. Before considering what will happen as a result of this force, we shall find it an advantage to introduce the concept of potential difference. Discussion of the consequences of (3.6) is resumed in Secs 3.7 and 3.8.

3.4 Electric potential difference

Concept In Sec. 2.7, we explained why it is useful to express the interactions between charges in terms of energy as well as force, and formulae were derived for the potential energy of a system of charges. We now look at the advantages of making a similar transition from the **force** per unit charge (electric field) to the **potential energy** per unit charge, which we call the **electric potential**.

Why should this be something worth doing? One answer lies in some general results obtained in Sec. 2.7. We saw there that, when one charge is moved in the electric field of another by an external force, the work done by this force was equal to the increase in potential energy and had two important properties: (a) it was independent of the path taken and (b) it reversed in sign when the direction of the path was reversed. Because of superposition, these properties will hold with any system of charges and make the electrostatic **E**-field an example of what is called a *conservative field* [see Eq. (3.26)].

In the situation just described the potential energy is a property of the whole system including the sources of **E** as well as the moved charge. However, we can adopt the idea of a test charge Q_t as in Sec. 3.1 and use it to explore the changes in potential energy as it moves in the region of fixed sources, say along a path like AB of Fig. 3.6. We may then reasonably talk of the potential energy of the test charge rather than of the whole system. If we form the work done *per unit charge* in moving Q_t, we shall have a quantity that depends only on the initial and final positions and on the sources of the field, and is thus no longer associated with any particular moving charge.

This new quantity is known as the **potential difference** or **p.d.** Unlike **E**, it is a scalar quantity, with magnitude only: this is yet another useful and important property because scalars are much easier to deal with than vectors.

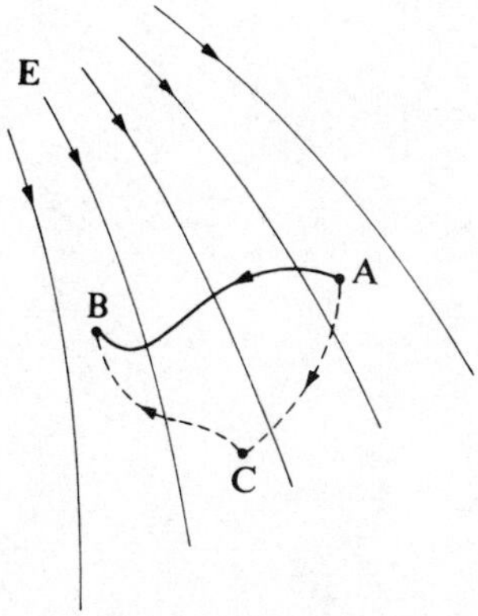

Figure 3.6 A path AB for the definition of potential difference. Because the work done in taking a test charge from A to B is independent of the path, $W_{AB} = W_{AC} + W_{CB}$.

Definition A formal definition of potential difference is required on the lines discussed. We start with a region in which there is an electrostatic field and consider two points A and B joined by any path (Fig. 3.6). A small test charge Q_t is moved along the path *from* A *to* B with constant speed by an external force which has to do work W_{AB}. The increase in potential energy of the system is then also $U_{AB} = W_{AB}$ because there is no increase in kinetic energy. The **potential difference** between B and A is defined by

$$\boxed{V_{AB} = W_{AB}/Q_t \qquad \text{(definition of } V_{AB}\text{)}} \tag{3.7}$$

or, in words, the work done per unit charge by an external force in taking a positive test charge from A to B. The point B is at the higher potential when positive work is done. *Thus we expect the potential to rise for displacements in opposition to* **E** *and to fall for displacements along* **E**.

Strictly, only differences of potential can be defined, as we should expect since they are potential energies per unit charge. We can, however, always choose a zero at some point A, and the potential difference between B and A becomes the **potential** at B with respect to a zero at A:

$$V_B = W_{AB}/Q_t \qquad \text{(zero at A)} \tag{3.8}$$

The zero should always be specified in this way when using the term *potential.*

The potential difference V_{AB} defined by (3.7) is independent of a zero. Suppose in Fig. 3.6 we take an arbitrary point C as a zero. The potentials at A and B are then $V_A = W_{CA}/Q_t$ and $V_B = W_{CB}/Q_t$ respectively. Because of path independence, $W_{AC} + W_{CB} = W_{AB}$, while $W_{AC} = -W_{CA}$. Using all these equations we have that the potential difference between A and B with C as zero is $V_B - V_A = (W_{CB} - W_{CA})Q_t = (W_{AC} + W_{CB})Q_t = W_{AB}/Q_t$ which, by (3.7), is V_{AB} independently of C.

To summarize: in defining a **potential** at any point in an electrostatic field we must choose a zero and use (3.8); but if a **potential difference** is required, (3.7) can be used independently of a zero. The V's in all cases are path-independent, and the potential of (3.8) is a scalar field quantity.

The **SI unit** of potential difference will be the $\mathrm{J\,C^{-1}}$ (joule per coulomb) known as the **volt**, symbol V. Thus the potential difference between two points is 1 V when 1 J of work is done per coulomb of charge taken from one point to the other.

Charge at a potential V We ask, as we did with **E** in Sec. 3.3, what happens if an actual charge is moved from A to B? Equation (3.7) suggests that an external force would need to do work $W_{AB} = QV_{AB}$ and that therefore this would be the increase U_{AB} in the potential energy of Q in the field. Thus, both W_{AB} and U_{AB} are given by

$$W_{AB} = U_{AB} = QV_{AB} = (V_B - V_A) = U_B - U_A \tag{3.9}$$

If Q *falls* through a potential difference V_{AB} then there will be a corresponding *decrease* QV_{AB} in its potential energy and $U_B < U_A$. Provided we choose zeros at the same point for U and V, we could use (3.9) in the form:

$$\text{Potential energy of } Q \text{ at potential } V,\ U = QV \qquad (\text{zero for } U \text{ where } V = 0) \tag{3.10}$$

If Q wanders through a field so that V varies, its potential energy also varies and presumably other forms of energy will change in amount so as to conserve the total. No more than this can be said about the energy balance until the exact situation is specified.

A warning similar to that in Sec. 3.3 should be sounded here. Equation (3.9) is not just a rearrangement of (3.7), in which Q_t is a test charge. The same remarks and conditions apply to the use of (3.9) and (3.10) as to (3.6). However, (3.10) will apply exactly for a vanishingly small charge. Sometimes we wish to use a theoretical model in which infinitesimal elements of charge dQ are moved through a potential difference V. The work done dW and the increment in potential energy dU are then given without qualification by $V\,dQ$. Hence

Work done dW in carrying dQ through $V=$
increment in potential energy $dU = V\,dQ$ (3.11)

3.5 Finding V due to charges

In this section we shall be using two mathematical ideas not encountered so far. One is the **scalar product** of two vectors (Appendix B.2). This involves a concise notation used when one vector is to be resolved in the direction of the other and then multiplied by the magnitude of the latter. This is precisely what is done when calculating the work done by a force $\mathbf{F}$ in moving a particle and causing a displacement $\mathbf{L}$. If θ is the angle between $\mathbf{F}$ and $\mathbf{L}$, it is only the $F\cos\theta$ component that does the work $FL\cos\theta$, and this latter quantity is written $\mathbf{F}\cdot\mathbf{L}$.

The second idea is that of a **line integral** (Appendix A.4). This is not very different from an ordinary integral $\int y\,dx$, regarded as the sum of products $y\,dx$ taken from one value of x to another. The line integral is a similar sum taken along a path in space and is another concept useful in connection with work. If a force $\mathbf{F}$ moves a particle along a path $\mathbf{L}$ through successive small displacements $d\mathbf{L}$, then the total work will be the integral of $F\,dL\cos\theta$ from one end of the path to the other. In vector notation, this can be written as $\int \mathbf{F}\cdot d\mathbf{L}$.

***V* in terms of electrostatic E-field** To calculate a potential difference $V_B - V_A$ between two points A and B in an electrostatic field we must find the work done by an external force in moving a test charge Q_t from A to B. In Fig. 3.7 the electric field at some element of the path $d\mathbf{L}$ is $\mathbf{E}$, so that an external force $\mathbf{F} = -Q_t\mathbf{E}$ is required to move Q_t with a constant speed. In moving the charge over the element $d\mathbf{L}$, the work done will be $F\,dL\cos\theta'$ or $\mathbf{F}\cdot d\mathbf{L}$ as discussed above. The work done over the whole path is then

$$W_{AB} = \int_A^B \mathbf{F}\cdot d\mathbf{L} = \int_A^B -Q_t\mathbf{E}\cdot d\mathbf{L}$$

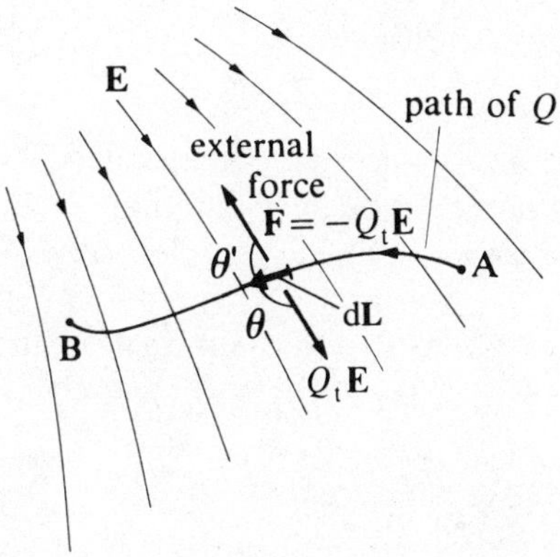

Figure 3.7 Calculation of potential difference V in terms of the **E**-field. The work done by the external force in taking a test charge along the increment of path **dL** is **F**·**dL**.

It follows from the definition of V in (3.7) that the potential difference between A and B is

$$V_B - V_A = \int_A^B -\mathbf{E}\cdot d\mathbf{L} = \int_A^B -E \cos\theta \, dL \tag{3.12}$$

where θ is the angle between **E** and d**L** as in Fig. 3.7. Moreover, if A is chosen as the zero of potential, the potential at B is

$$V_B = \int_A^B -\mathbf{E}\cdot d\mathbf{L} \qquad \text{(zero at A)} \tag{3.13}$$

Equations (3.12) and (3.13) enable V to be determined if **E** is known, but they should not be applied without thought—the path of integration may cover several regions over which **E** is given by different expressions [see Eqs (4.8) and (4.9)]. **For electrostatic fields the potential difference does not depend on the path, which is therefore usually chosen to make the integration as simple as possible.** The integrals in (3.12) and (3.13) are line integrals as discussed above.

***V* due to point charges** To evaluate the potential at a point P a distance r_P from a single point charge Q we could choose a path along a line of force so that the angle between **E** and d**L** is always π. Instead we choose a more general path to demonstrate the path-independence of V *without appealing to the result of Sec. 2.7.* Since **E** at any point is given by $Q\hat{\mathbf{r}}/(4\pi\varepsilon_0 r^2)$, Eq. (3.13) gives

$$V_P = \int_\infty^{r_P} -\frac{Q\hat{\mathbf{r}}\cdot d\mathbf{L}}{4\pi\varepsilon_0 r^2}$$

Figure 3.8 shows that the scalar product $\hat{\mathbf{r}}\cdot d\mathbf{L}$ is always equal to $dL\cos\theta$ and therefore to dr, i.e. that only radial displacements from Q contribute to the work and potential energy and not transverse ones. Hence the potential at P is

$$V_P = \int_\infty^{r_P} -\frac{Q\,dr}{4\pi\varepsilon_0 r^2}$$

or

point charge:
$$V_P = \frac{Q}{4\pi\varepsilon_0 r_P} \qquad \text{(zero at } \infty) \tag{3.14}$$

irrespective of the path taken. [Had we assumed path-independence, we could have integrated along the line of force through P so that in Eq. (3.12) $E = Q/(4\pi\varepsilon_0 r^2)$, $dL = -dr$ and $\cos\theta = \cos\pi = -1$. This gives the same integral as in (3.14).]

The potential difference between two points P_1 and P_2 distances r_1 and r_2 from Q will be

$$V_1 - V_2 = \frac{Q}{4\pi\varepsilon_0}\left(\frac{1}{r_1} - \frac{1}{r_2}\right) \tag{3.15}$$

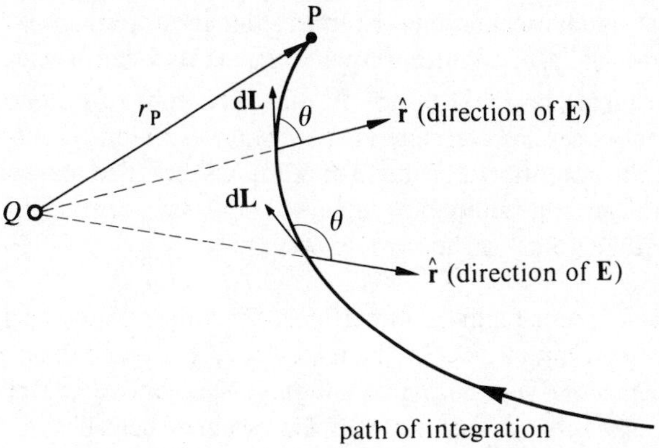

Figure 3.8 Finding the potential at P due to a charge Q. At all elements of the path $\mathbf{dL}\cdot\hat{\mathbf{r}} = dL\cos\theta = dr$, the increment in the radial distance from Q. Note that when θ is obtuse, $\cos\theta$ is negative and $dL\cos\theta$ will still be in the same direction as $\hat{\mathbf{r}}$.

For a collection of discrete charges, the potentials at a point due to the individual charges may be added algebraically since they are scalar quantities and the electric fields superpose. Thus, at a point a distance r_1 from Q_1, r_2 from Q_2, etc., set of point charges:

$$V = Q_1/(4\pi\varepsilon_0 r_1) + Q_2/(4\pi\varepsilon_0 r_2) + \cdots = \frac{1}{4\pi\varepsilon_0}\sum_i \frac{Q_i}{r_i} \qquad \text{(zero at } \infty) \quad (3.16)$$

Worked example In the arrangement shown in Fig. 3.3d, the potential at P is simply $2Q/(4\pi\varepsilon_0 r)$ or $Q/(2\pi\varepsilon_0 r)$. Whether this is a useful form of the potential depends on the context of the problem. If the variation along the y direction is required, for example, it might be better to express V in terms of y as $Q/[2\pi\varepsilon_0(y^2 + a^2/4)^{1/2}]$.

V due to continuous distributions of charge Suppose charge is distributed over a volume τ with a density ρ at any point within the volume. The potential at any field point P due to an element of the charge $dQ = \rho\, d\tau$ is given by Eq. (3.14), i.e. $\rho\, d\tau/(4\pi\varepsilon_0 r)$. To obtain the total potential at P this expression is integrated over the whole volume τ. The same method is adopted for surface and line distributions of charge with densities σ and λ, respectively. Thus the potentials at a field point P due to volume, surface and line densities of charge may be written charge distributions:

$$V_\sigma = \int_\tau \frac{\rho\, d\tau}{4\pi\varepsilon_0 r}; \qquad V_\sigma = \int_S \frac{\sigma\, dS}{4\pi\varepsilon_0 r}; \qquad V_\lambda = \int_L \frac{\lambda\, dL}{4\pi\varepsilon_0 r} \qquad (3.17)$$

all for a zero at infinity.

Worked example Suppose we wish to find the potential at P in Fig. 3.4 due to the disc of charge. If we divide the disc into annular rings as we did for the calculation of E, the potential due to each ring at P is given by Eq. (3.14) as $2\pi\sigma x\, dx/(4\pi\varepsilon_0 r)$. Since this is a scalar, the contributions from the elementary rings simply add together, which means that we have to integrate from $x=0$ to $x=b$. Expressing r as $(a^2+x^2)^{1/2}$, the integration with respect to x can be carried out and yields V in the form $[\sigma/(2\pi\varepsilon_0)](\sqrt{a^2+b^2}-a)$. The zero of potential is at infinity. Note that since this expression tends to infinity as b tends to infinity, we cannot use it to find the potential at P due to an infinite sheet. The reason for this is that we have used a point at infinity as our zero. The answer to such a problem lies in choosing another point as the zero of potential.

Obtaining *V* from E For some highly symmetrical distributions it is easier to find E and use $V=\int -\mathbf{E}\cdot d\mathbf{L}$ than to use (3.17). However, *care must be taken in using infinity as a zero when the distribution is not finite* (see above example). An important special case is that of a uniform field. If such a field has a constant magnitude E in the constant direction x, say, then Eq. (3.13) integrates at once to give

uniform E:
$$V=C-Ex \tag{3.18}$$

where C is a constant depending on the point chosen for the zero of V.

3.6 Potential gradient and E-field

Potential gradient At the beginning of the previous section we obtained a general expression for V in terms of **E**, as a line integral. We now turn to the inverse problem of finding **E** in terms of V and it comes as no surprise to find the relationship a differential one.

For a small element of the path $d\mathbf{L}$ in Fig. 3.7, we found that the work done per unit positive charge by an external force was $-\mathbf{E}\cdot d\mathbf{L}$. This is an increment of the potential difference dV so that

$$dV=-\mathbf{E}\cdot d\mathbf{L} \tag{3.19}$$

However, the scalar product is equal to $E_L\,dL$, where E_L is the component of **E** in the direction L. Hence

$$\boxed{E_L=-\partial V/\partial L} \tag{3.20}$$

where partial derivatives occur because V in general depends on three coordinates (the partial derivative means that the differentiation is carried out keeping all variables constant other than the one under consideration: see Appendix A.3).

In words, Eq. (3.20) states that *the resolved part of* **E** *in any direction is equal to the negative rate of change of V in that direction*, i.e. is equal to the negative potential gradient $-\partial V/\partial L$. In cases where V can be expressed in terms of one coordinate only, r say, a graph of V against r can be plotted in what is called a **potential diagram**. The slope of the graph at any point, dV/dr, gives the value of $-E$ at the point.

However, V is in general a function of more than one variable. If it is expressed in terms of Cartesian coordinates x, y, z, then

$$\boxed{E_x=-\frac{\partial V}{\partial x}; \qquad E_y=-\frac{\partial V}{\partial y}; \qquad E_z=-\frac{\partial V}{\partial z}} \tag{3.21}$$

If, instead, V can be specified in terms of plane polar coordinates (r,θ) only, then the radial and tangential components of **E** are

$$E_r = -\frac{\partial V}{\partial r}; \qquad E_\theta = -\frac{1}{r}\frac{\partial V}{\partial \theta} \qquad (3.22)$$

The expressions (3.20) to (3.22) can be used to obtain **E** from V when it is easier to calculate the latter independently. An example illustrating the use of (3.21) is given below, while (3.22) is used in finding the fields due to an electric dipole (Sec. 3.9).

The mathematical process of partial differentiation is discussed in Appendix A.3. Polar coordinates are explained in Appendix A.2. We shall see in Sec. 4.4 that (3.21) and (3.22) can be concisely written as $\mathbf{E} = -\textbf{grad}\, V$ or $\mathbf{E} = -\nabla V$, where ∇ is a vector operator. It is not, however, necessary to use ∇ at this stage.

Worked example If V is known to vary in a limited region of space according to the expression $V = K(x^2 - 2y^2)$ where K is a constant, then $E_x = -\partial V/\partial x = -2Kx$ and $Ey = 4Ky$.

Unit of E Electric fields are measured by finding potential gradients rather than forces on charges, so that the SI unit for **E** is most conveniently expressed in $\mathrm{V\,m^{-1}}$ rather than the equivalent $\mathrm{N\,C^{-1}}$. In practice, E-fields may be quoted in $\mathrm{V\,cm^{-1}}$, $\mathrm{kV\,mm^{-1}}$, etc., depending on their magnitudes. In Sec. 3.2, we showed that the E-field at 1 cm from a charge of 1 μC was $9 \times 10^7\ \mathrm{V\,m^{-1}}$ or $90\ \mathrm{kV\,mm^{-1}}$. Our everyday knowledge of voltage tells us that this is a strong field by normal laboratory standards and again shows how large even a net charge of 1 μC is. As another numerical example, the uniform field between parallel plates separated by 2 cm as in Fig. 3.9b is $2\ \mathrm{V\,cm^{-1}}$ or $200\ \mathrm{V\,m^{-1}}$ if a potential difference of 4 V is maintained across the plates.

Equipotentials and lines of force Equipotential surfaces are those connecting points at the same potential. In a diagram we can only show two-dimensional cross-sections and Fig. 3.9 shows equipotentials drawn at equal intervals of V for a point charge and a uniform field.

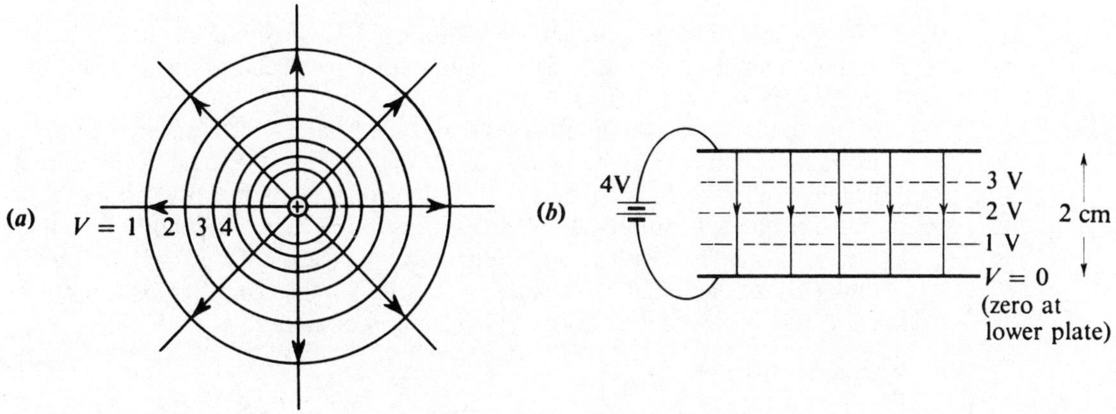

Figure 3.9 Lines of **E** and cross-sections through equipotential surfaces for (a) a point charge, (b) a uniform field.

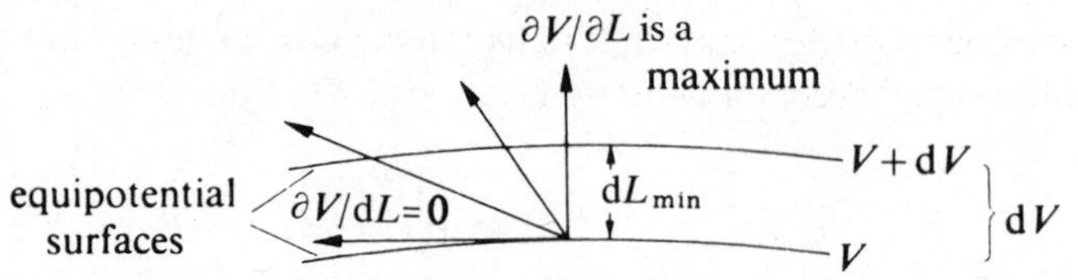

Figure 3.10 dV is the same for all paths between the surfaces, but dL is a minimum for the direction normal to both. Hence the potential gradient $\partial V/\partial L$ is a maximum in this direction.

When the lines of **E** are plotted on these diagrams, it becomes clear that they are everywhere at right angles to the equipotentials. More accurately, we ought to say that the **E** lines are *normal* to the equipotentials since the latter are really surfaces. It is fairly easy to see that this must be a general property by the following argument: in a direction tangential to an equipotential surface there is no change in V. It follows from (3.20) that there is no component of **E** in that direction, so that there can only be a component normal to the surface. In other words, the **lines of E are always normal to equipotentials.**

Another important property can be derived from Fig. 3.10, where we have drawn two equipotentials differing in potential by dV. This means that when we travel from one surface to the other, by any path of length dL, the potential difference is always dV. However, the shortest path is that along the normal to both surfaces, denoted by dL_{min}. It follows that the rate of change of V, dV/dL, is greatest along this direction, which is also that of **E**. Furthermore, if equipotentials are drawn at equal intervals of V as they are in Fig. 3.9a, the above result indicates that **more closely spaced equipotentials will imply a greater E-field.**

Potential diagrams The equipotentials are analogous to the contours on a map, where the lines of greatest slope are at right angles to the contours while the steepness of the slope is indicated by the closeness of the contour lines. We can carry the analogy still further by drawing a cross-section through Fig. 3.9a as if it were a hill, with the result shown in Fig. 3.11. Here, the curve of V against r which appears as the profile of the hill is a *potential diagram*. We shall have occasion later to use such diagrams in which the vertical axis may represent either potential as in Fig. 3.11 or potential energy (see Sec. 4.2 and Appendix A.6).

3.7 Action of E-fields on charged particles (*in vacuo*)

We now return to the problem we left in Sec. 3.3: what happens to charges in regions where E-fields exist? In this section we deal with charges in free space and in the next with conductors and insulators.

In free space, we assume either (1) that the E-field in which Q is placed can be maintained constant by its sources or (2) that Q is small enough to produce a negligible effect on **E**. In either case, the value of **E** is unaffected by the presence of Q and the force is simply Q**E**. What happens to the charge depends upon other forces. We deal first with an application in which Q**E** is balanced by mechanical forces in the determination of electronic charge. We then consider cases in which there are no other forces, so that the equation of motion is

motion of Q in **E**: $$Q\mathbf{E}=m\mathbf{a} \qquad (3.23)$$

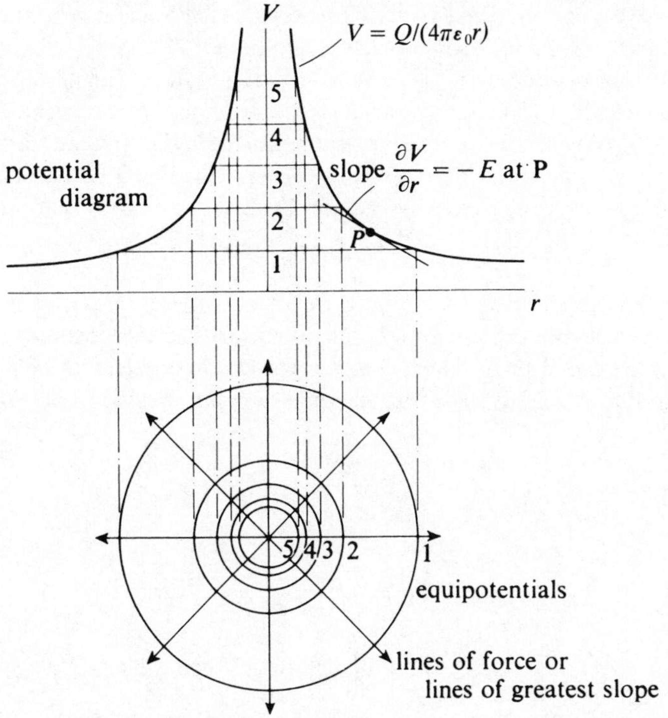

Figure 3.11 Potential diagram for a positive charge. The corresponding equipotentials drawn in the lower diagram reveal their analogy to contours of equal height on a map.

The acceleration **a** is thus $Q\mathbf{E}/m$, where m is the mass of the particle carrying the charge. If **E** is known as a function of position, the path of the particle can be calculated using standard methods of kinematics.

Determination of the electronic charge Methods for the direct determination of e depend upon measurements of the force exerted by an applied electric field. Classical experiments (like those of Millikan and of Hopper and Laby, 1941) used oil drops falling through air with their terminal velocity so that the resultant force on them was zero. If uncharged, the only forces would be those due to gravity (downwards) and the viscosity of air (upwards), giving a particular value for the terminal velocity. Ionization of the air causes some drops to take up an integral number n of charges e and thus to experience an additional force $ne\mathbf{E}$: this gives the drop a different terminal velocity. Observation of the two terminal velocities for a number of different drops gave a series of values for ne, from which e could be calculated.

The most accurate value for e to date is that given by Cohen and Taylor (1986):

$$e = (1.602\,177\,33 \pm 0.000\,000\,49) \times 10^{-19}\,\mathrm{C}$$

while experiments have also shown that the charge on the electron and proton do not differ in magnitude by more than 1 part in 10^{20}. Comment C3.2

Energy of accelerated charges: the electron-volt We now turn to the motion of charged particles in the *absence* of other forces.

It is reasonable to neglect the force due to the weight of particles when they are

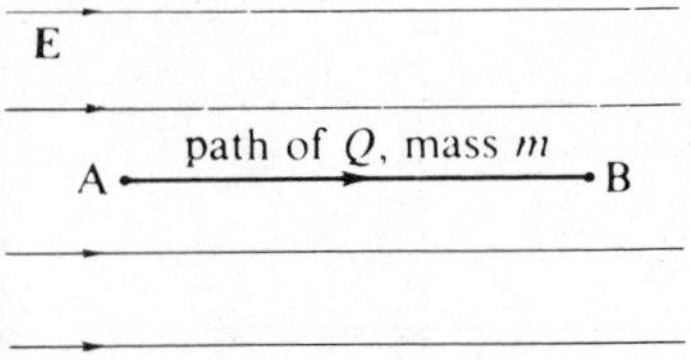

Figure 3.12 Acceleration of a charged particle in an **E**-field.

no more massive than ions. Even in moderate electric fields it is easy to show that the force eE exceeds mg for ions by factors of about 10^9.

Firstly, we consider the motion between two points in an **E**-field in terms of potential. According to (3.23), a positively charged particle accelerates in the direction of **E** and in doing so moves from a region of higher potential to one of lower (a negative charge will do the opposite). Thus in Fig. 3.12 the change in potential energy in moving from A to B under the action of **E** only is, from (3.10),

$$U_{AB} = U_B - U_A = Q(V_B - V_A) \qquad (3.24)$$

This change is compensated by a change in kinetic energy K as we now see. Because the path is always parallel to **E**, the potential difference is simply the line integral of $-E$ from B to A. Thus, using dr to denote an element of the path directed from B to A, and ds to denote an equal element directed from A to B, we have

$$\begin{aligned} U_B - U_A = Q(V_B - V_A) = Q\int_B^A -E\,\mathrm{d}r &= \int_B^A +QE\,\mathrm{d}s \\ &= \int_B^A ma\,\mathrm{d}s \qquad \text{by (3.23)} \\ &= \int_B^A mv\,\mathrm{d}v \qquad \text{using } a = v\,\mathrm{d}v/\mathrm{d}s \\ &= \tfrac{1}{2}mv_A^2 - \tfrac{1}{2}mv_B^2 \qquad (3.25) \end{aligned}$$

Hence

$$U_B - U_A = K_A - K_B \qquad (3.26)$$

It follows that $U_A + K_A$ is equal to $U_B + K_B$, i.e. the total mechanical energy is the same at A and B and is thus conserved as a constant of the motion, a result which would be expected because Coulomb forces are central.

It is a general result from mechanics that motion of a particle under a central force (i.e. directed towards a fixed centre) is conservative, i.e. the total mechanical energy, or the sum of kinetic and potential energy, is conserved.

Although (3.25) was derived for a charge moving parallel to **E**, it is a general result. If the charge possesses a velocity component *perpendicular* to **E**, u say, it would remain unchanged since there is no force in the direction. The contribution $\frac{1}{2}mu^2$ to the kinetic energy merely adds on to both terms on the right-hand side of (3.25). Since $(u^2 + v_A^2)$ and $(u^2 + v_B^2)$ are the squares of the resultant velocities at the two endpoints, it is clear the (3.25) will apply to *any* trajectory between any two points, A and B, where v_A and v_B are then the resultant velocities and not just the components parallel to **E**. (If the velocities v are not small compared with c, the velocity of light, then relativistic mechanics must be applied and the mass m cannot be taken as constant in deriving (3.25). Nevertheless, the final result is unaffected because the force $Q\mathbf{E}$ is replaced by $\mathrm{d}(m\mathbf{v})/\mathrm{d}t$ and this leads to the correct relativistic expression for K instead of the classical $\frac{1}{2}mv^2$. Thus $U + K$ is still a constant of the motion.)

In atomic and nuclear physics, charged particles of high kinetic energies are produced by acceleration in electric fields. Rather than quote the *gain* in kinetic energy in joules of such particles, it is more convenient to use as a unit the equal *loss* in potential energy of an electron falling *in vacuo* through a potential difference of one volt, known as the **electron-volt**, symbol eV. For higher energies 1 MeV and 1 GeV (see notes on prefixes inside the front cover) are often used as units, although none of these is an SI unit. The latter remains the joule, related to the eV by $1\,\text{eV} = 1.6 \times 10^{-19}\,\text{J}$.

Trajectories in uniform E-fields A charged particle in a uniform **E**-field will move in exactly the same way as a projectile does in a uniform gravitational field: the path will in general be a parabola (remember that in the case of the charged particle, we are neglecting the force due to its weight, which is reasonable for electrons and ions).

If the initial velocity of the charged particle is **parallel** to **E** (Fig. 3.13a), acceleration occurs without deflection and the increase in kinetic energy will be, by (3.26):

$$\Delta K = \tfrac{1}{2}mv_2^2 - \tfrac{1}{2}mv_1^2 = Q(V_1 - V_2) \quad (3.27)$$

The signs look after themselves in this equation: e.g. for a positive Q, acceleration will occur if V_2 is less than V_1, thus correctly making ΔK positive.

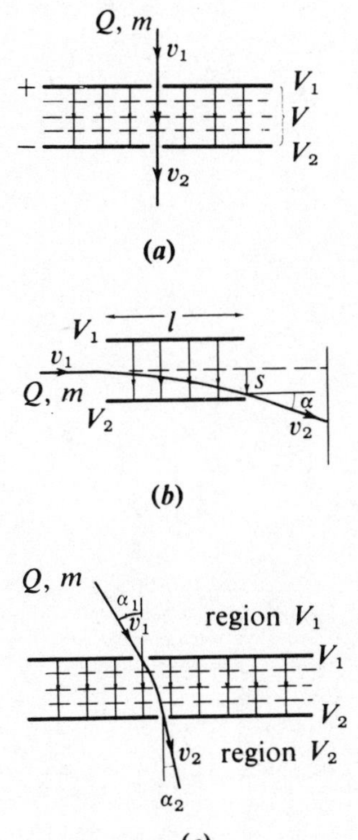

Figure 3.13 Positive charged particles in uniform **E**-fields (a) projected parallel to **E**, (b) projected perpendicular to **E**, (c) projected at an angle of incidence α_1 to **E**.

General applications This result can be used:

1. to measure the kinetic energy of charges emitted from a source by retarding them with an opposing field until $v_2 = 0$, when the *initial* kinetic energy is given by $Q(V_2 - V_1)$. For example, if a beam of monovalent ions emitted from a source to a collector is just cut off by an opposing potential of 1000 V, the initial kinetic energy is 1 keV.
2. to produce a beam of charges of approximately known and uniform energy from a source emitting them with negligible velocity: in this case, $v_1 = 0$ and the *final* energy is $Q(V_1 - V_2)$. As an example, an accelerating potential of 10 000 V will produce a 10 keV beam of electrons or monovalent ions.

Application to ion propulsion for artificial satellites The particles accelerated by a field have gained kinetic energy, but have also gained linear momentum. The conservation of momentum is a universal law, so that there must be a reaction on the source of the field which creates an equal momentum in the opposite direction. This is completely negligible in terrestrial installations, but can be used in space as a form of 'jet engine' using charged particles as the 'fuel'. Possible applications include the maintenance of artificial satellites in orbit and control of their attitude, and even interplanetary transfer.

A typical arrangement is shown in Fig. 3.14. There are two problems to be overcome. The first is the fact that the emission of positive ions, say, leaves the satellite charged negatively and this would ultimately destroy the accelerating field. The difficulty is overcome by having secondary sources emitting negative ions to maintain the neutrality. These secondary sources also enable the second problem to be solved: the ejected positive ions form a space charge at the rear

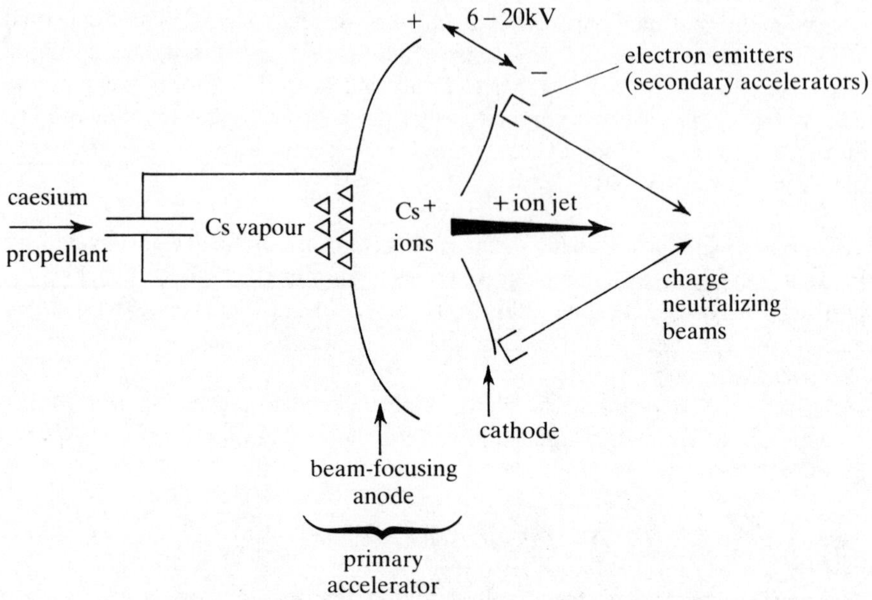

Figure 3.14 Principle of ion propulsion for spacecraft (adapted, with permission, from McGraw-Hill, 1987).

which repels further emission and splays out the beam. The secondary beams are so directed as to neutralize the space charge. The reader is referred to McGraw-Hill (1987) for further information.

When the initial velocity v_1 is **perpendicular** to **E** (Fig. 3.13b) a particle spends a time l/v_1 undergoing a transverse acceleration QE/m. The final velocity is thus the resultant of v_1 perpendicular to **E** and QEl/m_1 parallel to **E** so that

$$\tan\alpha = QEl/mv_1^2 = Q\frac{l}{d}\frac{V}{2K} \tag{3.28}$$

where K is the initial kinetic energy and the field is produced by a potential difference V between plates a distance d apart. The linear deflection in the field is given by

$$s = \frac{1}{2}\frac{QE}{m}\frac{l^2}{v_1^2} = Q\frac{l^2}{d}\frac{V}{4K} \tag{3.29}$$

General applications Transverse electric fields were used by Thomson (1897, 1899) to determine e/m for cathode rays and thermionically emitted charges, although a beam of known v_1 was required. Present-day applications include:

1. the deflection of electron beams in cathode-ray tubes by an amount proportional to V, an externally applied voltage. The deflecting plates have to be mounted inside the tube.
2. the electrostatic separation of particles with different energies from a single source. Those with the same K will suffer the same deflection provided their charges are the same. If, for example, a beam already has a well-defined

momentum from a magnetic analyser (Sec. 7.5), electrostatic deflection enables a beam to be picked out having a K and hence a velocity that are also well defined.

For **oblique incidence** (Fig. 3.13c) the particle enters the field from an equipotential region V_1 with velocity v_1 and leaves with velocity v_2 in a region with potential V_2. The following apply:

$$\tfrac{1}{2}mv_1^2 + QV_1 = \tfrac{1}{2}mv_2^2 + QV_2 \qquad \text{from (3.25)} \tag{3.30}$$

$$v_1 \sin\alpha_1 = v_2 \sin\alpha_2 \tag{3.31}$$

If the charges are originally produced by acceleration from rest at a point where the potential is taken as zero, then the total energy on both sides of (3.30) is zero and $v_1^2 = (2Q/m)V_1$ and $v_2^2 = (2Q/m)V_2$. Using this together with (3.31), we obtain the following expression:

$$\frac{\sin\alpha_1}{\sin\alpha_2} = \frac{v_2}{v_1} = \sqrt{\frac{V_2}{V_1}} \tag{3.32}$$

Trajectories in non-uniform E-fields If the thickness of the uniform field in Fig. 3.13c is very small, the path of the particle is clearly refracted in the same manner as a ray of light at the interface between two media. Equation (3.32) shows that the quantity corresponding to refractive index is v or $V^{1/2}$. Non-uniform fields are analogous to optical media with continuously variable refractive indices and the surfaces labelled V_1 and V_2 in the figure can be considered as two typical equipotentials whose separation is small enough for the field between them to be considered approximately uniform. Equation (3.32) thus gives the relation between the direction of the particle (defined by α), its velocity and the potential for the path in any electric field. Beams of charged particles can therefore be deviated in ways analogous to the refraction of light.

Applications to electron optics When the trajectories in non-uniform **E**-fields are those of electrons, we have a branch of **electron optics**. Note, however, that whereas positive charges increase their velocity as V decreases, electrons will suffer an increase in v as V increases (in contrast to light). Figure 3.15a illustrates the difference in behaviour of positive and negative charges.

The focusing properties of fields, which we shall encounter several times, are important for two reasons. The first is that beams of charge of a single sign are naturally defocusing because the electrostatic repulsion exceeds the magnetic attraction (Problem 8.4), and the second is that a slit used to define a beam has a finite width so that even with what is effectively a point source, the beam is naturally divergent. Focusing with electrostatic lenses is common in cathode-ray tubes and a typical form is shown in Fig. 3.15b and c. The two cylindrical electrodes produce equipotential surfaces and lines of force as shown and, even though one half is divergent, the whole is convergent because the electrons spend more time in that part of the field with the lower potential. In a cathode-ray tube the focusing is controlled by altering the relative potentials of the two cylinders.

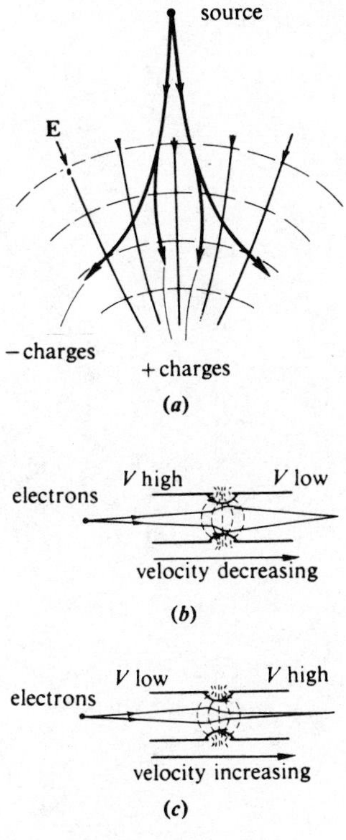

Figure 3.15 Charged particles in non-uniform **E**-fields. (a) Showing the different behaviour of positive and negative charges in a convergent field; (b) and (c) an electron lens. The broken lines are equipotentials.

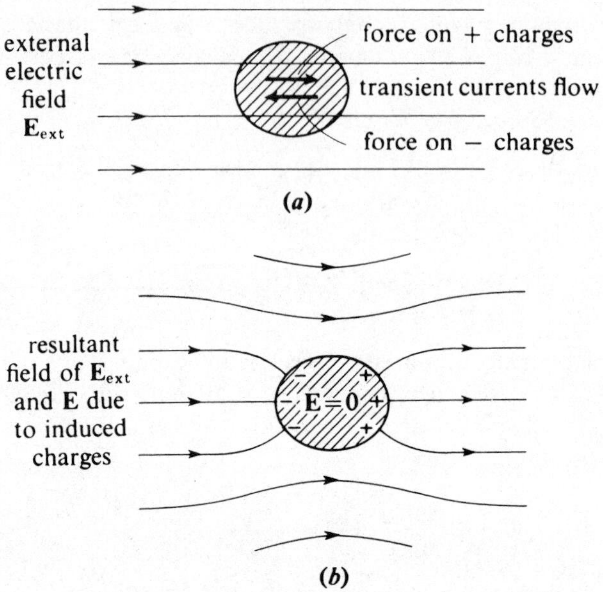

Figure 3.16 An isolated conductor in an electric field. (a) Initial state when first placed in the field; (b) final state: electrostatic induction has occurred.

3.8 Action of E-fields on conductors and insulators

Conductors We know already that a conductor contains charges free to move within it so that, if an electric field is established in conducting material, charges will move as long as the field exists.

If the conductor is isolated, Fig. 3.16 illustrates what will happen when it is placed in a field. As soon as **E** is established, one part of the conductor is at a higher potential than another and any free positive charges will move in the direction of **E** (and any negative charges in the opposite direction). Whatever the sign of the moving charges, the result is the same: charges reach the surface of the conductor and can go no further. They collect and produce a field within the conductor which opposes the applied field. This process continues until, within the material, there is no resultant field. Static conditions will then again prevail.

We have thus accounted for electrostatic induction mentioned in Sec. 1.1 and in addition have shown that **a conductor carrying only static charge can have no electric field within its material**, and hence that throughout the volume of such a conductor there is no potential difference. **A conductor carrying static charge is an equipotential volume and its surface an equipotential surface.** A similar argument shows that the electric field from the surface is at right angles to it at all points.

(It should be noted here that the **E**-field being discussed is a macroscopic or smoothed-out mean value. On an atomic scale the microscopic **E**-field will fluctuate violently from one point to another. This section has shown that over finite times and distances a charge in a conductor experiences an average field of zero.)

If any number of conductors at different potentials are connected by conducting wires, the potential gradients in the wires will cause currents to flow until the potentials are the same. Thus, **connecting together charged conductors will equalize their potentials, provided no batteries or other sources equivalent to batteries are present** (Fig. 3.17).

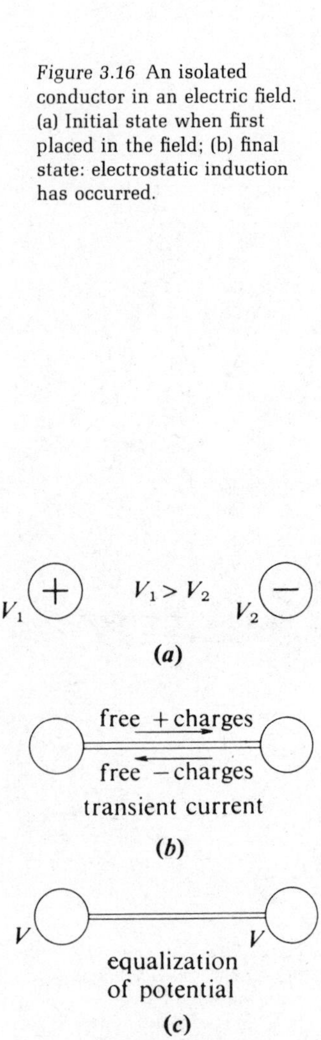

Figure 3.17 Interconnection of conductors equalizes their potentials with the flow of a transient current.

Sources of electromotance and of steady currents While the potentials of the two conductors of Fig. 3.17 are becoming equal, a transient current flows in the connecting wire, as we saw in Sec. 1.3. If a device could be arranged to take the charge from one conductor back to the other as fast as it arrives, the electrostatic field and potential difference between them would be maintained and a steady current set up in the connecting wire (Fig. 3.18). Such a device, known as a **source of electromotance** or as a **generator**, we shall examine in more detail in Sec. 6.1, merely noting for the moment that a voltaic cell is one example of such a source, that these sources cause charge to flow round a complete circuit or closed path and that, although the potential energy lost by a charge Q in the wire is $Q(V_1 - V_2)$ as before, no kinetic energy is now gained but heat is produced instead.

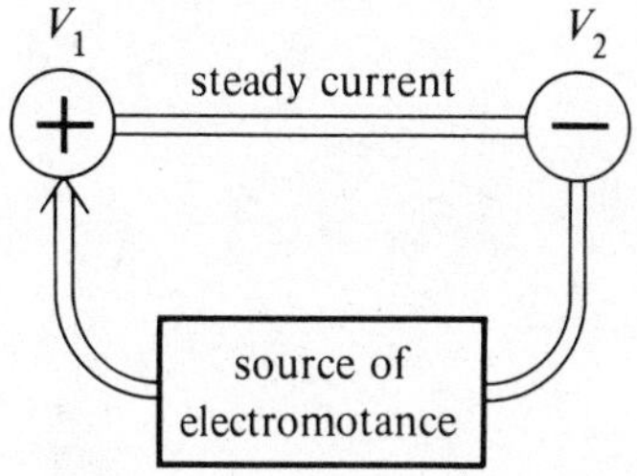

Figure 3.18 Action of a source of electromotance (e.m.f.). By conveying charges through itself, it maintains the potential difference $V_1 - V_2$ and a steady current in the wire. The arrow indicates the direction in which positive charges move.

Insulators A perfect insulator would contain no free charges, but real insulators inevitably contain a few and these will move under the action of an electric field just as in a conductor. However, for a good insulator the number is so small that it would take weeks to establish an appreciable amount of electrostatic induction. It therefore rarely makes sense to talk of 'the potential' of an insulator since it does not become an equipotential volume in a reasonable time.

If the few free charges are ignored, the only effect of placing an insulator in an electric field is the immediate production of a small and limited movement of + and − bound charges in opposite directions (we know that this occurs, from Sec. 1.1). This displacement of charge, which disappears on removal of the field, is known as **polarization**. While polarization may seem a much less dramatic effect than electrostatic induction, the presence of dielectrics has a marked effect on electric fields. We shall meet an example in connection with capacitors in Chapter 5, but postpone a detailed treatment until Chapter 11.

Applications to electrostatic separation The different behaviour of conductors and insulators in an electrostatic field can be used in the practical separation of particles in mineralogy or industrial processes. In **triboelectric separation**, use is made of the fact that the charging of insulators by friction may produce different levels of positive or negative charging (this goes right back to the original electrostatic observations of Sec. 1.1). A rotating drum or chute produces the friction and charges of opposite sign are separated by an electric field. In **induction separators**, conducting particles are charged by induction and removed by applying an electric field, while insulator particles are unaffected. See Cross (1987) for further details.

3.9 The electric dipole

Both in theory and practice, there are many examples of systems which are electrically neutral overall but still produce an electric field and are affected by an electric field. Molecules and thunderclouds are examples at the extreme ends of a range that includes ordinary insulators as well as some extraordinary ones (Sec. 11.10).

The simplest system of this type is the **electric dipole**, which consists of two equal and opposite charges $\pm Q$ separated by a distance l. The properties of many systems can be explained in terms of such dipoles, so that it is important to explore the properties of the dipoles themselves.

It might be asked how a dipole can exist at all in view of the attractive force between its charges. The answer lies in the existence of other forces that produce the displacement of charges in the first place. Examples in atomic systems are illustrated in Fig. 11.10.

The ideal dipole An ideal dipole is one in which l is negligible compared with the distance of other charges and with the distance of points where we wish to calculate fields. At the same time, Q is correspondingly large so that the product Ql is still finite.

We first calculate the potential and **E**-field produced by an ideal dipole at a field point, and define a quantity called the dipole moment. Then we place a dipole in an **E**-field and find the forces and torques exerted on it, and what its potential energy is. Finally, we look at the interaction between two dipoles, one in the **E**-field of the other.

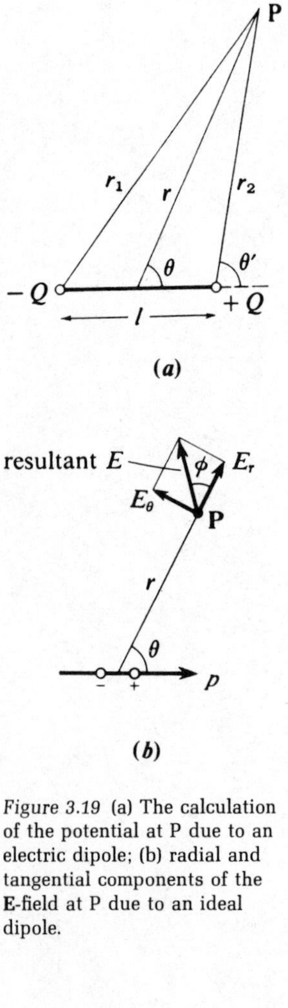

Figure 3.19 (a) The calculation of the potential at P due to an electric dipole; (b) radial and tangential components of the **E**-field at P due to an ideal dipole.

Potential and E-field due to an ideal dipole The method adopted is as follows: taking a non-ideal dipole as in Fig. 3.19a, we find the potential V_P at a point P having polar coordinates (r, θ) with respect to the centre. We then proceed to the limit in which l becomes negligible in comparison with r, giving V_P due to an ideal dipole. The components of **E** are then obtained from V_P by taking its gradient using Eqs (3.22).

From Eq. (3.16)

$$\begin{aligned} V_P &= Q/4\pi\varepsilon_0 r_2 - Q/4\pi\varepsilon_0 r_1 \qquad \text{(zero at } \infty) \\ &= Q(r_1 - r_2)/4\pi\varepsilon_0 r_1 r_2 \end{aligned} \tag{3.33}$$

But $r_1^2 = r_2^2 + l^2 + 2r_2 l \cos\theta'$ or $r_1^2 - r_2^2 = l(l + 2r_2\cos\theta')$ and, using this to substitute for $(r_1 - r_2)$ in (3.33), we obtain

$$V_P = \frac{Ql(l + 2r_2\cos\theta')}{4\pi\varepsilon_0 r_1 r_2 (r_1 + r_2)}$$

This is exact, but for an ideal dipole $(l \ll r) r_1 \to r, r_2 \to r$ and $\theta' \to \theta$, and in the limit

$$V_P = \frac{Ql\cos\theta}{4\pi\varepsilon_0 r^2} \tag{3.34}$$

[A more elegant method of deriving this expression is as follows. Let the potential at P due to $+Q$ alone by $V_Q = Q/(4\pi\varepsilon_0 r)$. The potential at P due to the $-Q$ is then $-V_Q - \mathrm{d}V_Q$, where $\mathrm{d}V_Q$ is the increment in V_Q produced by displaying Q a distance l along x, where $x = r\cos\theta$. The resultant potential at P due to both charges is thus simply $-\mathrm{d}V_Q$. Hence

$$V_P = -\mathrm{d}V_Q = -l\,\partial[(Q/(4\pi\varepsilon_0 r)]/\partial x = -\frac{Ql}{4\pi\varepsilon_0}\frac{\partial(1/r)}{\partial x} = \frac{Ql}{4\pi\varepsilon_0 r^2}\frac{\partial r}{\partial x} \tag{3.35}$$

This is the same as (3.34) because, as shown in Eq. (A.12) of Appendix A.3, $\partial r/\partial x = x/r = \cos\theta$.]

The quantity Ql is the electric moment of the dipole or its **electric dipole moment**, denoted by p. If a positive *direction* is associated with p along the line of l

from $-Q$ to $+Q$, then we shall see that it behaves as a vector quantity in that it can be resolved into components whose fields superpose. Hence we have

$$\mathbf{p}=Q\mathbf{l} \qquad \text{(definition of } \mathbf{p}\text{)} \tag{3.36}$$

where $\mathbf{l}$ is the vector displacement of $+Q$ from $-Q$. The **SI unit** of $\mathbf{p}$ is the C m.

Equation (3.34) can now be written

dipole:
$$V_P=\frac{p\cos\theta}{4\pi\varepsilon_0 r^2} \tag{3.37}$$

If we multiply numerator and denominator by r and use the scalar product $\mathbf{p}\cdot\mathbf{r}=pr\cos\theta$, then

dipole:
$$V_P=\frac{\mathbf{p}\cdot\mathbf{r}}{4\pi\varepsilon_0 r^3}=\frac{\mathbf{p}\cdot\hat{\mathbf{r}}}{4\pi\varepsilon_0 r^2} \tag{3.38}$$

where $\hat{\mathbf{r}}$ is a unit vector along $\mathbf{r}$. From Eq. (3.22) applied to (3.37), the components of $\mathbf{E}$ are

dipole:
$$E_r=\frac{2p\cos\theta}{4\pi\varepsilon_0 r^3}; \qquad E_\theta=\frac{p\sin\theta}{4\pi\varepsilon_0 r^3} \tag{3.39}$$

as in Fig. 3.19b. The resultant **E**-field has a magnitude $p(1+3\cos^2\theta)^{1/2}/(4\pi\varepsilon_0 r^3)$ at an angle ϕ to r such that $\tan\phi=\frac{1}{2}\tan\theta$.

The **E**-fields of dipoles are shown in Fig. 3.20a and b, the former also including equipotential surfaces. Note particularly that the plane bisecting the line joining the charges and perpendicular to it has $V=0$ (zero at ∞). Equations (3.39) show that $\mathbf{p}$ can be resolved into $p\cos\theta$ along $\mathbf{r}$ (end-on position) and $p\sin\theta$ perpendicular to $\mathbf{r}$ (broadside position): see Problem 3.5.

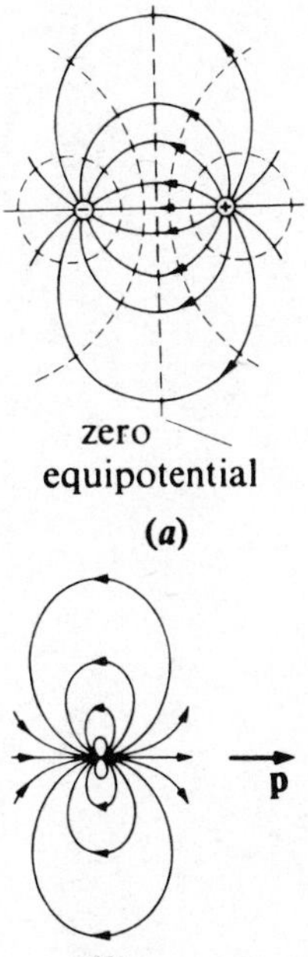

Figure 3.20 (a) The **E**-field (full lines) and equipotentials (broken lines) of an electric dipole; (b) the **E**-field of an ideal dipole with moment **p** in the direction shown but situated at the centre of the system of lines of force. Note that these figures are two-dimensional cross-sections: the complete system of **E**-lines and equipotential surfaces are obtained by rotating the diagrams about the central horizontal line which is parallel to **p**.

Dipole in a uniform E-field In a uniform field **E**, the forces on the charges of the dipole are both $Q\mathbf{E}$ but are in opposite directions (Fig. 3.21), thus constituting a couple of moment $QEl\sin\theta$. The angle θ is that between **p** and **E**, and so the torque is

$$T_\theta=pE\sin\theta$$

in a sense to decrease θ.

A couple or torque not only has a magnitude equal to the product of the forces and the distance between them but a direction normal to the plane containing the forces and the distance (compare this with the idea of area as a vector in Sec. 1.6). If the torque is given this direction, it is found to behave like a vector quantity, but one that is a little different from vectors we have met so far. It is called an **axial vector** since its direction is along an axis of rotation rather than along a direction specified by a single quantity like force or velocity.

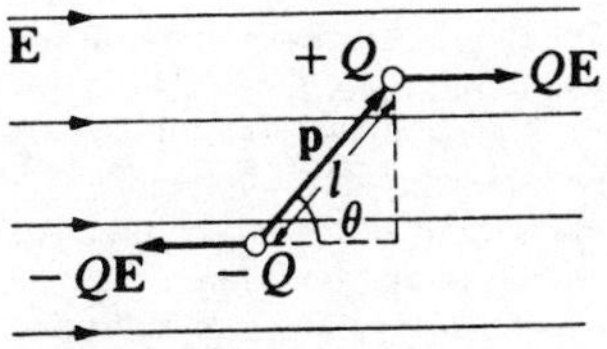

Figure 3.21 The torque on an electric dipole in a uniform **E**-field.

More precisely, we need to choose a positive direction and to do that we need the concept of the **right-handed screw**. Nearly all real screws have a right-handed thread: this means that if we look at it as we screw it into a piece of wood, say, we have to rotate it clockwise for it to move away from us into the wood. The same applies to the direction of a torque and other axial vectors: its positive direction is that in which a right-handed screw would progress if it were rotated in the same sense as the torque. In Fig. 3.21, this direction is into the plane of the page.

This can be used to write Eq. (3.40) in vector form using the idea of the **vector product** defined in Appendix B.5, which is useful in all cases where moments occur and will be vital when we come to magnetic forces that have a more complicated three-dimensional arrangement than electric forces. The vector product of two vectors **A** and **B** with an angle θ between them ($\theta \leq 180°$) is written $\mathbf{A} \times \mathbf{B}$: it is a vector with magnitude $AB \sin\theta$ and direction perpendicular to the plane containing **A** and **B**. The positive direction is that in which a right-handed screw would proceed if rotated from **A** to **B** through θ. Applying this to the vectors **p** and **E** in Fig. 3.21, we see that the torque **T** of Eq. (3.40) can be written as $\mathbf{p} \times \mathbf{E}$.

Since the total force is zero, an electric dipole of moment **p** in a uniform field **E** experiences a total force and torque given by

dipole in uniform **E**: $$\boxed{\mathbf{F}=0; \quad \mathbf{T}=\mathbf{p}\times\mathbf{E}} \tag{3.41}$$

Dipole in a non-uniform E-field (may be omitted in a first course) The general case here is complex, so let us begin with a dipole lying along the x direction in a field also in the x direction but not uniform (Fig. 3.22a). If the field at $-Q$ has the magnitude E_x, then for a dipole of length $\mathrm{d}x$ the field at $+Q$ will be $E_x+\mathrm{d}E_x$ or $E_x+(\mathrm{d}E_x/\mathrm{d}x)\,\mathrm{d}x$. The forces are now not equal and opposite but have a resultant $Q\,\mathrm{d}x(\mathrm{d}E_x/\mathrm{d}x)$. Since $Q\,\mathrm{d}x$ is the dipole moment p, the force exerted by a non-uniform field on a dipole lying along the field is

$$F_x = p\frac{\mathrm{d}E_x}{\mathrm{d}x} \tag{3.42}$$

In general, the displacement of $+Q$ from $-Q$ will not lie along x but will be, say, $\mathrm{d}l$ with components $\mathrm{d}l_x$, $\mathrm{d}l_y$ and $\mathrm{d}l_z$. Moreover, the field will also have components E_x, E_y and E_z, all of which may vary. Consider the force in the x direction, F_x, which must be due to variations in E_x. If we use E_x to denote the x component of **E** at $-Q$, then that at $+Q$ will be $E_x+(\partial E_x/\partial x)\,\mathrm{d}x + (\partial E_x/\partial y)\,\mathrm{d}y + (\partial E_x/\partial z)\,\mathrm{d}z$ and, although the QE_x's cancel, there is a resultant force on a general dipole in a non-uniform electric field whose x component is

dipole in non-uniform **E**: $$F_x = p_x\left(\frac{\partial E_x}{\partial x}\right) + p_y\left(\frac{\partial E_x}{\partial y}\right) + p_z\left(\frac{\partial E_x}{\partial z}\right) \tag{3.43}$$

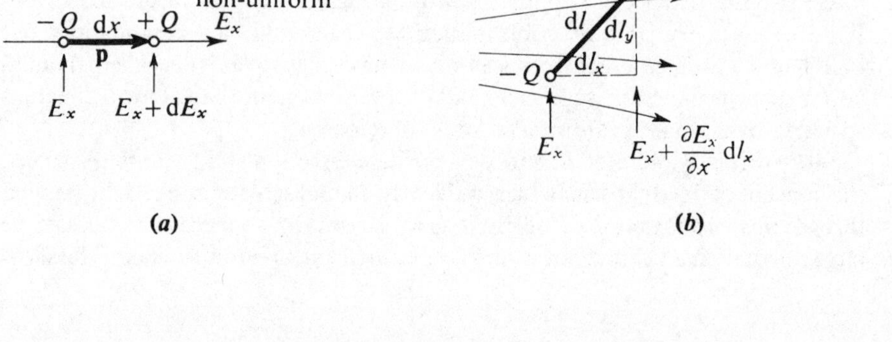

Figure 3.22 Electric dipoles in non-uniform **E**-fields. (a) Forces when **p** is in the same direction as **E**; (b) a more general situation: the dipole is in the xy plane. More generally still, there will be a displacement $\mathrm{d}l_z$, which will add a fourth term to the x-field at $+Q$.

because $Q\,dl_x = p_x$, the x component of $\mathbf{p}$, etc. Expressions for F_y and F_z are similar to (3.43).

The torque on an ideal dipole in a non-uniform field is still given by $T_\theta = pE\sin\theta$ because both forces are still $Q\mathbf{E}$, neglecting terms of a smaller order of magnitude.

Potential energy of a dipole in an E-field Suppose a dipole of moment $\mathbf{p}$ makes an angle θ with an electric field $\mathbf{E}$ as in Fig. 3.23. To calculate its potential energy, a zero position must be chosen. The natural choice would seem to be when $\theta = 0$, but it is more usual to take $\theta = 90°$ for the following reason: since the equipotentials are at right angles to $\mathbf{E}$, the charges comprising the dipole lie on the same equipotential and any work done in establishing the dipole in the field is not counted (the work done in bringing up $+Q$ is equal and opposite to that done in bringing up $-Q$).

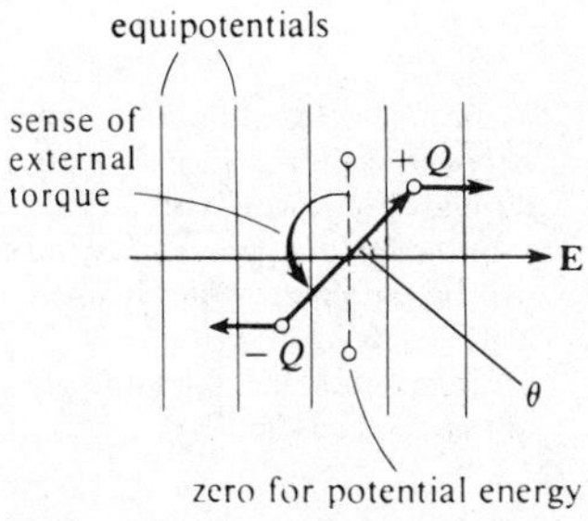

Figure 3.23 Finding the potential energy of a dipole in an **E**-field by calculating the work done in rotating it from a standard position to its final position.

Starting then from $\theta = 90°$, we now have to calculate the work done by external forces in rotating the dipole to its final position. By Eq. (3.40), the external torque necessary has a moment $pE\sin\alpha$ for a general angle α and will do work $pE\sin\alpha\,d\alpha$ in rotating the dipole through a small angle $d\alpha$. The work done in rotating it from $90°$ to θ is

$$W = \int_{\pi/2}^{\theta} pE\sin\alpha\,d\alpha = [-pE\cos\alpha]_{\pi/2}^{\theta}$$

and the potential energy of a dipole $\mathbf{p}$ in a field $\mathbf{E}$ is thus

dipole: $$U = -pE\cos\theta = -\mathbf{p}\cdot\mathbf{E} \qquad (3.44)$$

the negative sign occurring because work is done *on* the external torque.

In Sec. 2.7 we saw that there was a general relationship between forces and potential energy. F_L, the component of the force on an object in any direction $\mathbf{L}$, is given by $-\partial U/\partial L$ ($E_L = -dV/\partial L$ is a special case when the object is a unit test charge). By an exactly analogous argument, the torque T_θ about any fixed axis is given by $-\partial U/\partial\theta$, where θ is the angular displacement about the axis. Thus in general

dipole in $\mathbf{E}$: $$F_L = -\partial U/\partial L; \qquad T_\theta = -\partial U/\partial\theta \qquad (3.45)$$

It is easily seen that the force and torque on a dipole in a uniform **E**-field are given by applying (3.45) to (3.44) with p and E constant.

The rest of this chapter is concerned with more detail of charge distributions and is not essential to the general development of electromagnetism. It can therefore be omitted in a first course without fear of missing any vital results.

Mutual action between dipoles The general formulae for the forces and torques between two ideal dipoles are complex and we shall limit the discussion to coplanar dipoles as in Fig. 3.24a. The two moments are $\mathbf{p}_1$ and $\mathbf{p}_2$ making angles of θ_1 and θ_2, respectively, with the line joining them. The distance apart is r. The first step is to obtain the potential energy U of the two by regarding $\mathbf{p}_2$ as in the field $\mathbf{E}$ of $\mathbf{p}_1$. For this, use an origin at $\mathbf{p}_1$ and x and y axes along and perpendicular to r, respectively. From (3.43), $U = -\mathbf{p}\cdot\mathbf{E}$ and this can be written as

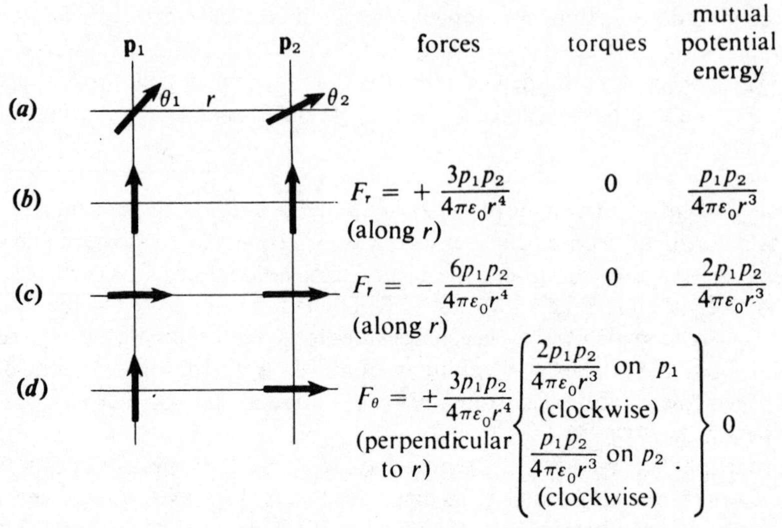

Figure 3.24 The mutual action between coplanar ideal dipoles. The forces and torques for the general case (a) can be obtained by applying (3.45) to (3.46). The arrangements of (b), (c) and (d) are special cases of (a).

$-p_xE_x - p_yE_y$ using Eq. (B.4) of Appendix B. In this expression, p_x and p_y are the components of p_2 ($p_2 \cos\theta_2$, $p_2 \sin\theta_2$), while E_x is the radial component of the field due to p_1 [i.e. $(2p_1 \cos\theta_1)/(4\pi\varepsilon_0 r_3)$ by (3.39)]. Similarly, E_y is the transverse component $-(p_1 \sin\theta_1)/(4\pi\varepsilon_0 r^3)$. Gathering all this together:

$$U = [p_1p_2/(4\pi\varepsilon_0 r^3)](\sin\theta_1 \sin\theta_2 - 2\cos\theta_1 \cos\theta_2) \qquad (3.46)$$

The results quoted in the diagram for the special cases of Fig. 3.24b, c and d can easily be derived from this by using (3.45) and inserting the appropriate values for θ_1 and θ_2. Note that the torques $-\partial U/\partial\theta_1$ and $-\partial U/\partial\theta_2$ may not sum to zero and must be balanced by that due to the F_θ's, i.e. rF_θ. Thus, in general, F_θ's are given by $-(1/r)(\partial U/\partial\theta_1 + \partial U/\partial\theta_2)$.

The resultant interaction in the general arrangement of Fig. 3.24a can also be obtained from (3.46) by using (3.45). An important feature of all these formulae is the way in which the forces and potentials vary with r. In order to emphasize this point, Fig. 3.25 lists the principal feature of the expressions for **F** and U in all possible interactions between charges and dipoles.

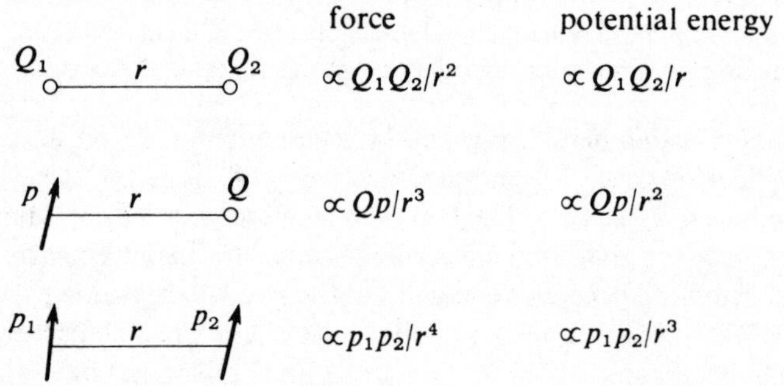

Figure 3.25 Summary of interactions between charges and dipoles, and the dependence upon r.

3.10 The quadrupole and general arrangements of charge

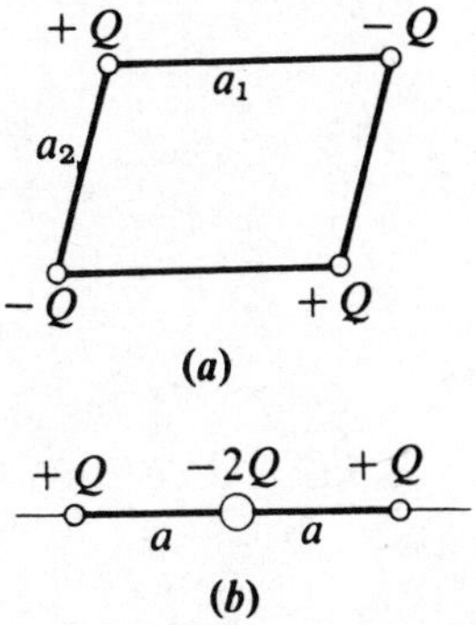

Figure 3.26 (a) A general quadrupole; (b) a positive linear quadrupole.

At very large distances, any collection of charges occupying a finite volume acts as if it were a point charge of ΣQ, where the sum is algebraic: the **E**-field falls off as the inverse square of the distance. If it happens that $\Sigma Q=0$, we now see that **E** will not necessarily be zero because the system could be a dipole whose field falls off as $1/r^3$. This $1/r^3$ field *may* be present when $\Sigma Q \neq 0$, but it falls off so much more rapidly than the inverse square that it can often be neglected.

Suppose now that we go one stage further and consider systems for which not only does $\Sigma Q=0$ (a scalar sum) but $\Sigma \mathbf{p}=0$ as well (a vector sum). Some arrangements for which this is true are shown in Fig. 3.26 and are known as **quadrupoles**. In particular, Fig. 3.26(b) is a **linear quadrupole**. General quadrupoles are dealt with in more advanced treatises and we consider here only the linear variety so as to exhibit the way in which **E** and V now behave.

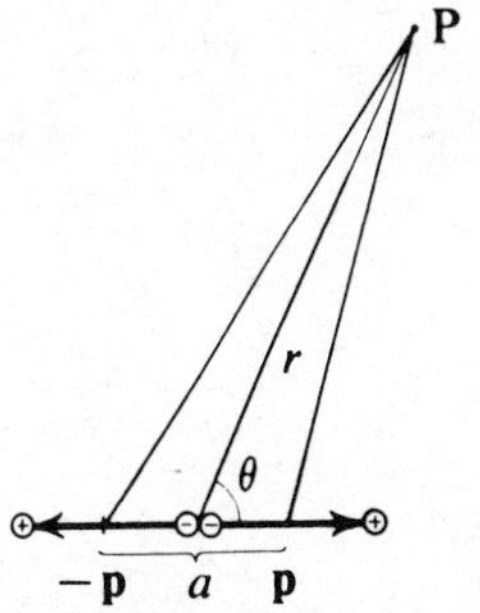

Figure 3.27 Calculation of the potential due to a linear quadrupole treated as two opposite dipoles in line.

Potential due to a linear quadrupole This can be obtained in an elementary way as in Problem 3.28, but since we need the expressions for an ideal quadrupole we shall use the method which led to (3.35). Regard the quadrupole as derived from two equal but opposite dipoles of moment **p** placed in line as in Fig. 3.27 and separated by a distance a. We need the increment in $(p\cos\theta)/4\pi\varepsilon_0 r^2$ for a small displacement a of **p** in the direction of x. Thus as in Eq. (3.35):

$$V_{\mathrm{P}}=-a\frac{\partial(p\cos\theta/4\pi\varepsilon_0 r^2)}{\partial x}=\frac{-pa}{4\pi\varepsilon_0}\frac{\partial(\cos\theta/r^2)}{\partial x}$$

Using $\partial\theta/\partial x=-(\sin\theta)/r$ and $\partial r/\partial x=x/r=\cos\theta$ (see Appendix A.3), this yields

$$V_{\mathrm{P}}=pa(3\cos^2\theta-1)/(4\pi\varepsilon_0 r^3)$$

The **quadrupole moment** q is defined as $pa(=Qa^2$ for the linear quadrupole) and thus

linear quadrupole: $$V_{\mathrm{P}}=\frac{q}{4\pi\varepsilon_0 r^3}(3\cos^2\theta-1) \qquad (3.47)$$

E will therefore decrease as $1/r^4$.

General arrangements of charge We can conceive of arrays for which $\Sigma Q=0$, $\Sigma\mathbf{p}=0$ and $\Sigma q=0$, forming **octopoles**, whose fields we should expect to fall off as $1/r^5$. For instance, alternate charges of $+Q$ and $-Q$ at the corners of a parallelepiped would possess such a field. The term **multipole** is used to describe the special arrangements we have been looking at (charge, dipole, etc.) and Table 3.1 summarizes some of their properties. Note particularly the last column of the table which shows that a charge may possess energy even when $\mathbf{E}=0$, that a dipole requires at least a non-zero **E**, while a quadrupole requires a *non-uniform* **E**.

It is not difficult to show that a *general* set of charges, whether discrete or continuous, produces a potential V and a field **E** at distant points which are the sum of an infinite series of terms. The first term of the series has $V\propto r^{-1}$, $\mathbf{E}\propto r^{-2}$ as does the first row of the table; the second term has $V\propto r^{-2}$ and is a dipole term; the third is a quadrupole term; and so on. The mathematical reason for this is that

Table 3.1. Multipoles

	Potential V	Field $\mathbf{E}$	Potential energy in electric field
Charge, Q	$\propto Q/r$	$\propto Q/r^2$	$\propto QV$
Dipole, p	$\propto p/r^2$	$\propto p/r^3$	$\propto pE$ or $p\partial V/\partial x$
Quadrupole, q	$\propto q/r^3$	$\propto q/r^4$	$\propto q\partial E/\partial x$ or $q\partial^2 V/\partial x^2$, etc.

the simple form $V = Q/(4\pi\varepsilon_0 r)$ cannot be used for more than one charge, because they cannot all be placed at the origin. If the field point P at which we want V has a vector position $\mathbf{r}$, then a typical charge in the set must be given a different position $\mathbf{r}_i$. The potential is then the sum of terms like $Q/(4\pi\varepsilon_0)(|\mathbf{r}-\mathbf{r}_i|)$. If θ is the angle between $\mathbf{r}$ and $\mathbf{r}_i$, V can be written

set of charges: $$V = \sum_i \left(\frac{Q}{4\pi\varepsilon_0 r}\right)\left(1 - 2\frac{r_i}{r}\cos\theta + \frac{r_i^2}{r^2}\right)^{-1/2}$$

and it is in the binomial expansion of the factor in parentheses that the multipole terms arise.

In most cases of interest one term is so predominant (the one falling off least rapidly with distance) that it is the only one we need consider. Sometimes, however, the small terms can be significant. As an example, an atomic nucleus with complete spherical symmetry would produce only an inverse square **E**-field around it, as we shall see in Sec. 4.2. Any departure from the spherical symmetry introduces extra small multipole terms in **E** that would have observable effects on spectra. Such effects can thus be used to gain information about nuclear structure.

3.11 Conclusion

Our initial incentive for introducing the electric field **E** and electric potential V was to make it easier to deal with problems involving the interactions of charges. Although *in principle* Coulomb's law is sufficient, calculations with **E** and V as intermediate steps are much simpler, even at the price of having to get used to new mathematical ideas.

However, it has become evident that **E** and V are more powerful than mere aids to calculation. The fact that they are field quantities, with values at every point over extended regions of space, puts them in a very important class of physical magnitudes. Field theories permeate almost every branch of physics, and the **E**-field and V-field are typical examples of vector and scalar fields, respectively.

In the next chapter, we explore the field properties of **E** and V more thoroughly, not only because they are important in their own right but also because the methods adopted form a pattern that will be repeated for other fields later in the development of the subject.

3.12 Revision summary

Electric field strength, **E**, at a point is defined as the force per unit positive test charge at the point

$$\mathbf{E} = \mathbf{F}/Q_t \tag{3.1}$$

where Q_t must be stationary and small enough, if necessary, to produce negligible effects. The **SI unit** for **E** is the *volt per metre*, $\mathrm{V\,m^{-1}}$. The **E**-field defined in this way

- is a **vector** field with magnitude and direction at every point in a region. Hence it possesses, in general, three components which may be specified in one of several coordinate systems, e.g. E_x, E_y, E_z in Cartesians. Each component may vary in space;
- is capable of being represented in direction by lines of force;
- gives the force which would be exerted on a charge Q at any point by

$$\mathbf{F} = Q\mathbf{E} \tag{3.6}$$

  where we assume for the moment that Q may be moving.

Objects giving rise to **E** are known as **sources of E**. From Chapter 2 we know that one type of source will be electric charges themselves, and the definition of **E** then allows Coulomb's law to be split into two parts:

$$\mathbf{F} = \frac{Q_1 Q_2}{4\pi\varepsilon_0 r^2}\hat{\mathbf{r}}$$

$$\swarrow \qquad \searrow$$

$$\mathbf{F} = Q\mathbf{E} \qquad \mathbf{E}_{\text{pt charge}} = \frac{Q}{4\pi\varepsilon_0 r^2}\hat{\mathbf{r}} \tag{3.3}$$

The **E**'s from a collection of charges may be added vectorially to give a resultant because the *principle of superposition* applies to Coulomb forces. Electric fields due to charges are known as *electrostatic fields* and they have properties (following from Coulomb's law) that distinguish them from other **E**-fields:

- $\mathbf{E}_{\text{pt charge}}$ is a central field,
- $\mathbf{E}_{\text{pt charge}}$ is an inverse square field.

Electric potential, V, arises from the concept of potential difference between two points, V_{AB}. This is defined for an electrostatic field as the work done per unit positive test charge in taking charge from one point to the other. Thus if the work is W_{AB} then

$$V_B - V_A = V_{AB} = W_{AB}/Q_t \tag{3.7}$$

The **SI unit** of potential is the *volt*, V. By using a zero chosen arbitrarily we can set up a potential V at every point in a region. The following properties are consequences of the definitions of V and **E**, and of Coulomb's law:

- V forms a **scalar** field with a magnitude at every point in a region. Hence it is in general a function of three variables like x, y, z.
- V is capable of representation by equipotential surfaces drawn through points at the same potential.
- V is given in terms of **E** by $\int -\mathbf{E}\cdot d\mathbf{L}$ (3.13), where the line integral extends from the zero to the field point at which V is required.

- V is path-independent.
- V due to a point charge Q is $Q/(4\pi\varepsilon_0 r)$ at a field point a distance r from the charge (3.14).
- The potential energy of a charge Q at a point where the potential is V is given by $U=QV$ (3.10).

Relationship between E and V is the same as that between force **F** and potential energy U. **E** is the negative gradient of V so that

- in Cartesians

$$E_x=-\partial V/\partial x, \qquad E_y=-\partial V/\partial y, \qquad E_z=-\partial V/\partial z, \tag{3.21}$$

- in plane polars

$$E_r=-\partial V/\partial r, \qquad E_\theta=-(1/r)\partial V/\partial \theta, \tag{3.22}$$

- lines of **E** are perpendicular to equipotential surfaces.

Calculation of E and V due to charges When the distribution of charge is given, this chapter yields two basic methods for finding **E** and V:

- Find V first by using $V=Q/4\pi\varepsilon_0 r$ for a point charge and using superposition, i.e. for a collection of point charges:

$$V=\sum_i Q_i/4\pi\varepsilon_0 r_i \tag{3.16}$$

  or for continuous distributions:

$$V_\rho=\int_\tau \frac{\rho\,\mathrm{d}\tau}{4\pi\varepsilon_0 r}; \qquad V_\sigma=\int_S \frac{\sigma\,\mathrm{d}S}{4\pi\varepsilon_0 r}; \qquad V_\lambda=\int_L \frac{\lambda\,\mathrm{d}L}{4\pi\varepsilon_0 r} \tag{3.17}$$

  Then calculate **E** from $E_x=-\partial V/\partial x$, etc. This method is generally the simpler because calculation of V involves scalar addition and of **E**, differentiation.
- Find **E** first by using $\mathbf{E}=Q\hat{\mathbf{r}}/(4\pi\varepsilon_0 r^2)$ for a point charge, and then using superposition. Then calculate V from $V=\int -\mathbf{E}\cdot\mathrm{d}\mathbf{L}$.

Further methods for finding **E** and V will be encountered in the next chapter.

Charges in free space A charge Q carried by a particle of mass m and placed in an electric field moves so that its total mechanical energy $U+\frac{1}{2}mv^2$ is constant, where $U=QV$. Kinetic energies are often given in electron-volts (eV) where $1\,\mathrm{eV}=1.6\times10^{-19}\,\mathrm{C}$.

Charges in conductors The properties of **E** and V allow us to deduce that a conductor carrying only *static* charges

- has zero **E**-field in the conducting material,
- is an equipotential volume and its surface is an equipotential surface. It follows that **E**-lines leave such a conducting surface along the normal.

Connecting together two charged conductors by a conducting path causes charge to flow from one to the other until their potentials are equal.

An electric dipole consists of two equal and opposite charges $\pm Q$ separated by a distance **l**. The dipole moment is defined as

$$\mathbf{p}=Q\mathbf{l} \tag{3.36}$$

An ideal dipole is one for which **l** is negligible compared with the distance **r** of a field point. For such a dipole the potential at a field point is

$$V=\frac{p\cos\theta}{4\pi\varepsilon_0 r^2}=\frac{\mathbf{p}\cdot\mathbf{r}}{4\pi\varepsilon_0 r^3} \tag{3.37}$$

and the components of its E-field are

$$E_r=\frac{2p\cos\theta}{4\pi\varepsilon_0 r^3};\qquad E_\theta=\frac{p\sin\theta}{4\pi\varepsilon_0 r^3} \tag{3.39}$$

The important feature of these expressions is the dependence on distance: an inverse square for the potential and an inverse cube for the fields.

An ideal dipole situated in an E-field experiences a couple

$$\mathbf{T}=\mathbf{p}\times\mathbf{E} \tag{3.41}$$

however **E** may vary. The force on such a dipole is zero if **E** is uniform, but for a non-uniform **E** is given by

$$F_x=p_x\left(\frac{\partial E_x}{\partial x}\right)+p_y\left(\frac{\partial E_x}{\partial y}\right)+p_z\left(\frac{\partial E_x}{\partial z}\right) \tag{3.43}$$

with similar expressions for F_y and F_z formed by a cyclic permutation of x, y and z. The potential energy of the dipole in the E-field is

$$U=-\mathbf{p}\cdot\mathbf{E}\qquad(\text{zero when }\mathbf{p}\perp\mathbf{E}) \tag{3.44}$$

Commentary

C3.1 On Definitions A clear distinction should be drawn between an equation representing a *law* and one representing a *definition.* The latter does not guarantee anything about the behaviour of the quantity defined when conditions are varied: that is a matter for experimental observations, often summarized in a law. Many defining equations take the form $Z=y/x$ or $Z=\mathrm{d}y/\mathrm{d}x$, where x and y are already well-defined quantities. It *may* be true that x and y are proportional under certain conditions, but this linearity must not be taken for granted (see, for instance, the discussion of $C=Q/V$ in Sec. 5.1 and of $R=V/I$ in Sec. 6.3).

What makes some derived quantities useful enough to define, while many other conceivable ones are not? One reason is that, like **E**, they may illuminate a subject by giving us another way of looking at it or they may make it easier to perform calculations. Another is that, like R, they may have for at least a good range of materials and conditions, a value which characterizes a substance or a system.

On the particular definition of **E** given in (3.2), some authors write the test charge as δQ, the force as $\delta\mathbf{F}$ and the limit as $\mathrm{d}\mathbf{F}/\mathrm{d}Q$, which appears logical. It is in fact not a correct definition of **E** as an example in Sec. 4.8 shows. The $\delta\mathbf{F}$ is *not* $\mathbf{F}(Q+\delta Q)-\mathbf{F}(Q)$, but is $\mathbf{F}(\delta Q)$, a very different quantity.

C3.2 On the Equality of Positive and Negative Elementary Charges Stover, Moran and Trischka (1967), among others, have repeated the Millikan experiment, not with the object of determining e, but rather to establish that its magnitude for the electron e_e was equal to that for the proton e_p. Their apparatus was far more sensitive, however, in that their 'oil-drop' was a minute iron spheroid which could be held stationary by magnetic fields: a vertical field to hold the spheroid up against gravity, and a variable horizontal field adjusted to balance a horizontal electric force. In this way they were able to show that any net charges which collected on the spheroid were integral multiples of a common unit (e) *to within a factor 0.2e.*

Their spheroids were of such a size that they contained about 2×10^{18} electrons and the same number of protons in the nuclei. If the difference in the magnitudes of e_e and e_p amounted to a fraction f of e, then a spheroid with an equal number of each would have an excess charge of $2 \times 10^{18} fe$. If this amounted in all to $(n+\delta)e$, where n is an integer and δ a fraction, then n extra electrons or ions would attach themselves to the spheroid to compensate and a few more than n would give the net charges measured in the experiment. However, δe could never be compensated in this way. The fact that $\delta \leq 0.2$ must mean that $f \leq 10^{-19}$.

Other methods of establishing this result do so by finding whether atoms and molecules are electrically neutral. Shull, Billman and Wedgwood (1967) used the absence of deflection in an electric field. King (1960) and Hillas and Cranshaw (1957) allowed a gas cleared of ions to flow out of a metallic chamber whose potential would clearly change if there were a small residual charge on each escaping molecule. Because the various substances used in all these experiments contained different proportions of neutrons and protons, they enable us to say not only that e_e and e_p do not differ by more than 1 part in 10^{20}, but that the neutron charge is less than about $2 \times 10^{-22} e$, i.e. less than 3×10^{-41} C—neutrality indeed!

C3.3 On the Importance of Comment C3.2 and on Quarks Why are experiments like those just described being carried out at such levels of precision? The original importance in the 1950s stemmed from the possible contribution of a charge excess on 'neutral' atoms to the recession of the galaxies through electrostatic repulsion or as a possible contribution to the earth's magnetic field. More recently it has become important in the physics of elementary particles. The electron and proton belong to classes of particles known, respectively, as leptons (light particles) and baryons (heavy particles), each being the lightest charged member of its class, and stable (although see the comment about the proton on page 14). The stability against decay is explained in terms of three fundamental conservation laws, those of charge, of number of leptons and of number of baryons. If e_p, the baryonic charge, and e_e, the leptonic charge, were slightly different, the conservation of charge alone would be sufficient to guarantee conservation of baryons and leptons. So far this reduction in the number of fundamental conservation laws is clearly not possible.

Another reason for experiments of the type described in comment C3.2 is the search for free **quarks**. The quark model, involving entities carrying charges of $\pm e/3$ and $\pm 2e/3$, is now well established and accounts admirably for the properties and behaviour of elementary particles [see, for example, Close (1983)]. Quark triplets are bound together in baryons, while mesons are formed from quark–antiquark pairs. There is experimental evidence (Sec. 4.2) that quarks exist within nucleons, but the many searches for free quarks by methods similar to those described in comment C3.2 have not shown convincingly that they can exist independently. A possible theory accounting for the non-existence of free quarks suggests that the attractive force between them *increases* with distance. An extensive review of work in the field up to the date of his paper is given by L. W. Jones (1977).

In spite of the reports of success in quark searches by Fairbank and colleagues (LaRue, Philips and Fairbank, 1981), it is clear that free quarks are, at best, extremely rare. Stover's spheroids, for example, must have contained less than 1 quark in 10^{18} nucleons and Boyd *et al.* (1979) report fewer than 1 quark in 4×10^{14} helium atoms. The latest reported work in this field (Smith *et al.*, 1987) found no evidence for a fractional charge and was able to provide an explanation for Fairbank's result.

Finally, the neutrality of atoms, in spite of the various velocities of the charges within them, demonstrates the **invariance** of charge, i.e. the magnitude is independent of the velocities either of the charges themselves or of any observers.

Problems

Section 3.1

3.1 Find the position of the neutral point in Fig. 3.1b if the charges are a distance a apart.

3.2 Sketch the lines of **E** due to charges $+2Q$ and $+Q$ separated by a distance a. Calculate the position of the neutral point.

Section 3.2

3.3 Three point charges $+Q$ are situated at three corners of a square of side a. Find the E-field at the fourth corner.

3.4 Estimate the electric field strength in $\mathrm{V\,m^{-1}}$ at distances of 0.5 nm and 50 pm (5×10^{-11} m) from a proton. (0.5 nm is the order of magnitude of the distance between atoms in solids; 50 pm is the approximate radius of the hydrogen atom.)

3.5 Point charges $+Q$ and $-Q$ are separated by a distance $2l$, and O is the point midway between them. Find the E-field at P, a distance r from O, (a) when OP is along the line joining the charges and $r > l$ and (b) when OP is perpendicular to the line joining the charges. Show that when $r \gg l$ both fields are nearly proportional to Ql/r^3. (The charges form a *dipole* and P is in the *end-on* and *broadside* positions, respectively—see Sec. 3.9.)

3.6 A thin rod of length $2l$ is uniformly charged with λ per unit length. Find the E-field at a point on the line of the rod, beyond its ends and a distance a from its midpoint.

3.7 A charge Q is spread uniformly round a thin wire bent into a circle O radius a. P is a point on its axis at a distance r from the wire itself and θ is the angle between r and the axis. Show that the E-field at P has a magnitude $Q \cos\theta/(4\pi\varepsilon_0 r^2)$. Show from that result that the work done in taking unit charge from O to infinity is $Q/(4\pi\varepsilon_0 a)$.

3.8 A charge Q is distributed uniformly over the surface of a hemisphere of radius r. What is the E-field at the centre of curvature?

3.9 A large horizontal plane sheet with a uniform surface density of charge σ has a circular hole of radius b cut in it. If the charge on the remainder of the sheet stays uniformly distributed, find the E-field at a point a distance a vertically above the centre of the hole. (Long method: integration. Short method: superposition.)

3.10 Show that the equations of lines of **E** in the xy plane are given by the solutions of $\mathrm{d}y/\mathrm{d}x = E_y/E_x$; and in the $r\theta$ plane by the solutions of $\mathrm{d}r/\mathrm{d}\theta = rE_r/E_\theta$. Find the equations for a point charge at the origin in both Cartesian and polar coordinates and check that they give the lines of Fig. 3.1a.

3.11 Sketch the E-lines in the xy plane if $E_x = 2y$ and $E_y = 2x$. (Use the method of Problem 3.10.)

Section 3.4

3.12 Use Eqs (3.11) and (2.12) to show that the potential energy of a finite collection of charges can be expressed as $\Sigma_i \frac{1}{2} Q_i V_i$, where V_i is the potential at the ith charge due to all the other charges.

Section 3.5

3.13 Three charges $+Q$ are situated at three corners of a square of side a. Find the potential at the fourth corner. What would be the potential energy of a charge $+Q$ placed at the fourth corner? What would be the potential energy of the system of four charges?

3.14 Find the potential (zero at infinity) at the point on the line of the charged rod of Problem 3.6.

3.15* A charge is distributed uniformly with density σ over the surface of a hollow conducting sphere of radius a. Show by direct integration that the potential at any point inside it is $a\sigma/\varepsilon_0$ (zero at infinity) and that this is the potential of the sphere itself.

Section 3.6

3.16 Show that the E-field of Problem 3.6 can be deduced from the potential calculated in Problem 3.14.

3.17 Find the components of **E** if (a) $V=3xy$ or (b) $V=r^2 \cos\theta$.

3.18 If $V=3xy$, show that the E-lines are of the same pattern as those of Problem 3.11. What are the equations of the equipotentials for $V=\pm 3$, $V=\pm 6$? Sketch them in relation to the E-field.

3.19 According to the calculation in Sec. 3.2, the **E**-field on the axis of a circular sheet of charge is $[1-a/(a^2+b^2)^{1/2}]\sigma/2\varepsilon_0$. Derive this by evaluating the potential at the point and finding its gradient.

Section 3.7

3.20 An electron starts from the cathode of a vacuum tube with negligible velocity. What are its kinetic energy and velocity at the anode if the potential difference between electrodes is 100 V? How many electrons per unit volume at the anode are there if 1 mA is collected by every square cm of electrode? Explain how it is that the current in such an electron beam is constant when v in Eq. (1.12) is increasing across the tube.

3.21 Find the relation between velocity and potential difference using classical mechanics for (a) electrons and (b) protons accelerated *in vacuo*. For what voltages do the velocities become one-tenth of the velocity of light?

3.22 The particle in Fig. 3.13a starts from the upper plate at time $t=0$ with negligible velocity. The plates are separated by a distance d, and $V=V_0 \sin\omega t$, the first half-cycle acting so as to accelerate the particle. Find how the distance travelled varies with t and sketch the variation. Is the particle ever recaptured by the upper plate? Does it ever reverse its direction of motion?

3.23 How long does an electron take to travel between two parallel plates separated by 1 cm between which a potential difference of 100 V is maintained? Assume that the initial velocity at one plate is zero. (This is the **transit time** between electrodes.)

3.24 The electron beam in a cathode-ray tube is deflected by plates 2 cm long and 5 mm apart. Estimate the linear deflection of the beam at a screen 20 cm away from the plates when 200 V is applied across them. The anode potential of the tube is 500 V.

3.25 The potentials across the x and y plates of a cathode-ray oscilloscope produce deflections given, respectively, by $A \sin\omega t$ and $A \cos\omega t$. Show that the trace is a circle of radius A. What is the effect on the trace of applying at the same time a modulation to the accelerating potential so that it varies by $(1+\sin\omega t)$ times its original constant value?

Section 3.9

3.26 Show that the equations of the **E**-lines of a dipole are $(\sin^2\theta)/r=\text{constant}$ and of equipotentials are $(\cos\theta)/r^2=\text{constant}$.

3.27 Find the forces and torques in cases (c) and (d) of Fig. 3.24 without using the general argument in the text. Why, in case (d), does the whole system of two dipoles not rotate under the action of the combined torques? Or perhaps it does?

Section 3.10

3.28 Take the linear quadrupole of Fig. 3.26b and place it along the x axis with its centre at the origin. Find the potential at a distance r along the x axis by elementary methods. Show that when $a \ll r$, the potential is consistent with Eq. (3.47).

3.29* Show that point charges $+Q$ situated at $(\pm a, 0)$ in the xy plane give an E-field at large distances the same as that of a charge and a positive quadrupole at the origin. Hence argue that an atomic nucleus in the shape of a spheroid elongated in the x direction (prolate) would possess a positive quadrupole moment, that one contracted in the x direction (oblate) would possess a negative quadrupole moment and that neither would possess an electric dipole moment.

4

Further properties of electrostatic fields

Our concern in the first three chapters has been to establish some basic laws and concepts. We are now in a position to use these in a further development of electrostatics with two aims in mind: to be able to tackle more complex problems and at the same time to put the laws into forms allowing a powerful theoretical treatment. Such forms allow us to link electrostatics more easily with other branches of the subject later in the book and point the way to suitable generalizations. No new laws are introduced in this chapter, and it is essentially a set of deductions from Coulomb's law.

Gauss's law, developed in Sec. 4.1 in integral form, results from the dependence of the Coulomb force on the inverse square of the distance. We use it to deduce some very general results and, in Sec. 4.2, to examine the fields of some highly symmetrical distributions of charge. The **circuital law** expresses another important property of **E** and is developed in Sec. 4.3. It results from the central nature of the Coulomb force.

In the subsequent four sections we show how to take the laws relating to paths, surfaces and volumes (integral forms) and convert them into differential forms applying at a point. These sections could perhaps be studied in outline only for a first-level course, but the methods and results are very important for later work.

Finally, we look at a more general type of problem in which the charge distribution is *not* specified from the start as we have assumed in most cases so far (Sec. 4.8).

4.1 Gauss's law for E (integral form)

Concept of flux In Sec. 3.1, it was promised that the idea of using the density of lines of force as a measure of **E** would be developed more precisely. In elementary courses this is sometimes done by letting the magnitude of **E** at any point equal the number of lines of force per unit area crossing an area perpendicular to **E** at that point. Taking now a point charge Q at the centre of a sphere of radius r, we find the value of E at all points on the surface to be $Q/(4\pi\varepsilon_0 r^2)$ and, since the area of the surface is $4\pi r^2$, the convention adopted means that Q/ε_0 lines of force leave a charge Q. This is then used to generalize the situation to any number of charges.

In more advanced work, the idea of individual lines of force crossing an area is replaced by the concept of the **flux** of the field across the area. Although the concept is discussed in a general way in Appendix B.3, its first appearance in this book warrants additional explanation here. We have already seen (page 17) that an area can be represented vectorially by giving it a direction normal to its plane: for example, the area of magnitude S in Fig. 4.1a is represented by the vector **S**. If another vector (such as a field) has a value **A** over the whole of **S**, the **flux of A over S** is obtained by multiplying the component of **A** normal to **S** by the magnitude of S, i.e. $AS \cos\theta$ in Fig. 4.1a. Remembering the notation for a scalar product from the last chapter, we see that this can be written as $\mathbf{A}\cdot\mathbf{S}$. If **A** should vary over **S**, then we have to divide **S** into elements d**S** small enough for **A** to be constant over it, so that the flux of **A** over d**S** is $\mathbf{A}\cdot d\mathbf{S}$. These elementary fluxes are then added over the large surface (i.e. integrated) to give the expression $\int \mathbf{A}\cdot d\mathbf{S}$ (the integral is a surface integral discussed in Appendix A.5, but at this stage can be taken simply as a sum over **S**). (If **A** is a quantity representing the motion of something, then the flux really does give the flow across the area in terms of an amount per unit time. In the case of electric fields, no motion is involved but the word flux is still retained.)

A final point about flux concerns its sign. For a general area this can be chosen arbitrarily, but when a closed surface is involved it is conventional to take the *outward* direction as positive. This is a sign convention we shall always adopt.

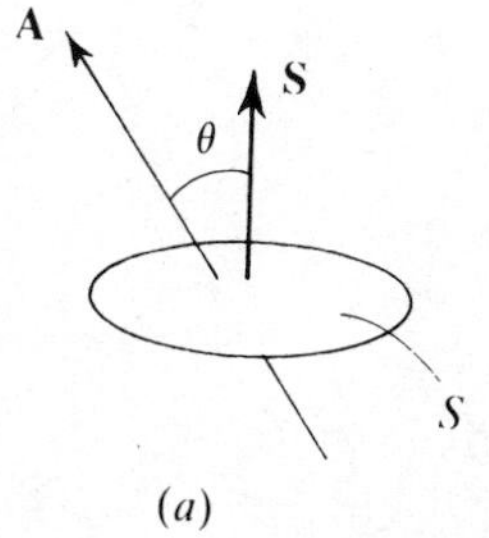

Figure 4.1 (a) The flux of **A** over **S** is $AS\cos\theta = \mathbf{A}\cdot\mathbf{S}$;

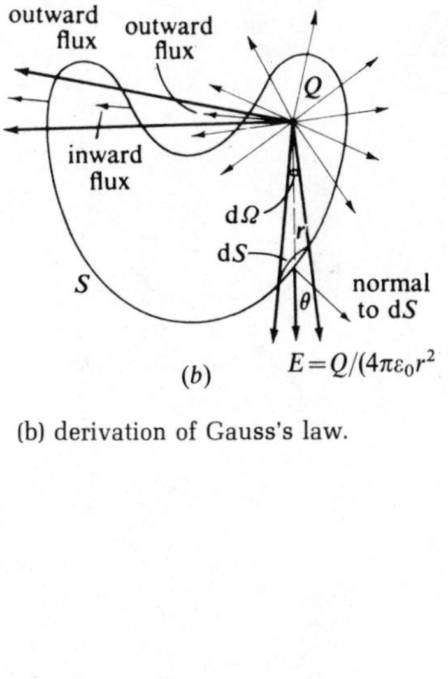

(b) derivation of Gauss's law.

Derivation of Gauss's law: special case Returning now to the special case of a point charge Q, let us imagine a spherical surface of radius r centred on Q over which we calculate the flux of **E**. Since **E** is everywhere normal to the surface already and has a constant magnitude $Q/(4\pi\varepsilon_0 r^2)$ over it, the flux of **E** over the whole surface of area $4\pi r^2$ is simply Q/ε_0. Since this is independent of r, the same flux is obtained for any such spherical surface and gives rise to the idea that Q is the source of a total electric flux of Q/ε_0. This is in fact a special case of Gauss's law. We now go on to prove the law for any shape of surface enclosing any number of charges at any locations.

Derivation of Gauss's law: general proof Consider a point charge Q (Fig. 4.1b) located inside an imaginary closed surface S, known as a **Gaussian surface**. Let a general element dS of this surface be a distance r from Q, so that the electric field **E** at dS is $Q/(4\pi\varepsilon_0 r^2)$ in the direction of r. The flux of **E** across dS (i.e. $\mathbf{E}\cdot d\mathbf{S}$) is outwards and is thus given by

$$\text{Outward flux of } \mathbf{E} \text{ across } d\mathbf{S}\ (\mathbf{E}\cdot d\mathbf{S}) = \frac{Q\, dS \cos\theta}{4\pi\varepsilon_0 r^2}$$

However, the quantity $(\mathrm{d}S\cos\theta)/r^2$ is one we have already encountered in connection with the Cavendish experiment (page 31): it is the element of solid angle, say $\mathrm{d}\Omega$, subtended at Q by $\mathrm{d}S$ (see also Appendix A.1). We therefore have

$$\text{Outward flux of } \mathbf{E} \text{ across } \mathrm{d}\mathbf{S}\ (\mathbf{E}\cdot\mathrm{d}\mathbf{S}) = \frac{Q\,\mathrm{d}\Omega}{4\pi\varepsilon_0}$$

The total flux of **E** over the whole surface S is the sum of these elementary fluxes. This will be simply $Q/(4\pi\varepsilon_0)$ multiplied by the complete solid angle at Q, i.e. 4π. Hence,

$$\text{Total outward flux of } \mathbf{E} \text{ across } S = Q/\varepsilon_0 \tag{4.1}$$

where the sign of the charge is incorporated in Q and inward flux is negative.

Any charges outside S contribute as much inward as outward flux, while the shape of S makes no difference to these results, as can be seen from Fig. 4.1b. Moreover, because electric fields superpose, a number of charges inside S give their own fluxes each obeying (4.1); these fluxes add, so that the right-hand side of (4.1) becomes $\Sigma Q/\varepsilon_0$, where the summation sign means the algebraic sum of all the charges within S. Finally, the total outward flux can be written mathematically as the surface integral of $\mathbf{E}\cdot\mathrm{d}\mathbf{S}$, written $\oint \mathbf{E}\cdot\mathrm{d}\mathbf{S}$, where the circle over the integral sign indicates that it applies to a *closed surface.*

We therefore arrive at the important general law expressed in the form:

Gauss's law for **E**:
$$\oint_S \mathbf{E}\cdot\mathrm{d}\mathbf{S} = \Sigma Q/\varepsilon_0 \tag{4.2}$$

or, in words, **the outward flux of E over any closed surface S is equal to the algebraic sum of the charges enclosed in S divided by ε_0**. This is **Gauss's law**, sometimes also referred to as **Gauss's theorem** (though not to be confused with a theorem of the same name in vector analysis). It is important to realize that, once **E** is defined, the law is a deduction from Coulomb's law and the principle of superposition.

What is the significance of Gauss's law? Its importance is threefold:

1. It gives the total charge in a region if **E** is known over the whole of its bounding surface. This is the inverse of the process used in the last chapter, where **E** was determined from known charges. Some general results following from this aspect of the law are indicated below.
2. It enables **E** to be calculated for highly symmetrical charge distributions, such as those discussed in Sec. 4.2 below.
3. It can be put into a form which applies at any *point* in a region rather than over a surface, a form that allows a powerful development of the subject later. This we shall call the *differential form* of the law (Sec. 4.5), as opposed to the form in Eq. (4.2) above, which we call the *integral form.*

General deductions from Gauss's law We first look generally at the **distribution of charge** over a conductor (Fig. 4.2). It has already been established in Sec. 3.8 that *in the material* of a conductor carrying static charges only, **E** is everywhere zero. Any Gaussian surface drawn entirely within conducting material can therefore have no flux of **E** crossing it and must contain zero resultant charge. A conductor with no interior surface (Fig. 4.2a) therefore *carries all its excess charge on the*

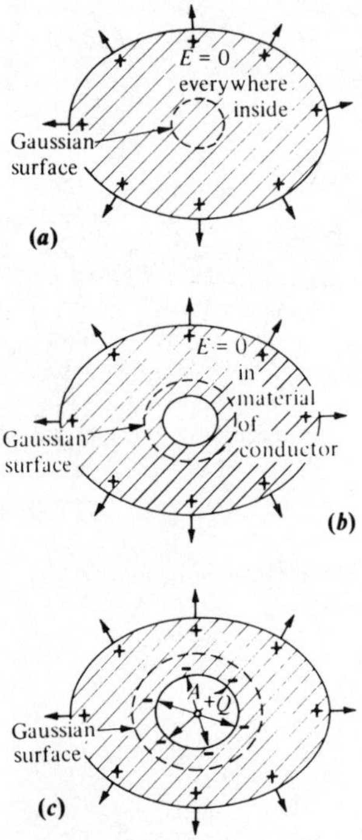

Figure 4.2 Distribution of charge on conductors deduced from Gauss's law: (a) net charge on surface, (b) net charge on outer surface unless (c) a charge is placed in the hollow.

surface. A hollow one, too, as in Fig. 4.2b, carries all its excess charge on the *outer* surface, unless (Fig. 4.2c) there is a charge $+Q$ on a body A placed in the hollow. In that case a charge of $-Q$ is induced on the inner surface. If the conductor in Fig. 4.2c was uncharged before the introduction of A into the hollow, then the charge induced on the *outer* surface must be $+Q$ whatever the position of A. If A is a conductor and is allowed to touch the inside, the $-Q$ on the inner surface and $+Q$ on A neutralize each other, leaving the outer $+Q$ unaffected.

Next we examine the **E**-field in a hollow conductor. The inner surface S of a charged hollow conductor as in Fig. 4.3a is an equipotential surface as we know from Sec. 3.8. Suppose that it is possible for an **E**-field to exist in the hollow with no charge placed in it. Then there would be points inside at potentials different from that of S. If all points just inside S are at a higher potential, then an equipotential surface like S' exists and lines of force between S and S' are all outwards: the flux of **E** over S' is not zero and a charge exists within it, which is self-contradictory. If all points just inside S are at a lower potential, a similar argument applies. If some are lower and some are higher, there exists a surface like S'' at the same potential as S dividing the hollow into two regions to which the same arguments apply since both are bounded by equipotentials. We can only conclude that **E** *is zero at all points within a hollow conductor not containing charge* and that the interior is an equipotential volume at the same potential as the conductor: all this is irrespective of happenings outside. Electrostatic screening (see below) is based on this property.

Comment C4.1

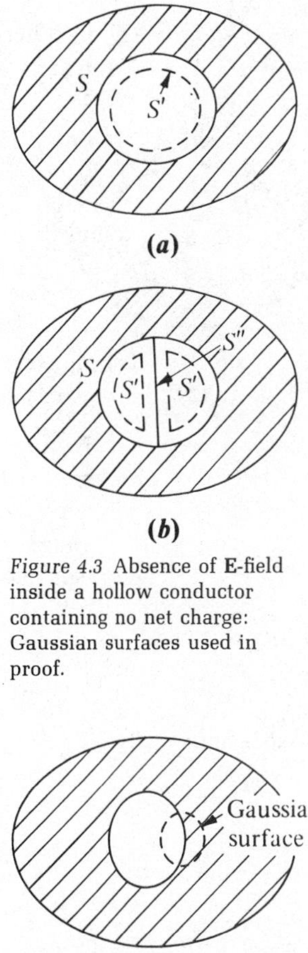

Figure 4.3 Absence of **E**-field inside a hollow conductor containing no net charge: Gaussian surfaces used in proof.

One last deduction. The fact that there is no *resultant* charge on the inner surface of the conductor in Fig. 4.2b does not guarantee the absence of equal and opposite charges. But *any* Gaussian surface chosen as in Fig. 4.4 to intersect part of the inner surface has $\mathbf{E}=0$ at all points and can contain no resultant charge.

To summarize: no electrostatic **E**-field or net electric charge can exist within a conductor even when it is hollow (unless a net charge is introduced into the hollow) and even when there are external electric fields. Any net charge given to a conductor is located entirely on its outer surface.

Experimental tests and applications That charge resides only on the outer surface of conductors was suggested in 1729 by Stephen Gray who showed that a solid and hollow cube of the same size produced the same electrical effects at distant points when charged in the same way: the absence of the inside made no difference.

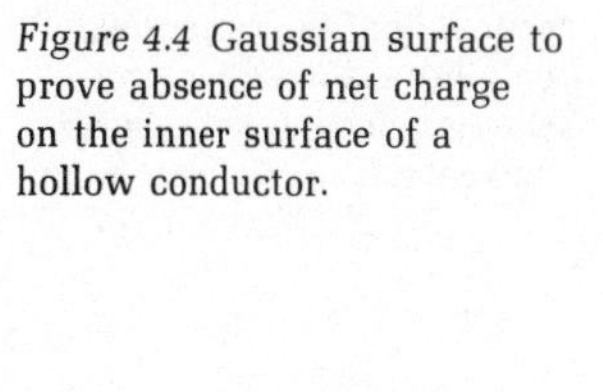

Figure 4.4 Gaussian surface to prove absence of net charge on the inner surface of a hollow conductor.

The results deduced above have two important applications whose success verifies the validity of our arguments. Firstly, a hollow conductor can be given an indefinitely large charge by conveying it to the interior in successive small amounts (Fig. 4.5). Secondly, any apparatus placed inside a hollow conductor whose potential is fixed is unaffected by external electric fields: this is the principle of **electrostatic shielding or screening.**

Any small apertures in the hollow conductor have little effect on these results and the reader should now be able to explain the operation of the Faraday cup in Sec. 1.1.

A practical application is the production of an equipotential region shielded from the effect of external electric fields by enclosing it in what is known as a **Faraday cage**. This consists of conducting sheets or wire mesh surrounding the region involved, welded or otherwise connected in such a way as to provide a continuous conducting path throughout the structure. Apparatus and people inside the cage are protected from electric fields generated externally. Cars and aircraft constructed from metallic panels form such a cage and protect individuals and equipment inside them from the effects of lightning discharges.

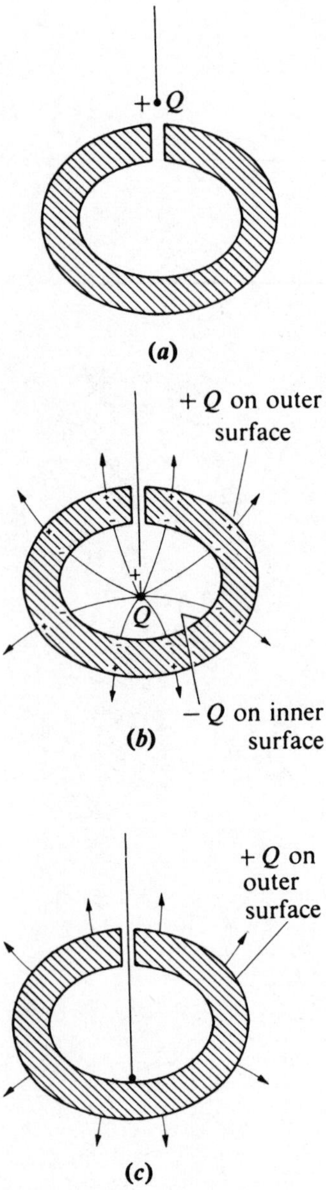

Figure 4.5 Charging a hollow conductor. The carrier need only bring successive small charges since it is discharged on contact.

4.2 E-fields due to symmetrical charge distributions

The use of Gauss's law to determine **E**-fields due to charge distributions is limited to cases where there is a high degree of symmetry: spheres, infinite cylinders and infinite planes. Although such cases would at first sight appear to be so unrealistic as to be hardly worth consideration, we shall see that there are both theoretical and practical applications of the results. In particular, the fields due to effectively infinite distributions are very good approximations when one or two dimensions in a practical situation are much greater than any other (e.g. a cylindrical conductor like a cable with a length much greater than its diameter).

Spheres of charge: spherical shells A conducting sphere of radius a carries all its charge Q on its surface and by its symmetry will constitute a spherical shell of uniform density of charge $\sigma = Q(4\pi a^2)$. Suppose we want to find the electric field strength at an external point a distance r from its centre. By the spherical symmetry of the problem if the field at P (Fig. 4.6) is E, then it must be radially outwards and have the same magnitude at all points on the surface of a sphere S of radius r concentric with the conductor. If S is chosen as a Gaussian surface, E is everywhere normal to it and has the same magnitude over it, so that the outward flux over S is $4\pi r^2 E$. By Gauss's law this is equal to Q/ε_0 and so

spherical shell, $r \geq a$: $$E = Q/(4\pi\varepsilon_0 r^2) \tag{4.3}$$

The potential will therefore be

spherical shell, $r \geq a$: $$V = Q/(4\pi\varepsilon_0 r) \qquad \text{(zero at } \infty) \tag{4.4}$$

Equations (4.3) and (4.4) show that the **E**-*field and potential outside a uniform spherical shell of charge are as if the whole charge were concentrated at the centre.*

Inside the sphere, $\mathbf{E} = 0$ of course, and the potential is therefore the same as that at the surface, i.e.

spherical shell, $r \leq a$: $$V = Q/(4\pi\varepsilon_0 a) \qquad \text{(zero at } \infty) \tag{4.5}$$

Application *discharge from points* (*corona discharge*) If the surface density of charge on a conducting sphere is σ, the electric field *just outside* the surface is

$$E_{\text{surface}} = \frac{Q}{4\pi\varepsilon_0 a^2} = \frac{\sigma}{\varepsilon_0} \tag{4.6}$$

while the potential is

$$V_{\text{surface}} = \frac{Q}{4\pi\varepsilon_0 a} = aE_{\text{surface}} = \frac{a\sigma}{\varepsilon_0} \tag{4.7}$$

Equation (4.7) shows that, if we have a series of spheres of varying radii but all charged to the same potential, the surface densities of charge and the electric fields just outside the surfaces are inversely proportional to a, the radii of curvature. We might expect the same to happen on a single conductor whose surface (Fig. 4.7) has various convex radii of curvature. The spheres of which the surfaces at points A, B and C are a part are all at the same potential, and we expect the surface density to be highest at A (this is not intended as a proof, but merely as a suggestion). In general, therefore, charge should collect more densely where the convex radius of curvature is smaller, and particularly at points. The electric field strength is correspondingly greater and at sharp corners may be large enough to cause the ionization of the air around them: ions of opposite sign are then attracted to the conductor and discharge it, while those of the same sign are repelled and cause a detectable wind. Discharge from points is often accompanied by a bluish glow and is known as **corona discharge**.

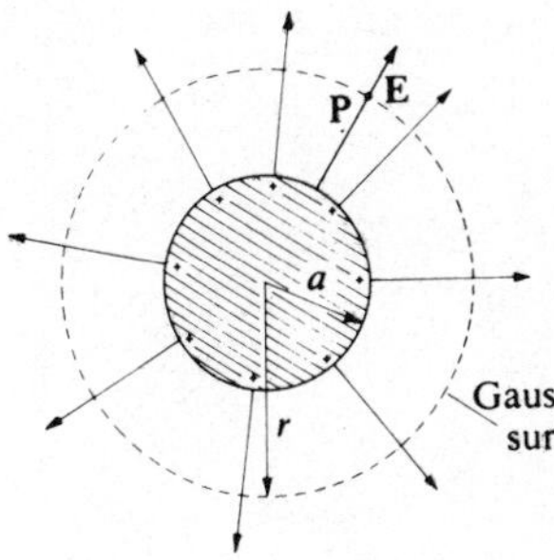

Figure 4.6 The **E**-field due to a charged spherical conductor.

Such discharge is to be avoided for obvious reasons with high-voltage apparatus, whose external surfaces should be free of sharp corners and dust particles. If these precautions are not taken, the large potential gradient between the apparatus and the surroundings is concentrated in the region around the corners or particles and the resultant ionization of the air can produce a conducting path which would allow a discharge to anybody nearby.

On the other hand, corona discharge can be useful as in a lightning conductor (providing a low resistance path for the lightning) and in charging belts or plates as in the Van de Graaff and other machines (page 121).

At extremely sharp metallic points at very high voltages, the **E**-field can be strong enough to cause free electrons or even ions to be emitted from the metal surface itself. This is an effect known as **field emission** (see Sec. 5.7).

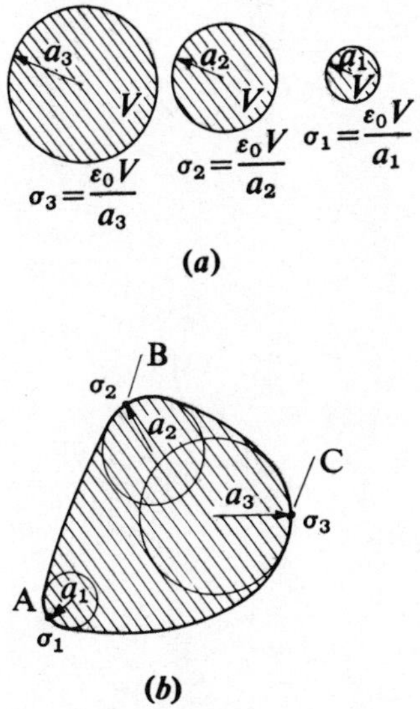

Figure 4.7 (a) Conducting spheres all at the same potential have surface charge densities $\sigma \propto 1/a$; (b) a conductor has an equipotential surface: the surface charge density at various locations on the surface will be approximately proportional to $1/a$, where a is the radius of curvature at a given location. A sharp point (small a) will have a high σ and thus an intense **E**-field just outside.

Spheres of charge: spherical volumes If we now take a spherical volume of radius a carrying a total charge Q distributed uniformly throughout it, the same argument as in the case of the shell will lead to (4.3) and (4.4) for the electric field and potential at points *outside*. At points *inside*, take a Gaussian surface S of radius r as in Fig. 4.8 over which E is the same at all points: the outward flux is then still $4\pi r^2 E$, but the enclosed charge this time is Qr^3/a^3 and so

spherical volume, $r \leq a$: $$E = \frac{Qr}{4\pi\varepsilon_0 a^3} \tag{4.8}$$

To evaluate V, we use Eq. (3.13), taking a zero at infinity and remembering that the path for $r > a$ has E given by (4.3), while for $r < a$, E is given by (4.8). Thus, using x as the variable distance from the centre,

$$V = \int_{\infty}^{a} -\frac{Q\,dx}{4\pi\varepsilon_0 x^2} + \int_{a}^{r} -\frac{Qx\,dx}{4\pi\varepsilon_0 a^3}$$

i.e.

spherical volume, $r \leq a$:

$$V = \frac{Q}{8\pi\varepsilon_0 a}\left(3 - \frac{r^2}{a^2}\right) \quad \text{(zero at } \infty) \tag{4.9}$$

Figure 4.9 compares the variations of E and V for a point charge, a spherical shell and a spherical volume of charge.

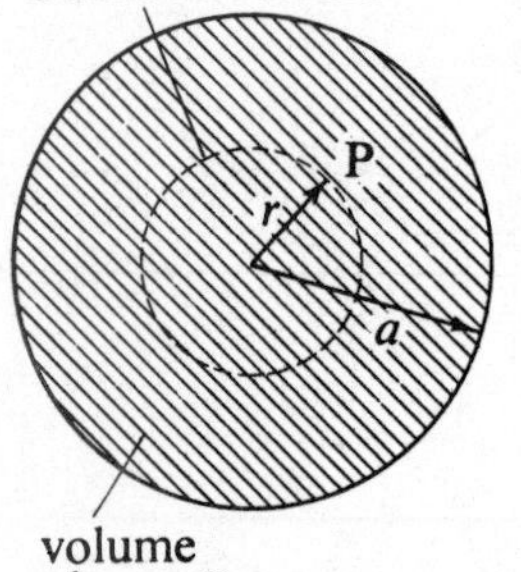

Figure 4.8 Calculating the **E**-field inside a spherical volume of charge (note, not a conductor).

Application *the nuclear atom* The scattering of an α-particle by a nearly spherical nucleus is produced by the Coulomb repulsion between the two positive charges. So, if we imagine Fig. 4.9a to be a potential diagram with energy plotted vertically, the barrier presented to an incident α-particle by a given charge on the sphere is clearly raised when a becomes smaller. Historically, the difference between scattering by a sphere with $a = 10^{-8}$ cm (Thomson's model of the atom, 1907) and that by a sphere with $a < 10^{-12}$ cm (Rutherford's nuclear model, 1911) produced evidence that the latter model was correct. With much higher energies now available to the bombarding particles, it has become possible to penetrate some of the charge in the nuclei themselves and obtain information about its distribution. In the past few years, the scattering of fast electrons by nucleons has revealed the existence within them of concentrated charges with properties that would be possessed by quarks (e.g. the constituent elements of a proton with charges of $2e/3$ and $-e/3$).

The results of scattering experiments also give information about the correctness of Coulomb's law at atomic distances. Experiments with electrons and muons have shown that the law may be applied without detectable error down to 10^{-14} cm, so that the assumption in Sec. 2.5 is amply justified.

E-fields within charge distributions We have been rather glib in accepting without question the concept of an **E**-field within a volume of charge as in Eq. (4.8). Since the point P is right in the middle of a continuous charge distribution, it is at zero distance from some charge. So why does **E** not become infinite using Coulomb's law with $r = 0$? The same question arises with V, yet Eq. (4.9) seems to show that it, too, remains finite.

The answer lies in our model of a charge distributed *continuously* with a density ρ. We know from (4.5) that V inside a *spherical shell* of uniform density ρ would be $4\pi r^2 \rho \, \mathrm{d}r/(4\pi\varepsilon_0 r)$ or $\rho r \, \mathrm{d}r/\varepsilon_0$. As such a shell becomes smaller and smaller, r tends to zero and hence so does V. Moreover, we know that **E** inside such as shell is zero. Thus, if we regard the continuous distribution as a set of shells converging on to the field point, it is clear that **E** will be zero at the centre and that V will converge rather than diverge. The immediately adjacent charges of a continuous distribution do *not* therefore cause **E** and V to become infinite.

On the atomic scale, of course, our continuous distribution will not be a valid model at all and the **E**-field inside materials will then fluctuate enormously from one point to another. This is usually called the **microscopic** **E**-field to distinguish it from the **macroscopic** **E**-field of the smoothed-out model.

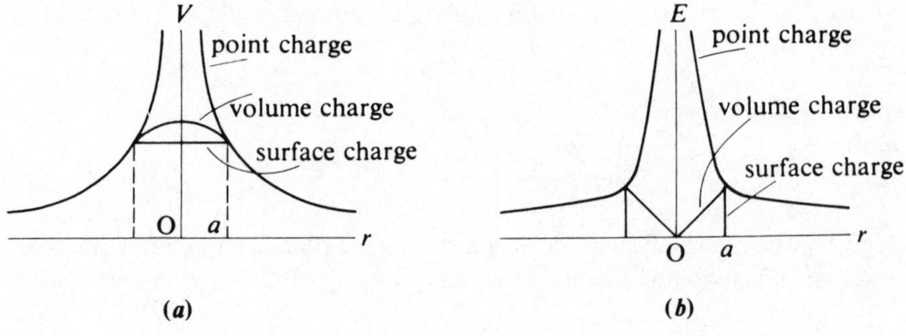

Figure 4.9 Comparison of (a) potential diagrams and (b) variations of **E** with distance, for a point charge and for surface and volume distributions of the same radius, all containing the same total charge.

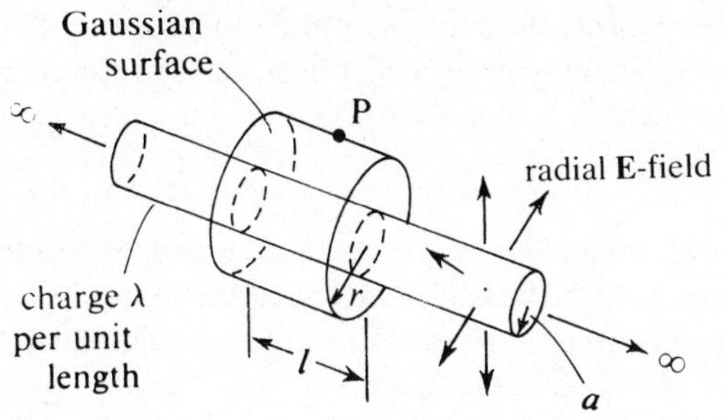

Figure 4.10 Calculating the E-field due to a charged conducting cylinder of infinite length.

Cylinders of charge An infinite cylindrical conductor of radius a carries a charge λ per unit length distributed uniformly over its surface, thus forming a cylindrical shell. To determine the electric field E at a point P a distance r from the axis, choose a Gaussian surface as in Fig. 4.10. Since the system has cylindrical symmetry, the field will be radially outwards and will have the same magnitude at all points on the curved surface of radius r. The outward flux of E is thus $2\pi rlE$ (none over the ends), while the enclosed charge is λl. By Gauss's law

cylindrical shell, $r \geq a$: $$E = \lambda/(2\pi\varepsilon_0 r) \tag{4.10}$$

which would not change if the charge were distributed along the axis. In practice we encounter only finite cylinders but we expect (4.10) to be the more accurate as the ratio of length to diameter becomes greater, provided P is far from the ends compared with its distance from the surface.

The potential V is evaluated by integrating $-\mathbf{E}\cdot d\mathbf{L}$ using (4.10) and yielding a logarithm. The zero for V cannot be chosen at infinity because the logarithm becomes infinite, but any other point may be chosen. Taking the cylinder itself, and using x as the variable distance from the axis:

$$V = \int_a^r -[\lambda/(2\pi\varepsilon_0 x)]\,dx = [\lambda/(2\pi\varepsilon_0)]\log_e(a/r) \qquad \text{(zero at cylinder)}$$

or, in general,

cylindrical shell, $r \geq a$: $$V = C - [\lambda/(2\pi\varepsilon_0)]\log_e r \tag{4.11}$$

where C is a constant determined by the zero of V.

If the cylindrical region has charge distributed throughout its volume, as might be the case with a beam of electrons, the same results are obtained for the outside field, but the inside field ($r < a$) is not now zero (see Problem 4.10).

Planes of charge We require the electric field strength **E** at a point P outside an infinite conducting plane carrying a charge σ per unit area distributed uniformly over the surface. Symmetry shows (a) that the direction of **E** will be normal to the plane surface and (b) that the magnitude of **E** will be the same at all points which are the same distance from the plane as P. The Gaussian surface chosen (Fig. 4.11)

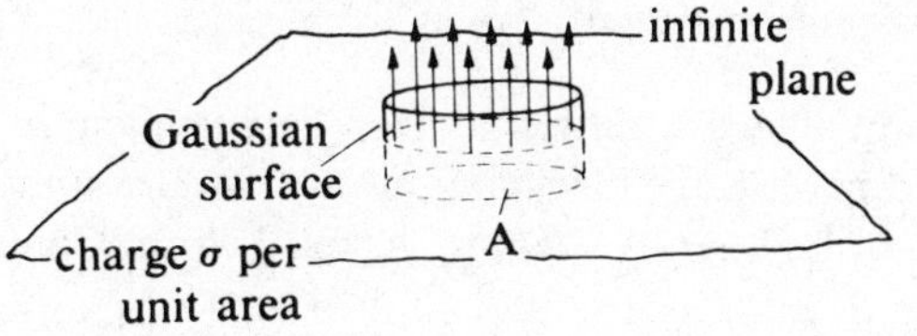

Figure 4.11 Calculating the E-field due to a charged conducting plane of infinite area.

is a cylinder with its lower end inside the conductor and its upper end at the same distance as P. The only flux over this surface is EA over the upper end and since the enclosed charge is σA, Gauss's law gives

$$\text{plane charged conductor:} \qquad E=\sigma/\varepsilon_0 \tag{4.12}$$

which is a uniform field as predicted by Eq. (3.5). Once again, we do not meet infinite planes, but we expect (4.12) to be sufficiently accurate for points very near a surface and some distance from any edges. The reader should check that for such points even (4.3) and (4.10) yield (4.12).

As with the infinite cylinder, the potential V cannot be evaluated with infinity as the zero. Any other point can be used and integration of $-\mathbf{E}\cdot d\mathbf{L}$ using (4.12) yields

$$\text{plane charged conductor:} \quad V=C-Ex=C-\sigma x/\varepsilon_0 \tag{4.13}$$

where C is a constant determined by the zero of V and x is the distance from the plane.

In this example, as in the others of this section, we have assumed that the conductors are isolated in space and that the lines of **E** go off to infinity. This implies that these lines eventually end on equal negative charges. In practice, our formulae are approximate, not only because real conductors are finite, but also because the "surroundings" are at a finite distance. This is a problem discussed in the next chapter.

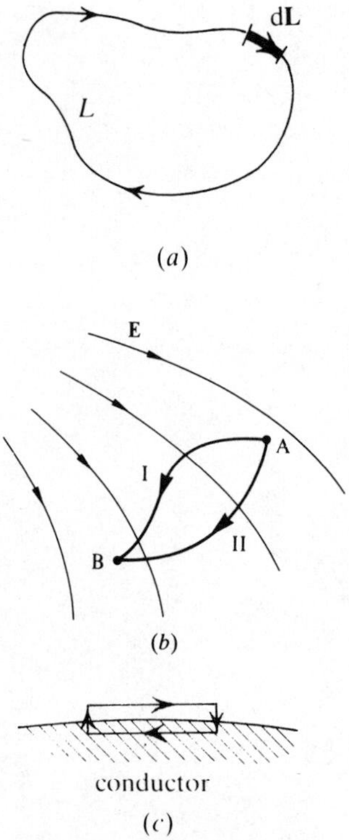

Figure 4.12 (a) A closed path L for finding whether lines of **E** can form closed loops; (b) the line integral of **E** from A to B is the same by paths I and II. Reversing II forms a closed path; (c) a closed path for examining the **E**-field near the surface of a charged conductor.

4.3 Circuital law for E (integral form)

Concept of circulation Gauss's law is concerned with the flux of E-fields and its origins—with where lines of **E** begin and end. The question it answers is: how much flux is generated in a given volume by the charges contained in it? The inverse square law, as we have seen, gives a simple answer.

We now turn to another aspect of E-fields, not concerned with the origin or destination of the lines of **E** but with their ability to close on themselves. The question we now pose is: can lines of **E** form a closed loop which shows no reversal of direction as we move round it? If such an **E** line existed (Fig. 4.12a), then as we move through successive vector elements $d\mathbf{L}$ of the path, the directions of **E** and $d\mathbf{L}$ would always be the same and the scalar product $\mathbf{E}\cdot d\mathbf{L}$ would always be positive. Summing these products round the whole path L would give the line integral of **E** round L, a quantity known as its **circulation** and written as $\oint \mathbf{E}\cdot d\mathbf{L}$. The circle on the integral sign is an indication that the line integral is to be taken round a closed path.

The original question can now be rephrased to ask whether the circulation of **E** due to charges can be non-zero. This is a question raised in connection with many vector fields, particularly those connected with the motion of fluids. Its importance there lies in the fact that a non-zero circulation indicates that the flow can in principle give rise to eddies or vortices. With E-fields, no motion is involved, but the property is no less important, as we now see.

Derivation of the circuital law The circulation of an E-field is given by the line integral of **E** round a closed path, L. To calculate this, we can use some general properties of the line integral between two points A and B from Chapter 3.

Consider two *different* paths I and II between A and B in an E-field (Fig. 4.12b), paths which will eventually be combined into a single loop. The potential difference between A and B is given by the negative line integral between A and B [Eq. (3.12)] and we know that this is (a) independent of the path taken and (b) reverses its sign when the path is reversed. In the figure, therefore, property (a) shows that the line integrals of **E** from A to B are the same by both paths, so that

$$\int_{\mathrm{I}\,A}^{B} \mathbf{E}\cdot d\mathbf{L} = \int_{\mathrm{II}\,A}^{B} \mathbf{E}\cdot d\mathbf{L}$$

If path II is reversed, the integral is taken from B to A and property (b) shows that its sign changes. We then have

$$\int_{\mathrm{I}\,A}^{B} \mathbf{E}\cdot d\mathbf{L} + \int_{\mathrm{II}\,B}^{A} \mathbf{E}\cdot d\mathbf{L} = 0$$

However, the left-hand side is now the line integral of **E** around a closed path starting and finishing at A, and so we can write:

circuital law for **E**:
$$\boxed{\oint_L \mathbf{E}\cdot d\mathbf{L} = 0} \tag{4.14}$$

We have in no way restricted our choice of path so that this applies to any closed path in any electrostatic field and is called the **circuital law** for E. In words, it states that **the circulation of E round any closed path in an electrostatic field is zero.** Another verbal form, because of the meaning of the line integral, is that **the work done in taking a test charge round any closed path in an electrostatic field is zero.**

It should be clear that the circuital law follows from the path-independence of V, which is itself a consequence of the central nature of the E-field due to charges. Thus

$$\text{Central E-field} \Leftrightarrow \text{path-independence of } V \Leftrightarrow \oint_L \mathbf{E}\cdot d\mathbf{L} = 0 \tag{4.15}$$

Applications of the circuital law The zero result means that the law cannot be used to solve any problems relating to specific charge distributions and its importance lies rather in its general nature and its contribution to further developments. As with Gauss's law, there is a differential form applying at a point (Sec. 4.6).

There are, however, two general deductions which can be made. The first concerns the E-field near the surface of a charged conductor. Consider a closed path chosen as in Fig. 4.12c, where the longer sections are parallel to the surface and the shorter sections are perpendicular to it. The contribution of the ends of $\oint \mathbf{E}\cdot d\mathbf{L}$ can be made as small as we wish by decreasing their length without limit and can be made negligible: they are likely to cancel in any case, since the normal component of the E-field sufficiently close to the surface is given by σ/ε_0 in both cases. The contribution of the path inside the conductor is zero since $\mathbf{E}=0$, and because $\oint \mathbf{E}\cdot d\mathbf{L}$ is zero for the whole path, it follows that the tangential component of **E** outside the conductor must also be zero, that is $E_{\tan}=0$. **The E-field just outside a charged conductor is always normal to its surface.**

The second general deduction from the circuital law is that it implies that lines of **E** due to charges cannot close on themselves to form loops without change of the field direction at any point. A little thought shows that such a closed **E** line would violate (4.14).

However, there must exist **E**-fields for which (4.14) is not true; otherwise it would be impossible to maintain a steady current in a closed circuit. All that equation (4.14) tells us is that such currents cannot be produced by electrostatic fields alone, but must arise from other sources of **E**-fields. We see, therefore, that electrostatic fields are a special sort of **E**-field to which (4.14) applies. Where we wish to distinguish **E**-fields due to charges from others, we shall use the symbol $\mathbf{E}_Q$. In Chapter 6, we shall identify the circulation of **E** in non-electrostatic fields with the **electromotance** or **e.m.f.** Since $\oint \mathbf{E}\cdot d\mathbf{L}$ is not zero for these, an attempt to define a V in the usual way will make it depend on the path: this explains why the definition of V in Chapter 3 was confined to electrostatic fields, as are all the results in this chapter.

Although the results derived in the next four numbered sections are important for later developments, they involve a further set of mathematical concepts. The reader may wish to omit them during a first course in the subject since the results are not used until much later.

4.4 The gradient and the ∇ operator

Before deriving the differential forms of Gauss's law and the circuital law, we take another look at the relation between **E** and V as developed in Chapter 3. The first thing to notice is that the two forms

$$V=\int_A^B -\mathbf{E}\cdot d\mathbf{L} \qquad \text{and} \qquad E_s=-\partial V/\partial L$$

are derivable from each other. The difference between them is that the first is an *integral* form applying to *a finite length of path*, while the second is a differential form applying *at a point*.

We now examine further the differential form. In Cartesian coordinates

$$E_x=-\partial V/\partial x; \qquad E_y=-\partial V/\partial y; \qquad E_z=-\partial V/\partial z$$

where we accept $\partial/\partial x$, etc., as differentiating operators. Using unit vectors, **E** can be written as

$$\mathbf{E}=\mathbf{i}E_x+\mathbf{j}E_y+\mathbf{k}E_z=-\left(\mathbf{i}\frac{\partial}{\partial x}+\mathbf{j}\frac{\partial}{\partial y}+\mathbf{k}\frac{\partial}{\partial z}\right)V \tag{4.16}$$

The expression in parentheses is also a differentiating operator and is denoted by the symbol **∇**, usually called 'del'. When **∇** operates on a scalar field quantity like V it produces a vector field like **E**, whose magnitude and direction at any point are those of the gradient of V, of its greatest rate of change. Consequently, when **∇** operates on a scalar it is often called the *gradient* and written **grad**. Thus (4.16) becomes

$$\boxed{\mathbf{E}=-\nabla V=-\mathbf{grad}\,V} \tag{4.17}$$

Although we have defined ∇ in terms of Cartesian coordinates by

$$\nabla \equiv \mathbf{i}\frac{\partial}{\partial x}+\mathbf{j}\frac{\partial}{\partial y}+\mathbf{k}\frac{\partial}{\partial z} \tag{4.18}$$

its components will vary according to the coordinate system chosen. Further information about ∇ is to be found in Appendix B.8 and the form of the gradient in other systems is given in Appendix B.12.

4.5 Differential form of Gauss's law: divergence

Gauss's law as developed in Sec. 4.1 applies to a finite volume and therefore takes an integral form. We now apply it to an elementary volume in x, y, z space, and shrink the volume so that in the limit we obtain the form of the law which applies at a point.

The volume element is a rectangular parallelepiped with faces parallel to the coordinate planes and with edges of lengths dx, dy and dz (Fig. 4.13). The mean volume density of charge within the parallelepiped is denoted by ρ. Let the electric field strength at A have components E_x, E_y and E_z and consider the outward flux over the shaded surfaces (due only to the x components of **E** because E_y and E_z are parallel to them). At B, the x field is $E_x+\mathrm{d}E_x$ or $E_x+(\partial E_x/\partial x)\,\mathrm{d}x$ and it should be clear that the x field at any point on the right-hand shaded surface exceeds that at the corresponding point on the left-hand surface by $(\partial E_x/\partial x)\,\mathrm{d}x$. The *resultant* outward flux from the two shaded surfaces is thus $(\partial E_x/\partial x)\,\mathrm{d}x\,\mathrm{d}y\,\mathrm{d}z$.

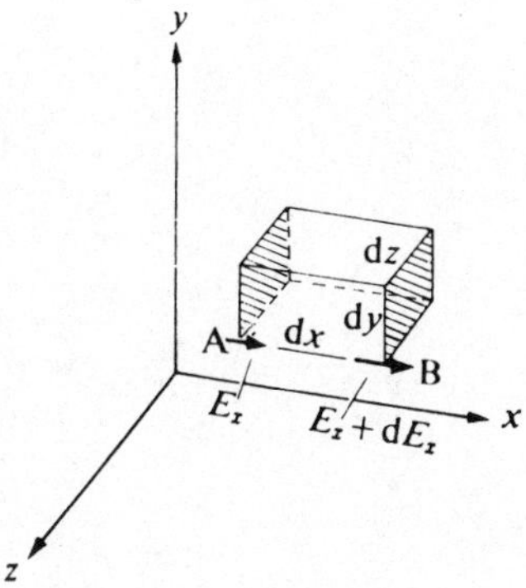

Figure 4.13 Applying Gauss's law to a volume element in Cartesian coordinates. In the limit, we obtain the outward flux per unit volume at a point, called the divergence of **E**.

Similar arguments applied to the other faces in pairs give a total outward flux of $(\partial E_x/\partial x+\partial E_y+\partial E_z/\partial z)\,\mathrm{d}x\,\mathrm{d}y\,\mathrm{d}z$, which must be equal to the enclosed charge, $\rho\,\mathrm{d}x\,\mathrm{d}y\,\mathrm{d}z$, divided by ε_0. If the volume is allowed to shrink to A, then in the limit ρ will be the charge density at A and thus in general

$$\frac{\partial E_x}{\partial x}+\frac{\partial E_y}{\partial y}+\frac{\partial E_z}{\partial z}=\frac{\rho}{\varepsilon_0} \tag{4.19}$$

at a point.

The method of calculation shows that the left-hand side is the **flux per unit volume** diverging from the point, a quantity known as the *divergence* of **E** and written div **E** (Appendix B.9). Another method of writing the divergence uses the ∇ operator whose Cartesian form was given in (4.18). In general, the scalar product of two vectors $\mathbf{A}\cdot\mathbf{B}$ can be written as the sum of products of corresponding Cartesian components $A_xB_x+A_yB_y+A_zB_z$. Treating ∇ and **E** as the two vectors it is clear that the left-hand side of (4.19) is equivalent to $\nabla\cdot\mathbf{E}$ so that

$$\operatorname{div}\mathbf{E}\equiv\nabla\cdot\mathbf{E} \tag{4.20}$$

Both forms are used. Gauss's law in differential form can thus be expressed by

$$\boxed{\operatorname{div}\mathbf{E}\equiv\nabla\cdot\mathbf{E}=\frac{\rho}{\varepsilon_0}} \tag{4.21}$$

A more general discussion of divergence is to be found in Appendix B.9 and its forms in other coordinate systems are given in Appendix B.12.

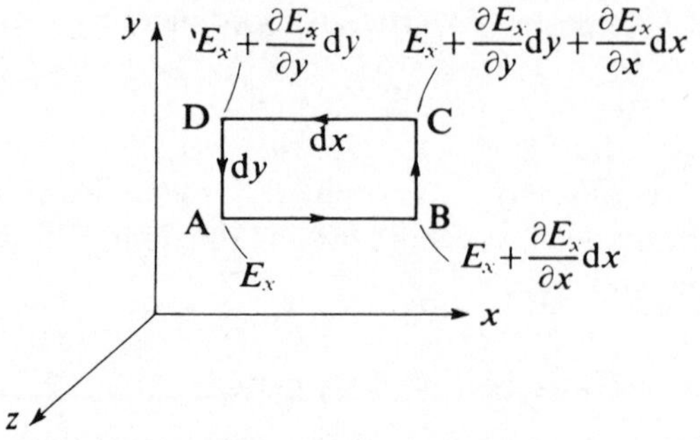

Figure 4.14 Applying the circuital law to a small closed path in the xy plane: for clarity only the x components of E are shown, but the y components are also involved in this calculation. In the limit, we obtain the circulation per unit area perpendicular to the z axis, called the z component of the **curl** of E.

4.6 Differential form of the circuital law: curl

Now let us apply the general circuital law (4.14) to a path enclosing an elementary area in Cartesian coordinates. The area in Fig. 4.14 is parallel to the *xy* plane, has sides d*y* and d*x* as shown and is traversed from A in a clockwise sense when looking along the positive *z* axis (this is related to the general right-hand screw rule of Appendix A.2). The electric field at A has components E_x, E_y and E_z. The contributions to $\oint \mathbf{E}\cdot\mathrm{d}\mathbf{L}$ from AB and CD result only from the *x* components whose values are shown in the figure at the corners. It is clear that the field along CD always exceeds that at the same distance along AB by $(\partial E_x/\partial y)\mathrm{d}y$. Hence, because AB and CD are traversed in opposite directions, their total contribution to $\oint \mathbf{E}\cdot\mathrm{d}\mathbf{L}$ is $-(\partial E_x/\partial y)\,\mathrm{d}y\,\mathrm{d}x$. Similarly the contribution from BC and DA is $(\partial E_y/\partial x)\mathrm{d}x\,\mathrm{d}y$, and so, from (4.14),

$$\partial E_y/\partial x - \partial E_x/\mathrm{d}y = 0$$

and similarly (4.22)

$$\partial E_z/\partial y - \partial E_y/\partial z = 0$$

$$\partial E_x/\partial z - \partial E_z/\partial x = 0$$

These equations could have been more simply derived from $E_y = -\partial V/\partial y$ and $E_x = -\partial V/\partial x$ by using the fact that $\partial^2 V/\partial x\,\partial y = \partial^2 V/\partial y\,\partial x$, but the method we have used is more general and works in cases when the right-hand sides are not zero.

The method of calculation shows that the left-hand sides of these equations are the **circulations per unit area** at a point, the first equation referring to an area perpendicular to the *z* axis, and so on. The circulation per unit area thus has three components and is a vector known as the curl of **E** and written **curl E**. Equations (4.22) can therefore be succinctly written in the form

$$\mathbf{curl}\,\mathbf{E} = 0 \tag{4.23}$$

and the zero **curl** is related to the other properties of **E** and *V* which occur in (4.15).

The differential operator **curl**, like div, can be expressed in terms of the∇ operator. This time the Cartesian form of a general vector product $\mathbf{A}\times\mathbf{B}$

(Appendix B.5) shows that **curl E** is equivalent to $\nabla \times \mathbf{E}$. The differential form of the circuital law for **E** is thus

$$\boxed{\mathbf{curl\,E} \equiv \nabla \times \mathbf{E} = 0} \tag{4.24}$$

Full use of the curl operator is postponed until Chapter 8.

4.7 Poisson's and Laplace's equations

The differential form of Gauss's law, (4.19), can be expressed in terms of V by substituting $E_x = -\partial V/\partial x$, etc., so that it becomes

$$\frac{\partial^2 V}{\partial x^2} + \frac{\partial^2 V}{\partial y^2} + \frac{\partial^2 V}{\partial z^2} = -\frac{\rho}{\varepsilon_0} \tag{4.25}$$

An alternative way of looking at this is to substitute (4.17) into (4.21) giving $\nabla \cdot \nabla V = -\rho/\varepsilon_0$ or

Poisson's equation:
$$\boxed{\nabla^2 V = -\frac{\rho}{\varepsilon_0}} \tag{4.26}$$

where ∇^2 is called the **Laplacian** (Appendix B.8) and has in Cartesian coordinates the form $(\partial^2/\partial x^2 + \partial^2/\partial y^2 + \partial^2/\partial z^2)$. At points where no charges exist, Poisson's equation becomes

Laplace's equation:
$$\boxed{\nabla^2 V = 0} \tag{4.27}$$

Expressions for ∇^2 in polar coordinates are given in Appendix B.12.

Laplace's equation occurs in many branches of physics and engineering (for instance, in problems of heat flow, of celestial mechanics, of hydrodynamics, as well as in several areas of electromagnetism) although the quantity represented by V must obviously be different for each case. The equation is a partial differential equation with an infinite number of possible solutions, each giving the variation of V as a function of position. With so many possibilities, an obvious question to ask is: how does one ever arrive at *the* solution to a problem? And, further, is there indeed only *one* solution?

In practice, the number of solutions applicable to a particular problem is limited by the **boundary conditions**, i.e. by the values that V or its derivatives are specified to take at particular points or over particular surfaces. For instance, although $V = kx$ is obviously a solution of Laplace's equation, it could not apply to the potential from a spherical region containing charge because we require $V \to 0$ as the distance from the charge becomes very great. Another common condition is that V shall have a specified constant value over the surfaces of conductors and this limits the possible solutions considerably. Given such boundary conditions, experience often tells us where to *start* looking for solutions.

Once V has been determined, then **E** is obtained from $E_x = -\partial V/\partial x$, etc., as usual. So solving Laplace's or Poisson's equation can be a very general method for solving electrostatic problems. Since other potentials obey Laplace's equation, we postpone a general discussion of solutions until Sec. 13.4.

Uniqueness We now ask whether there is only *one* solution to a problem. The answer is that, provided the conditions are sufficiently well defined, Poisson's and Laplace's equations do have unique solutions. We shall look at a general theorem in Chapter 13 and here we illustrate the approach by a simple but important case.

Suppose the potential V is required over a region of space bounded by surfaces over which it is fixed at specified values. We allow charge density to be present in some parts of the region so that V satisfies Poisson's equation. The general argument is to assume that it is possible to find *two* solutions, V_1 and V_2, and then to show that $V_1 = V_2$ everywhere.

First we note that both V_1 and V_2 satisfy Poisson, so that $\nabla^2 V_1 = -\rho/\varepsilon_0$ and $\nabla^2 V_2 = -\rho/\varepsilon_0$. Consider now the potential $V_1 - V_2 = V_3$, say. Subtracting Poisson's equations shows that $\nabla^2 V_3 = 0$ everywhere in the region. In addition, since V_1 and V_2 must have the same specified values at the boundaries, V_3 must be zero over them. Hence V_3 has the following properties: it obeys *Laplace's* equation throughout the region and has the value zero over the boundaries.

In Sec. 4.1 we found that if a region contained no charge and was bounded by an equipotential surface, there could be no **E**-field within it, and V was constant throughout it (the region was a hollow in a conductor). Now our V_3 is the potential in just such a region only *its* value is *zero* at the boundaries. It must therefore be zero everywhere and it follows that $V_1 = V_2$. There is thus only one solution V to the original problem. An important consequence is that **E** is also unique and therefore, since $E = \sigma/\varepsilon_0$ near the surface of any conductor, the surface charge density is also unique: only one distribution will satisfy the original conditions.

Uniqueness is a powerful weapon since it is sometimes easy to find solutions by mere inspection, by simple trials or by analogy, and we now know that a solution found in any of these ways is the only one (see Sec. 13.4 for examples).

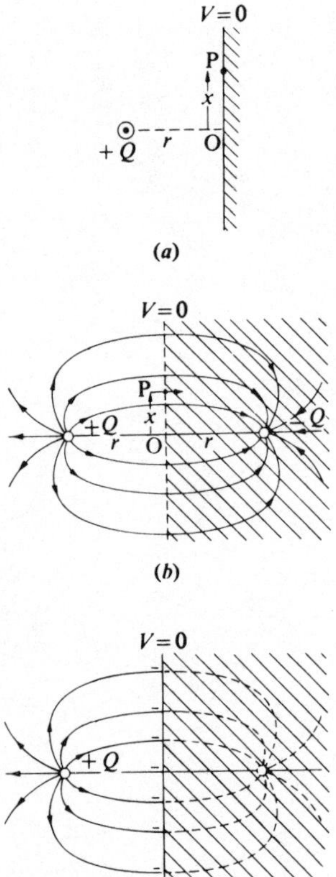

Figure 4.15 (a) A point charge and an infinite conducting plane; (b) a point charge and an image $-Q$ in the plane; (c) lines of **E** (full lines) due to a point charge and conducting plane: the broken lines do not exist in the real problem of diagram (a).

4.8 The method of images

In all the problems solved so far we have either been *given* a charge distribution or have been able to *assume* a uniform density because of the high degree of symmetry present. All we had then to do was to calculate V and **E**. However, in many problems the charge distribution on conducting surfaces will be neither given nor of high enough symmetry to make many assumptions. For such cases the surface density σ becomes another unknown to be determined as well as **E** and V. In this section we look at a method of solution, the method of images, that works when conducting surfaces have the same shape as the equipotentials in an already-solved problem.

Consider the situation in Fig. 4.15a. Q is a point charge at a perpendicular distance r from an infinite plane conducting surface which we shall take as a zero of potential. There is axial symmetry about the line QO, but this only means that the surface density and field at P will be the same as those at any other point on the surface at the same distance x from O as P. We can get no further by methods used so far and call on the uniqueness theorem to proceed.

Look at the lines of **E** and the equipotentials in Fig. 4.15b, which are a repetition of those in Fig. 3.20a with distances r and x added. Compare the region to the left of the $V=0$ equipotential with that in Fig. 4.15a: the regions contain the same charge distribution, i.e. the point charge $+Q$, and are bounded by identical conditions, i.e. $V=0$ over the plane. By uniqueness, V and **E** must be the same in the two cases, and since they are easily found for Fig. 4.15b, the problem of Fig. 4.15a is also solved, as shown in Fig. 4.15c. Note that to the *right* of the $V=0$ equipotential the solutions are quite different.

Thus to solve the given problem of Fig. 4.15a we replace the plane by a charge $-Q$ known as the *image charge* because it is at the same point as the image of $+Q$ would be in a plane mirror through OP. The **E**-field at P in Fig. 4.15b is $2Qr/[4\pi\varepsilon_0(r^2+x^2)^{3/2}]$. Outside any conducting surface with a density of charge σ, we have that $E=\alpha/\varepsilon_0$ from (4.12), and so the surface density of charge at P is $\varepsilon_0\ E$ or

$$\sigma_P = -2Qr/4\pi(r^2+x^2)^{3/2}$$

Moreover, the force between Q and the plane will be $Q^2/(16\pi\varepsilon_0 r^2)$, as it would be between $+Q$ and $-Q$. [Notice that defining the **E**-field as $d\mathbf{F}/dQ$ yields the incorrect result referred to in Comment C3.1: an uncharged plane would give a field of $Q/(8\pi\varepsilon_0 r^2)$.]

Other image problems The general method of images is appropriate when any conducting surfaces present have the same shape as equipotentials from specified distributions of charge. Thus we know that two *point* charges of opposite sign give equipotentials that can be (a) plane, as above, or (b) spheres by choosing magnitudes and distances appropriately (see Problem 4.24). Two *line* charges of equal and opposite densities give equipotentials that can be (a) plane, midway between them, or (b) cylindrical in general. It follows that many problems involving plane, cylindrical, or spherical conducting surfaces can be solved using the method of images in a way similar to the example above, although it is fair to say that the algebra is usually more complicated and it is sometimes necessary to superpose several image systems. Note that the 'image system' is not necessarily the same as that in the equivalent optical system, as Problem 4.24 shows. For an example outside electrostatics, see Sec. 13.4.

4.9 Conclusion

This chapter has been devoted mainly to the development of new expressions for the properties of **E**-fields that are a consequence of Coulomb's law. The mathematical forms of these expressions are vital for future work, but they seem at first sight to be far removed from the basic law of Chapter 2. For that reason, the derivations used in the text have been given in forms that attempt to maintain contact with the physics of the situations. This has made them more elaborate than is strictly necessary, and in this section we provide more concise derivations suitable for the advanced reader.

These derivations concern the transformation of Gauss's law and the circuital law from their integral to their differential forms. The methods are mathematically neater but physically less revealing.

Gauss's law The right-hand side of the integral form of this law in Eq. (4.2), $\Sigma Q/\varepsilon_0$, can be written as a volume integral $\int \rho\, d\tau/\varepsilon_0$ over the whole volume τ bounded by the closed surface S, so that

$$\oint_S \mathbf{E}\cdot d\mathbf{S} = \int_\tau \rho\, d\tau/\varepsilon_0 \tag{4.28}$$

There is a general theorem in vector analysis, known as *Gauss's divergence theorem*, which converts the surface integral of a vector over a closed surface into a volume integral of its divergence over the enclosed volume [see Appendix B.10, Eq. (B.17)]. Applied to **E**, this produces the identity:

$$\oint_S \mathbf{E}\cdot d\mathbf{S} \equiv \int_\tau \operatorname{div}\mathbf{E}\, d\tau \tag{4.29}$$

where S bounds τ. Using this in (4.28) converts Gauss's electrostatic law to the form:

$$\int_\tau \operatorname{div}\mathbf{E}\, d\tau = \int_\tau \rho\, d\tau/\varepsilon_0 \tag{4.30}$$

Since this applies to *any* volume τ, the integrands must be equal, so that

$$\operatorname{div}\mathbf{E} \equiv \nabla\cdot\mathbf{E} = \rho/\varepsilon_0 \tag{4.31}$$

as before.

Circuital law In this case, there is only the line integral of **E** round the closed path L to consider. Again, there is a general theorem in vector analysis, known as *Stokes's theorem*, which converts the line integral of a vector round a closed path into a surface integral of its curl over the enclosed area [Appendix B.10, Eq. (B.18)]. Applied to **E**, this produces the identity

$$\oint_L \mathbf{E}\cdot d\mathbf{L} \equiv \int_S (\mathbf{curl\, E})\cdot d\mathbf{S} \tag{4.32}$$

where L bounds S. Using this in (4.14), the circuital law becomes

$$\int_S (\mathbf{curl\, E})\cdot d\mathbf{S} = 0 \tag{4.33}$$

Since this is true for *any* surface area S, the integrand is zero and

$$\mathbf{curl\, E} \equiv \nabla\times\mathbf{E} = 0 \tag{4.34}$$

as before.

4.10 Revision summary

The E-field introduced in Chapter 3 has been used to express the laws of electrostatics in further ways through *Gauss's law* and the *circuital law*. Both of these have integral forms, applying to surfaces and paths, respectively, and differential forms, applying at a point. The potential V of Chapter 3 was also, it will be remembered, related to **E** through both an integral and differential form and these now lead to the very important Poisson's and Laplace's equations. We exhibit all these results in tabular form, using the ∇ operator which is given in Cartesians by

$$\nabla \equiv \mathbf{i}\frac{\partial}{\partial x}+\mathbf{j}\frac{\partial}{\partial y}+\mathbf{k}\frac{\partial}{\partial z} \tag{4.18}$$

	Integral forms		Differential forms	
A Definitions of E, V	$\int_A^B -\mathbf{E}\cdot d\mathbf{L}=V_{AB}$	(3.13)	Gradient: $\mathbf{E}=-\nabla V$	(4.17)
			$\nabla^2 V=-\rho/\varepsilon_0$	(4.26)
B Gauss's law	$\oint_S \mathbf{E}\cdot d\mathbf{S}=\frac{\Sigma Q}{\varepsilon_0}$	(4.2)	Divergence: $\nabla\cdot\mathbf{E}=\frac{\rho}{\varepsilon_0}$	(4.21)
C Circuital law	$\oint_L \mathbf{E}\cdot d\mathbf{L}=0$	(4.14)	Curl: $\nabla\mathrm{x}\mathbf{E}=0$	(4.24)

The forms taken by the operators gradient, divergence, curl and ∇^2 in various coordinate systems are to be found when needed in Appendices B.8, B.9 and B.12.

General properties of E and *V* due to charges:

From **A** above: E lines are perpendicular to equipotential V surfaces.
From **B** above: flux of E starts from positive charges and ends on negative charges. In general, charges are *sources* of E (negative charges are sometimes called *sinks*).
From **C** above: E lines due to charges cannot close on themselves. We say that such an E-field has no *vortices*.

Using these properties, plus the fact that conducting surfaces are equipotentials, it is often possible to sketch lines of E and of constant V on a diagram to a fair degree of accuracy in electrostatic situations without calculation.

Solving problems To determine E and V we now have the following methods:

- Determination of E or V using Coulomb's law and superposition as in Chapter 3. Then determine E from V or vice versa using **A** above.
- Determination of E through Gauss's law in integral form. A method only useful with highly symmetrical charge distributions. It yields, for instance:

  outside a charged sphere: $E=Q/(4\pi\varepsilon_0 r^2)$ (4.3)

  outside a charged cylinder: $E=\lambda/(2\pi\varepsilon_0 r)$ (4.10)

  outside a charged plane: $E=\sigma/\varepsilon_0$ (4.12)

- Determination of E, V and surface densities of charge on conductors by the method of images. For problems where the conducting surfaces can be made to coincide with equipotentials of charge distributions in already-solved problems.

- Determination of V by solving Poisson's or Laplace's equation either analytically or numerically. See Sec. 13.4.
- Sketching lines of **E** and equipotentials using the general properties listed above. Capable of fair accuracy and useful where the conducting surfaces have complicated shapes.
- The uniqueness theorem ensures that, however a solution is obtained, it is the only solution.

Commentary

C4.1 On E-fields within Hollow Charge Distributions Suppose electric charge is sprayed in any fashion over the surface of a hollow *insulator*. We may take a closed Gaussian surface inside the hollow and, because it encloses no charge, the total outward flux of **E** over it is zero. Some would then argue that because the Gaussian surface can be chosen anywhere within the hollow, **E** must be zero everywhere inside. This conclusion is clearly absurd since our original charge could have been sprayed on to a very small area of the insulator where it would have stayed and produced a $1/r^2$ field around it. The argument is clearly faulty, yet an exactly similar one is sometimes used to 'prove' the absence of an **E**-field inside a hollow *conductor*: it is certainly a simpler proof than the one we have used in Sec. 4.1!

The fallacy lies in the assumption that if the total *flux* of **E** over any surface in the region is zero, the value of **E** *at every point* on the surface is zero. One has only to take a region containing no charge but just a uniform **E**-field to see the error. The difficulty arises with hollow distributions of charge because we are used to the idea that **E** *is* zero inside for some highly symmetrical cases. However, we should remember that the only surface distributions inside which **E** is everywhere zero are those which can exist in equilibrium on a conducting surface (including of course uniform spherical and cylindrical surface distributions). For further discussion see Ehrlichson (1970).

C4.2 On the Inverse Square Law We can return at this stage to the question raised at the end of Sec. 2.3: what is the importance to modern physics of the inverse square part of Coulomb's law that warrants the experiments now being carried out to establish it with high precision? In Chapter 2 we discussed departures from the law in terms of the value q by which the exponent differs from 2, but an alternative form of departure has more physical significance. This is one that gives the potential

$$V = \frac{Qe^{-kr}}{4\pi\varepsilon_0 r} \tag{4.35}$$

where k is a constant having the dimensions of an inverse length and whose value would be zero if the inverse square law were exact. From the discussion in Sec. 2.5 we see that the deviations from inverse square will be negligible at distances much less than $1/k$ but that the field will fall off rapidly at distances much larger than $1/k$. Thus, $1/k$ is a measure of the *range* of such a field.

The form of (4.35) is suggested by analogy with the Yukawa potential for strong nuclear interactions. The Yukawa theory is that the strong forces between nucleons are brought about by the exchange of what are called virtual pions: particles with a predicted mass of $hk/(2\pi c)$, where h is Planck's constant, c the velocity of light, and $1/k$ the range of nuclear forces. Since this range is approximately known, the mass can be calculated and turns out to be identical with that of the free pion as measured by the usual techniques.

If the Coulomb force is due to a similar exchange, the particle concerned would be the photon. Thus the departure from Coulomb's law in terms of k and the possible range $1/k$ of the Coulomb force are related to the upper limit for the mass of the photon through the relation $m = hk/(2\pi c)$. Experiments have shown that the photon mass must be less than 2×10^{-44} kg and that the range of the Coulomb force must be at least 10^7 m. A complete set of references is given in Goldhaber and Nieto (1971).

C4.3 On Gravitationally Induced E-fields Can the electrons in a stationary piece of metal sink under gravity so that they have a greater concentration at the bottom than at the top? If so, they would set up an E-field tending to oppose further fall. Equilibrium would be established when the E-field thus set up produced a force $e\mathbf{E}$ on an electron exactly equal to $m\mathbf{g}$, assuming that **g** is the same for a single electron as for bulk matter. The expected field would thus be mg/e, about 5.6×10^{-11} V/m. This field would still exist just outside the metal and would therefore be expected to occur down the axis of a hollow vertical metal tube. Witteborn and Fairbank (1967) measured the rate of fall of some free electrons down the axis of such a tube and showed that the field mg/e did exist: **g** is thus the same for free electrons as for bulk matter. Theory and experiment are not yet in accord, however, because the non-rigidity of the positive-ion metallic lattice should produce an additional effect not observed (Strnad, 1971).

A similar field would be set up in a rotating cylinder. It would be radial and of magnitude $mr\omega^2/e$ at a distance r from the axis, ω being the angular velocity. This effect is closely related to the electromechanical experiments of Sec. 1.5.

Problems

Section 4.1

4.1 A point charge Q is placed at the centre of a cube. What is the flux of **E** over each face? The charge moves to a corner of the cube. What is now the flux over each face?

4.2 Estimate the total charge on the earth given that the mean fair-weather atmospheric E-field near the surface is $100\ \mathrm{V\,m^{-1}}$. (Radius of the earth is about 6400 km.)

4.3 Show that the flux of **E** over the base of a cone of semi-vertical angle θ with a charge Q at the apex is $Q(1-\cos\theta)/(2\varepsilon_0)$.

4.4 The potential in a region of space is given by $V = -kx^3$, where k is a constant. Find the total charge contained in a cube of side a with one corner at the origin and three edges along the positive x, y and z directions.

Section 4.2

4.5 A conducting sphere 18 cm in diameter is charged to 9 kV above earth potential and isolated. Find the charge on the sphere, the E-field just outside it and the field and potential 18 cm from its centre.

4.6 If air ionizes at a potential gradient of $30\ \mathrm{kV\,cm^{-1}}$, what is the greatest charge which can be carried by a sphere of 18 cm diameter without the occurrence of corona?

4.7 If some apparatus is to be operated at 90 kV, what is the smallest external radius of curvature which should be permitted? (Dielectric strength of air = $30\ \mathrm{kV\,cm^{-1}}$.) If smaller radii are unavoidable, what is a possible solution to the difficulty?

4.8 A 5 MeV α-particle makes a head-on collision with a gold atom. What is the distance of closest approach? (Charge on α-particle $= 2e$; charge on gold nucleus $= 79e$.)

4.9 Find the **E**-field at a distance r from the centre of each of the spherically symmetrical charge distributions of Problem 1.14.

4.10 Assuming that a beam of electrons can be treated as a charge distributed uniformly over a cylindrical region of radius a and of infinite length, find the **E**-field at a distance r from the axis for $r > a$ and $r < a$, taking the line density of charge as λ. If there are n electrons per unit volume in the beam, show that the force on an electron at a distance r from the axis is $ne^2r/(2\varepsilon_0)$ radially outwards, e being the electronic charge.

4.11 A sheet of charge, surface density σ, gives an **E**-field of $\sigma/(2\varepsilon_0)$ on both sides. If, however, a thin sheet of metal is charged with σ per unit area, Eq. (4.12) seems to show that the field outside will be σ/ε_0. Resolve the paradox.

4.12 An infinite plane sheet of charge gives an **E**-field of $\sigma/(2\varepsilon_0)$ at a point P a distance a from it. Show that half the field is contributed by charge whose distance from P is less than $2a$ and that in general all but f per cent of the field is contributed by charge whose distance from P is less than $100a/f$.

4.13 A conductor carries its charge in a thin sheet on its surface, and the **E**-field just inside and just outside the sheet should be $\sigma/(2\varepsilon_0)$ as in Fig. 3.5a. How do you reconcile this with the zero field which in fact occurs inside conducting material and the field of σ/ε_0 which occurs just outside?

Section 4.3

4.14 The absence of an **E**-field inside a hollow charged conductor might be justified by a method different from that of Sec. 4.1. Consider the possible existence of electric flux inside such a conductor which (a) begins and ends on the conductor, (b) begins on the conductor and ends in space, (c) closes on itself or (d) does something other than (a), (b) or (c).

Section 4.5

4.15 Show that an **E**-field *in vacuo* whose lines of force are straight lines all parallel to the x axis cannot vary in the x direction and is thus a uniform field. (Note that along a *single* straight line of force the **E**-field is not necessarily constant in magnitude—see Fig. 3.20, for example.)

4.16 Solve Problem 4.4 by using the differential form of Gauss's law to determine the volume charge density ρ and hence the total charge in the cube by integration.

Section 4.6

4.17 Can any distribution of charge be found which could give rise to **E**-fields with the following components? If so, find the potential:

(a) $E_x = -ky$, $E_y = -kx$, $E_z = 0$.

(b) $E_x = -kx$, $E_y = -ky$, $E_z = -kz$.

(c) $E_x = k(x^2 + y^2)$, $E_y = 2kxy$, $E_z = 2kyz$.

Section 4.7

4.18 Take those parts of Problem 4.17 to which the answer is 'yes' and find what volume charge density is needed (if any) to give the potential.

4.19 The volume charge density in a region of space varies in the x direction only and is given by $-aV^{1/2}$, where a is a constant and V is the potential. If V and $\mathrm{d}V/\mathrm{d}x$ are both zero when $x = 0$, find how the charge density varies with x.

4.20 Poisson's equation in cylindrical coordinates is

$$\frac{1}{r}\partial V/\partial r + \partial^2 V/\partial r^2 + \frac{1}{r^2}\partial^2 V/\partial \theta^2 + \partial^2 V/\partial z^2 = -\rho/\varepsilon_0$$

For constant ρ, show that if cylindrical symmetry is complete (no variation with θ) and all cross-sections perpendicular to the z axis are identical (no variation with z), V must be of the form $k_1 \log_e r + k_2 r^2 + k_3$, where k_1, k_2 and k_3 are constants. Apply this to the solution of Problem 4.10.

Section 4.8

4.21 Two very large conducting planes intersect at right angles and a point charge Q is situated in the corner a perpendicular distance a from both. Find the set of image charges making *both* planes zero equipotentials and find the force on Q.

4.22 Show, by integration over the surface, that the charge induced on the plane in Fig. 4.15 is $-Q$.

4.23 A simple pendulum has a mass m at the end of its string of length l. The mass possesses a charge Q and swings a distance a above a large horizontal conducting plane at zero potential. Find the period of oscillations with amplitude much smaller than a or l, neglecting all damping.

4.24* A point charge Q is a distance d from the centre O of an earthed conducting sphere of radius a. Show that a charge $-Qa/d$ at a distance a^2/d from O along OQ is a satisfactory image charge. Find the ratio of the maximum to the minimum surface density of charge induced on the sphere.

4.25 An electric dipole of moment **p** is a perpendicular distance r from a semi-infinite conducting plane. If **p** is perpendicular to the plane, find the force on the dipole.

4.26* A thundercloud may be idealized as a vertical dipole with its positive charge uppermost. Consider the changes in potential gradient at a point A on the earth's surface as such a cloud approaches at a constant height from a great distance to a point directly over A. Assume that the earth's surface can be treated as a plane conductor. For what horizontal distance R is the field at A zero if the height of the cloud is H?

5

Capacitance and electric energy

The methods and results developed in the previous two chapters can now be used to investigate more practical situations than most of those encountered so far. The important concept of **capacitance** as a quantity relating charge and potential is introduced for an isolated conductor (Sec. 5.1) and for a pair of conductors forming an ideal capacitor (Sec. 5.2) when unaffected by other conductors. We see how such capacitors combine into networks (Sec. 5.3), how real capacitors are constructed and how far they approximate to the ideal (Sec. 5.4).

Capacitors can be regarded as means of storing **electric energy** and in Sec. 5.5 we continue the study of such energy begun in Chapter 2. Expressions are first derived in terms of the charges themselves and then, in Sec. 5.6, we look at an alternative form in terms of the **E**-field. The expressions for energy lead naturally to an examination of the forces and torques between charged conductors and their application in certain instruments (Sec. 5.7). Finally, we look briefly at some applications of electrostatics.

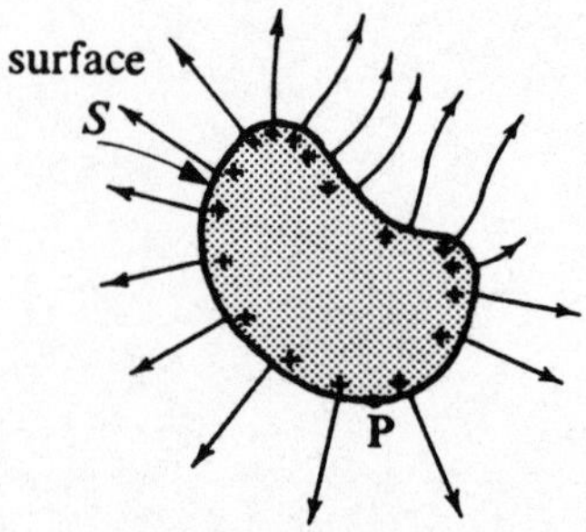

Figure 5.1 A single charged conductor remote from other bodies. The flux of **E** ends on surroundings effectively at infinity.

5.1 Capacitance of a conductor

Concept and definition When an uncharged finite conductor remote from other bodies is given a charge Q (Fig. 5.1), its potential with respect to a zero at infinity rises from zero to a certain value V. We might think of Q as the amount of charge the conductor can hold at the potential V, i.e. its *capacity* for carrying charge, and we intuitively expect it to increase with the physical size of the conductor.

This property can be specified more precisely by defining the **capacitance** C of the conductor as the ratio Q/V, i.e.

$$C=Q/V \qquad \text{(definition of } C\text{)} \tag{5.1}$$

The definition enables C to be calculated for some particular conductors: e.g. Eq. (4.5) shows that for a conducting sphere of radius a,

$$\text{conducting sphere:} \qquad C=4\pi\varepsilon_0 a \tag{5.2}$$

although for other shapes an accurate calculation is not easy to carry out.

The **SI unit** of capacitance is the CV^{-1} or **farad**, symbol F, although the submultiples 1 μF and 1 pF (10^{-12} F) are more convenient in practice. The definition (5.1) means that a capacitance will always be of the form of ε_0 multiplied by an expression with the dimensions of length, like (5.2). It follows that the dimensions of ε_0 are those of capacitance per unit length, i.e. $\mathrm{F\,m^{-1}}$ in SI units. Using (5.2), a conducting sphere of radius 10 cm will have a capacitance of $0.1/(9\times 10^9)$F or approximately 11 pF.

The proportionality of Q and V The definition (5.1) does not guarantee that C is constant for any given conductor, even though (5.2) shows that it will be for a sphere. We now prove that Q and V are proportional for a conductor of *any* shape or size and thus that C will depend only on the geometry.

This proof is given for completeness: it is not an essential step in the understanding and use of capacitance as a property of conducting systems.

First we recall the uniqueness theorem of Sec. 4.7. This shows that when the potential of a conductor such as that of Fig. 5.1 is specified, there is only one distribution of charge over its surface that can give that potential. Thus, for a given V, the surface density of charge σ is determined at every point. Now consider the conductor first with a potential V_1 and then with V_2. Let the surface density at any point in the first case be σ_1 and in the second, σ_2. The potentials are then given by (3.17) as

$$V_1=\int\sigma_1\,\mathrm{d}S/(4\pi\varepsilon_0 r) \qquad \text{and} \qquad V_2=\int\sigma_2\,\mathrm{d}S/(4\pi\varepsilon_0 r) \tag{5.3}$$

Suppose that $V_2/V_1=n$; then

$$V_2=nV_1=\int n\sigma_1/(4\pi\varepsilon_0 r) \tag{5.4}$$

Equations (5.3) and (5.4) show that V_2 is produced by a distribution σ_2 and also by $n\sigma_1$, so that the uniqueness theorem gives $n\sigma_1=\sigma_2$. If we now evaluate the total charge on the conductor in the second case, we have

$$Q_2=\int\sigma_2\,\mathrm{d}S=\int n\sigma_1=nQ_1 \tag{5.5}$$

So $Q_2/Q_1=n=V_2/V_1$, i.e. Q and V are proportional. The capacitance C is thus a constant quantity for a conductor of given shape and size when remote from other bodies.

5.2 Ideal capacitors and calculation of capacitance

Capacitance of an isolated pair of conductors In practice, conductors are not isolated as in Sec. 5.1 but are influenced by other bodies which could, of course, be of either conducting or insulating material. Charges on any one of the bodies will affect the distribution on all the others, influencing the potentials of conductors and the polarization of insulators, and a very complicated set of relationships will arise. As usual, we begin with simple situations and shall assume for most of this chapter that non-conducting bodies are not present: we shall deal briefly with their effect in Sec. 5.4 and leave a detailed treatment until Chapter 11.

In this section we consider the important case illustrated in Fig. 5.2, where only one other conductor B is near enough to influence the original conductor A. All the flux of **E** from A therefore ends on B. The application of Gauss's law to any surfaces completely enclosing A or B shows that the charges on the two conductors must be equal and opposite, as if A were charged from B. Such an arrangement is known as a **capacitor**, but a somewhat idealized one in that it is assumed that no other bodies affect the field. If the conductors are initially uncharged, the transfer of a charge Q from A to B will produce a potential difference $V_A - V_B$ between the two conductors, and the capacitance of the capacitor is defined by

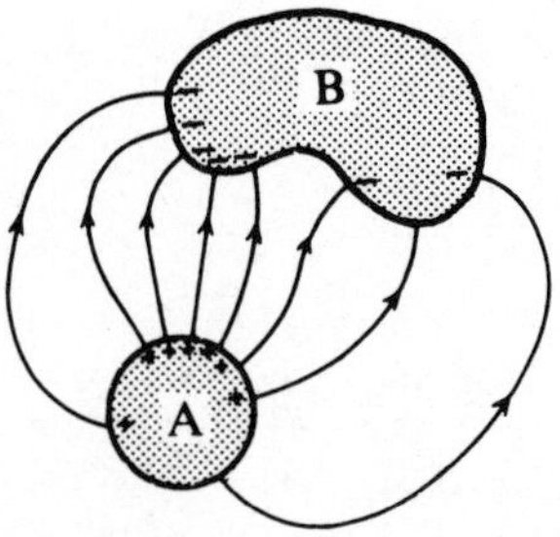

Figure 5.2 An ideal capacitor. All the flux of **E** from A ends on B.

$$C = \frac{Q}{V_A - V_B} \qquad \text{(definition of } C\text{)} \tag{5.6}$$

Definition (5.1) can be regarded as a special case of (5.6) in which B surrounds A and is a large distance away.

The proportionality of Q and $V_A - V_B$ in the absence of insulators now follows as in the previous section. The only difference is that the integrals for V must be taken over both surfaces. It is otherwise just as easy to show that first V_A and then V_B must both be proportional to Q. This argument rests very much on that part of Coulomb's law denoted by $F \propto Q_1 Q_2$ because it is this which yields the numerator of the expression $\int \sigma\, dS/(4\pi\varepsilon_0 r)$ for potential. An experiment in which Q on a capacitor is measured by the throw on a galvanometer used ballistically and $V_A - V_B$ by a potentiometer (Sec. 6.9) can be performed in the laboratory. Their proportionality, and thus the constancy of C, is verified and provides a highly sensitive test of the law of Sec. 2.2. (Note that the experiment would not be valid using an electrometer which is calibrated on the assumption that $Q \propto V$.)

Calculation of capacitance In principle, the value of C for a given pair of conductors can be determined using (5.6), i.e. by imagining a charge Q transferred from A to B and calculating the resultant potential difference. In practice it is only possible to carry out accurate theoretical calculations if the geometry is simple. To illustrate the method, consider two concentric spherical conducting shells of radii a and b as in Fig. 5.3. If a charge Q is transferred from the outer conductor to the inner, we can assume that the resultant charge densities will be uniform by symmetry and uniqueness. In that case, the **E**-field and potential in the space between the conductors is, by Gauss's law, the same as if all the $+Q$ on the inner sphere were concentrated at its centre (Sec. 4.2), while the $-Q$ produces no **E**-

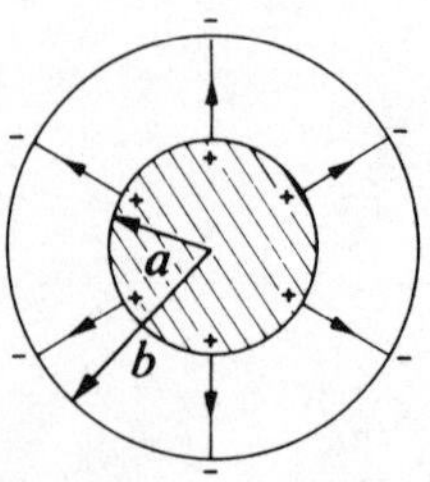

Figure 5.3 A spherical capacitor.

field inside itself. Hence the potential difference between inner and outer is $Q/(4\pi\varepsilon_0 a) - Q/(4\pi\varepsilon_0 b)$ or $Q(b-a)/(4\pi\varepsilon_0 ab)$. Using (5.6), we have

spherical capacitor: $$C = \frac{4\pi\varepsilon_0 ab}{b-a} \qquad (5.7)$$

The earth and the ionosphere can be regarded as a giant spherical capacitor (Comment C5.2).

Equation (5.7) tends to (5.2) as b tends to infinity and this illustrates the point that the surroundings in Sec. 5.1 cannot really be ignored since there must be a destination for the flux of **E** from the apparently isolated conductor.

Capacitor geometry An ideal capacitor is best realized by enclosing one conductor within another as in the spherical capacitor of Fig. 5.3, but this arrangement is inconvenient because of the inaccessibility of the inner conductor. The shapes in common use are the parallel plate and the cylindrical (Fig. 5.4), in both of which there will be two effects occurring at the edges. One is a possible leakage of electric flux to surrounding objects (see 'stray capacitance' in Sec. 5.4). The other is the distortion of the E-field from its ideal configuration called the **edge effect** or **fringing** (Fig. 5.4b). Both effects will become less important as the spacing between the conductors decreases in comparison with other linear dimensions and in this way the ideal situation can be approached very closely. In certain uses the edge effects can also be virtually eliminated by using guard-rings, illustrated later (Fig. 5.18b), or guard-cylinders. We shall carry out the calculations for the two ideal geometries as if each capacitor were part of one with infinite extent.

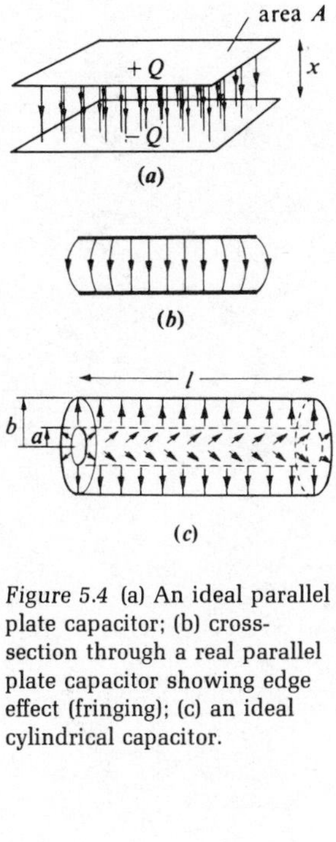

Figure 5.4 (a) An ideal parallel plate capacitor; (b) cross-section through a real parallel plate capacitor showing edge effect (fringing); (c) an ideal cylindrical capacitor.

In Fig. 5.4a, the transference of charge Q from one plate to the other produces uniform surface densities of charge $\sigma = Q/A$, positive on one plate and negative on the other. By (4.12), this produces a uniform E-field of magnitude σ/ε_0 or $Q/(\varepsilon_0 A)$. The potential difference is thus Ex or $Qx/(\varepsilon_0 A)$ and so

parallel-plate capacitor: $$\boxed{C = \varepsilon_0 A/x} \qquad (5.8)$$

In Fig. 5.4c, the transference of Q from the outer to the inner cylinder produces a charge per unit length of Q/l. Thus, by (4.10), there is an E-field of $Q/(2\pi\varepsilon_0 rl)$ at any distance r from the axis, the outer cylinder producing no field inside itself. The potential difference given by $\int -E\,\mathrm{d}r$ is thus the integral of $Q\,\mathrm{d}r/(2\pi\varepsilon_0 rl)$ between a and b, i.e. $(Q \log_e b/a)/(2\pi\varepsilon_0 l)$. Hence $C = Q/V$ is

cylindrical capacitor: $$C = 2\pi\varepsilon_0 l/\log_e(b/a) \qquad (5.9)$$

Distributed capacitance In a capacitor, the capacitance is deliberately localized within a relatively small volume, and it is then described as being **lumped**. In extended conductors, however, such as cables or transmission lines used to convey electric currents over large distances, the capacitance is **distributed continuously** and is an important factor in any electrical changes which occur. Transmission lines, treated in detail in Chapter 10, consist most often of two conductors arranged either as a coaxial cable (Fig. 5.5) or as a twin cable (Fig. 5.6).

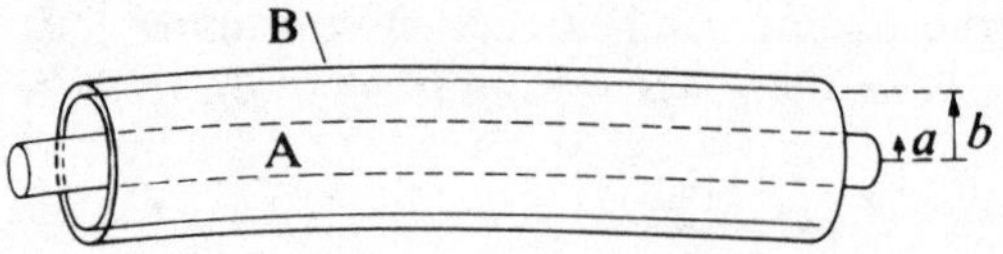

Figure 5.5 A section of coaxial cable.

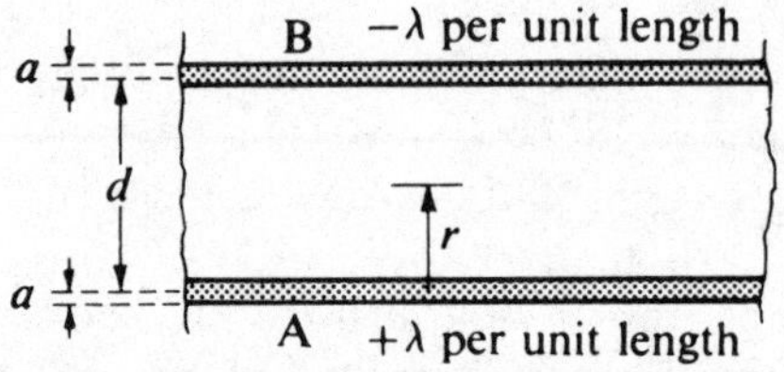

Figure 5.6 A section of twin cable.

The **coaxial cable** forms an extended cylindrical capacitor and we expect (5.9) to apply quite accurately here in the form

$$\text{coaxial cable:} \qquad C = 2\pi\varepsilon_0/\log_e(b/a) \qquad \text{per unit length} \tag{5.10}$$

The capacitance of **twin cable** can be most easily found if the radius of the wires (Fig. 5.6) is small compared with their distance apart because the non-uniformity of the distribution of charge over the conductors can be neglected. If a charge λ per unit length is transferred from one to the other, the **E**-field at a distance r from one wire has a magnitude $\lambda/(2\pi\varepsilon_0 r) + \lambda/[2\pi\varepsilon_0(d-r)]$. The first term is the field of the lower wire and the second that of the upper. The potential difference

$$V_A - V_B = \frac{\lambda}{2\pi\varepsilon_0}\int_{d-a}^{a}\left(-\frac{1}{r} - \frac{1}{d-r}\right)\mathrm{d}r$$

$$= \frac{\lambda}{\pi\varepsilon_0}\log_e\frac{d-a}{a}$$

For twin cable in which $d \gg a$, this becomes $[\lambda/(2\pi\varepsilon_0)]\log_e(d/a)$. Since the charge per unit length is λ,

$$\text{twin cable:} \qquad C = \frac{\pi\varepsilon_0}{\log_e(d/a)} \qquad \text{per unit length} \tag{5.11}$$

This result can also be used to find the capacitance between a single elevated horizontal cable, like a telegraph wire, and the earth, as in Problem 5.6.

Making an estimate of capacitance The ability to obtain an order of magnitude for a physical quantity is extremely useful because rough estimates are often sufficient to begin practical design work. This is true of capacitance: several of the above formulae, although developed for ideal cases, often give close approximations to the real values. This is particularly true of the parallel plate formula when the two conductors are very close together and of the formulae for cables when they are very long. The reason why good estimates are obtained in these cases is that the charge distribution over the conductors is in practice very nearly uniform. We now look at an example in which the charge is not at all uniform, so that none of the formulae developed so far can be applied.

Worked example *A thin conducting disc* of radius a would have the greatest charge density at the edges where the curvature is large. The distribution will in fact be intermediate between a uniform one and one in which *all* the charge is concentrated round the rim. A very simple calculation of the potential at the centre of the disc (and therefore of the disc as a whole) can be carried out for these two extremes.

For a uniform charge density σ, a typical annulus of radius r (Fig. 5.7a) carries a charge $\sigma\, 2\pi r\, \mathrm{d}r$ which produces a potential $\sigma\, \mathrm{d}r/(2\varepsilon_0)$ at the centre. Integration of this from 0 to a gives the potential of the whole disc as $\sigma a/(2\varepsilon_0)$. Since the charge is $\pi a^2 \sigma$, the capacitance would be $2\pi\varepsilon_0 a$. In a similar way, if the charge were all round the edge (Fig. 5.7b) the capacitance would be $4\pi\varepsilon_0 a$. The true value should lie between the two we have obtained, i.e. between about $6\varepsilon_0 a$ and $12\varepsilon_0 a$, so that we have at least an order of magnitude (accurate calculation gives $8\varepsilon_0 a$).

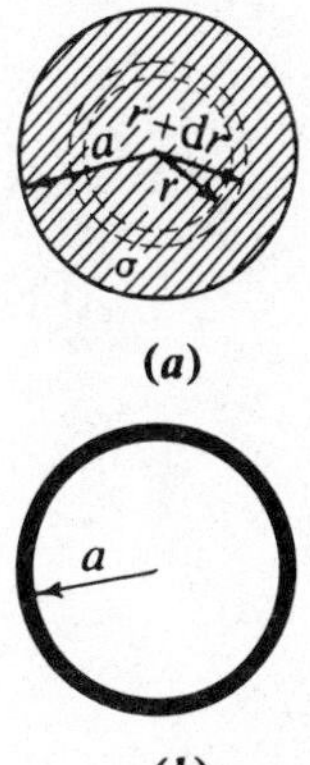

Figure 5.7 Estimating the capacitance of a disc. The true charge distribution lies between the uniform distribution (a) and one in which it is concentrated round the edge (b).

5.3 Combinations of capacitors

Equivalent capacitance We now wish to look at the connection of several ideal capacitors by conducting wires into a network. A simple example is shown in Fig. 5.8, where a capacitor is represented diagrammatically by a pair of thick parallel lines (crossed by an arrow if adjustable). It is useful to be able to calculate the **equivalent capacitance** between two points such as A and B, defined once again by $Q/(V_A - V_B)$ where $V_A - V_B$ is the potential difference produced by the transference of Q from B to A. This is a slight generalization of (5.6) and means that we are asking for the capacitance of that single capacitor which, if connected between A and B, would produce the same effect as the original network.

To obtain the equivalent capacitance we use two principles already established: the conservation of charge and the path-independence of potential difference. First we establish two useful formulae for special cases.

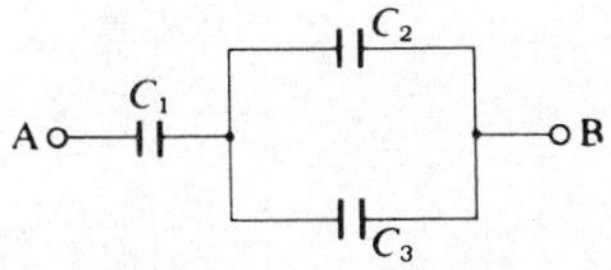

Figure 5.8 A simple capacitor network.

Capacitors in series A set of capacitors as in Fig. 5.9 are in series. A charge Q transferred from one end to the other produces charges $\pm Q$ on the plates as shown, using the conservation of charge. The potential differences are therefore $V_1 = Q/C_1$, $V_2 = Q/C_2$, $V_3 = Q/C_3$. The path-independence of potential difference shows that the total V across AB is the sum of these three. The equivalent capacitance C_{eq} is thus given by

$$1/C_{eq} = V/Q = (V_1 + V_2 + V_3)/Q = 1/C_1 + 1/C_2 + C_3$$

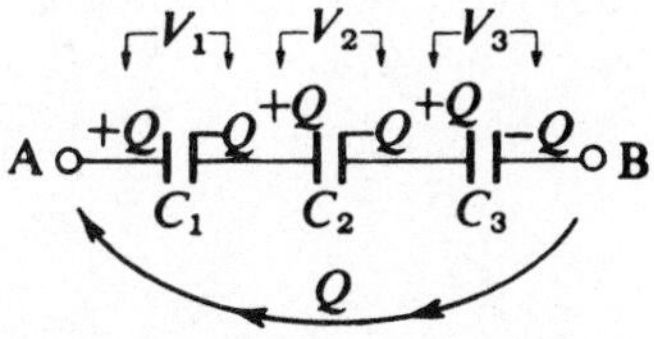

Figure 5.9 Capacitors in series (common Q).

In general, therefore:

capacitors in series:
$$\frac{1}{C_{eq}} = \sum_i \frac{1}{C_i} \tag{5.12}$$

Note particularly that these capacitors act as a *potential divider*, with voltages each inversely proportional to their capacitance but with common charges.

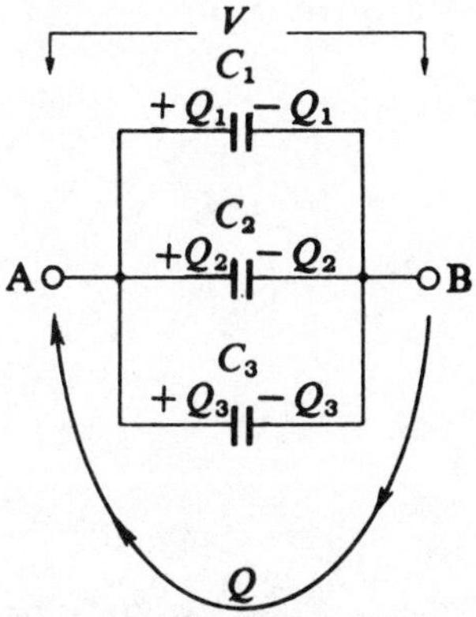

Figure 5.10 Capacitors in parallel (common V).

Capacitors in parallel In Fig. 5.10 it is the charge Q which divides into Q_1, Q_2 and Q_3 such that $Q = Q_1 + Q_2 + Q_3$ by the conservation of charge. On the other hand, all capacitors now have the same V (path-independence of potential difference).

The equivalent capacitance is therefore

$$C_{eq} = Q/V = (Q_1 + Q_2 + Q_3)/V = C_1 + C_2 + C_3$$

or in general

capacitors in parallel: $$\boxed{C_{eq} = \sum_i C_i} \qquad (5.13)$$

General networks Many networks can be treated as a collection of series and parallel combinations and the equivalent capacitance found by using (5.12) and (5.13). For instance, the network of Fig. 5.8 is equivalent to $C_1(C_2+C_3)/(C_1+C_2+C_3)$. However, a network like that of Fig. 5.11 cannot be treated in that way, for no two capacitors are in series (common Q) or in parallel (common V). A general method based on our two fundamental principles must be used.

Worked example Taking the network of Fig. 5.11, we again transfer Q from B to A. Let Q_1 and Q_2 be the charges on the capacitors shown and apply the conservation of charge to the junctions A, B, D, and E, giving the charges also shown in the figure. To find Q_1 and Q_2, apply the path-independence of V in the form: potential difference round a closed path is zero. This gives

$$Q_1/C - (Q - Q_1)/C + (Q_1 - Q_2)/C = 0 \quad \text{round DAE}$$
$$Q_2/2C - (Q_1 - Q_2)/C - (Q - Q_2)/C = 0 \quad \text{round BDE}$$

Solved for Q_1 and Q_2, these equations give $Q_1 = 7Q/13$ and $Q_2 = 8Q/13$. The potential difference between A and B via D is thus $11Q/13C$ and the equivalent capacitance Q/V is $13C/11$.

A further method for reducing networks is given in Problem 5.12.

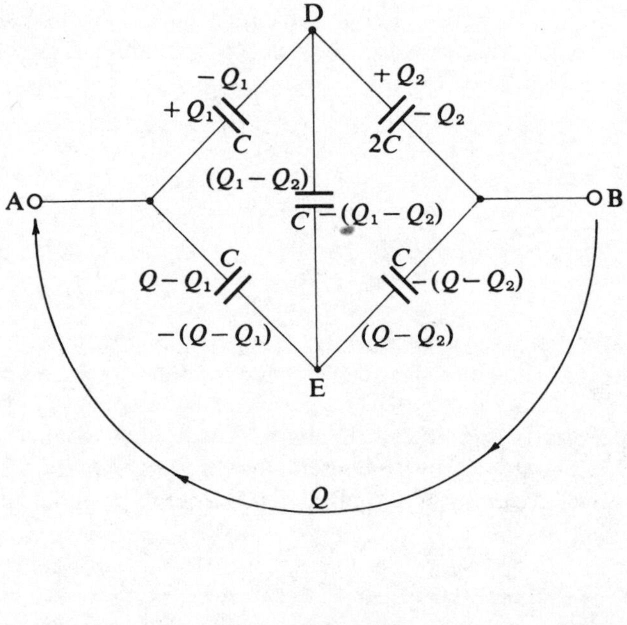

Figure 5.11 A network which cannot be broken into series and parallel combinations.

5.4 Capacitors in practice

Capacitor applications Capacitors form one of the most widely used components in electrical and electronic circuits, but the great variety of these circuits means that there is an equally great diversity in capacitor design. The following list gives some indication of the range of applications, together with references to their treatment later in this book, although such a list clearly cannot be exhaustive:

1. Standardizing methods (Appendix C) and measuring circuits and instruments (e.g. bridges, electrometers).
2. Storage of electric energy (Sec. 5.5).
3. Circuits using the ability of capacitors to block direct current and provide a low-impedance path for alternating current depending on its frequency: filters (Sec. 10.9), bypass and noise-suppression circuits, smoothing circuits (e.g. in power supplies), coupling capacitors.
4. Tuning circuits (Sec. 10.6).
5. Wave-shaping and timing circuits (Sec. 10.12).

Dielectrics The space between the plates of a real capacitor is rarely a vacuum, as we have assumed, but is usually filled with an insulator or **dielectric**. Dielectric materials are dealt with fully in Chapter 11, but they have several important effects which must be outlined here since they affect the design of capacitors.

First, dielectrics increase the capacitance. If C_0 is the capacitance of a capacitor *in vacuo*, and C_m that of the same capacitor with a homogeneous insulating material filling the whole of the region in which an electric field exists, then

$$C_m/C_0 = \varepsilon_r \qquad \text{(definition of } \varepsilon_r\text{)} \tag{5.14}$$

defines the **relative permittivity** or **dielectric constant** of the material. Thus, for a parallel plate capacitor whose plates are separated by an insulator of relative permittivity ε_r, the capacitance is $\varepsilon_r\varepsilon_0 A/x$. For some materials ε_r is a function of the potential difference and C_m, unlike C_0, is not then dependent only on the geometry. The definition (5.14) means that the factor ε_0 should be replaced by $\varepsilon_r\varepsilon_0$ in all the expressions for capacitance in terms of the geometry of the capacitor.

The fact that C_m is always greater than C_0 is explained in terms of the polarization of the dielectric in the **E**-field between the plates (end of Sec. 3.8). The negative charges produced by the polarization will be close to the positive charge on the capacitor plate and will thus reduce the **E**-field that it generates. The same effect occurs at the other plate and the total reduction in **E** causes a similar reduction in V. Since the true Q on the plates has remained constant, $C = Q/V$ will increase. For further details see Chapter 11.

The second effect produced by an insulator is a limitation on the potential which can be applied across the plates. At some potential gradient known as the **dielectric strength** of the material it breaks down and conducts.

For the moment, the following orders of magnitude for the two properties of dielectrics will suffice: common gases at STP have ε_r between 1 and 1.001 and dielectric strengths of about 3 kV mm^{-1}, while common solid insulators have ε_r between 2 and 10 and dielectric strengths from a few to several hundred kilovolts per millimetre (see Sec. 11.10).

Because ε_r may vary with applied potential difference, with pressure, temperature, humidity, frequency and with age, real capacitors may be constructed to possess a wide variety of characteristics determined almost entirely by the dielectric. They might also, of course, exhibit unwanted variations in properties. Briefly, the types of capacitor manufactured are designed to give the required capacitance in the smallest volume subject to the following:

1. Whether the need is for a fixed or continuously variable capacitance.
2. The largest potential difference that needs to be applied across the plates without breakdown (the working voltage).
3. The required stability under varying physical conditions and with time.
4. The maximum energy consumption (dielectric loss) and maximum current which can be permitted (no dielectric being perfect).
5. Cost.

For small distances between the plates, capacitances are always given effectively by the parallel plate formula $\varepsilon_r\varepsilon_0 A/x$, so that small x (obtained by using thin films of sufficient dielectric strength) and large ε_r both contribute to compactness and enable A to be reduced.

Types of capacitor ***Air and vacuum capacitors*** are bulky and are only used as primary standards or where a continuous variation of C is required. As primary standards, such capacitors are invaluable because C depends only on ε_0 and certain dimensions and is thus *calculable* (see Appendix C). Continuously variable capacitors for tuning circuits can be of the parallel plate type with a movable set of plates overlapping a fixed set by an amount determined by the rotation of a calibrated head; or of the cylindrical type, with a central conductor whose length of overlap with the outer is controlled by a micrometer screw head. Typical capacitances are from a few picofarads to about 0.01 μF.

Solid and plastic dielectric capacitors have a much larger capacitance-to-volume ratio because their plates can be so much closer and the very high ε_r of some materials enhances the ratio even more. For relatively stable secondary standards, sheets of tinfoil and **mica** only a few thousandths of an inch thick are interleaved, rigidly clamped, heated to drive off moisture and then encased in wax. Mica sprayed with silver is also used in small capacitors. For less exacting requirements **paper**, impregnated with paraffin wax, interleaved with aluminium or tinfoil and rolled into a small volume, was long the main dielectric for general use, but **plastics** have replaced paper and metallized films the foils. **Ceramic** capacitors are now in widespread use because they are cheap and compact: the ceramics may be ferroelectrics such as barium titanate with ε_r up to many thousands in the form of discs or tubes which are coated on two opposite surfaces with silver to form the electrodes. These are used as components in miniaturized or hybrid circuits (the latter consisting of combinations of integrated circuits and lumped components). At microwave frequencies, capacitance can be provided by sections of **transmission line** (Sec. 10.10).

In monolithic (i.e. completely integrated) circuits, the function of capacitors is performed by the capacitance across a *p-n* junction. Such junctions show a variation of capacitance with applied voltage, and this can be used to produce non-linear lumped circuit components known as **varactors** or **varactor diodes**.

Electrolytic capacitors are produced by the electrolysis of certain salts between aluminium or tantalum foils, causing an insulating film of thickness about 0.1 μm to be formed on one foil. Used as a dielectric, this film enables capacitances of up to several thousand microfarads to be obtained in a reasonable volume although with a relatively low working voltage. Since quite a small potential difference applied in the reverse direction would destroy the film, the main uses for these capacitors are those in which unidirectional voltages are encountered.

A survey of capacitor technology is to be found in Fairs (1974), while Trotter (1988) gives an up-to-date descriptive treatment.

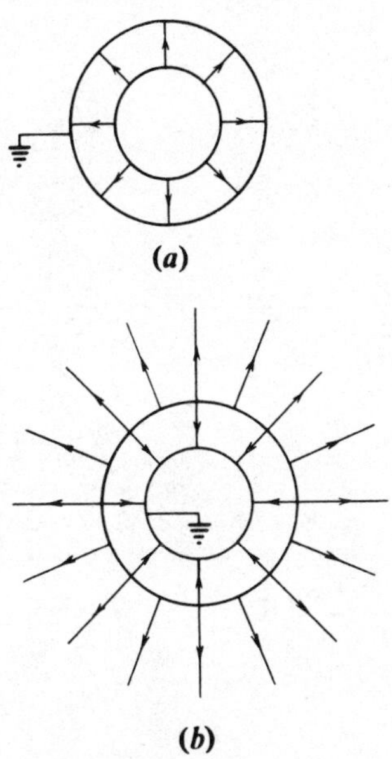

Figure 5.12 A spherical capacitor with (a) outer conductor earthed, (b) inner conductor earthed.

Stray capacitance and screening If one plate of a capacitor completely encloses another and is earthed (i.e. connected to the surroundings at constant potential) as in Fig. 5.12a, the capacitance is just that between the inner and outer plates: the capacitor is *screened* from external effects. If, however, the inner plate is earthed (Fig. 5.12b), there is an additional capacitance between the outer conductor and the surroundings in parallel with the original one (why in parallel?). The extra C is known as a *stray capacitance* and could be eliminated in this case by using the connection of Fig. 5.12a (see Problem 5.13).

A practical capacitor often does not have one plate screening another and strays will occur between *both* plates and earth as in Fig. 5.13a. They can be appreciable fractions of C and they make the actual capacitance between A and B indeterminate. A complicating factor in variable capacitors is the change in capacitance between the movable plate and earth as the hand approaches the knob to turn it, and as the knob is turned. To avoid this, a metal case is provided to which plate A is connected as in Fig. 5.13b. C_1 is now eliminated and C_2 is quite definite so that the actual capacitance is $C+C_2$ and this is the value quoted by the makers. If C is variable, the movable plate is connected to A.

All electrical apparatus with metallic sections will have stray capacitance to the surroundings and between components, so that variable and undesirable effects will be introduced when alternating current is used. Connecting wires are a great source of trouble in this way and are therefore usually screened by a flexible outer cylindrical sheath of braided copper connected to earth.

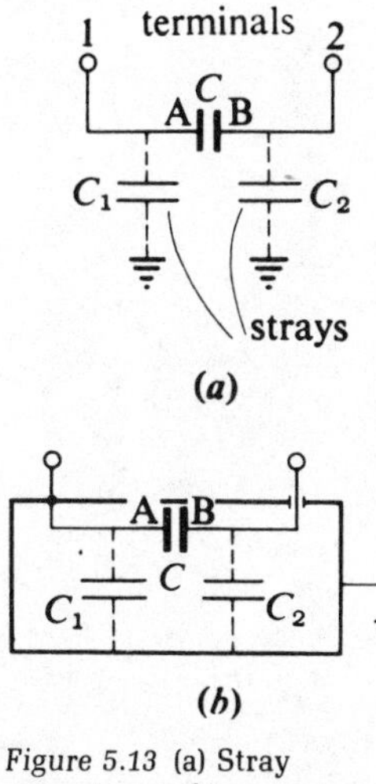

Figure 5.13 (a) Stray capacitance; (b) screening.

5.5 Electric energy 1: in terms of Q, V and C

We now return to a consideration of the energy due to electrostatic forces, a subject we left in Chapter 2. As in Sec. 2.7, the method of finding the potential energy of a system of charges is to calculate the work done in assembling them. We generalize the result of Sec. 2.7 below and then apply it to conductors. First, however, we consider the energy stored in an ideal capacitor of only two charged conductors.

Work done in charging a capacitor or conductor Whenever an agency such as a voltaic cell charges a capacitor, it does work in transferring the charge. Let us take a capacitor with capacitance C which is charged by a cell as in Fig. 5.14 to a final potential difference V and charge $\pm Q$. The figure itself shows an intermediate stage of the charging process when the charge already transferred is q and the resultant potential difference is v, and, of course, $q=vC$. A further infinitesimal

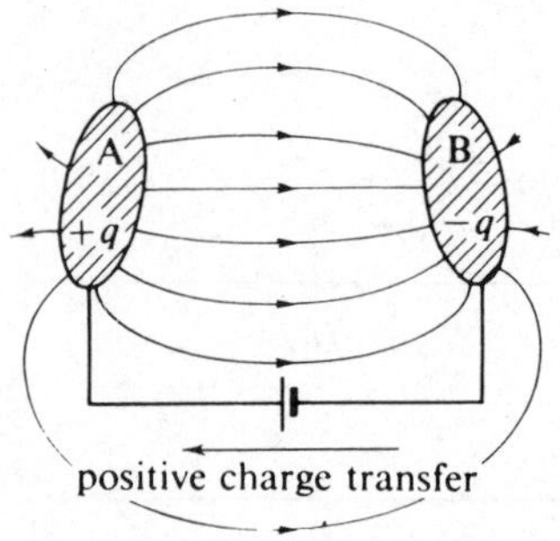

Figure 5.14 A capacitor AB charged by a cell has charges conveyed from one plate to the other against the electrostatic field.

charge dq transferred by the cell requires an amount of work $dW = v\,dq$. The total work done in the complete charging process is thus the integral of $v\,dq$ from $q=0$ to $q=Q$. Using $v=q/C$, this gives $Q^2/2C$. The work done in charging a capacitor is therefore

capacitor or conductor: $W=\frac{1}{2}Q^2/C=\frac{1}{2}CV^2=\frac{1}{2}QV$

A single conductor remote from its surroundings can be regarded as a special case and the same expression can be used with the understanding that V is the potential (zero at infinity) of the conductor.

Potential energy of a charged capacitor Because W represents work done in moving charges to their final positions, it also gives the **potential energy** U of a charged capacitor using a zero when it is uncharged, so that

capacitor or conductor: $$U=W=\tfrac{1}{2}Q^2/C=\tfrac{1}{2}CV^2=\tfrac{1}{2}QV \qquad (5.15)$$

Another way of regarding the expression is to write it as $\frac{1}{2}Q(V_A-V_B)$. If the charges are assembled, not by transference but by being separately brought from infinity in small increments, the work done would be $\frac{1}{2}QV_A$ for the charge on one plate and $-\frac{1}{2}QV_B$ for the charge on the other. We wish to generalize these expressions.

Electric energy of charged conductors We start from Eq. (2.12), which gave the value of U for three *charges* (not conductors at that stage) assembled from infinity:

$$U=Q_1Q_2/(4\pi\varepsilon_0 r_{12})+Q_2Q_3/(4\pi\varepsilon_0 r_{23})+Q_3Q_1/(4\pi\varepsilon_0 r_{31}) \qquad (5.16)$$

and this could be written as

$$U=\tfrac{1}{2}Q_1\left(\frac{Q_2}{4\pi\varepsilon_0 r_{12}}+\frac{Q_3}{4\pi\varepsilon_0 r_{31}}\right)+\tfrac{1}{2}Q_2\left(\frac{Q_3}{4\pi\varepsilon_0 r_{23}}+\frac{Q_1}{4\pi\varepsilon_0 r_{12}}\right)$$
$$+\tfrac{1}{2}Q_3\left(\frac{Q_1}{4\pi\varepsilon_0 r_{31}}+\frac{Q_2}{4\pi\varepsilon_0 r_{23}}\right) \qquad (5.17)$$

in which each of the terms of (5.16) has been split into two halves. Equation (5.17) can be written as

$$U=\tfrac{1}{2}Q_1V_1+\tfrac{1}{2}Q_2V_2+\tfrac{1}{2}Q_3V_3 \qquad (5.18)$$

in which V_1 is the potential at the point occupied by Q_1 due to all the *other charges* (i.e. except Q_1), and similarly for Q_2 and Q_3. The same argument can be applied to any number of charges and (5.18) can be generalized to

set of charges: $$U=\sum_i \tfrac{1}{2}Q_iV_i \qquad (5.19)$$

Suppose now that the charges are distributed over the surfaces of N *conductors* A, B, C, etc., so that all those charges on conductor A occupy points at which the potential is V_A and can thus be taken out of the sum in (5.19) to form a term $\frac{1}{2}Q_AV_A$ where Q_A is the total charge on A. Repeated for all the conductors, (5.19) becomes set of conductors:

$$\boxed{U = \sum_{A=1}^{N} \tfrac{1}{2} Q_A V_A} \tag{5.20}$$

This gives $\frac{1}{2}Q(V_A - V_B)$ for an ideal capacitor.

In all cases the zero of U is taken when all the charges are remote from each other at infinity or when all conductors are uncharged.

We shall refer to U in the above expressions as **electric energy** and denote it by U_E: it is potential energy stored in the system because of the positions of the charges and the electric forces between them, and is energy *internal* to the system. (As we have seen, the potential energy of Q in an *external* field is QV not $\frac{1}{2}QV$.) When the forces are allowed to move the charges, for instance through conducting wires, the electric energy decreases at the expense of heat energy and becomes zero when the conductors equalize their potentials. Alternatively, the forces may move the conductors as a whole when the electric energy decreases due to the external work done (Sec. 5.7).

5.6 Electric energy 2: in terms of E

In many problems involving systems of conductors, the **E**-field is a more useful quantity than the individual charges and potentials. Moreover, the methods for solution often give the results in terms of field quantities rather than the Q and V of the separate conducting elements. It would therefore be useful to have an expression for U_E in terms of **E** instead of Q and V as above. This ought to be possible since the electric energy of a system arises because of the work done in gradually assembling charges against the electric field already produced. However, a derivation which attempts to be very general from the beginning is either very cumbersome or involves complex mathematical operations, so we begin with a particular case and generalize from it.

Consider a single charged conductor as in Fig. 5.15 carrying a charge Q and hence at a potential V (zero at infinity). The electric energy of this system is

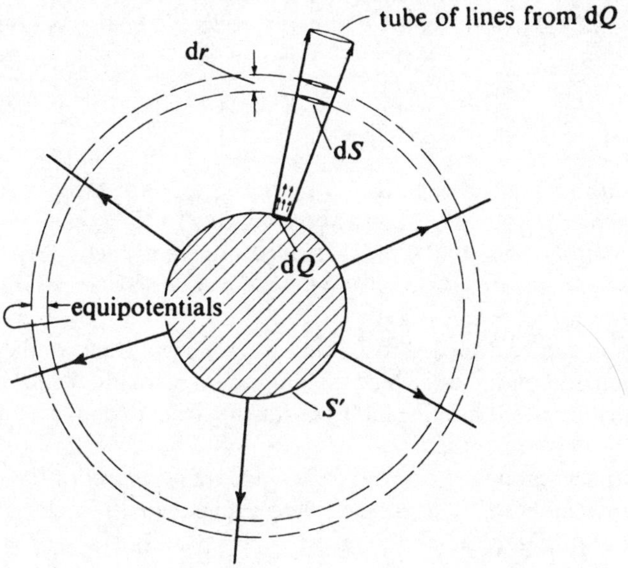

Figure 5.15 Calculation of the energy of a charged conductor in terms of **E**.

internal since Q has been brought to the conductor by doing work against its own field. The electric energy U_E is thus $\frac{1}{2}QV$ as in the previous section.

A small area of the surface carrying a charge dQ will have an electric energy $dU_E = \frac{1}{2}V\,dQ$ associated with it. The lines of **E** starting at dQ form a tube diverging as shown and intersecting equipotential surfaces as they do so. Two such equipotential surfaces differing in potential by dV and at a distance r are drawn in the figure. The tube of **E** lines from dQ carves out a small volume between the equipotentials which has an area dS and a thickness dr, and the electric field at the volume has a magnitude E and is, of course, normal to the equipotentials.

Gauss's law is now applied to the volume bounded by (a) the conducting surface, (b) the walls of the tube of lines from dQ and (c) the area dS. The only outward flux of **E** arises from (c) since **E** is parallel to the walls and is zero in the conductor, and the value of this outward flux is simply $E\,dS$. The enclosed charge is dQ, so that $E\,dS = dQ/\varepsilon_0$. The electric energy $dU_E = \frac{1}{2}V\,dQ$ associated with dQ can therefore be written as $\frac{1}{2}V(\varepsilon_0 E\,dS)$.

However, V is given by

$$V = \int_{\infty}^{S'} -E\,dr = \int_{S'}^{\infty} E\,dr$$

and is the sum of all the elementary potential differences stretching from S' to infinity. Hence

$$dU_E = \tfrac{1}{2}\varepsilon_0 E\,dS \int_{S'}^{\infty} E\,dr$$

Because $dQ = \varepsilon_0 E\,dS$ applies to all the equipotentials cut by the tube, it can be taken inside the integral and hence

$$dU_E = \int_{S'}^{\infty} \tfrac{1}{2}\varepsilon_0 E^2\,dS\,dr$$

The same expression applies to every tube starting from a dQ and the total energy is obtained by integrating over the other two coordinates represented by dS. If we denote the volume element $dS\,dr$ by $d\tau$, then the electric energy is given by

electric energy:
$$U_E = \int_{\text{space}} \tfrac{1}{2}\varepsilon_0 E^2\,d\tau \qquad (5.21)$$

where the volume integration extends over the whole of space, including now the inside of the conductor where $E = 0$ and no contribution to U_E is made. *We have, in short, assumed that the energy* $\frac{1}{2}QV$ *can be divided up so that* $\frac{1}{2}\,dQ\,dV$ *is associated with the volume element* $d\tau$; *shown that* $\frac{1}{2}\,dQ\,dV$ *can be written as* $\frac{1}{2}\varepsilon_0 E^2$; *and integrated over space to obtain (5.21).*

This derivation can be extended to an ideal capacitor in which all the tubes of lines end on another conductor instead of going to infinity, and thence to sets of charged conductors by considering all tubes starting from dQ on one conductor and ending at $-dQ$ on another.

That the two expressions (5.20) and (5.21) give the same result for the total energy can be exemplified by a parallel plate capacitor with a charge Q and capacitance $C = \varepsilon_0 A/x$. Equation (5.21) gives U_E as $\frac{1}{2}Q^2x/(\varepsilon_0 A)$ since $E = Q/(\varepsilon_0 A)$

everywhere between the plates and the volume is xA. This is the same as $\frac{1}{2}Q^2/C$ or $\frac{1}{2}QV$.

Equation (5.21) can be interpreted by saying that the energy per unit volume, or the **energy density**, is $\frac{1}{2}\varepsilon_0 E^2$, but we shall not take the view that this implies any storage of energy in space since we must always use the integrated expression giving the total energy of the system *as a whole*: there is little point in arguing about the location of the energy.

5.7 Forces and torques between charged conductors

The expressions (5.20) and (5.21) for the internal potential energy of charged conductors and capacitors can be used to obtain the forces and torques acting on any particular conductor. The method (of allowing the conductor to move under the force or torque) is similar to that used in Sec. 3.6 to obtain **E** from V and leads to a similar result [compare (3.20) with (5.22)]. Here, however, the movement may take place *either* with the conductors isolated so that Q is constant *or* with them connected to sources of voltage such that V is held constant. Both cases are considered.

Changes at constant Q Consider the two conductors of a capacitor as in Fig. 5.16 where the internal forces of attraction are indicated by the broken arrows and the external forces (keeping the plates apart and in equilibrium) by full arrows. Conductor B has an internal force **F** and an external force $\mathbf{G} = -\mathbf{F}$ acting on it. For any displacement d**L** of B the increment of internal energy is $\mathrm{d}U_\mathrm{E}$ and the work done *on* the capacitor *by* **G** is $\mathbf{G}\cdot\mathrm{d}\mathbf{L}$. Hence

$$\mathrm{d}U_\mathrm{E} = \mathbf{G}\cdot\mathrm{d}\mathbf{L} = -\mathbf{F}\cdot\mathrm{d}\mathbf{L} = -F_L\,\mathrm{d}L$$

or

$$\text{force on conductor:} \qquad \boxed{F_L = -\left(\frac{\partial U_\mathrm{E}}{\partial L}\right)_Q} \qquad (5.22)$$

giving the component of the force on B in any direction **L**. Because the charges are constant

$$\mathrm{d}U_\mathrm{E} = \tfrac{1}{2}Q\,\mathrm{d}V = \tfrac{1}{2}Q^2\,\mathrm{d}(1/C) = -\tfrac{1}{2}V^2\,\mathrm{d}C$$

so that, from (5.22),

$$F_L = \tfrac{1}{2}V^2\frac{\partial C}{\partial L} \qquad (5.23)$$

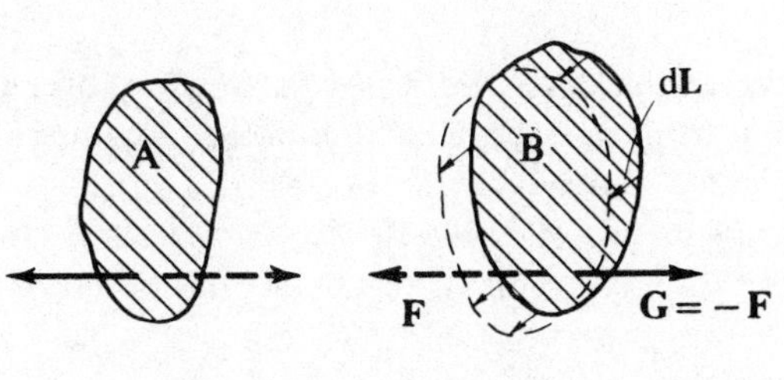

Figure 5.16 Forces between charged conductors. The internal forces are drawn with broken lines (**F**). The external forces (**G**) drawn with full lines keep the system in equilibrium. Conductor B is displaced through d**L** by the internal force.

Similarly, the torque about any axis is given by

torque on conductor:
$$T_\theta = -\left(\frac{\partial U_E}{\partial \theta}\right)_Q \tag{5.24}$$

which gives, for one conductor of a capacitor

$$T_\theta = \tfrac{1}{2}V^2\frac{\partial C}{\partial \theta} \tag{5.25}$$

Worked example *parallel plate capacitor* Using the notation of Fig. 5.4a, we have $C=\varepsilon_0 A/x$. If σ is the surface charge density on the plates, the force is given by (5.23) as

$$F_x = \tfrac{1}{2}V^2\frac{\mathrm{d}C}{\mathrm{d}x} = -\tfrac{1}{2}\varepsilon_0 A\frac{V^2}{x^2}$$

$$= -\tfrac{1}{2}\varepsilon_0 E^2 A \qquad \text{or} \qquad -\sigma^2 A/2\varepsilon_0 \tag{5.26}$$

The sign indicates that F_x acts to decrease x: it is an attraction.

Changes at constant V Suppose now that the same capacitor is maintained by a source of electromotance at a constant potential difference V and that conductor B again undergoes the displacement $\mathbf{dL}$ of Fig. 5.16. The increment of potential energy, $\mathrm{d}U_E$, is this time equal to $\frac{1}{2}V\mathrm{d}Q$ and not to $\frac{1}{2}Q\,\mathrm{d}V$. The source of electromotance has to transfer a charge $\mathrm{d}Q$ to keep V constant, and in doing so does work $V\mathrm{d}Q$ and thus loses this amount of energy. The increment in the energy of the whole system (capacitor + source) is thus $-V\mathrm{d}Q+\frac{1}{2}V\mathrm{d}Q$ or $-\frac{1}{2}V\mathrm{d}Q$. This equals $-\mathrm{d}U_E$. As before, the work done *on* the capacitor is $\mathbf{G}\cdot\mathbf{dL}$ or $-\mathbf{F}\cdot\mathbf{dL}$, and so:

$$-\mathrm{d}U_E = \mathbf{G}\cdot\mathbf{dL} = -\mathbf{F}\cdot\mathbf{dL} = -F_L\mathrm{d}L$$

or

force on conductor:
$$F_L = +\left(\frac{\partial U_E}{\partial L}\right)_V \tag{5.27}$$

Because V is constant, this gives $F_L=\frac{1}{2}V^2(\partial C/\partial L)$ as in (5.23): naturally, however the *changes* are made, the forces in a static situation are the same.

Notice that changes at constant V mean that, of the energy supplied by the source of electromotance, half goes to increasing the electric energy U_E *and half goes in external work.*

Force on the surface of a charged conductor Of the several methods of calculating this force, we choose two: one to show some detail of how the force arises and the other to illustrate the use of U_E again.

Method 1 Figure 5.17 represents a part of the surface of a charged conductor where the surface density of charge is σ. This charge resides in a thin surface layer

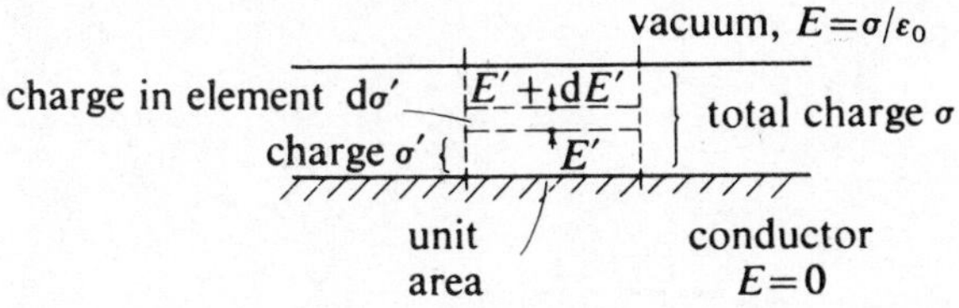

Figure 5.17 Detail of the surface of a charged conductor.

of unspecified thickness. Consider the unit area of the layer containing a total charge σ, spread in fact throughout a cylinder, at the bottom of which $E=0$ and at the top of which $E=\sigma/\varepsilon_0$. Since E must increase through the layer let its magnitude at some level be E' and the charge in the cylinder up to this level be σ'. A small element of the cylinder containing a charge $d\sigma'$ thus experiences an outward force $E'\,d\sigma'$. If, however, the electric field increases across the element to $E'+dE'$ the application of Gauss's law to the same element gives $dE'=d\sigma'/\varepsilon_0$ and the force on the element is thus $\varepsilon_0 E'\,dE'$. On the whole cylinder the force is

$$\int_0^E \varepsilon_0 E'\,dE' \qquad \text{which is } \tfrac{1}{2}\varepsilon_0 E^2 \text{ or } \sigma^2/2\varepsilon_0$$

This force is strictly exerted on the charges in the surface, but because emission of charge does not normally occur, there must be a balancing non-electrostatic force exerted by the conductor on the charges. This causes the force to be communicated to the conductor itself by Newton's third law so that

$$\text{Outward pressure on the surface of a charged conductor} = \frac{\varepsilon_0 E^2}{2} = \frac{\sigma^2}{2\varepsilon_0} \tag{5.28}$$

At sharp points, where σ is very large, the force *can* be great enough to cause charges to leave the surface and *field emission* is then said to occur. This effect forms the basis of field-electron and field-ion microscopes, in which electrons and ions respectively emitted from the tip of a pointed metallic specimen under a high voltage can be made to produce a visual image on a fluorescent screen showing the atomic arrangement in the metal (see, for example, Atkins, 1986, p. 769).

Method 2 Using (5.21) we can also find the force on the surface by allowing a virtual displacement dx of area A to occur under the action of the outward force F_x. By equating the work done by the internal force, $F_x\,dx$, to the decrease of energy $U_E A\,dx$ because of the disappearance of a volume $A\,dx$ in which E existed, (5.28) is again obtained.

Practical applications *electrostatic measuring instruments* These can be constructed so that one of two adjacent conductors is fixed, while the other can move under the action of a force or torque given by (5.23) or (5.25). An opposing force or torque (due to a spring or gravity) is brought into play and the resultant equilibrium position depends on the potential difference. Except for specialized uses mentioned below, instruments of this type have become obsolete with the advent of electronic measurement.

Electrostatic instruments have a very small inherent capacitance of only a few

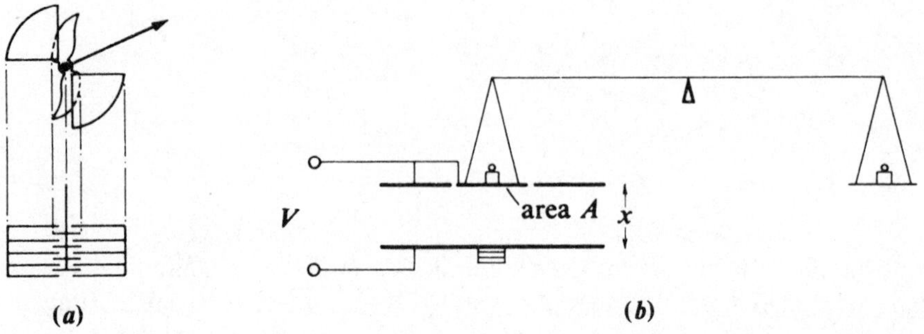

Figure 5.18 (a) An electrostatic voltmeter; (b) an attracted disc electrometer. The plates in (b) are shown in cross-section. The disc of area A is surrounded by an annular guard-ring under which the edge effects of Fig. 5.14b occur. There is a highly uniform field between the disc proper and the lower plate.

picofarads. This makes them especially useful for high voltages, which only produce moderate forces and torques, and for detection of small charges, which still produce a high V because $V=Q/C$. The dependence of the action on V^2 also makes the polarity of V immaterial so that alternating voltages may be used. The **Braun electrometer** is constructed like the electroscope of Fig. 1.1 but has a light aluminium pointer and a scale. It can be used for detection of charge or for rough measurements of potential from 500 V to 100 kV for elementary work. Large alternating voltages (several kilovolts and over) can be measured by the **electrostatic voltmeter** (Fig. 5.18a) in which the torque on the vane, $\frac{1}{2}V^2\,dC/d\theta$, is opposed by that due to a spring of torsional constant c. At equilibrium, $\theta=\frac{1}{2}(V^2/c)\,dC/d\theta$. The way in which θ varies with V depends on the shape of the vane which determines $dC/d\theta$. The UK National Physical Laboratory has used an electrostatic voltmeter as a transfer standard d.c. to a.c. (see Appendix C).

A particular type of voltmeter known as the **attracted disc electrometer** (Fig. 5.18b) does not need calibration if ε_0 is known, because the force is accurately $\varepsilon_0 AV^2/(2x^2)$ by Eq. (5.26) owing to the presence of guard-rings. The guarded disc is balanced with $V=0$ and a weight mg on it. This weight is then removed and the disc rebalanced by applying V and adjusting the distance x. The method is tedious and not suited to routine measurement: it has been used at the US National Bureau of Standards for measuring alternating potential differences of up to 250 kV in order to check the accuracy of methods normally used.

More recently (Sloggett, Clothier and Ricketts, 1986), the Australian National Measurement Laboratory has designed a similar electrometer with a mercury surface replacing the lower plate in order to provide a transfer standard between mechanical units and the volt. The attractive force given by (5.26) lifts the liquid through a distance h against gravity and is therefore equal to $hA\rho g$, where ρ is the relative density of mercury. The voltage is then given by $x(2\rho gh/\varepsilon_0)^{1/2}$ and is thus expressed in terms of mechanical quantities.

5.8 Applied electrostatics

Comment C5.1

Hazards and applications of electrification Unbalanced static charges are produced haphazardly in many natural and industrial processes that have only recently been thoroughly investigated. Such charges are mostly dissipated steadily by conduction through neighbouring material or by corona discharge from points. There are instances, however, when neither of these processes is sufficient to prevent a large accumulation of charge. For instance, the use of highly insulating artificial fibres in modern fabrics reduces dissipation through clothing and floor-covering. There then comes a point at which catastrophic breakdown occurs, resulting in a spark discharge. Sprays and dusts (because of their large surface area) and roller-driven belts are particularly prone to static electrification, with consequent hazards when inflammable materials are present. In these cases, additional means of dissipation are provided, usually in the form of air-ionizers (points, radioactive sources), humidifiers or special conducting dressings. On the other hand, controlled electrification of dusts and sprays has technical applications. For instance, a dust may be precipitated by first charging the particles by passing them over wires at a high potential and then collecting them on earthed plates. An article describing general applications of electrostatics including methods of photocopying is that of Moore (1972).

Practical applications *electrostatic generators* Controlled electrification also finds an application in electrostatic generators, which convert mechanical energy into electric energy. They fall broadly into two classes, both generating small currents at high voltages, in contrast with electromagnetic generators (Chapter 9) which operate at a low potential but can produce large currents.

The first class uses an **insulator** in the form of a moving belt to carry charge from a source A to a collector B connected to the inside of a conductor whose potential rises to a value determined by the amount of dissipation. The source consists of a comb of points or a sharp-edged blade maintained at a high positive potential so that corona discharge takes place, and charge is sprayed on the belt. At the collector a further comb B becomes charged by induction, further corona occurs and positive charge is transferred to the outside of the metallic enclosure, the belt being discharged.

Some machines in this class are designed to produce a high potential, others to produce large power. The Van de Graaff generator of Fig. 5.19 is typical of the high-voltage type and here the distance AB needs to be large so that the potential gradient in the surrounding gas can be kept down. High pressures may be used to prevent breakdown in the gas. Such instruments are used to produce beams of ions with only a small spread of energies. The ion source at S sends positive ions accelerating down the tube T to the target at earth potential. The supply of charge by the belt is balanced by the currents formed by the ions and any leakage currents in the ambient gas. Machines producing ions at voltages up to 20 MV are in use and the energy can be at least doubled by using a **tandem generator**. In this, negative ions are accelerated towards the high-potential conductor, stripped

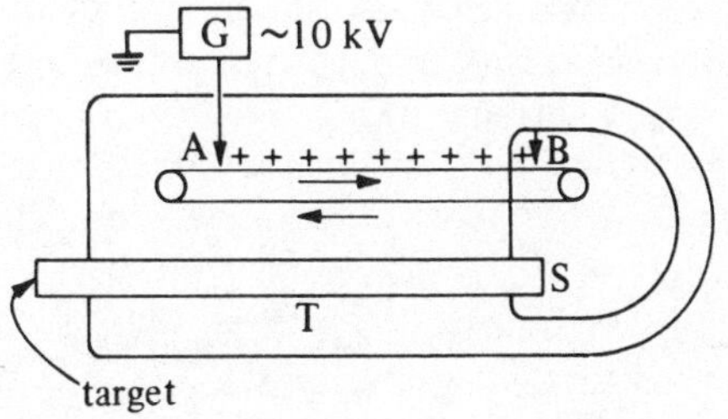

Figure 5.19 The Van de Graaff generator.

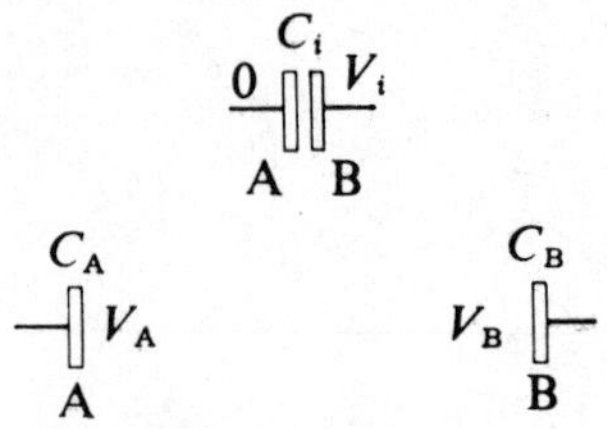

Figure 5.20 The principle of the condensing electroscope.

of some electrons within it so that they become positive ions and then accelerated again but this time away from the high-potential electrode (Rose and Wittkower, 1970). Tandem accelerators in which heavy ions are stripped to yield ionic charges of up to 10*e* can now (1989) produce beams with energies of over 300 MeV from a machine operating at 20 MV.

The *power generators* of this class use a much smaller belt length AB but have a much larger sprayed width so that currents can be much higher. These machines have been greatly developed by Felici at Grenoble and are used industrially.

A second class of machine, still being developed, uses a **conducting** carrier of charge and is based on the principle of the condensing electroscope. In Fig. 5.20, two conductors A and B are very close and have a capacitance C_i between them. A is earthed and B is raised to a small potential V_i so that the magnitude of the charge on both is given by $Q = V_iC_i$. They are then isolated and separated by such a large distance that they both function almost as isolated conductors with capacitances C_A and C_B with respect to the surroundings. Because they retain the same charge, the final potential of *B* is V_iC_i/C_B and of A is V_iC_i/C_A. Since C_i is much greater than either C_A or C_B, the final potential of both conductors is much higher than the original, one being negative and the other positive. The condensing electroscope itself, the electrophorus and charging by induction are all examples of the process described in elementary texts, although the importance of the increased potential is often not pointed out.

Continuous generators using the principle have been developed at Grenoble. A rotor has vanes which pass between a pair of plates at a slightly higher potential and each vane is earthed as it passes, the connection being broken almost at once. When the vane has moved so far that the capacitance is a minimum, and the potential therefore a maximum, the charge is collected. See Bright and Makin (1969).

The reader is referred to Moore (1973) and Cross (1987) for greater detail about electrostatic applications and hazards, while Walker (1978) discusses bioelectricity and provides references to further reading on the subject.

5.9 Conclusion

Once again, most of the results in this chapter have been consequences of Coulomb's law. The only exception lies in the introduction of dielectric materials whose properties have to be established by further experiments. Readers who wish to go on immediately to a study of these may proceed directly to Chapter 11, the initial section of which can be understood without trouble at this stage. In this book we delay the full treatment so that we can use more sophisticated arguments, and for the time being we leave electrostatics proper. In the next chapter we look at what happens when an E-field in a conductor is not allowed to decay to zero by a temporary movement of charge, but is maintained by a source of electromotance to produce a steady current.

5.10 Revision summary

Capacitance is defined for a pair of conductors by the relation

$$C = Q/V \tag{5.6}$$

where V is the potential difference produced between them when a charge Q is transferred from one to the other. No other conductors should be close enough to affect the electric field. Such a pair is known as an *ideal capacitor*. The **SI unit** of capacitance is the *farad*, F, equivalent to the $\mathrm{C\,V^{-1}}$.

The following properties of, and values for, C can be established using electrostatic laws:

- C is a constant for a given pair of conductors *in vacuo* at fixed positions.
- An isolated conductor is a special case of an ideal capacitor in which a second conductor surrounds it but is large enough to be considered located at infinity.
- Values of C for particular ideal capacitors are:

  isolated sphere, radius a:

  $$C = 4\pi\varepsilon_0 a \tag{5.2}$$

  spherical capacitor, inner radius a, outer radius b:

  $$C = 4\pi\varepsilon_0 ab/(b-a) \tag{5.7}$$

  cylindrical capacitor or coaxial cable, inner radius a, outer radius b, length l:

  $$C = 2\pi\varepsilon_0 l/\log_e(b/a) \tag{5.9}$$

  parallel plate capacitor, area A, separation of plates x:

  $$C = \varepsilon_0 A/x \tag{5.8}$$

  parallel twin cables, separation $d \gg$ radius a, length l:

  $$C = \pi\varepsilon_0 l/\log_e(d/a) \tag{5.11}$$

Networks of capacitors Any network of capacitors connected between two terminals A and B can be replaced by a single equivalent capacitor C_{eq}. The value of C_{eq} can be obtained by (a) imagining the transfer of charge Q from B to A, (b) finding the resultant charges on all capacitors in terms of as few unknowns as possible using conservation of charge, (c) equating to zero the sum of the potential differences Q/C round a number of closed paths, yielding sufficient equations to obtain the unknown charges on all capacitors and hence the potential differences across them, (d) finding the total potential difference V_{eq} by any path from A to B and (e) using $C_{eq} = Q/V_{eq}$.

Special cases can often be used to reduce networks without that method, e.g.

capacitors in series:

$$1/C_{eq} = \sum_i 1/C_i \tag{5.12}$$

capacitors in parallel:

$$C_{eq} = \sum_i C_i \tag{5.13}$$

Real capacitors approximate to parallel plate or cylindrical ideal capacitors and the formulae for C given above often yield good estimates. In practical circuits there are stray capacitances to other conductors and these can be represented by a network of ideal capacitors.

Dielectrics If the whole of the space in which an electric field exists in a charged capacitor is filled with a honogeneous insulator (a dielectric), the capacitance increased from its vacuum value C_0 to a new value C_m. The ratio is known as the *relative permittivity* ε_r of the dielectric:

$$\varepsilon_r = C_m/C_0 \tag{5.14}$$

All expressions for capacitance in terms of dimensions should therefore include an ε_r in the numerator when a dielectric fills the space between the plates.

Dielectrics will break down at a potential gradient or electric field known as their *dielectric strength.*

Work and energy The work done in charging a conductor or a capacitor is $\frac{1}{2}QV$, an expression also written as $\frac{1}{2}CV^2$ or $\frac{1}{2}Q^2/C$. This is also the electric potential energy U_E stored in a single capacitor. The total energy stored in a number of conductors carrying charges Q_i and at potentials V_i is

$$U_E = \sum_i \tfrac{1}{2}Q_iV_i \tag{5.19}$$

The same energy can be expressed in terms of the electric field as

$$U_E = \int \tfrac{1}{2}\varepsilon_0 E^2 \, d\tau \tag{5.21}$$

where the integration extends over the whole of space. The E-field thus behaves as if it stored energy per unit volume equal to $u_E = \frac{1}{2}\varepsilon_0 E^2$.

Forces and torques The force F_L in any direction $\hat{\mathbf{L}}$ on a single conductor in a collection (e.g. one plate of a capacitor) is given by

$$F_L = -\left(\frac{\partial U_E}{\partial L}\right)_Q = +\left(\frac{\partial U_E}{\partial L}\right)_V \qquad (5.22)\ (5.27)$$

while torques T_θ about an axis are given by similar expressions involving $\partial U_E/\partial\theta$. In terms of capacitance, the same force and torque are given by

$$F_L = \tfrac{1}{2}V^2\,\partial C/\partial L; \qquad T_\theta = \tfrac{1}{2}V^2\,\partial C/\partial\theta \qquad (5.23)\ (5.25)$$

When movement of a conductor takes place at constant potentials maintained by sources of electromotance, there is both an increase in U_E, ΔU_E, and an equal amount of external work done, ΔW. The total energy supplied by the sources is thus split equally between ΔU_E and ΔW.

Finally, there is an outward force on the surface of a conductor carrying a surface charge density σ which is given by $\sigma^2/(2\varepsilon_0)$ per unit area (5.28).

Commentary

C5.1 On Static Electrification Investigations of static electrification in solids and liquids have long been stimulated by the industrial processes which generate unwanted charge or which make use of charged materials. More recently, the desire to understand natural phenomena such as the electrification of thunder-clouds has led to much work in connection with water and ice. The mechanisms involved are complex and many are not fully understood in detail. A broad picture is now emerging, however, and a brief summary is given here.

The transfer of charge between *solid and solid* can take place through:

1. Contact. Here the transfer is of electrons when metals or semiconductors are involved. The mechanism is the same as that which produces contact potentials (see any solid-state text). With good insulators, there is still some controversy over the relative contributions from electron and ion transfer.
2. Deformation. When a substance is deformed, piezoelectric effects may occur (Sec. 11.10) or dislocations may move, carrying charge with them to various parts of the surface.
3. Cleavage. Cleaved solids reveal new surfaces together with their usually imperfect atomic arrangements. The imperfections can leave the pieces oppositely charged and this mechanism is often responsible for the charging of dusts.
4. Friction. Local temperature gradients will be different in two substances rubbed asymmetrically against each other, and this can cause electron or ion transfer. If abrasion takes place as well, actual material is transferred.

In *fluids*, the transfer of charge can take place by:

1. Flow. Generation of charge in pipes and filters seems to be due mainly to adsorption of ions from the fluid on to the solid surfaces. The source of a major hazard in filling tankers.
2. Fragmentation. Splashes, sprays, and drops may all give rise to charged droplets. This is particularly evident where an **E**-field already exists because charging by electrostatic induction then takes place as in Fig. 1.3.

For detailed discussion and description of experiments, see Moor(1973) and Institute of Physics (1967, 1971, 1975, 1979, 1983).

C5.2 On the Terrestrial Electric Field An electric field of great interest is that of the earth. It is generally directed downwards during fine weather with an **E**-field at the surface of about 100 V m^{-1}. This indicates that the earth has a resultant negative charge, and since it is known that the ionosphere at some distance above the surface has a resultant positive charge, the system forms a giant spherical capacitor (Stow, 1969). The capacitor is continuously discharging through fair-weather currents and various forms of precipitation (rain, snow). How then is the charge maintained?

The principal mechanism is through the agency of thunderstorms. Thunderclouds generally form with a massive negative charge centred below a positive charge. Lightning discharges and induced point discharges from leaves and grass below a thundercloud both transfer enormous quantities of negative charge to the earth, and similar amounts of positive charge are conducted to the ionosphere from the top of the cloud. Stow gives the overall balance in current densities to the earth as follows:

Fair-weather conduction:	$+90\,\mathrm{C\,km^{-3}\,y^{-1}}$
Precipitation:	$+30\,\mathrm{C\,km^{-3}\,y^{-1}}$
Lightning:	$-20\,\mathrm{C\,km^{-3}\,y^{-1}}$
Point discharge:	$-100\,\mathrm{C\,km^{-3}\,y^{-1}}$

The electrification of thunderclouds is clearly the crucial process maintaining the terrestrial field and many possible mechanisms have been suggested to account for the charge separation occurring in opposition to the preexisting field. As a result of investigations during the past decades, a convincing theory has developed that the principal contributions come from the splintering of supercooled droplets, and from the collision between small ice particles and much larger soft hail pellets. In both cases, temperature gradients are responsible for the migration of ions from one particle to the other and subsequent gravitational settlement of heavier negative particles through the air produces the observed charge separation. See Malan (1963), Latham (1966), Stow (1969), Walker (1978) and Williams (1988).

The earth's electric field can be measured directly by a rotating vane field mill (Cross, 1987). The field is large enough to provide voltages that could be tapped for useful purposes such as electrolysis, as was discovered as long ago as 1831 (Shoop, 1988), but it never seems to have been exploited in this way.

Problems

Section 5.2

5.1 Estimate the capacitance of (a) a 5p or 25¢ coin, (b) a person, considered as an isolated conducting cylinder of length 2 m and diameter 30 cm, (c) a straight copper wire of length 10 cm, diameter 1 mm and situated 1 cm away from and parallel to a plane earthed chassis, and (d) the earth.

5.2 The voltage between the plates of a cylindrical capacitor is fixed at V and the outer radius at r_2. Find the electric field E_1 just outside the inner conductor of radius r_1 in terms of V, r_1 and r_2. Find the value of r_1 which makes E_1 a minimum, neglecting end effects.

5.3* Using the method at the end of Sec. 5.2, find the limits between which the capacitance of a conducting cylinder of length $2l$ and radius a must lie. To what value will the capacitance per unit length at the centre tend as l increases relative to a?

5.4* A twin transmission line consists of two thin conducting strips of width $2a$ with their flat surfaces facing each other and separated by a distance x. Find the capacitances per unit length on the extreme assumptions of uniform distribution of charge and distribution only along the edges.

5.5* Find whether edge effects increase or decrease the capacitance of a parallel plate capacitor compared with its ideal value (use Problem 5.4).

5.6 A single horizontal telegraph wire of radius a is at a height h above the earth's surface, which behaves as a plane conductor. Find the capacitance per unit length between the wire and earth.

Section 5.3

5.7 Capacitors A and B of 0.5 and 0.2 μF, respectively, are connected in series across a 140 V d.c. supply. What are the charges and voltages for each capacitor? Repeat with A and B in parallel.

5.8 A 0.1 μF capacitor is charged to a voltage of 2 V and isolated. It is then connected across an uncharged 0.3 μF capacitor. What is now the voltage across the two capacitors and what are their charges?

5.9 Four points A, B, C, D are at the corners of a square. Capacitors of 1 μF are connected between each pair of points, including the diagonally opposite ones. Find the equivalent capacitance between the two adjacent points on one side of the square.

5.10 Two parallel metallic plates of area A are placed a distance a apart and both are earthed. A similar plate is placed between the first two, parallel to them and at a distance x from one of them. A charge Q is given to the central plate. What is its voltage and what charges are induced on the two outer plates?

5.11 Use superposition and Problem 5.10 to determine the charges induced on two earthed parallel plates when a point charge Q is placed between them. The plates are separated by a and Q is x from one of them. If the plates are connected externally by a wire and if Q moves a distance y towards one plate, show that the charge passing through the wire is Qy/a.

5.12 The arrangement of Fig. 5.21a is to be equivalent to that of Fig. 5.21b when the points A, B and C are connected into any network. Show that, for this to be so, $C_A = (C_1C_2 + C_2C_3 + C_3C_1)/C_1$ with similar expressions for C_B and C_C. (Use $K_A = 1/C_A$, etc.) Show also that $C_1 = C_BC_C/(C_A + C_B + C_C)$, etc., by using $C_1C_A = C_2C_B = C_3C_C = k$, say. Solve the network of Fig. 5.11 by reducing one of the deltas to a Y (the delta–Y transformation or $\pi - T$ transformation: see also Secs 6.6 and 10.4).

5.13 If the inner sphere of a spherical capacitor is earthed instead of the outer, show that the total capacitance is $4\pi\varepsilon_0 b^2/(b-a)$ where $a<b$. If a charge Q is given to the outer sphere from surroundings at earth potential, what proportions reside on the outer and inner surfaces of the outer sphere?

Section 5.4

5.14 Estimate the capacitance of the following: (a) an air capacitor of 16 plates each 6×4 cm separated by 2 mm, alternate plates being connected together, (b) a rolled paper capacitor with strips of tinfoil 3×40 cm as the plates, the paper being 0.02 mm thick and having $\varepsilon_r = 2$, and (c) a ceramic capacitor with plates 1 cm square and a dielectric of thickness 0.5 mm and $\varepsilon_r = 3000$.

5.15 If the largest safe potential gradient which can be allowed in air is $3\,\text{kV}\,\text{mm}^{-1}$, what is the greatest capacitance which can be obtained in a volume $10 \times 8 \times 6$ cm using air as the dielectric and having a working potential of 600 V?

5.16 Estimate whether corona discharge is likely to take place from the surface of an oil drop of radius 5μm carrying about 100 electronic charges.

Section 5.5

5.17 In Problem 5.7, compare the energy stored in the capacitors with that supplied by the source.

5.18 Eight identical spherical drops of mercury charged to 12 V above earth potential are made to coalesce into a single spherical drop. What is the new potential and how has the internal electric energy of the system changed?

5.19 A capacitor of capacitance C_1 and potential difference V is connected across an uncharged capacitor of capacitance C_2. Show that the new potential difference is $VC_1/(C_1 + C_2)$ and that the loss in electric energy is $\frac{1}{2}C_1C_2V^2/(C_1 + C_2)$.

Section 5.6

5.20 Show that the electric energy of a charged conducting sphere of radius a carrying a charge Q is $Q^2/(8\pi\varepsilon_0 a)$. Carry out the calculation by two methods, using Eq. (5.20) first and then (5.21).

5.21 Show that the electric energy of a spherical region of space of radius a filled with a charge Q spread uniformly throughout it is $3Q^2/(20\pi\varepsilon_0 a)$ (nuclear model).

Section 5.7

5.22 Find the mechanical work needed to double the separation of the plates of a parallel plate capacitor *in vacuo* if a cell maintains them at a constant potential difference V and the area and original separation are A and x, respectively.

5.23 The central conductor of a cylindrical capacitor has only half its length overlapping the outer. What is the force on the inner conductor if the potential difference is V and the radii are a and b?

5.24 An electrometer measures potential but has an effective capacitance C_i across its input terminals due to its internal circuits. If it is to be used to measure the *potential* across an external charged capacitor C_e, generalize the result of Problem 5.8 to see what the desirable relation between C_i and C_e should be. If the electrometer is to be used to measure the *charge* on C_e rather than its potential, what is now the desirable relation?

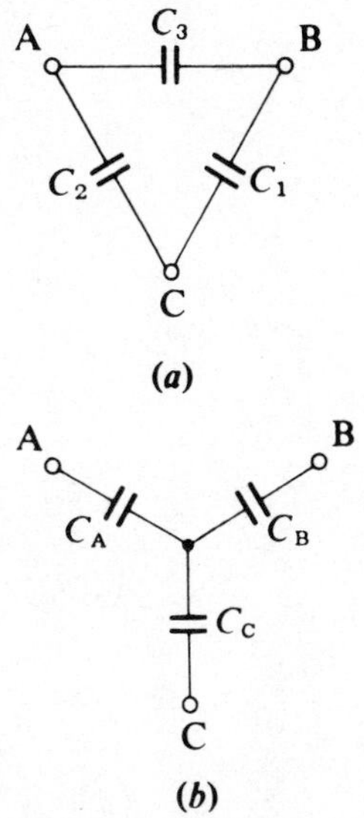

Figure 5.21 The delta–Y (or delta–star) transformation.

6

Steady electric currents

In Chapter 1 we saw that temporary currents could be produced in conductors by electrostatic fields, but that the maintenance of a *steady* current required a complete circuit incorporating an **electromotance** or **electromotive force** (e.m.f.). In the first two sections of this chapter we try to make the concept of electromotance more precise before looking at various aspects of steady current flow itself.

We then look at a class of problem in which the existence of fields can be ignored, and in which the systems to be dealt with can be represented by **networks**. The networks consist of connected **elements**, each with two terminals and each characterized solely by the relation between the **voltage** V across the terminals and the **current** I flowing between them. For steady currents, real networks can often be sufficiently well represented by combinations of resistors (Sec. 6.3) and voltage or current generators (Sec. 6.4). In Secs 6.5, 6.6 and 6.7, we show how problems involving such networks can be solved by the application of laws already established in previous chapters but expressed in more convenient forms. We also see how to apply the laws to a case where the current varies slowly—the charge and discharge of a capacitor. Section 6.9 deals briefly with d.c. measurements.

Leaving networks, we return to more general aspects of current flow in Sec. 6.10. We first define **resistivity** and **conductivity**, which are properties of a material rather than a system, and these are then used to examine currents flowing in extended media and not just in filamentary wires (Sec. 6.11). The chapter ends with a brief look at some atomic aspects of conduction.

6.1 Electromotance or e.m.f.: field aspect

Development of the concept A steady current consists of charges moving with a constant average speed. Unless these charges move in a closed path, they will accumulate somewhere, producing an increasing electrostatic field and thus changing the current. It is therefore an essential feature of a *steady* current that a closed path or *circuit* is necessary.

It is true that we talk later of currents in wires of infinite length, but this should be recognized as a conceptual device to obtain results for very long wires by simple methods.

Let us look, therefore, at what is needed to *produce* and *maintain* such a current. It is clear that, if we begin with stationary charges (whether *in vacuo* or on a conductor), some force $\mathbf{F}$ will be required to produce motion in the first place so as to start the current, and that $\mathbf{F}$ will do work. Such a force could be provided by an electrostatic field $\mathbf{E}_Q$ as in Fig. 3.17. There, a wire between two charged conductors at different potentials carries a current generated by a force $\mathbf{F}=e\mathbf{E}_Q$ acting on each electron in the wire. The current is only transient, however, because there is no circuit and $\mathbf{E}_Q$ eventually falls to zero. If there *were* a circuit, what of the force needed to *maintain* a current—and would an $\mathbf{E}_Q$ be sufficient for that in view of the circuital law?

At first sight, we might think that an $\mathbf{E}_Q$ *would* be sufficient. After all, if a steady current flowing in a closed path *in vacuo* has already been started, it is undoubtedly true that no extra input of energy is needed provided it does no work. For example, a beam of electrons following a circular path around a small positively charged sphere would be analogous to a satellite moving in a circular orbit around the earth: once started, its motion is maintained without further input of energy, provided it encounters no resistance and provided radiation is neglected (Chapter 14).

However, a current of this sort is not useful. In practice, a steady current will need an input of energy to maintain it because of two separate factors: firstly, it may flow in a conductor where there is effectively a resistance to motion; secondly, it may be required to do external work (as in an electric motor). In neither case will the current remain steady unless an equivalent amount of energy is forthcoming from the forces maintaining the current.

It is perfectly clear that an electrostatic field cannot provide any resultant input of energy to a charge moving round a complete circuit L because of the general condition (4.14):

$$\oint_L \mathbf{E}_Q \cdot d\mathbf{L} = 0 \tag{6.1}$$

It follows that $\mathbf{E}_Q$ cannot sustain a *useful* steady current.

In order, therefore, to maintain such a current there must be, *for at least part of the circuit*, non-electrostatic forces acting on the charge carriers. These forces must have a non-zero line integral round a closed path L, so that they can do non-zero work on charges in moving them round L. Put more formally, if dW is the work done by a force $\mathbf{F}$ in moving a charge Q through an element $d\mathbf{L}$ of the closed path L, then we require that the quantity $\oint dW/Q$ or $\oint \mathbf{F}\cdot d\mathbf{L}/Q$ shall not be zero.

Definition of electromotance The input of energy or work done per unit charge provides a definition of the **electromotance, electromotive force** *or* **e.m.f.** of the source of $\mathbf{F}$. To make the definition very general, the charge involved is a test charge Q_t as defined in Sec. 3.1:

$$\mathscr{E} = \oint_L \frac{dW}{Q_t} = \oint_L \frac{\mathbf{F}\cdot d\mathbf{L}}{Q_t} \qquad \text{(definition of } \mathscr{E}\text{)} \tag{6.2}$$

In this form, the definition can be applied to *any* closed path L, even when no conductors or charges are present. It therefore enables us to find the electromotance for any closed path in any field.

The definition (6.2) shows that $\mathscr{E}$ is equal to the work done per unit positive charge in taking a charge completely round L, so that its **SI unit** is the same as that of electric potential: the **volt**. (Note that $\mathscr{E}$ is not a force, in spite of one of its names.)

Types of electromotance We now ask what *can* contribute to $\mathscr{E}$ since all we have shown so far is that electrostatic fields cannot do so because of (6.1). First take the case of a closed path L completely within a homogeneous conductor, of the same material throughout and at the same temperature, pressure, etc., everywhere. Then one possible type of contribution to $\mathscr{E}$ is from electromagnetic induction, the subject of Chapter 9. This leads to a type of E-field, which we denote by $\mathbf{E}_M$, having a non-zero line integral round a closed path. If $\mathbf{E}_M$ exists at all points round L, as it may, it produces an electromotance which is said to be **non-localized**. An example would be the $\mathscr{E}$ produced in a loop of copper wire if a bar magnet were pushed through it (see Fig. 9.1a).

Comment C6.4

A second type of force may contribute to the electromotance when there are non-homogeneities such as boundaries between dissimilar materials or gradients of temperature, concentration, etc. These are often **localized** in only a part of the circuit. Chemical cells are an obvious example, and even the Van de Graaff generator could be considered as belonging to this category. For these forces we shall write $\mathbf{F}/Q$ as $\mathbf{e}$ rather than $\mathbf{E}$ to indicate that they are not necessarily electric fields under the definition (3.1)—they cannot exist *in vacuo* but require the presence of material. In general, therefore,

Comment C6.1

$$\mathscr{E} = \oint_L (\mathbf{E}_Q + \mathbf{E}_M + \mathbf{e})\cdot d\mathbf{L} \tag{6.3}$$

from (6.2), although we should note that if we wish to calculate $\mathscr{E}$ by adding elements from small parts of the path L, we do not need to include $\mathbf{E}_Q$ because of (6.1) so that

$$d\mathscr{E} = (\mathbf{E}_M + \mathbf{e})\cdot d\mathbf{L} \tag{6.4}$$

Work, energy and power If an electromotance $\mathscr{E}$ causes a total charge Q to move once round L, the charge crossing any cross-section of the circuit is also Q. From the definition (6.2), the work done must be $Q\mathscr{E}$ and the rate of working or power must be $I\mathscr{E}$. Hence

Work done by a source $\mathscr{E}$ supplying $Q = Q\mathscr{E}$
Rate of working of a source $\mathscr{E}$ supplying $I = I\mathscr{E}$

For steady currents, this work represents a transfer of energy from the sources of electromotance to the conductor as heat or to any external system on which work is being done. If the current is *in vacuo* and does no work, it cannot remain steady under the action of an electromotance and the energy appears as increased kinetic energy of the charges themselves (see the betatron in Sec. 9.12).

6.2 Electromotance or e.m.f.: network aspect

Localized electromotance, relation to potential difference The account given in Sec. 6.1 is rather abstract and the definition of $\mathscr{E}$ very general. We need to relate the concept to real situations and to the reader's experience of actual sources of $\mathscr{E}$, which in practice are often located between two terminals. The source itself can be extremely complex and not available for detailed examination, particularly if it is an electronic circuit. It is therefore unrealistic to think of applying the definition (6.2) because of the difficulty of tracing the path L within the source. How, then, can we proceed? In particular, how in practice shall we *measure* $\mathscr{E}$ for such a source?

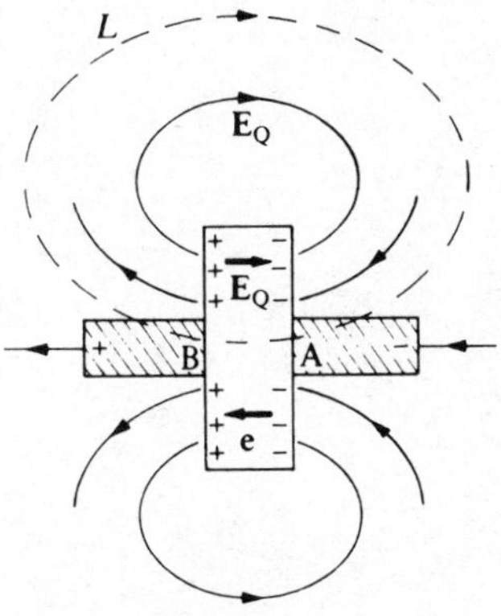

Figure 6.1 A source of electromotance on open circuit. The centre portion represents the region in which the non-electrostatic field **e** is located. The shaded region outside AB represents conductors leading to the terminals. $\mathbf{E}_Q$ outside the source would be very small.

In order to see what to do, we look at an oversimplified picture of a localized source as illustrated in Fig. 6.1. Inside the source there exists a non-electrostatic field of the type denoted by **e** in Sec. 6.1. We are not concerned here with the origin of **e**, but with its effect. As soon as the source is constructed or put into operation (but still on open circuit), the field **e** acts on charges free to move within the source itself and separates positive and negative charges as shown. These charges can move no further than the boundaries of the source or the terminals, where they accumulate and produce a growing electrostatic field $\mathbf{E}_Q$. *Inside* the source, $\mathbf{E}_Q$ acts in opposition to **e**, so the charges will only continue to accumulate until $\mathbf{E}_Q$ and **e** are equal and opposite, i.e., until $(\mathbf{E}_Q + \mathbf{e}) = 0$. No external conducting path is connected at this stage and the source is said to be on *open circuit* (supplying no current). The conductors connected to A and B form terminals to which external connections can be made. The material of the terminals form equipotential volumes with small surface charge densities.

Comment C6.2

Now let us find $\mathscr{E}$ from (6.3) with $\mathbf{E}_M = 0$, using the path L shown as a broken line in Fig. 6.1, with the sense indicated by the arrow. That part of the path between A and B *inside* the source contributes nothing to the line integral because $(\mathbf{E}_Q + \mathbf{E}) = 0$. That part of L *outside* the source has $\mathbf{e} = 0$, so all that is left is

$$\mathscr{E} = \int_B^A \mathbf{E}_Q \cdot d\mathbf{L} = \int_A^B -\mathbf{E}_Q \cdot d\mathbf{L} = V_B - V_A$$

the last step following from Eq. (3.12). Thus *the electromotance is equal to the potential difference between the terminals on open circuit.*

Now we do not pretend that any source can be described as simply as that of Fig. 6.1, although a chemical cell comes close to it. However, we can use the open-circuit potential difference as a definition of $\mathscr{E}$ without any assumptions about the internal structure, and can be confident that it is consistent with (6.2). So in network theory, where our concern will be with localized sources, we shall define $\mathscr{E}$ using

$$\boxed{\mathscr{E} = V_B - V_A \qquad \text{(open circuit)} \qquad \text{(network definition of } \mathscr{E}\text{)}} \qquad (6.6)$$

This is a quantity that can be measured by the potentiometer of Sec. 6.9.

Closed circuit When the terminals A and B of a source are connected externally by a conducting path, the field $\mathbf{E}_Q$ exists along this path and causes a current to flow. The charges on the terminals are replenished by processes within the source, processes we have denoted by **e**. Provided **e** remains constant throughout this, the potential difference between the terminals is maintained steady, as is the current in the circuit. However, because **e** now has to do work in moving charges within the source, $\mathbf{e} + \mathbf{E}_Q > 0$ and hence $\mathbf{e} > -\mathbf{E}_Q$. It follows that

$$\mathscr{E} > V_B - V_A \qquad \text{(closed circuit)}$$

provided there are no other sources of electromotance in the external part of the circuit.

Any conducting system connected externally between the terminals so that it only consumes energy is known as a **load** (see Fig. 6.3). From the results of Chapter 3, we know that the following expressions will apply to a source and load where there is a potential difference V between the terminals:

Work done by the source in delivering a charge $Q = QV$
Power consumed (rate of energy consumption) from the source in supplying current I to a load $= IV$ (6.7)

Combining these results with (6.5) shows that the consumptions of energy and power *within* a source of electromotance are $Q(\mathscr{E} - V)$ and $I(\mathscr{E} - V)$, respectively.

Active and passive components Any two-terminal component which can supply power continuously to a load is described as a **generator** and is said to be **active**. A component which can only consume or store energy is **passive**. For steady currents, the two types can be differentiated by finding the potential difference between the terminals on open circuit: if the result is zero, the component is passive.

6.3 Resistance and conductance

Concept and definitions For the next seven sections we shall be considering networks of components with two terminals like the localized electromotance of the last section. First, we look at **passive** components for which the important property is their resistance to the flow of steady current.

When the steady current flowing through a two-terminal passive component is plotted against the potential difference between its terminals, the graph obtained is known as its *characteristic* (sometimes *static* characteristic, to indicate that it is obtained under steady conditions). The ratio of the potential difference V to the current I defines the **resistance** R of the component and its reciprocal G, the **conductance**:

$$R = V/I \qquad \text{(definition of } R\text{)}$$
$$G = 1/R = I/V \qquad \text{(definition of } G\text{)} \qquad (6.8)$$

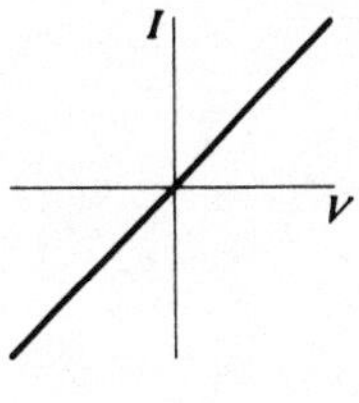

(a)

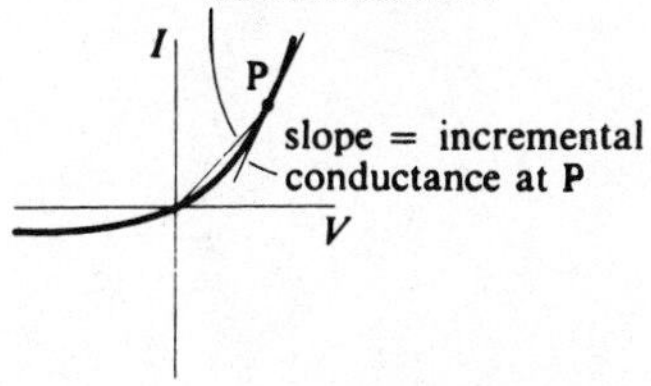

(b)

Figure 6.2 Static current–voltage characteristics of (a) a metal, (b) a non-linear device usable as a rectifier.

The **SI unit** of resistance, the V A^{-1}, is known as the **ohm** (symbol Ω) and the unit of conductance as the **siemens** (symbol S).

The characteristics of homogeneous conductors such as metals and alloys are linear as in Fig. 6.2a, provided conditions such as temperature are unvarying, a property known as **Ohm's law**. Such materials are said to be **ohmic** or **linear** and they clearly have a resistance independent of the current or potential difference. Chrystal (1876) showed that the conductors of iron, platinum and German silver with cross-sections of 1 cm^2 and resistances of 1 Ω for infinitesimally small currents did not change in resistance by as much as 1 part in 10^{12} when the current increased to 1 A, the temperature being held constant. Since then, there is ample evidence that little change occurs even when current densities of up to 10^5 A cm^{-2} are used.

When a component possesses a non-linear characteristic, for instance as in Fig. 6.2b, the **incremental resistance** dV/dI is sometimes more important than the resistance V/I. This occurs when the component is used with changing V and I, for the changes ΔV and ΔI are small enough, $\Delta V/\Delta I$ is given sufficiently accurately by dV/dI. Under these conditions the component may be treated as a linear resistance of that value for the purposes of analysis. For large changes in V and I, the non-linearity cannot be ignored, but it may serve a useful purpose as in the rectification of alternating current. These matters will not concern us for the moment since we are dealing with steady currents.

A **resistor** is a component in an electrical network used to control current: it is represented in a diagram of the network by the symbol as at r in Fig. 6.3 on page 134, crossed by an arrow if adjustable. Both the use of the term 'resistor' and the network symbol imply linearity, unless it is specifically described as non-linear in some way.

Production of heat In Sec. 3.7 we saw that a charge Q falling through a potential difference V loses potential energy QV. In a conductor carrying a steady current no kinetic energy is gained overall by the charges so that all the energy lost must be given to the conducting medium and appear eventually as heat. Thus, QV is the total heat produced in a conductor through which a total charge Q falls through V, and the rate of production of heat if the current is steady at I is IV. Thus

$$\text{Rate of production of heat in a conductor} = IV = RI^2 = GV^2 \qquad (6.9)$$

in watts when R is in ohms and I in amps. For ohmic conductors, R is constant and Joule's law (1.2) is now seen as a consequence of our choice of the magnetic effect for the measurement of current, of the conservation of energy and of Ohm's law. Alternatively, we can look upon Joule's law and Ohm's law as jointly confirming that electric energy of charges (QV) must be included in the general law of conservation of energy.

Internal or output resistance The definition of resistance in (6.8) does *not* apply to **active** components. For simple sources, such as chemical cells, the potential difference V between the terminals *decreases* with the current I supplied. For sufficiently small currents the variation is often such that $\mathrm{d}V/\mathrm{d}I$ is constant, and we are led to define an *internal* or *output* resistance of the source by

$$\text{internal resistance:} \qquad r = -\mathrm{d}V/\mathrm{d}I \qquad \text{(definition of } r\text{)} \tag{6.10}$$

If the source is used under conditions for which r is indeed independent of V or I, then the relationship between V and I from (6.10) is $V = -rI + \text{constant}$. The constant is given by the value of V when $I = 0$, i.e. by the open-circuit terminal voltage which (6.6) shows to be $\mathscr{E}$ (although see the cautionary remarks in the next section). Thus

$$\text{constant } r: \qquad V = \mathscr{E} - rI \tag{6.11}$$

It must be emphasized that (6.11) may only be a reasonable description of actual sources for a limited range of V and I, and that outside this range $\mathscr{E}$ and r are not necessarily constants independent of I or V. Note also that I may be negative, for instance in a cell being charged from another source, and then $\mathscr{E}$ will be smaller than V.

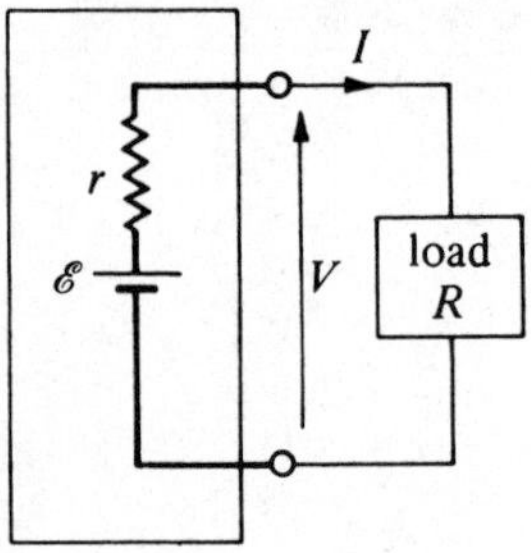

(*a*)

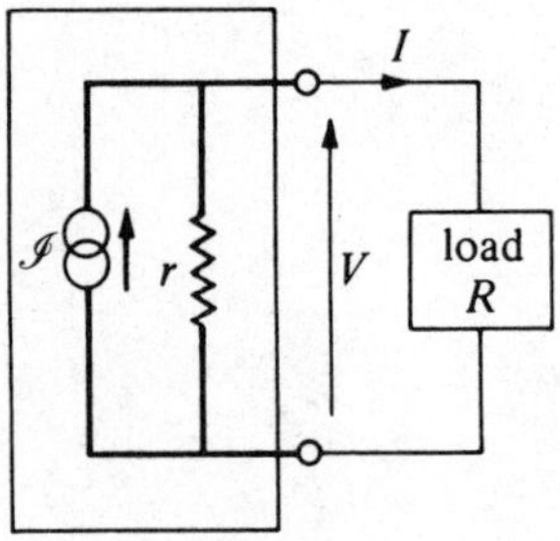

(*b*)

Figure 6.3 Equivalent networks for a source with internal resistance as (a) a voltage generator, (b) a current generator.

6.4 Network elements and terminology

Voltage generators The variation of V with I in many generators is sufficiently well described by (6.11) above. Such a generator can be replaced by an ideal source, with voltage $\mathscr{E}$ but no internal resistance, in series with a resistor of value r, because the potential difference V between the terminals is also $\mathscr{E} - rI$ for any I. The combination, shown boxed in Fig. 6.3a, is said to be an **equivalent network** to the real one because it has the same relationship between V and I at its terminals. This concept of equivalence is used frequently in subsequent sections for reducing complicated networks.

The ideal source is called a **voltage generator** or **voltage source** and $\mathscr{E}$, its open-circuit voltage, is independent of any external connections. $\mathscr{E}$ is also called the electromotance of the source, but *in network theory neither the use of the symbol* $\mathscr{E}$ *nor the use of the terms 'electromotance' or 'e.m.f.' must be taken to imply that we have any knowledge of the internal structure of the source.* In other words, we are concerned with $\mathscr{E}$ only as the open-circuit voltage and must not use the definition (6.2) in network theory. This point needs emphasizing now that many sources are complex electronic circuits.

Circuit symbols for voltage generators vary somewhat and we shall continue to use that of Fig. 6.3a, originally meant to indicate a chemical cell. Other symbols include a circle with the positive side marked or with an arrow pointing in the direction from negative to positive terminal.

Current generators An alternative way of regarding a source is to use the concept of a **current generator** in parallel with a resistor r as in Fig. 6.3b. In this case the current from the generator is independent of external connections and has the value $\mathscr{I} = \mathscr{E}/r$. The boxed networks in Fig. 6.3a and b are both equivalent to the same source because they both yield $V = \mathscr{E} - rI$ at the terminals. A source will be more aptly represented by a current generator if its output resistance is much higher than that of any load. $\mathscr{I}$ is called the short-circuit current of the source, i.e. the current when $V=0$, although in practice most sources would cease to obey (6.11) under short-circuit conditions.

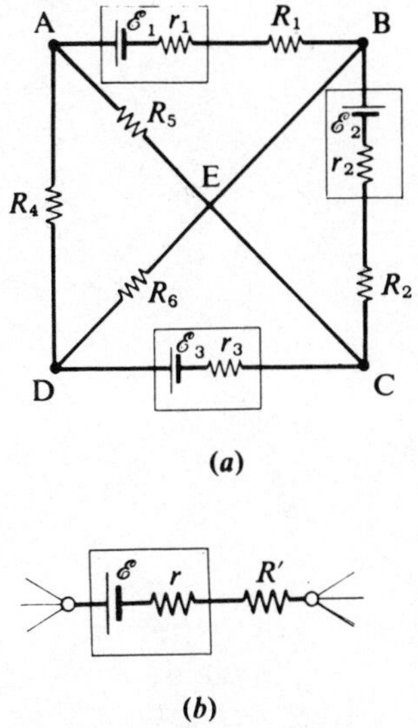

Figure 6.4 (a) Illustrating a general network: junctions or nodes are indicated by dots as at A, B, C, D (E is not a junction). AB, BC, etc., are branches. Circuits such as ABCA, ABCDA are meshes or loops. (b) A general branch.

General networks and their terminology In order to calculate the currents and potentials in an actual network of two-terminal components, we try to replace it by an idealized network which serves as a mathematical model of the real one and which is represented by a circuit diagram. Each component is replaced by an **element** with an I–V relationship at its terminals which matches that of the component as closely as possible. Thus any passive component with a sufficiently constant resistance can be replaced by a resistor for which $V = RI$. The voltage generator (electromotance or e.m.f.), the current generator and the resistor are the three elements from which any linear steady current network may be constructed. We shall tend to work mostly in terms of voltage rather than current generators since they are likely to be more familiar to the reader.

Some terminology adopted in describing networks is illustrated in Fig. 6.4a. The points to which three or more terminals are connected are known as **junctions** or **nodes** and should be represented in a circuit diagram by dots. The point E is simply a crossing of connections *in the diagram* and is *not* a junction unless a dot is drawn. The collection of components between two nodes is known as a **branch**. Any set of branches forming a closed path is a **mesh** or **loop**.

6.5 Solving network problems: Kirchhoff's laws

The first and obvious step in dealing with any network is to draw a circuit diagram corresponding to the given network of components. Any *given* values should be inserted in the diagram, and *unknown* I's and V's indicated by labelled arrows—for currents, *on* the lines representing connecting wires and for voltages, *across* the elements. Any negative results mean that the current or voltage is in the opposite direction to the arrow. The scene is now set for the application of laws yielding equations between the known and unknown quantities, enabling the unknown to be determined.

Kirchhoff's laws: proof The fundamental laws to be applied, known as Kirchhoff's laws, are based on two principles already established—and in fact applied in Chapter 5 to reduce capacitor networks. The principles are those of *the conservation of charge* and *the path-independence of potential difference.*

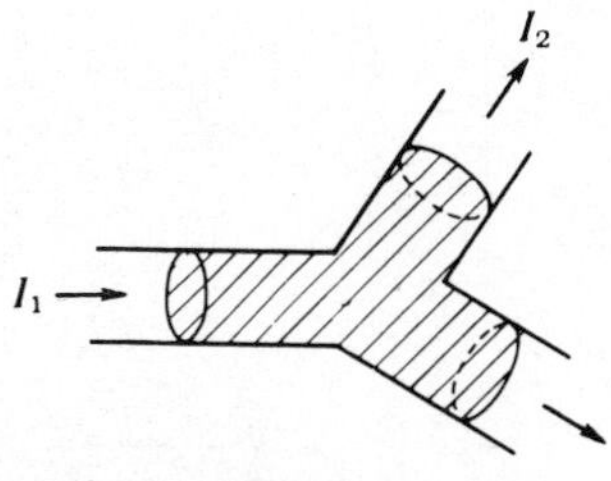

Figure 6.5 Steady currents at a junction. The conservation of charge in the shaded volume, coupled with $I = dQ/dt$, gives $I_2 + I_3 - I_1 = 0$ (Kirchhoff's first law)

The first of these is applied to a junction as in Fig. 6.5. Conditions are steady and so no charge can accumulate within the shaded volume. It follows that in any time interval the charge entering the volume must equal the charge leaving it. Since $dQ/dt = I$, the total current entering the junction must equal that leaving it, or

$$\text{K1 (steady currents):} \quad \boxed{\Sigma I = 0 \quad \text{at any junction}} \tag{6.12}$$

in which a negative sign is allocated to a current entering the junction. This convention is adopted so that the current leaving the shaded volume in Fig. 6.5 shall be positive, in line with the sign allocated to outward normals in Appendix B.3.

The second general principle, the path-independence of V, is expressed by

K2: $\Sigma V=0$ round any closed path (6.13)

Some branches of the network will only contain resistors for which the V in K2 is simply written as RI, but *in general* a branch will contain a source $\mathscr{E}$ with internal resistance r and other resistances totalling R' as in Fig. 6.4b. We then modify K2 as follows: the current I in a general branch is counted as positive when flowing through the source as if supplied by it (i.e. right to left in Fig. 6.4b). The potential difference across the branch is then $\mathscr{E}-rI-R'I$ or $\mathscr{E}-RI$, where R is the total resistance in the branch. Round a mesh, therefore, K2 will give that $\Sigma V=\Sigma(\mathscr{E}-RI)=0$ or

K2′: $\Sigma\mathscr{E}=\Sigma RI$ round any mesh (6.14)

In this equation, the sums are algebraic and signs are allocated to $\mathscr{E}$ and I by choosing positive senses round each mesh.

Kirchhoff's laws: application and worked example Slavish application of the laws is usually unwieldy and the following hints may be found useful. We use the network of Fig. 6.6 as an example in which all the currents are required (for the moment, ignore the arrows labelled 'mesh 1, x', 'mesh 2, y').

Avoid the use of subscripts in numerical problems: they are useful in theoretical work [cf. Eq. (6.19)] but the use, say, of α, β, etc., for currents saves time and prevents confusion. Secondly, never write down explicit equations for unknown currents using Kirchhoff's first law, but reduce the number of unknowns by applying the law directly in a circuit diagram: thus if α is the current from A in Fig. 6.6 and β is the current from B then that in the 10 Ω resistor is $\alpha+\beta$ and a third unknown is avoided. Thirdly, apply the second law to as many *independent* meshes as there are unknowns: thus

$$3.5-7=\alpha-\beta \quad \text{in mesh 1}$$
$$7=\beta+10(\alpha+\beta) \quad \text{in mesh 2}$$

from which α is -1.5 A, β is 2 A and $(\alpha+\beta)$ is $+0.5$ A.

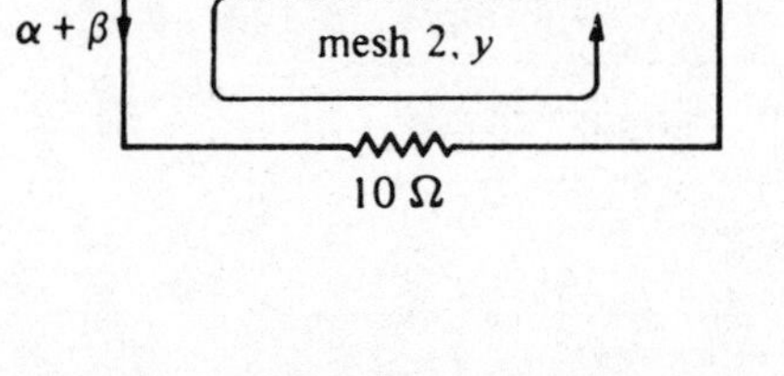

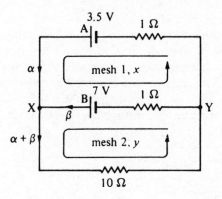

Figure 6.6 A network to illustrate the application of Kirchhoff's laws and other circuit theorems.

Other methods K1 and K2 (or K2′) are sufficient to solve any linear network problem, but there are a number of methods that reduce the labour considerably for specific types of problem. We deal first, in the next section, with the reduction of passive networks before looking at more general theorems in Sec. 6.7.

6.6 Combinations of resistors

Equivalent resistance Resistors connected by wires of negligible resistance into a network with n external connections are said to form an n-terminal passive network. When there are only two terminals we can define an *equivalent resistance* by V/I, where V is the potential difference between the terminals when a current I flows in at one and out of the other. We use this idea in conjunction with K1 and K2 to derive some useful formulae.

Resistances in series and parallel It is often possible to divide a network into collections of series and parallel combinations. For resistors in *series* (Fig. 6.7a), K1 tells us that the current is common to all, while the total potential difference V is the sum $V_1+V_2+V_3$ or $I(R_1+R_2+R_3)$. Hence, because the equivalent resistance $R_{eq}=V/I$,

resistors in series: $$R_{eq}=\Sigma R_i \tag{6.15}$$

In *parallel*, as in Fig. 6.7b, K2 tells us that the potential difference V is common to all, while the currents add according to K1. Hence $I=I_1+I_2+I_3=V(G_1+G_2+{}_3)$ so that

resistors in parallel: $$G_{eq}=\Sigma G_i \quad \text{or} \quad \frac{1}{R_{eq}}=\Sigma\frac{1}{R_i} \tag{6.16}$$

A useful pair of simple results for *two* resistors in parallel, say P and Q, are (a) that the equivalent resistance is $PQ/(P+Q)$, i.e. product divided by sum, and (b) any current flowing into the pair divides between them in inverse ratio to the resistances, i.e. P carries $Q/(P+Q)$ of the total. These are worth remembering.

Delta–Y transformation This is useful in reducing networks which *cannot* be treated as collections of series and parallel combinations and is similar to that for

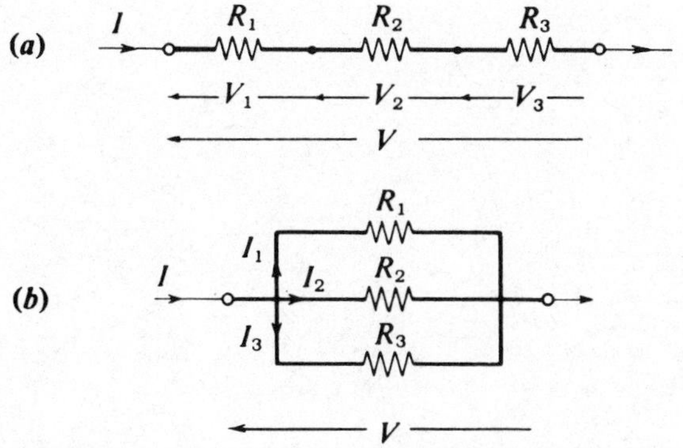

Figure 6.7 (a) Resistances in series (common current); (b) resistances in parallel (common voltage).

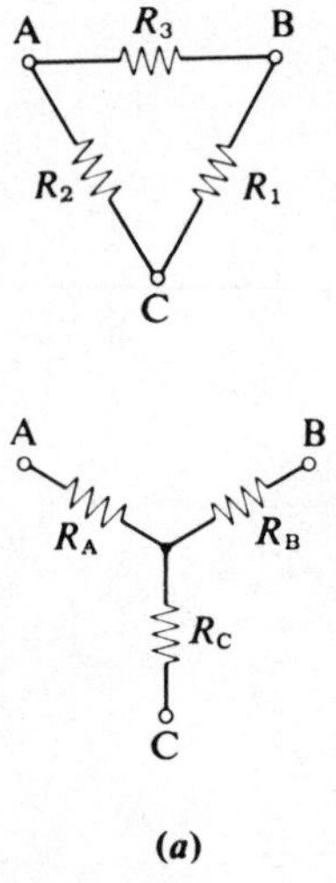

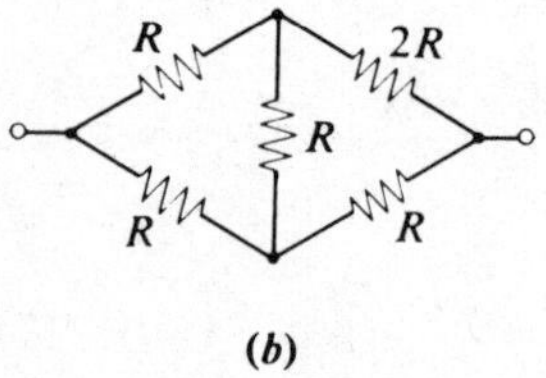

Figure 6.8 (a) The delta–Y transformation; (b) a network which canot be broken down into series and parallel combinations.

capacitors in Problem 5.12. In Fig. 6.8a, by equating the equivalent resistances between A and B, between B and C and between C and A, three relations are obtained which yield

$$R_A = R_2R_3/(R_1+R_2+R_3), \text{ etc.}$$
and
$$R_1 = (R_BR_C+R_CR_A+R_AR_B)/R_A, \text{ etc.} \tag{6.17}$$

for the equivalence of the delta and Y.

As an example, the reader should find the equivalent resistance of the network of Fig. 6.8b between the two external terminals using (6.17). The result could be checked by allowing I to enter and leave, determining the currents in the various branches using K1 and K2, finding the potential difference between the terminals and dividing it by I. The result by both methods should be $13R/11$.

Just as any two-terminal network can be replaced by an equivalent R, so a three-terminal network can be replaced by a delta or a Y.

6.7 General network theorems

Circuital currents A method that often shortens the use of K1 and K2′ is one using circuital currents or mesh currents. For instance, in Fig. 6.6, let the current in mesh 1 be x and that in mesh 2 be y so that the current originally denoted by β is now $(y-x)$. *This method ensures that Kirchhoff's first law is automatically satisfied*, and by applying the second law we obtain

$$3.5-7 = x-(y-x) \quad \text{in mesh 1}$$
$$-7 = (y-x)+10y \quad \text{in mesh 2}$$

giving $x = -1.5$ A and $y = +0.5$ A as before.

Superposition theorem In a general network, let the meshes be labelled 1, 2, 3, etc., and let us use mesh currents I_1, I_2, etc. (Fig. 6.9). Then the application of K2 to each mesh gives

$$\mathscr{E}_1 = R_{11}I_1 - R_{12}I_2 - R_{13}I_3 - \cdots$$
$$\mathscr{E}_2 = R_{22}I_2 - R_{12}I_1 - R_{23}I_3 - \cdots \text{ etc.} \tag{6.18}$$

where $\mathscr{E}_1$ is the *total* electromotance in mesh 1, etc., R_{12} is the resistance common to meshes 1 and 2, etc., and R_{11} is the *total* resistance round mesh 1, etc. Equations (6.18) can be solved for I_1, I_2, etc., to take the form

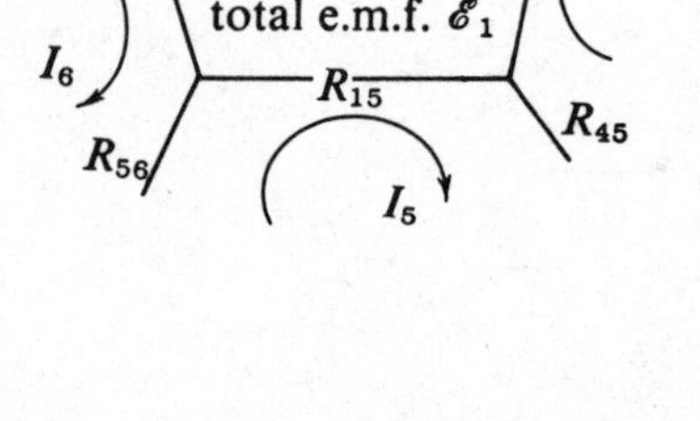

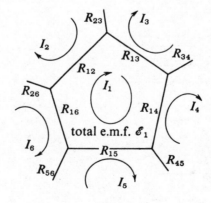

Figure 6.9 A mesh in a general network. R_{ii} stands for the total resistance in the ith mesh, e.g. $R_{11} = R_{12} + R_{13} + \cdots + R_{16}$.

$$I_1 = A_{11}\mathscr{E}_1 + A_{12}\mathscr{E}_2 + A_{13}\mathscr{E}_3 + \cdots \qquad (6.19)$$

with corresponding equations for I_2, etc., in which the A's are functions only of the resistances. Equations (6.19) show that the mesh currents, and hence the branch currents, superpose: put formally, if the current in a branch is I_1 due to an electromotance $\mathscr{E}_1$ acting alone at its proper place in the network and if the current in the same branch is I_2 due to $\mathscr{E}_2$ acting alone, the current in the branch when both $\mathscr{E}_1$ and $\mathscr{E}_2$ act is $I_1 + I_2$. It is understood that '$\mathscr{E}_1$ alone' means that all other electromotances are replaced by inert resistances equal to their internal resistance, i.e. that any voltage generators other than the one being considered are short-circuited and only the internal resistances left. Similarly, when current generators are present, the network can be solved for each generator separately, provided all the other current generators are, this time, open-circuited. Note that superposition depends on the linearity of (6.18) and therefore of the R's.

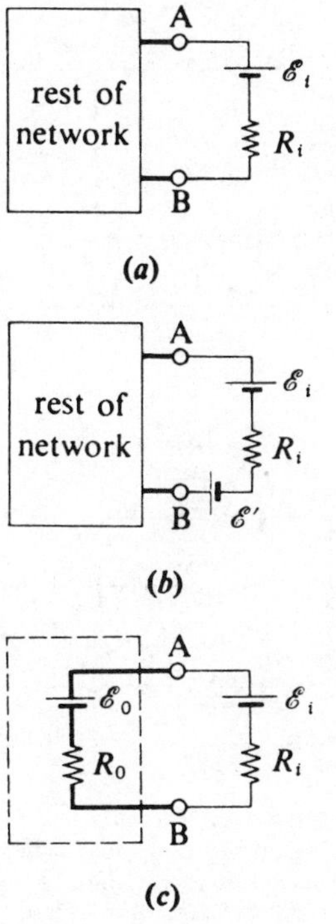

Figure 6.10 Thévenin's theorem. (a) The actual situation; (b) derivation of the theorem by the insertion of $\mathscr{E}'$; (c) the equivalent circuit.

Thévenin's theorem: statement This theorem enables us to obtain the current in any single branch of a network without calculating any of the other currents. If, for example, the branch concerned is the ith branch between nodes A and B as in Fig. 6.10a, it will in general contain an electromotance $\mathscr{E}_i$ and a total resistance R_i. Suppose the required current in this branch is I. Then it is always possible to find a circuit equivalent to the rest of the network as in Fig. 6.10c consisting of an electromotance $\mathscr{E}_0$ and a resistance R_0 which produces the same current I as the real circuit given simply by $I = (\mathscr{E}_0 - \mathscr{E}_i)/(R_i + R_0)$. The question is: what are the values of $\mathscr{E}_0$ and R_0?

Thévenin's theorem answers the question as follows. $\mathscr{E}_0$ is equal to the potential difference which appears across AB if the ith branch is open-circuited. R_0 is the resistance of the 'rest of the network' between A and B when all voltage generators in it have been short-circuited and all current generators open-circuited, leaving only their internal resistances.

Thévenin's theorem: proof The required current in Fig. 6.10a is I and is not yet known. Insert an electromotance $\mathscr{E}'$ into the ith branch as shown in Fig. 6.10b so as to oppose $\mathscr{E}_i$. By superposition, the new current in the ith branch will now be the sum of I and the current that flows if the only electromotance in the whole network is $\mathscr{E}'$. The latter current is $\mathscr{E}'/(R_i + R_0)$, so that

$$\text{New current} = I - \frac{\mathscr{E}'}{R_i + R_0} \qquad (6.20)$$

If, however, $\mathscr{E}'$ is so chosen that the new current is zero, then the 'rest of the network' across AB is effectively on open circuit. Under those conditions, the combined electromotance $\mathscr{E}_i - \mathscr{E}'$ must be equal to the open-circuit potential difference across AB, which is $\mathscr{E}_0$. In other words, the value of $\mathscr{E}'$ that reduces the current to zero is $\mathscr{E}_i - \mathscr{E}_0$. Hence, from (6.20):

$$0 = I - \frac{(\mathscr{E}_i - \mathscr{E}_0)}{(R_i + R_0)}$$

or

$$I = (\mathscr{E}_i - \mathscr{E}_0)/(R_i + R_0) \qquad (6.21)$$

and this is just the current which would be produced by the equivalent circuit of Fig. 6.10c. If the ith branch contains only a resistor R_i, then $I = \mathscr{E}_0/(R_i + R_0)$.

Thévenin's theorem: application From the proof, we see that the application of the theorem requires the following steps:

Step 1. Isolate the branch in which the current is required. Remove the circuit elements in the branch, leaving two terminals X and Y coming from the rest of the network.
Step 2. Determine $\mathscr{E}_0$ by calculating the potential difference across XY.
Step 3. Determine R_0 by replacing all sources by resistances equal to their internal resistances and calculating the equivalent resistance between XY.
Step 4. Replace the original branch across XY and replace the rest of the network by $\mathscr{E}_0$ and R_0 in series. This is now a simple series circuit in which I is given by total electromotance divided by total resistance.

Thévenin's theorem: worked examples We illustrate this important theorem with two examples using the network of Fig. 6.6. In the first example, the current in a branch containing an active element is calculated; in the second, we calculate the current in a branch containing only a resistor. It is not suggested that Thévenin's theorem should normally be used in networks as simple as this.

First, suppose that we require only the current through A. The rest of the network below A is replaced by a Thévenin source $\mathscr{E}_0$ equal to the voltage across XY when the branch containing A is open-circuited. This will be the voltage across 10 Ω produced by a current of 7/11 A, i.e. $\mathscr{E}_0 = 70/11$ V. The output resistance R_0 of the Thévenin source is that of the 1 and 10 Ω in parallel across XY, i.e. $R_0 = 10/11$ Ω. The equivalent network is thus as in Fig. 6.11a and a simple series-circuit calculation gives the current as −1.5 A as before.

To illustrate the theorem used in finding the current in a resistor, suppose we require that in the 10 Ω of Fig. 6.6. $\mathscr{E}_0$ is the voltage across XY when the 10 Ω branch is open-circuited. The current round mesh 1 would then be 1.75 A and the voltage across XY would be the 7 V of B less the drop of 1.75 V across the 1 Ω, i.e. $\mathscr{E}_0 = 5.25$ V. The output resistance is that of the two 1 Ω in parallel, i.e. $R_0 = 0.5$ Ω. The equivalent network is as in Fig. 6.11b and the current is 0.5 A as before.

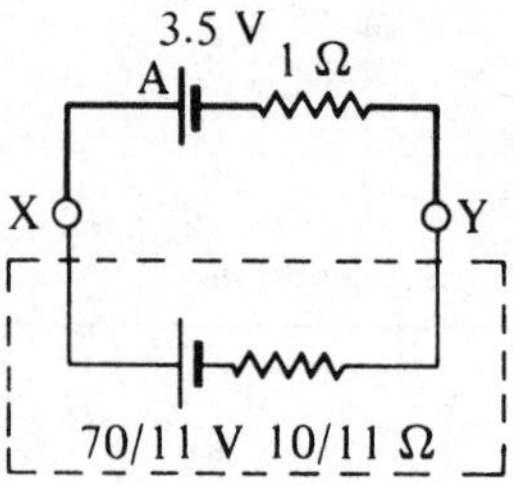

(*a*)

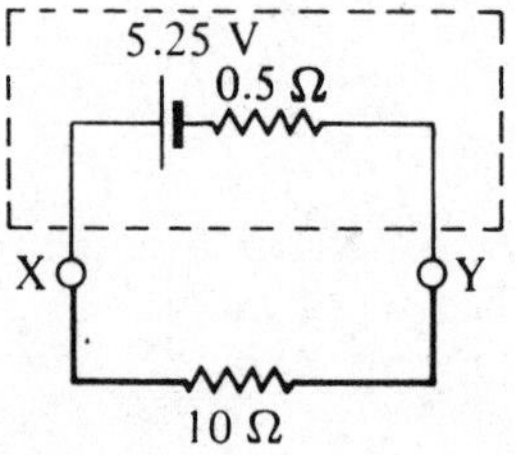

(*b*)

Figure 6.11 (a) Thévenin equivalent of Fig. 6.6 to obtain the current in A; (b) Thévenin equivalent of Fig. 6.6 to obtain the current in the 10 Ω resistor.

Norton's theorem In a similar way, a two-terminal network can also be replaced by an equivalent constant current generator $\mathscr{I}_0$, shunted by a parallel resistor R_0: this is *Norton's theorem.* The value of $\mathscr{I}_0$ is calculated by finding the short-circuit current at the terminals and the value of R_0 is the same as that in Thévenin's theorem. It is often useful in the reduction of a network to be able to change from Thévenin to Norton sources and vice versa, and this is the very simple conversion between Fig. 6.3a and b, where $\mathscr{E}_0 = R_0 \mathscr{I}_0$. Problem 6.18 illustrates this method.

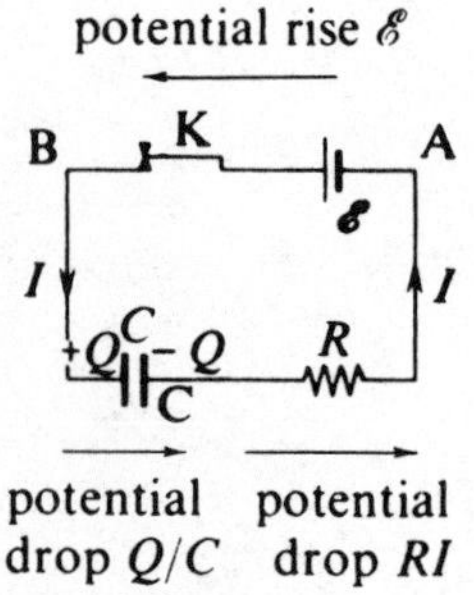

(*a*)

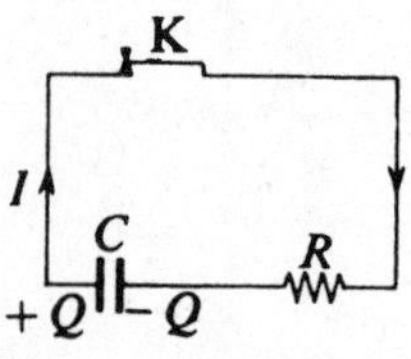

(*b*)

Figure 6.12 (a) Charge and (b) discharge of C through R.

Power transfer and matching We sometimes wish to ensure that the maximum possible amount of power is transferred from a source to a load. If the source is represented by its Thévenin equivalent as in Fig. 6.3a, the load of resistance R will carry a current $\mathscr{E}/(R+r)$, and will consume power $P = R\mathscr{E}^2/(R+r)^2$. For a given load, R, P is a maximum for $r=0$, but it is more common to find that r is fixed and that there is some choice of R. By equating $\mathrm{d}P/\mathrm{d}R$ to zero, the condition for maximum power transfer to the load is found to be $R=r$. The load is then said to be **matched** to the source.

6.8 Charging and discharging a capacitor through a resistor

While Kirchhoff's laws strictly refer to steady currents, we can assume that they apply at any instant in networks carrying time-varying currents, provided we watch for experimental contradiction of any deductions made. If the laws do apply, the currents are said to be **quasi-steady** (see Sec. 13.5).

An example is provided by the *charging* of a capacitor C through a resistor R which incorporates all the resistance in the circuit including the internal resistance of the source (Fig. 6.12a). Let the current at a time t after the switch K is closed by I in the direction shown and let the charge which has accumulated on C be $\pm Q$. Then, because the source of electromotance has no internal resistance, the rise in potential from A to B is $\mathscr{E}$, the fall from B to C is Q/C and the fall from C to A is RI. Thus, because $\Sigma V=0$ round the circuit, $\mathscr{E}-Q/C-RI=0$. However, $I=\mathrm{d}Q/\mathrm{d}t$, so that

$$R\frac{\mathrm{d}Q}{\mathrm{d}t}+\frac{Q}{C}=\mathscr{E} \tag{6.22}$$

Let $C\mathscr{E}$ be a new constant Q_0 so that (6.22) can be written

$$R\frac{\mathrm{d}(Q_0-Q)}{\mathrm{d}t}+\frac{(Q_0-Q)}{C}=0$$

or

$$\frac{\mathrm{d}(Q_0-Q)}{(Q_0-Q)}=-\frac{\mathrm{d}t}{RC}$$

Integrating,

$$\log_e(Q_0-Q)=-t/RC+\text{constant}$$

At $t=0$, $Q=0$ and the constant is thus $\log_e Q_0$. This finally gives

charging of capacitor: $$Q=Q_0(1-\mathrm{e}^{-t/RC}) \tag{6.23}$$

and

charging of capacitor: $$I=\mathrm{d}Q/\mathrm{d}t=\frac{Q_0}{CR}\mathrm{e}^{-t/RC} \tag{6.24}$$

Q_0 is thus identified with the charge on the capacitor after an infinite time, while the quantity RC, known as the **time constant** or the **relaxation time** of the circuit, gives the time for the charge to rise to $1-1/\mathrm{e}$ or about $\frac{2}{3}$ of its final value. It thus gives an indication of the rate of charge, as shown in Fig. 6.13a.

For the *discharge* of a capacitor with an initial charge Q_0, let the current and charge after a time t be I and $\pm Q$ as in Fig. 6.12b, where this time $I=-\mathrm{d}Q/\mathrm{d}t$ [see Eq. (1.7)]. $\Sigma V=0$ gives

$$R\frac{\mathrm{d}Q}{\mathrm{d}t}+\frac{Q}{C}=0$$

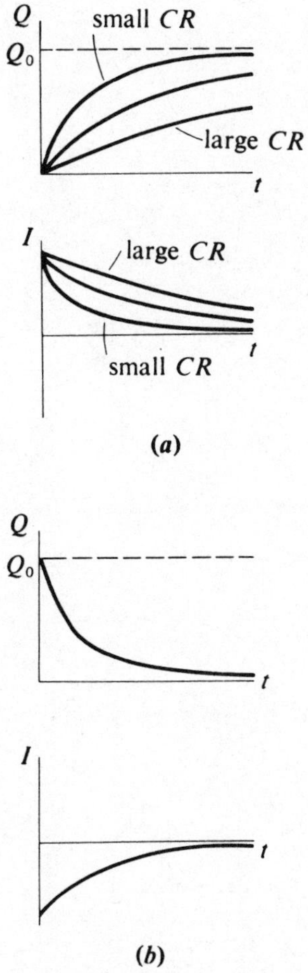

Figure 6.13 Growth and decay of charge and current in a CR circuit. (a) Charging a capacitor; (b) discharging a capacitor.

yielding, by a similar method to that used above,

discharge of capacitor: $$Q = Q_0 e^{-t/RC} \tag{6.25}$$

and

discharge of capacitor: $$I = \frac{Q_0}{RC} e^{-t/RC} \tag{6.26}$$

plotted as negative in Fig. 6.13b because it flows in the opposite direction to (6.24) as Fig. 6.12 shows.

The charge remaining on the capacitor after various times can be measured by an electrometer and a graph of $\log_e Q$ against t gives a straight line of slope $-1/RC$. In practice, the dielectric in the capacitor will have a small conductivity producing an effective leakage resistance in parallel with C and R: this can be determined by carrying out the experiment without R.

For high resistances, the experiment can be used to determine R in terms of a standard capacitor: for instance, with $C = 1\,\mu\text{F}$ and R about $50\,\text{M}\Omega$ the time constant is conveniently about 50 s.

Comment C6.3

Such experiments justify our assumption that steady current laws can be applied at least to slowly varying currents.

Practical applications of RC circuits The simple series RC circuit provides the basis for a number of widely used applications based on the shape of the voltage variations across C or R.

1. *Sweep and timer circuits.* The time base of a cathode-ray oscilloscope or the sweep circuit of a television tube need a voltage that repeatedly increases linearly with time and then flies back quickly to zero. The voltage across a charging capacitor, Q/C, has the shape shown in the upper set of curves in Fig. 6.13a, i.e. it varies as $(1 - e^{-t/RC})$. If the times involved are very short compared with RC, the approximation $e^{-t/RC} \approx 1 - t/RC$ shows that the voltage will nearly be proportional to t/RC. The linearity can be improved using auxiliary electronic circuits.

   The flyback is obtained by arranging for the capacitor to discharge at a predetermined voltage through a circuit with a much smaller value of R, i.e. with a much lower time constant. Repetition of the process yields a saw-tooth waveform that can be used either for voltage sweep circuits or timing circuits.

   High-voltage television and radar tubes need a linear current sweep rather than a voltage sweep since their electron beams are deflected magnetically. In these cases, the circuits are based on an LR combination rather than a CR combination (Sec. 9.5).

2. *Pulse circuits.* The way in which an RC combination can be used to generate a sequence of pulses is discussed in Sec. 10.12.

6.9 D.C. measurements

For measurements involving steady currents, a laboratory needs standards of resistance and potential to use for calibration. These standards will be resistors of various values (see Sec. 6.10) and either standard cells (Appendix C) or, more likely, stabilized electronic voltage sources.

Measurement of potential difference V All instruments or methods for measuring V between two points in a network must be connected *across* the points, and should ideally have an infinite resistance so as to leave unchanged the value of V being measured. In practice, it is sufficient if the instrumental resistance is much higher than that in the network across which the connection is made. Electronic voltmeters, often constructed with a digital display, have resistances up to $10^{14}\,\Omega$, in contrast with moving coil voltmeters with perhaps no more than a few thousand ohms. (Note that, although a voltmeter with a finite resistance still records the voltage actually presented across its terminals, the voltage after connection is not the same as before connection. The difference is sometimes called the *load error*.)

For reliable measurements, the potentiometer can be used (Fig. 6.14a). A steady current I flows through R, and the potential difference V to be measured is connected across R_1. The tappings A and B, or one of them, are adjusted until no current flows in the galvanometer G. Application of K2, $\Sigma V=0$, to the circuit ABV gives $V=R_1 I$. When V is replaced by a standard cell of electromotance $\mathscr{E}_s$, the balance point shifts so that the resistance between A and B is R_s. This time $\mathscr{E}_s=R_s I$. Hence, *provided I remains constant throughout*, $V=R_1\mathscr{E}_s/R_s$. An essential condition for the use of the method is therefore that the driving electromotance $\mathscr{E}$ and the resistances r and R shall be constant from the time the first balance point is obtained, although their values do *not* need to be known. Apart from that, it is also necessary that V and $\mathscr{E}_s$ shall be (a) smaller than $\mathscr{E}$ and (b) connected to A and B with the correct polarities.

The particular value of the method lies in the absence of current drawn from the source of potential difference. The potentiometer thereby acts like a voltmeter of infinite resistance and can measure electromotances. A further consequence, common to all null methods, is that the calibration of the galvanometer is of no importance.

In elementary work, R is often a uniform wire on which the balance point B is obtained using a sliding contact. The length AB is then proportional to the resistance R_1 between A and B. If the balance lengths corresponding to V and $\mathscr{E}_s$ are l_1 and l_s, then $V=l_1(\mathscr{E}_s/l_s)$. In modern potentiometers of high precision, R can be an elaborate arrangement of high-quality resistance coils and for more details and further references see Kite (1974).

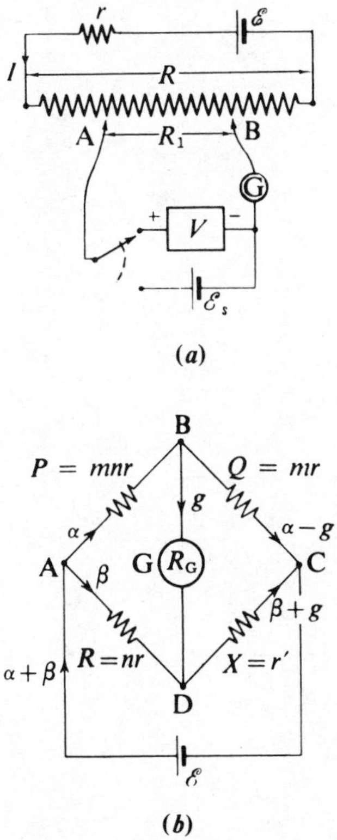

Figure 6.14 (a) The potentiometer; (b) the basic Wheatstone bridge.

Measurement of resistance For resistances from about $1\,\Omega$ to $1\,M\Omega$ and possibly higher, depending on the accuracy required, the Wheatstone bridge is the basic network for comparing an unknown with a standard resistance. The condition for zero current through the galvanometer in Fig. 6.14b is that B and D should be at the same potential so that $P\alpha=R\beta$ and $Q\alpha=X\beta$. Hence $P/Q=R/X$ independent of the electromotance $\mathscr{E}$. If the unknown is X, the absolute values of P and Q are not important but only their ratio, and this can be obtained either by using a uniform slide-wire so that the ratio is equal to that of the lengths of wire (the so-called metre bridge) or by using sets of resistors whose ratios are accurately known. The only standard needed is R.

Special methods are needed for high resistances (the bridge becomes insensitive —see Problem 6.24) and for low resistances (no provision is made for eliminating leads or contact resistances). For high resistance, a direct measurement with a meter of the current produced by a known voltage is often sufficient. For low resistances of $10^{-2}\,\Omega$ and below, the potentiometer above can be used to compare $V_1=RI$ across a standard R with $V_2=XI$ across an unknown X carrying the same current (R and X being in series). This method enables the current and potential terminals (see Fig. 6.16b) to be used, and the contact resistances in the latter carry no current at balance and thus do not affect the balance points.

Finally, one can make direct measurements of resistance R by *ohm-meter*. An electronic ohm-meter will consist of a highly stabilized current source supplying a predetermined current I which is independent of the load. Such a current is passed through the R connected across the terminals producing $V=RI$. The voltage is thus directly proportional to R and the output meter can be calculated linearly in ohms.

Measurements of current All instruments for measuring current must be connected so that the current flows through them. Although moving-coil ammeters often give sufficiently good values for most work, the most accurate and precise measurement of current involves passing it through a standard resistance R. The potential difference V across R is then measured by a potentiometer or electronic voltmeter, and $I=V/R$.

Electrometers which are electronic voltmeters with current, voltage, resistance and charge ranges are increasingly the first choice for making d.c. measurements.

6.10 Resistivity and conductivity

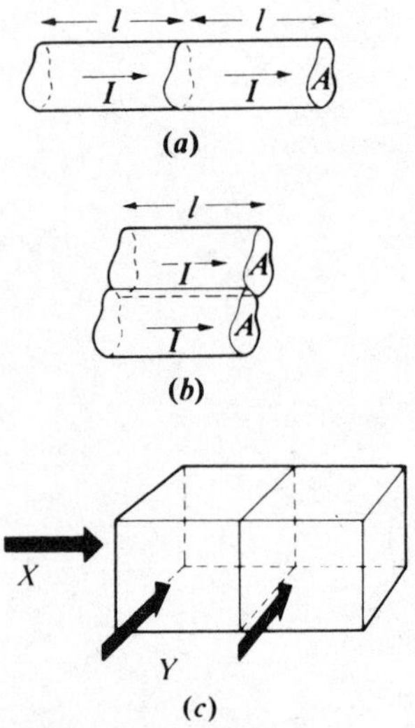

Figure 6.15 Resistance of cylinders is (a) proportional to length, (b) inversely proportional to cross-sectional area. The two cubes of (c) have twice the resistance of one for current in the direction X, but only half the resistance of one for current in direction Y.

Concepts and definitions The network theory of the previous eight sections has treated conductors either as two-terminal elements we called resistors or as connecting leads with negligible resistance. We now take a more detailed look at the flow of current in conductors because real conductors are *volumes* in which a current density exists at every point. Another, more practical, reason for examining flow in more detail is that we need to examine the materials of which resistors and leads are constructed and to form theories as to why different substances conduct in such different ways.

The resistance R of an ohmic conductor depends upon *geometry* (shape, size, direction of current) and on the *material* of which it is composed. The geometrical factors can be separated out by taking cylindrical elements of the conductor which carry current parallel to their axes. If the length along the axis of such an element is l and the cross-section A, then the formula for resistances in series (6.15) shows that two cylinders end-to-end as in Fig. 6.15a will have twice the resistance of one. In general, therefore, R is proportional to l for constant A. Similarly, the formula for resistances in parallel (6.16) shows that R is proportional to $1/A$ for constant l (Fig. 6.15b). In the most general case for any cylindrical conductor, therefore, R is proportional to l/A or

$$R=\rho\frac{l}{A} \qquad \text{(definition of } \rho) \tag{6.27}$$

where ρ is a constant for an ohmic conductor, known as its **resistivity**. The reciprocal is known as the **conductivity** σ, so that

$$\sigma = 1/\rho = \frac{l}{RA} \qquad \text{(definition of } \sigma) \tag{6.28}$$

While ρ and σ can still be defined in this way for non-ohmic materials, they will then be quantities which vary with the current passed. The **SI unit** for ρ is the $\Omega\,m$ and for σ the $S\,m^{-1}$.

The statement often encountered that the resistivity of a material is the resistance per unit cube is incorrect: it implies that the unit of ρ is the $\Omega\,m^{-3}$ which is clearly not in accord with (6.27), and it ignores any reference to the current flow. Even if the direction of this flow is specified as parallel to one pair of sides, the definition still implies that two unit cubes have a resistance twice that of one, whereas it may be halved (Fig. 6.15c). The correct definition (6.27) contains the understanding that l is a dimension parallel to the current and A an area perpendicular to it.

The conductivity of materials It is not difficult to distinguish a class of good conductors (ρ around $10^{-7}\,\Omega\,m$) whose conductivity is due to the presence of free electrons and one of good insulators (ρ upwards of $10^{11}\,\Omega\,m$) containing very few free charge carriers. However, in between these two extremes lies a continuous range of materials whose conductivity arises from the various types of carrier indicated in Table 6.1. The table also demonstrates the enormous range of resistivities exhibited by real substances. Note once again that only metals are known to be strictly linear or ohmic and to have a constant ρ under constant physical conditions. For other materials the values quoted are representative.

Variation of resistivity When the physical conditions of materials (e.g. temperature) are changed, then so is the resistivity. The resultant change in total resistance can be measured simply and accurately by incorporating it in one arm of a Wheatstone bridge. The out-of-balance current through the detector in the bridge arm BD (Fig. 6.14b) can thus be used to monitor or measure the change in the physical condition. As examples we consider temperature, magnetic field and mechanical strain.

The resistance of metals and alloys increases with rise in **temperature** according to the empirical law

$$R_{\theta_1} = R_\theta[1 + \alpha(\theta_1 - \theta) + \beta(\theta_1 - \theta)^2 + \cdots] \tag{6.29}$$

where θ and θ_1 are temperatures, and α, β, etc., are coefficients decreasing rapidly in magnitude from one to the next. Typical values are $0.004\,K^{-1}$ for α and $10^{-6}\,K^{-2}$ for β so that only α, the *temperature coefficient of resistance*, need normally be considered. Because α has a value close to 1/273 for many metals, a temperature scale furnished by the variation of resistance is very approximately the same as that on the ideal gas scale, and platinum resistance thermometers can be corrected to such a scale if α and β for Pt are known. Equation (6.29) also gives the variation of *resistivity* with temperature although the values of α, β, etc., are slightly different because of the changes in dimensions: the difference for moderate $(\theta_1 - \theta)$ is negligible.

At very low temperatures, usually below 10 K, the resistance of many metals

Table 6.1 Conductivity and resistivity of materials

Class	Conduction by	$\log_{10}\sigma$ (σ in S m^{-1})	$\log_{10}\rho$ (ρ in Ω m)	Examples (ρ in Ω m at 20°C)
		8	−8	Silver (1.6×10^{-8})
Metals and alloys	Free electrons			
		6	−6	
				Mercury (0.96×10^{-6})
		4	−4	
		2	−2	
Electrolytes	Ions in solution			Saturated NaCl solution (0.04)
		0	0	
Semi-conductors	Excitation of electrons and holes			Germanium (0.46) Silicon (2×10^3)
		−2	2	
		−4	4	Distilled water ($10^3 - 10^4$)
		−6	6	
Ionic solids	Ionic diffusion			Crystalline NaCl (10^7)
		−8	8	
				Wood ($10^8 - 10^{11}$)
		−10	10	
				Glass ($10^9 - 10^{14}$)
	(ceramics)	−12	12	
Insulators		−14	14	
				Mica (5×10^{14}) Polystyrene ($10^{15} - 10^{16}$)
	(thermoplastics)	−16	16	
		−18	18	PTFE ($10^{17} - 10^{18}$)

Comment C6.5

and alloys suddenly decreases to a negligible value, a phenomenon known as **superconductivity**. Semiconductors, insulators and electrolytes all have resistivities which *decrease* as temperature rises.

Bismuth shows a remarkably high variation of resistivity when a **magnetic field** is applied at right angles to the current flow, and this property has been used to measure the strength of such fields: other materials show extremely small changes by comparison. The phenomenon is known as **magnetoresistance** and is allied to the Hall effect (Sec. 7.6).

A stretched metal wire undergoes a measurable change in resistance which can be used as a measure of **mechanical strain**. A fine wire bent backwards and forwards until it covers an area of about the size of a postage stamp is sandwiched between two fine pieces of paper and the whole strain gauge thus formed is made to adhere to the surface whose strain is required. The gauge is strained by the same amount as the surface and the change in resistance is measured. Gauges of metal foil are now frequently used.

Resistors in practice **Fixed resistors of high stability** for use in measurements are usually wire or sheet of some quaternary alloy based on nickel and chromium (see Dix, 1975). These alloys have a relatively large resistivity with a very small temperature coefficient and a small thermal electromotance against copper. *Power resistors*, capable of dissipating up to several hundred watts, are wire-wound. **General purpose resistors**, whose value need not be known with precision, dissipate only a few watts and may be 'composition' (graphite, filler and resin baked into a rod), 'cracked-carbon' (carbon deposited at a high temperature on to a ceramic rod) or 'metal-film' in increasing order of stability. Resistors in hybrid circuits consist of a resistor paste (often of ruthenium oxide mixed with a glass frit) deposited on a substrate of alumina. In monolithic (completely integrated) electronic circuits, resistance is provided by *p*-type or *n*-type regions in slices of the basic semiconductor used, invariably silicon.

Standard resistors are constructed with wire, again of a quaternary alloy, sealed into a container often filled with oil whose temperature can be measured; they are fitted with thick copper terminals as in Fig. 6.16a. For standards of less than about 1 Ω the use of only two terminals makes the resistance indeterminate, not only because of leads (whose resistance can be allowed for) but principally because of **contact resistances** at the terminals themselves. These are due to contamination of the surface and can amount to 0.0001 Ω even when care is taken: this is a serious error particularly with low resistances and it is, moreover, variable. Consequently, four terminals are provided as in Fig. 6.16b, the resistance R being defined as the ratio of the potential difference between PP, the potential terminals, to the current between CC, the current terminals. It is clear that contact resistances at CC do not affect R and those at PP are arranged to produce zero or negligible effect by the circuit connections (see previous section).

Variable resistors are either (a) resistance boxes giving resistances in steps by removable plugs or by dials controlling a decade or (b) rheostats with sliding contacts. The latter can also be used to obtain variable potential differences from

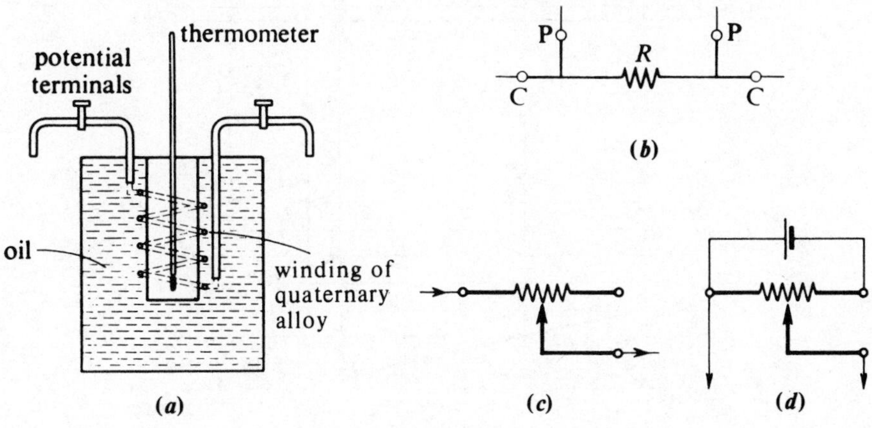

Figure 6.16 (a) Construction of a standard resistor; (b) current and potential terminals; (c) rheostat used as a variable resistor; (d) rheostat used as a potential divider.

a fixed source as shown in Fig. 6.16d, and it is then known as a potential divider: electronic engineers often refer to it confusingly as a potentiometer, a term we shall reserve for the measuring instrument described in Sec. 6.9.

Thermistors are resistors with a very high temperature coefficient of resistance, either negative or positive (NTC or PTC thermistors, respectively). The NTC type are constructed from ceramic materials, mostly metallic oxides, and the PTC type from barium titanate or silicon. The resistance of these devices may change by several orders of magnitude for temperature variations of 100°C, so that they form very effective sensors for changes in temperature.

Varistors are non-linear resistors whose dynamic resistance increases markedly with increasing voltage. They are particularly useful for suppressing large and abrupt surges of voltage that might otherwise damage equipment.

A survey of resistor technology is to be found in Fairs (1975).

6.11 Current flow in three dimensions

The one-dimensional flow of current in metallic wires is so simple in form that it is easy to forget that the wire is a volume and that three dimensions are involved. As well as that, we are familiar nowadays with many examples of currents that flow throughout three-dimensional regions in quite complicated patterns. If we are to deal with such cases we need to look at current flow more generally than hitherto. In this section we examine flow in extended conductors and ways of generalizing to three dimensions some of the laws already encountered in this chapter.

Distributed resistance We consider first, in slightly more detail, the ordinary flow of current in a long wire or cable. While it is often possible to neglect the resistance of leads to resistors or from a source of electromotance, the transmission of power to a load sometimes takes place through cables so long that the potential drop is

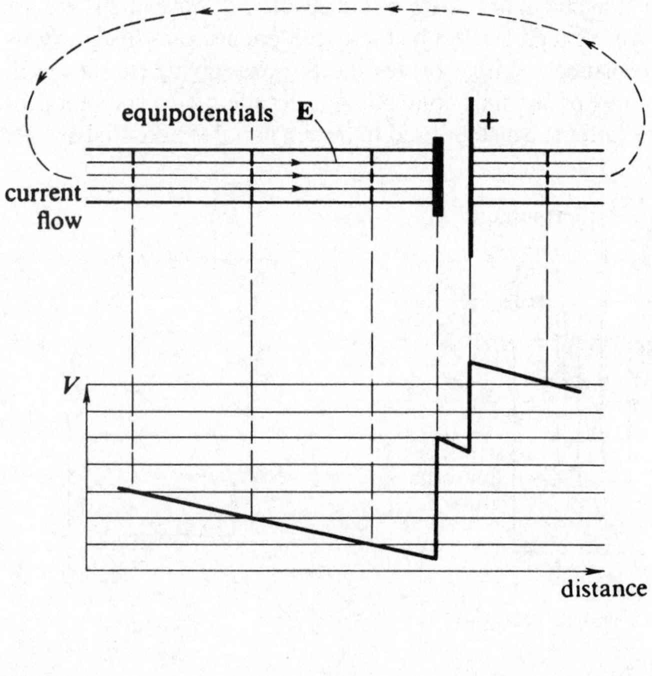

Figure 6.17 Equipotentials, lines of **E** and potential diagram for a steady current flowing in a uniform wire.

serious unless *either* the resistance per unit length *or* the current is kept small. For overhead cables in the national grid, the diameter cannot be indefinitely increased because of weight and cost and the loss is reduced by using high voltages (and hence smaller currents). For distribution within buildings, weight is not a limitation and larger diameter cables are used in preference to higher voltages for obvious reasons.

In a *straight, uniform current-carrying wire*, the potential drop along the length has a uniform gradient, and so there exists a uniform **E**-field along it. Equipotential surfaces are cross-sectional planes normal to the **E**-field, as usual. Figure 6.17 shows the complete picture for a steady current produced by a source of electromotance such as a chemical cell. When the wire is not straight, the situation is not basically changed because the current flow and **E**-field are still along the axis.

Comment C6.2

However, currents do not always flow along parallel lines in a conductor (see, for example, the situations in Fig. 6.18b and Problem 6.25). If the total resistance is required in such cases, the conductor must be divided into elementary cylinders with lengths $\mathrm{d}l$ along the current flow and areas $\mathrm{d}A$ perpendicular to it. The resistance of such an element is then $\mathrm{d}l/\sigma\,\mathrm{d}A$ by (6.27) and (6.28), and the total resistance is obtained by appropriate summation or integration of the elements in series or parallel.

Worked example | *current flow between concentric spheres* Suppose a potential difference is maintained between an inner sphere of radius a and an outer concentric sphere of radius b as in Fig. 6.18b. The space between them is filled with a uniform conductor of conductivity σ and we require the total resistance to the radial flow of current.

A typical spherical shell of radius r and thickness $\mathrm{d}r$ consists of elementary cylinders all in parallel, so that their *conductances* add. The conductance of one, with cross-section $\mathrm{d}A$, say, is $\sigma\,\mathrm{d}A/\mathrm{d}r$. Since σ and $\mathrm{d}r$ are common to them all, the total conductance of the shell is its total area $4\pi r^2$ multiplied by $\sigma/\mathrm{d}r$, i.e. its resistance is $\mathrm{d}r/(4\pi r^2\sigma)$. As all such shells are in series, the total resistance is the integral of $\mathrm{d}r/(4\pi r^2\sigma)$ between $r=a$ and $r=b$. That gives $(b-a)/(4\pi\sigma ab)$.

Ohm's law at a point The relation $V=RI$ includes Ohm's law if R is constant: it applies between *two* points along the length of a conducting wire or cable. We wish to find a form more general than this, one which will apply at any *single* point within a conductor.

To do this, consider a cylindrical element of a passive current-carrying conductor of conductivity σ as in Fig. 6.18a. The voltage across the ends is $-\mathrm{d}V$ (negative because the potential falls along the direction of positive current flow). Hence

$$-\mathrm{d}V=R\,\mathrm{d}I=\frac{\mathrm{d}l}{\sigma\,\mathrm{d}A}\mathrm{d}I$$

Since $-\mathrm{d}V/\mathrm{d}l$ is the electric field E_Q in the element and $\mathrm{d}I/\mathrm{d}A$ is the current density J, the equation may be written in the limit as $J=\sigma E_Q$ or, in vector form,

passive conductor: $$\boxed{\mathbf{J}=\sigma\mathbf{E}_Q} \qquad (6.30)$$

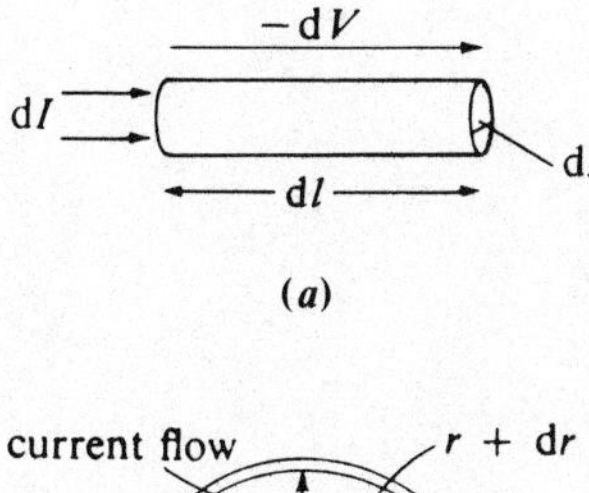

(*a*)

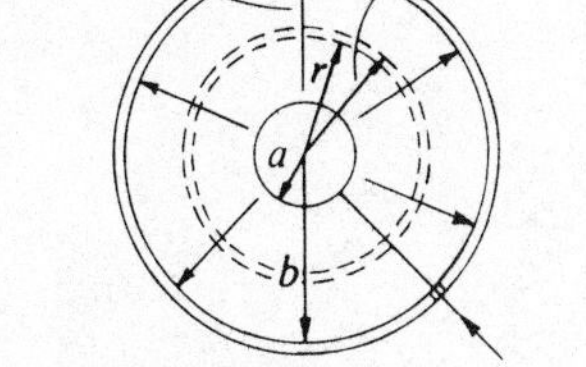

(*b*)

Figure 6.18 (a) An element of a current-carrying conductor; (b) radial current flow in a conductor between two concentric spherical surfaces.

Like $V=RI$, this is strictly a definition of σ at a point and an *ohmic conductor* is one for which σ is independent of the magnitudes of **J** and $\mathbf{E}_Q$ at constant temperature, etc. A conductor for which σ is the same at every point is *homogeneous*, and one for which its value at a point is independent of the direction of **J** or $\mathbf{E}_Q$ is *isotropic*. We shall not concern ourselves with non-homogeneous or anisotropic conductors.

What if the conductor is not passive, but contains seats of electromotance within it, accounted for by additional fields $\mathbf{E}_M$ and **e** as in Sec. 6.1? These extra fields will have the same effect as $\mathbf{E}_Q$ and produce the same current density, so that in general

active conductor: $$\boxed{\mathbf{J}=\sigma(\mathbf{E}_Q+\mathbf{E}_M+\mathbf{e})} \qquad (6.31)$$

When a current I is confined to a closed conducting path of cross-section A, J/σ can be written as $I/\sigma A$. For an element of the path length dl, (6.31) shows that $(\mathbf{E}_Q+\mathbf{E}_M+\mathbf{e})\cdot \mathrm{d}\mathbf{l}=I\,\mathrm{d}l/\alpha A$. If this is integrated round the complete path, definition (6.1) gives that

$$\mathscr{E}=RI \qquad (6.32)$$

where R is the total resistance in the circuit. This is identical to Kirchhoff's second law in the form (6.14).

Note on $J=nev$ and carrier mobility Equation (1.28), $J=nev$, is a very general one giving any current density in terms of its constituent moving charges. It is in no way superseded by (6.30) or (6.31), which apply to conductors only. In fact, the two equations can be combined in a very useful way, for if the carriers of current in a conductor all possess a single electronic charge and the same velocity, then $\sigma=J/E=nev/E$. The quantity v/E, the drift velocity per unit electric field, is called the *mobility* of the carrier:

$$\mu=v/E \qquad \text{(definition of } \mu) \qquad (6.33)$$

It follows that $\sigma=ne\mu$ or, if more than one carrier exists,

$$\sigma=\Sigma\, ne\mu \qquad (6.34)$$

Equation of continuity The conservation of charge leads to a very general property of current density summed up in what is called the equation of continuity. We look at this first for steady currents and then generalize it.

Consider any volume of a space bounded by a closed surface S which encloses a total net charge Q (the region does not necessarily contain any conducting material). If any currents that flow are steady, then no accumulation of charge can occur anywhere. It follows that Q must be constant and by the conservation of charge this means that no *resultant* current can flow out over S (we used this argument to establish Kirchhoff's first law). Since the outward current over S can be written as $\oint \mathbf{J}\cdot \mathrm{d}\mathbf{S}$, then for steady currents $\oint \mathbf{J}\cdot \mathrm{d}\mathbf{S}=0$.

Now suppose that currents are allowed to vary. The conservation of charge means that any net current that *does* leave S must lead to an equivalent reduction in Q, i.e. that

$$\oint_S \mathbf{J}\cdot d\mathbf{S} = -dQ/dt \tag{6.35}$$

which is more general than the result obtained for steady currents.

Equation (6.35) applies to a finite volume, just as Gauss's law for **E** did in Sec. 4.2, and it has the same form as Gauss's law with **J** replacing **E** and $-dQ/dt$ replacing Q/ε_0. Consequently, the same method as used in Sec. 4.5 can be used to obtain a version applying at a point. This will take the form $\operatorname{div}\mathbf{J} = -\partial\rho/\partial t$ or $\nabla\cdot\mathbf{J} = -\partial\rho/\partial t$ (where ρ is here the volume density of charge and not, it should be emphasized, the resistivity).

These equations are known collectively as the **equation of continuity** and are typical of equations governing conserved quantities encountered throughout physics. They can be summarized as follows:

general equation of continuity: $$\oint_S \mathbf{J}\cdot d\mathbf{S} = -dQ/dt;\quad \nabla\cdot\mathbf{J} = -\partial\rho/\partial t \tag{6.36}$$

for steady currents: $$\oint_S \mathbf{J}\cdot d\mathbf{S} = 0;\qquad \nabla\cdot\mathbf{J} = 0 \tag{6.37}$$

Quasi-steady conditions Kirchhoff's first law is a result of (6.37) and applies only under steady conditions. When currents vary, (6.36) applies and strictly speaking K1 is invalid. However, we shall see in Chapter 11 (Problem 11.16) that for good conductors any accumulated charges disappear so quickly that $\partial\rho/\partial t$ can be taken as zero after a very short time, provided the frequency of any time variations is not too great. These conditions—high conductivity and not too high a frequency—are known as **quasi-steady conditions**. Under them, K1 can still be applied even if currents and voltages change with time, as it was, for instance, in Sec. 6.8. Further discussion of this point is to be found in Sec. 13.5.

6.12 Conduction in solids: atomic aspects

Metals: the classical free-electron model (Drude–Lorentz theory) We have already seen that electrons are responsible for metallic conduction and we therefore adopt an atomic model in which the outermost (or valence) electrons of the metallic atoms are so weakly bound that they move freely among the framework formed by the ions and suffer collisions with them. This is a *free-electron* model where the paths of the electrons are similar to those of the molecules of a gas in the kinetic theory. Such an electron 'gas' would have a mean free path λ between collisions and a mean random velocity $\bar{c}$ so that the mean time between collisions is $\lambda/\bar{c}$.

Since $\bar{c}$ is random, it does not contribute to the electric current. Two possible effects can contribute to the general drift of electrons in a particular direction. One is *diffusion*, caused by the concentration n of electrons being greater in one region than another. If the gradient of the concentration, $\operatorname{grad} n$, is small, the current density **J** due to diffusion is given by $-De \operatorname{grad} n$, where D is the coefficient of diffusion. The other effect causing drift is the application of an electric field **E** giving a current density $\mathbf{J}=\sigma\mathbf{E}$ as in (6.30). In a metal, the drift velocities due to **E** are so much greater than diffusion velocities that the latter can be neglected and we shall consider only the conduction current.

The striking features of metallic conduction are the obedience to Ohm's law and to the Wiedemann–Franz law, the latter stating that the ratio of thermal to electrical conductivity is LT, where T is the absolute temperature and L a constant (the Lorenz number) which has almost the same value for all metals. Let us see how far classical theory will take us.

When an electric field **E** is applied the electrons accelerate in the direction of **E** with an acceleration $e\mathbf{E}/m$, thus gaining kinetic energy and a drift velocity in the same direction. We know that overall there is no increase in kinetic energy or drift velocity because a steady current flows, so we must make some assumption about the way this energy is lost. The simplest assumption to make is that at a collision an electron loses all the drift energy gained since the previous collision, the lost energy appearing as vibrations of the framework (heat). We also assume that the drift velocities are much smaller than $\bar{c}$ and that λ is unaffected by the application of **E**.

Since $e\mathbf{E}/m$ is constant when **E** is uniform, the drift velocity just before a collision is $e\mathbf{E}\lambda/m\bar{c}$ (acceleration × time) and, because the initial velocity is zero and the increase uniform, the mean drift velocity is

$$\mathbf{v}=e\mathbf{E}\lambda/2m\bar{c}$$

The current density **J** is given by $ne\mathbf{v}$, so that $\mathbf{J}=ne^2\mathbf{E}\lambda/2m\bar{c}$ and

$$\sigma=\frac{ne^2\lambda}{2m\bar{c}} \tag{6.38}$$

The kinetic theory of gases gives the thermal conductivity K as $\frac{1}{2}nk\lambda\bar{c}$, where k is the Boltzmann constant, and so $K/\sigma=km\bar{c}^2/e^2$. Now $\bar{c}^2$ is given by $8kT/\pi m$ according to classical kinetic theory, although many authors equate it to $3kT/m$ which is strictly $\overline{c^2}$. Hence we have

$$\frac{K}{\sigma}=C\left(\frac{k}{e}\right)^2 T \tag{6.39}$$

where C is either $8/\pi$ or 3. Equation (6.38) embodies Ohm's law since none of the quantities on the right would be expected to vary if physical conditions were held constant. Equation (6.39) predicts the Wiedmann–Franz law with a Lorenz number $C(k/e)^2$ of about $2.0\times10^{-8}\,\mathrm{V\,K^{-2}}$. Many metals have Lorenz numbers close to this (e.g. at 100°C, Na 2.19, Al 2.23, Ni 1.83), although others do show marked differences (e.g. W 3.20, Bi 2.89).

However, the classical theory runs into greater difficulties when examined in more detail. The value of λ given by (6.38) when n is assumed to be 1 per atom turns out to be several hundred times the interatomic spacing and does not show the predicted variation with temperature. Moreover, and much more significantly, the electron gas would be expected to contribute an extra $3R/2$ to the molar heat capacity of metals in addition to the $3R$ given by the Dulong–Petit law for all solids. In fact, metals differ very little in their molar heat capacities from insulators.

Metals: the quantum free-electron model The application of quantum mechanics instead of classical theory removes most of the difficulties mentioned. In the first place, the electrons in their wave-mechanical aspect show no interaction with a perfectly periodic framework of ions. Finite resistance to electron flow is only exhibited because of *thermal vibrations*, giving a resistivity ρ_T, and *defects*, giving a resistivity ρ_D. The total resistivity is

$$\rho = \rho_T + \rho_D \tag{6.40}$$

Only ρ_T varies with temperature, tending to zero as T tends to zero, while ρ_D remains constant.

Secondly, only a small fraction of the electrons takes part in the conduction process, whether of electricity or heat, and this accounts for the small contribution to the specific heat capacity. For further details the reader is referred to solid-state texts explaining the quantum-mechanical Fermi–Dirac distribution.

Solids in general: the band theory Semiconductors form an important class of solid differing from metals in having conductivities which *increase* with a rise in temperature and which are sensitive to minute amounts of impurity. These effects are explained in terms of the *band theory* of conduction in solids. This theory shows that the periodically varying potential in a crystalline solid causes the sharp energy levels of the individual atoms to spread out into energy bands each of N levels, where N is the total number of atoms in the specimen. The Pauli exclusion principle forbids any level to contain more than two electrons and this means that all the bands are completely full, except possibly the band with the highest energy. This is known as the **valence band** and contains the valence electrons. The (almost) empty band above it is known as the **conduction band**.

The application of an electric field can give electrons only a very small amount of energy, so that if a solid is to conduct there must be empty levels available immediately above those already occupied. Because there are no empty levels above any of the bands below the valence band, they cannot take part in conduction. In a metal, however, the valence and conduction bands overlap, so that electrons at the top of the former have the empty levels of the latter available to them. The small increase in energy provided by an applied electric field in producing a current can now be accepted by the electrons, which move to the slightly higher levels when taking part in conduction.

In a semiconductor or insulator, the valence and conduction bands are separated by an **energy gap** that is small in a semiconductor (for example 1.14 eV in silicon and 0.66 eV in germanium) and large in an insulator (for example 5.3 eV in diamond). Appreciable conduction can only take place if the number of electrons in the conduction band (and of holes in the valence band) is sufficient. In an intrinsic (pure) semiconductor, the conductivity is increased by any excitation that can take electrons across the energy gap (rise in temperature, bombardment by photons or particles with sufficient energy). Extrinsic semiconductors are formed by doping—adding small quantities of impurity atoms to provide further electrons or holes that can also take part in conduction. The reader is referred to texts on solid-state physics for further details (e.g. Guinier and Jullien, 1989).

6.13 Revision summary

Electromotance: field aspect An electromotance $\mathscr{E}$ round a closed path L is defined as the work done per unit positive test charge in taking a charge completely round L:

$$\mathscr{E} = \oint_L \frac{dW}{Q_t} = \oint_L \frac{\mathbf{F}\cdot d\mathbf{L}}{Q_t} \tag{6.2}$$

The force per unit charge $\mathbf{F}/Q_t$ may be an electrostatic field $\mathbf{E}_Q$, an electric field $\mathbf{E}_M$ produced by electromagnetic induction (Chapter 9) or a field $\mathbf{e}$ produced by other non-electrostatic sources (Comment C6.4). Hence:

$$\mathscr{E} = \oint_L (\mathbf{E}_Q + \mathbf{E}_M + \mathbf{e})\cdot d\mathbf{L} = \oint_L (\mathbf{E}_M + \mathbf{e})\cdot d\mathbf{L} \qquad (6.3)\ (6.4)$$

the last step being the result of $\oint \mathbf{E}_Q \cdot d\mathbf{L} = 0$. When $\mathscr{E}$ causes a total charge Q to move once round L, the work done is therefore $Q\mathscr{E}$ so that

Rate of working of an electromotance producing current $I = I\mathscr{E}$ (6.5)

Electromotance: network aspect The complete path L in the above definition is often not accessible, and practical sources of electromotance are located within a box with two terminals. The electromotance $\mathscr{E}$ of such a source is defined as the potential difference between the terminals on an open circuit (supplying no current). This definition is compatible with the field definition above but is more useful in indicating how $\mathscr{E}$ can be measured. In either case, the **SI unit** for $\mathscr{E}$ is the *volt*.

Network elements An electric network consists of *components* with terminals connected by conducting *leads*. In this chapter we have dealt with the flow of *steady currents* in networks of *two-terminal components*. Each component is characterized by the relation between the voltage V across its terminals and the current I flowing in or out of them. We have:

- **Passive components with resistance: resistors.** The resistance R between the terminals is here defined as the ratio V/I and the conductance G as $1/R$. For metals R is constant under constant physical conditions (Ohm's law) and they are said to be *linear* or *ohmic* conductors. A component with such a constant R is called a *resistor*. The **SI unit** for R is the *ohm*, Ω, and for G is the *siemens*, S.

- **Active components: sources of electromotance.** If a voltage V appears across the terminals when $I = 0$, then we have seen that this defines the electromotance $\mathscr{E}$ of the component. When such a component *does* supply a current I then the voltage V may vary, and we define an *internal* or *output resistance* by

  $$r = -dV/dI \tag{6.10}$$

  If the source is used under conditions such that $\mathscr{E}$ and r are constant, then

  $$\mathscr{E} = V + rI \tag{6.11}$$

  where V is the voltage between the terminals when I is supplied. This relation is not necessarily an adequate description of all actual sources.

Network elements are idealizations of these components. All passive elements are assumed absolutely linear and possess simply a constant resistance R. All active elements are assumed to be *constant voltage generators* $\mathscr{E}$ in series with an internal resistance r or *constant current generators* $\mathscr{I}(=\mathscr{E}/r)$ in parallel with r.

Solution of network problems Draw a circuit diagram in which each passive component is represented by a linear R, each active component by a constant voltage or constant current generator and all leads are assumed to be of negligible resistance, as a first approximation. Then apply laws derived from those of the E-fields in Chapters 3 and 4:

Conservation of charge → Kirchhoff's first law (K1): $\Sigma I=0$ at a junction
Path-independence of V → Kirchhoff's second law (K2): $\Sigma V=0$ or $\Sigma\mathscr{E}=\Sigma RI$ round a mesh

Reduction of passive networks can often be achieved by using equivalent resistances to a set in series

$$R_{eq}=\Sigma R_i \tag{6.15}$$

or in parallel

$$1/R_{eq}=\Sigma 1/R_i \tag{6.16}$$

For *general networks*, solutions can sometimes be found more quickly using either circuital currents or the superposition theorem. **Thévenin's theorem** is particularly important: if we imagine part of a network placed in a box with only two leads X and Y emerging to the rest of the network, the theorem tells us how to replace all the elements within the box by a single electromotance $\mathscr{E}_0$ in series with a single resistor R_0. To find $\mathscr{E}_0$ we imagine the network outside the box removed and just calculate the voltage between X and Y. To find R_0 we again imagine the network outside the box removed, we replace all the electromotances within the box by their internal resistances and calculate the resistance between X and Y.

Field properties The forces and fields contributing to $\mathscr{E}$ in any material cause currents to flow if the path consists entirely of conducting material. The resistivity ρ or conductivity σ of an element of any material is defined by

$$\rho=RA/l; \qquad \sigma=1/\rho \tag{6.27) (6.28}$$

where l is the length parallel to the current flow and A the area perpendicular to it. The current density **J** at any point is then related to the total field **E** by

$$\mathbf{J}=\sigma\mathbf{E} \tag{6.30}$$

and this could be regarded as an alternative definition of σ at a point. For ohmic materials σ is independent of **J** and **E**.

The current density **J** obeys an **equation of continuity** at any point

$$\nabla\cdot\mathbf{J}=-\partial\rho/\partial t \tag{6.36}$$

which becomes for steady currents

$$\nabla\cdot\mathbf{J}=0 \tag{6.37}$$

whose integral form is $\oint\mathbf{J}\cdot d\mathbf{S}=0$ for any closed surface S. Since $\mathbf{J}\cdot d\mathbf{S}$ is the *current* across any element of S, this also is a version of Kirchhoff's first law $\Sigma I=0$. When the current is not steady, K1 can still be applied in spite of the full equation of continuity, provided the conductivity is high and time variations are not too rapid (quasi-steady conditions).

Commentary

C6.1 On Electromotance and Potential Difference The definition of electromotance as a line integral round a closed path (Sec. 6.1) makes it a difficult concept at first. We generally find quantities easier to grasp if they are field quantities with a unique value at every point in a region of space, like V. However, V is single-valued precisely because $\mathbf{E}_Q$ has a zero line integral round a closed path. Suppose we try to define, by analogy with V, an electromotance say in a field $\mathbf{E}_M$ for which $\oint \mathbf{E}_M \cdot d\mathbf{L} \neq 0$. Then at a point B, we should have $\mathscr{E}_B = \int \mathbf{E}_M \cdot d\mathbf{L}$ between A and B, choosing the point A as the zero. Unfortunately, $\mathscr{E}_B$ has an infinite number of possible values depending on the path between A and B, and it cannot therefore have a unique value. (The same problem arises with the use of a magnetic scalar potential in the magnetic fields of currents—see Sec. 7.7.) We have to be content, therefore, with electromotance defined round a closed path: later, we shall see that it is related to the circulation or *curl* of the field concerned. For a further comment, see Comment C9.1.

C6.2 On Electric Fields in Current-Carrying Wires Granted that the current flowing in wires containing no electromotances are produced by electric fields due to charges, how is it that such a field can follow the tortuous meanderings of typical networks? Furthermore, the wires are often very far from the sources of electromotance: does the electrostatic field of the charged terminals (Fig. 6.1) really extend as far? If not, how is the field maintained at all?

Let us consider in detail the whole process of connecting up a simple circuit using a chemical cell as our source of electromotance. On open circuit, there are two double layers instead of the single one of Fig. 6.1, but the situation is basically as indicated there: there is in fact only a small electric field outside the cell. If we connect a short wire to each terminal as represented by the shaded areas in the figure, there is no current and each wire must form an equipotential volume. The extreme ends of the wires are thus at a potential difference equal to the electromotance in magnitude, a capacitance must exist between the wires and a small charge must be distributed over their surfaces with a slight concentration at the ends. This charge would be supplied by a temporary current from the cell when the wires were attached.

If we now connect the ends of the wires, the capacitance is reduced to zero and the ends of the wires are discharged, but just enough charge remains along the surfaces to produce the internal **E**-field and keep it parallel to the wire itself as in Fig. 6.17. Figure 6.19 shows diagrammatically (a) how a charge density decreasing slowly along the surface of a wire produces an internal **E**-field along the wire and (b) how a slight excess charge on one side can bend the field into the new direction. Such excess charges need only be extremely small and Rosser (1970) has shown that no more than an odd electron is needed to bend **E** round a 90° corner in a typical wire.

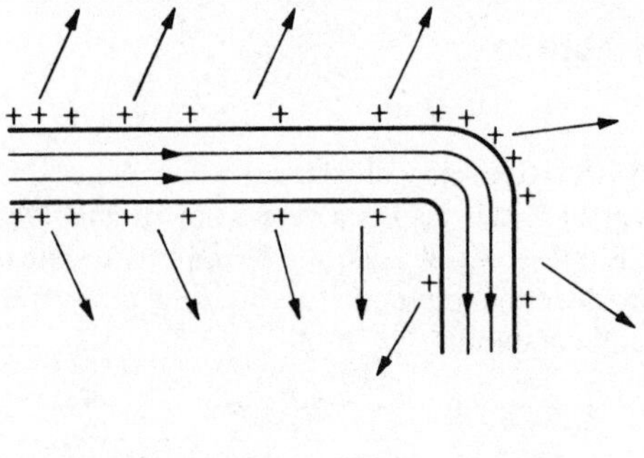

Figure 6.19 Lines of **E** and surface charges in a current-carrying wire.

Should not such surface charges produce *external* fields? They should, and do. Jefimenko (1962) shows how to demonstrate experimentally that the lines of **E** outside a current-carrying wire are in the direction shown in Fig. 6.19. The outside **E**-field has a longitudinal component equal to the field in the wire, but a radial component which is usually much greater. The demonstration of the external field shows that the explanation of the origin of the internal field is correct.

C6.3 On Currents 'through' Capacitors The statement is often made, somewhat loosely, that a varying current passes *through* a capacitor when clearly there is usually a good insulator between the plates stopping all but the most minute of leakage currents. How then does it happen that we can talk with any sense of the whole current in a network branch passing through the capacitor? The answer lies in the attitude of network theory. As with other circuit elements, the theory is concerned only with the current I in the leads to the capacitor and the potential difference V between them, and not with the processes occurring within the capacitor. *We* know that a current I flowing into one plate increases its positive charge, that this will induce an exactly equal negative charge on the other plate and that a current exactly equal to I therefore flows out of the second plate. No charge has passed across the space between the plates, yet the effect on a network in which the capacitor C is placed is as if I had passed right through it. Since $Q = VC$, $I = C\,\mathrm{d}V/\mathrm{d}t$ is the relation between I and V of sole interest to the network. This is the origin of the commonly heard statement that C will 'pass' a.c. but not d.c.

C6.4 On Sources of Electromotance We have not dealt with the internal mechanisms by which electromotance is produced inside a source. Such mechanisms can be regarded as means of converting other forms of energy into electrical energy, but in general the physics needed to understand them in detail is outside the scope of this book with one exception: the electromagnetically induced electromotance in Chapter 9. Apart from this we have:

1. *Chemical cells*, used widely as localized sources for small electrical and electronic equipment requiring steady voltages for their operation. A Weston cell (Appendix C) still forms the reference standard for voltage. Texts on electrochemistry should be consulted for details of the origin of chemical electromotance.
2. *Thermocouples*, widely used for measurement and control of temperature. Two dissimilar metal wires are connected to form a circuit with two junctions. If one junction is maintained at a different temperature from the other an electromotance is produced round the circuit. Many thermodynamic treatments of thermoelectricity are of doubtful validity and reliable accounts should be sought, for instance Zemansky and Dittman (1981), Scott (1966) and Chambers (1977).
3. Irradiated with light, *p-n junctions* form the basis of solar cells. Semiconducting materials can be produced in which there is a preponderance either of negative or positive current carriers, producing *n*-type or *p*-type regions, respectively. An interface between a *p*-type and *n*-type region in the same semiconductor forms a *p-n* junction. Irradiation of such a junction with light or nuclear particles provides energy for the release of further carriers in the junction and is the origin of the electromotances used in solar cells or particle detectors.

C6.5 On Superconductivity As mentioned in Sec. 6.12, normal resistance in metals would not occur in a perfectly periodic array of metals ions through which the electrons move. However, two sources of imperfection are always present: thermal vibrations, which diminish with lower temperatures but never disappear, and defects in the ionic structure due to impurities, vacancies, etc. In spite of this, a number of substances show a sudden decrease in resistivity to a negligible value as the temperature is reduced below a critical value T_c. This effect, first discovered in 1911 and known as **superconductivity**, is accompanied by the inability of a magnetic **B**-field to penetrate the material to a depth of more than about 1 μm. Any magnetic flux already present is reduced to zero, or more colloquially is said to be 'expelled', a property known as the **Meissner effect**. One readily observable consequence of this is the repulsion experienced by a magnet near the surface of a superconductor (see Sec. 13.4 for an explanation).

What might be called conventional superconductors are all metals or alloys with T_c near liquid helium temperatures (e.g. tin 3.7 K, mercury 4.1 K) or, at best, well below liquid nitrogen temperatures (e.g. Nb_3Ge 23 K). Although the superconductivity disappears in high enough magnetic fields, superconducting alloys are now extensively used in the construction of air-cored coils for strong magnetic fields at a great saving in the cost of the power normally wasted as heat. The saving is to some extent offset by the cost of producing liquid helium, so that the discovery of ceramic materials with T_c above liquid nitrogen temperatures in 1986–1987 was an important breakthrough. Progress since the initial discovery has been so rapid that any reported achievement at the time of writing (1989) is likely to be outdated. A T_c of 125 K for a mixed oxide of thallium, calcium, barium and copper has been achieved, and the aim is clearly to produce materials with T_c around room temperature.

The explanation of conventional superconductivity is provided by a theory of Bardeen, Cooper and Schrieffer (the BCS theory) who in 1957 put forward the idea, now corroborated, that *pairs* of electrons known as **Cooper pairs** showed an attraction for each other via the ionic framework, an attraction which produced superconductivity if it was greater than the Coulomb repulsion. If the electrons pair up in this way, they do so with equal and opposite spins and momenta when no current is flowing and are generally about 1 μm apart—a large distance on an atomic scale. It is the size of this distance that enables Cooper pairs to move without resistance since neither thermal vibrations nor defects are sufficient to break the pairing below T_c. It has also been shown that paired electrons move in such order through materials that they can be described by a de Broglie wavelength comparable with the sizes of the conductors. Large-scale quantum effects can be produced using this property. However, the high-temperature superconductivity of the new ceramics cannot be completely accounted for by the BCS theory and the basic mechanism of this is not yet clear, although electron pairing still seems to be involved (Gough, 1987). Mendelssohn (1966) gives a very readable account of the discovery of 'classical' superconductivity and Hazen (1988) is equally entertaining on the 1986 discoveries.

Problems

Section 6.1

6.1 A 45 ampere-hour battery will supply 45 A for 1 h, 2 A for 22.5 h, etc. If such a battery has an electromotance of 12 V, find (a) the total charge which can be supplied, (b) the total energy available, (c) how long the sidelights of a car could be operated if they consumed 24 W, (d) how long the battery could operate the starter motor taking 135 A. Assume that the potential difference at the terminals is 12 V and remains at this value until complete discharge.

Section 6.2

6.2 According to Fig. 3.5b and the argument leading to it, there is no field outside the plates. Yet the double layer of Fig. 6.1, which is similar in having equal and opposite charge densities, is drawn with a small field outside the source. Resolve the apparent contradiction.

Section 6.3

6.3 The relation between I and V for a certain rectifier is $I = I_0(e^{eV/kT} - 1)$, where I_0 is a constant and T is the temperature. Show that the incremental conductance of such a rectifier increases linearly with current.

6.4 A d.c. source has an electromotance of 120 V and a negligible internal resistance. If n cells, each of electromotance 2.1 V and internal resistance 0.1 Ω, are to be charged from this source with a charging current of 3 A, find the series resistance necessary. If $n = 20$, what proportion of energy delivered by the source is wasted as heat? Find also the potential difference across the 20 cells.

Section 6.4

6.5 The two representations of a source in Fig. 6.3 are equivalent if $\mathscr{I} = \mathscr{E}/r$. In terms of $\mathscr{E}$, r and R, find the power consumed in R and the power consumed in the source itself in the two cases. Comment on the results.

Section 6.5

6.6 If the sources A and B in Fig. 6.6 were both 7 V but all resistances were unchanged, what currents would be supplied by each?

6.7 In a Wheatstone bridge network, the galvanometer is replaced by a battery of the same electromotance $\mathscr{E}$ as the one already present. If the resistances of all the arms and of both batteries are equal to R, find the currents in the various branches.

6.8 A d.c. generator in series with a resistance R is connected in parallel with both a battery of electromotance 12 V and a resistive load. The electromotance of the generator may fluctuate, and R is of such a value that all the current to the load is supplied by the generator when its electromotance is 60 V. What fraction of the current does it supply when its electromotance drops to 50 V? Neglect the internal resistances of the generator and battery.

6.9 Three of the four arms of a Wheatstone bridge network are of 40 Ω, the fourth is 41 Ω and the resistance of the detector is 20 Ω. If the driving source has an electromotance of 2 V and negligible internal resistance, find the current through the detector using Kirchhoff's laws.

6.10 A current I divides between two resistances R_1 and R_2 in parallel. Show that the current through each resistor is correctly given by assuming only that the total rate of production of heat is a minimum together with Kirchhoff's first law.

Section 6.6

6.11 Two 10 Ω resistors are connected in series and a third resistor R is connected in parallel with one of them. What value of R makes the equivalent resistance of the whole combination equal to 18 Ω?

6.12 Show that the Wheatstone bridge network (Fig. 6.14b) with all resistances, including that of the galvanometer, equal to R presents a resistance R to the driving electromotance.

6.13 A skeleton cube is made of wires soldered together at the cube corners, the resistance of each wire being R. Calculate the equivalent resistance of the network between (a) diagonally opposite corners and (b) the two ends of one wire. Solve this classic problem either by sending a current I into the network and using Kirchhoff's laws and symmetry to find the currents in the wires, or by using the delta–Y transformation.

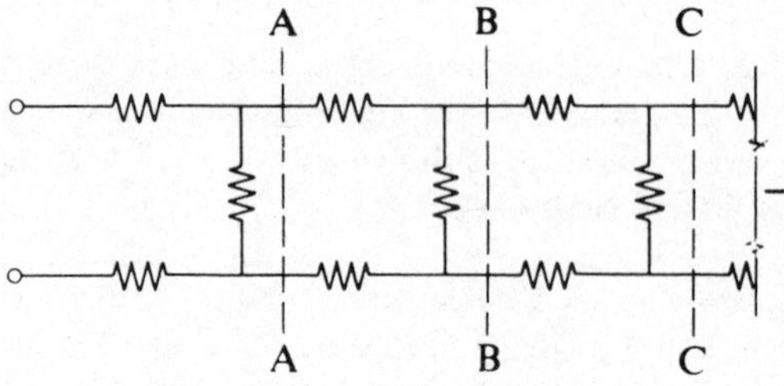

Figure 6.20 Ladder network for Problem 6.14.

6.14 In the ladder network of Fig. 6.20 all the resistances are equal to R. Find the resistance between the input terminals when the network is completed up to AA, up to BB, etc. What is the input resistance if the network is of infinite length?

6.15* The following problems have little value but might prove interesting as intellectual exercises.

(a) Is it possible to construct a network of resistors which *cannot* be reduced by using the series and parallel formulae together with the delta–Y or Y–delta transformation?

(b) An infinite triangular mesh of 1 Ω resistors is constructed in a plane. What is the equivalent resistance between two adjacent nodes?

(c) Nine resistors are of identical appearance but eight are of 1 Ω while the ninth is of 2 Ω. Can a two-terminal network be constructed out of these resistors of such a form that *one* measurement of resistance across it would enable the odd resistor to be unambiguously selected?

Section 6.7

6.16 Solve the problem of Fig. 6.6 by using the superposition theorem.

6.17 A source of electromotance 6 V and internal resistance 2 Ω has in parallel across its terminals (a) a 2 Ω resistor and (b) a series combination of a 1 and a 4 Ω resistor. A further 1 Ω resistor R is connected across the 4 Ω resistor. Find the current in R using Thévenin's theorem.

6.18 Solve the preceding problem by alternately converting from a Thévenin source to a Norton source and back again, each time incorporating any series or parallel resistors as convenient.

6.19 Because of the superposition theorem, Problem 6.17 can be solved in yet another way. Draw the network but let the voltage generator be simply $\mathscr{E}$ volts instead of 6 V. Now *assume* that the current in R is 1 A and work back through the network to find $\mathscr{E}$. Use superposition to find the current in R when $\mathscr{E}=6$.

6.20 The resistor R in Problem 6.17 is replaced by a current generator delivering 5 A towards the junction of the 1 and 4 Ω resistors. Find the current in the central 2 Ω resistor.

6.21 Find the current in Problem 6.9 using Thévenin's theorem.

6.22 Construct your own network problems and solve them by several methods so as to check the correctness of the answers.

Section 6.8

6.23 A capacitor of capacitance C_1 has a charge Q_0 on its plates. It is connected at time $t=0$ in series with a capacitor of capacitance C_2 and a resistor R by the closing of a switch. Find the time constant of the ensuing discharge.

Section 6.9

6.24 Using the notation of Fig. 6.14b, show that the current g is given by $\beta(r-r')/[mr+r')+R_G(1+1/n)]$. If the bridge is nearly balanced, let $r-r'=\mathrm{d}r$ and the out-of-balance current be $\mathrm{d}g$. Find the sensitivity of the bridge, defined as $r\mathrm{d}g/\mathrm{d}r$, and investigate how it depends on m, n and r.

Section 6.10

6.25 A copper wire of radius a forms the inner conductor of a coaxial cable, and an imperfect insulator of conductivity σ and external radius b fills the space between the inner and outer conductors. Find the resistance per unit length of the insulator for currents flowing in it parallel to the wire. Find also the conductance per unit length for currents flowing radially outwards from the wire. Why is the conductance rather than the resistance asked for in the second case?

6.26 A parallel plate capacitor has an imperfect insulator of conductivity σ and relative permittivity ε_r between its plates. What is the time constant of self-discharge of the capacitor? Use the data from Table 11.1 (page 299) to find the time constant if water is placed between the plates.

6.27 A twin cable carries current between A and B, 5 miles apart. It is known to have one fault, i.e. a breakdown in the insulation between the two conductors, but not necessarily a short-circuit. A potential difference of 200 V maintained across the ends at A produces a potential difference of 50 V across the ends at B, while 200 V across B produces 20 V across A. Locate the fault.

6.28 In the potential divider network of Fig. 6.16d, let the total resistance of AB be R, the resistance of AC be R_1 and of the load be R_L. Find the input resistance (resistance as seen from the input terminals) and its value for large and small loads.

Section 6.11

6.29 Show that the rate of production of heat per unit volume at a point in a conductor is $\mathbf{J}\cdot\mathbf{E}$ or $\sigma\mathbf{E}^2$, where $\mathbf{J}$ is the current density, $\mathbf{E}$ the electric field and σ the conductivity.

6.30 Calculate the mobility of electrons in copper assuming that each atom contributes one conduction electron.

6.31 A steady current flows across the plane boundary between two conductors of conductivities σ_1 and σ_2. Find a relation between the angles θ_1 and θ_2 which the current flow makes with the normal to the boundary on the two sides. What will be the effect if one conductor is a metal and the other an electrolyte?

6.32 A steady current flows across the plane boundary between two conductors of different conductivities. Assuming that Gauss's law holds in the form of Eq. (4.2), show that the boundary must have a steady surface charge density along it.

Section 6.12

6.33 Use Eq. (6.38) to determine λ for copper, assuming that $\bar{c}$ is given by $(8kT/\pi m)^{1/2}$.

7

The laws of magnetic interaction (*in vacuo*)

In Chapters 2–5 the electric forces between static charges *in vacuo* were thoroughly examined. In this chapter we begin a similar investigation of *magnetic* forces which were first mentioned in Sec. 1.2 as those occurring between currents. As far as possible, we wish to use the experience gained in handling the electric forces by adopting the same general approach as in Chapters 2 and 3. Let us first, therefore, summarize that approach in the following four stages:

- *Choice of basic element* (the point charge) to use in the statement of fundamental laws. Large systems can be dealt with by summing or integrating the effects of elements.
- *Establishment of the fundamental laws* (Coulomb's law, superposition).
- *Introduction of a field quantity to ease calculations* (the **E**-field). The effect of this stage is to split the fundamental law into two parts. Thus in the electric case:

  charge in **E**-field $\mathbf{F}=Q\mathbf{E}$ (from definition of **E**)
  E-field due to charge: $\mathbf{E}_\mathrm{Q}=Q\hat{\mathbf{r}}/(4\pi\varepsilon_0 r^2)$
- *Use of the last two equations to solve problems.*

When dealing with magnetic interactions we encounter two factors making the treatment inevitably more complex:

1. *The choice of basic element* is not straightforward, in that there are four distinct possibilities: the current element, the moving charge, the magnetic dipole or, in much older treatments, the monopole. Each has its advocates and any one could be used as a starting point. The most generally useful of them is undoubtedly the current element, already introduced in Sec. 1.7, and we adopt it here.
2. *The fundamental laws* cannot be so compactly expressed as Coulomb's law between charges, except in the case of the monopole, whose applications are limited. The reason for the complexity is that magnetic forces are not in general central forces and depend on the *directions* of currents as well as on their magnitudes and distance apart.

This last point leads us to introduce the field quantity, the **B**-field, as early as possible so that our fundamental law can be expressed in two relatively simple parts from the beginning. These parts are, firstly, the force on an element in a **B**-field and, secondly, the **B**-field arising from an element. As with the corresponding **E**-field formulae, these can then be used to solve simple problems. This treatment means, however, that this chapter is the magnetic parallel of *both* Chapters 2 and 3.

The general plan adopted in this chapter is to begin in Sec. 7.1 with a survey in which the various magnetic interactions are summarized and in which we can make some initial progress in introducing a field quantity. In Secs 7.2 to 7.4, the current element is then used as the basic element to define the **B**-field, to find the **B**-field due to currents and thus to express the fundamental law. Some simple problems are also solved at this stage. In subsequent sections, we see how the alternative approaches using moving charges (Secs 7.5 and 7.6), magnetic dipoles (Secs 7.8 and 7.9) and monopoles (Sec. 7.10) add to our understanding and provide their own methods for tackling specific problems. Among all these matters, we see how an attempt to introduce a magnetic potential by analogy with electric potential has only limited application, and we discuss why this should be so (Sec. 7.7).

7.1 Magnetic forces and torques – a survey

In Chapter 1 we used the adjective 'magnetic' to describe the forces between one electric current and another. The reader will, however, probably be familiar with forces between permanent magnets and between a magnet and a current. In this section we survey all three types of interaction and justify the use of 'magnetic' to describe them all.

Magnet–magnet interactions Permanent magnets are recognized by two effects known to the ancient Greeks, the *directive* and the *attractive*. The first effect is that, when permanent magnets are freely suspended so that they can oscillate about a vertical axis through their centre, they always turn and come to rest in the same approximately N–S direction at a given point on the earth's surface. One end is then called an N-seeking or N pole of the magnet and the other an S pole. The second effect is that they attract pieces of iron and certain other metals. In this, they are distinguishable from charged bodies (Sec. 1.1) whose attraction for other objects is independent of the material of which the objects are composed.

The poles of two magnets exert forces on each other summarized by the qualitative law 'like poles repel, unlike poles attract' (Gilbert, 1540–1603) and they behave in some respect like electric charges but with two important differences. One is that the poles can never be separated, N and S poles always being found in pairs of equal strength. The second difference is that there is no process which could be described as the conduction of poles (Sec. 7.10).

Permanent magnets are therefore always **magnetic dipoles**, behaving at least qualitatively like electric dipoles in exerting torques on each other. These torques reveal the presence of a **magnetic field** at any point in space whose direction is obtained by placing a small permanent magnet there and allowing it to come to rest in stable equilibrium. The required direction is that of the S-pole-to-N-pole axis. This is the 'compass-needle' method of plotting magnetic fields with which readers will be familiar. Magnetic lines of force plotted in this way have a *sense* indicated by an arrow giving the direction in which the N pole of the compass needle would point. Figure 7.1a and b gives examples of such plottings.

If we were to base our investigation of magnetism on the interactions between permanent magnets, we should be led to the idea that a small magnetic dipole would provide an element similar to the electric charge as a testing device, but similar to the electric dipole in its detailed behaviour. Quite crude experiments show results for small magnets like those of Sec. 3.9. We shall see in Sec. 7.8 that it is then possible to define a magnetic dipole moment, and hence to obtain the magnitude of a magnetic field as well as its direction. Early measurements of such magnitudes were in fact made by finding the torques on small suspended magnets, often using the period of angular oscillations (Problem 7.17).

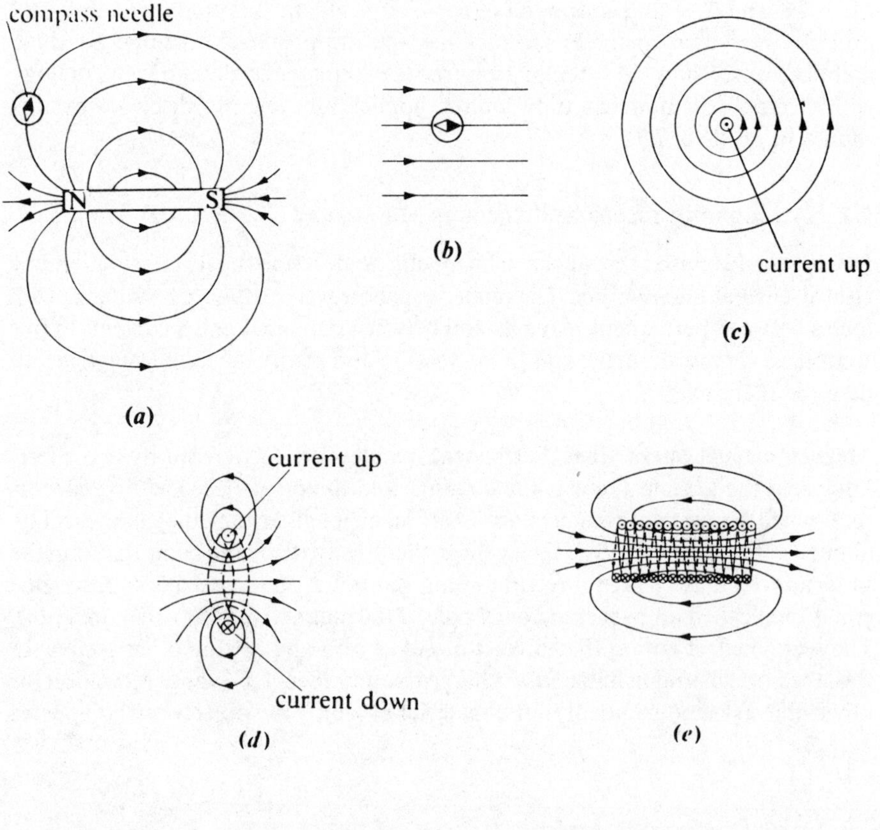

Figure 7.1 Magnetic field lines (a) outside a bar magnet, (b) over a small region of the earth's field, (c) outside a long straight current-carrying wire, (d) due to a current-carrying coil, (e) due to a current-carrying solenoid.

Current–magnet interactions In 1819, Oersted discovered that a horizontal current-carrying wire exerted a torque on a compass needle suspended above or below it when the needle was *parallel* to the wire (a perplexing result at that time). That was the earliest observation of an electric current as a further source of a magnetic field and Oersted deduced from his experiments that the shape of the field lines due to a long straight current is as shown in Fig. 7.1c. The reader should note the convention used for indicating the direction of an electric current perpendicular to the page. Other important examples of fields are shown in Fig. 7.1d and e. Because of Newton's law of action and reaction, we should expect a magnet to exert a force or torque on a nearby current. This is in fact observed and shows that the action between them is mutual.

The similarity between the field lines outside a bar magnet and a solenoid (Fig. 7.1a and e) suggests that a small solenoid would do as well as a small magnet for detecting and plotting magnetic fields. When the solenoid is small enough it becomes a single-turn current loop as in Fig. 7.2b, with a field indistinguishable from that of Fig. 7.2a. Moreover, the torques *on* the two are also similar, so that a small pivoted coil like that of Fig. 7.2c can be used just as well as a compass needle to plot the fields. Thus a small current loop behaves like a magnetic dipole, an analogy pursued in Sec. 7.8. We note for the moment only that *the positive direction of the axis of the current loop is obtained from the direction of current flow by the right-hand screw rule.*

(The right-hand screw rule will be used many times in this chapter, so that readers should make sure they understand it. The easiest way of recalling it is to realize that with a right-handed screw **a clockwise rotation produces motion away from the observer.**)

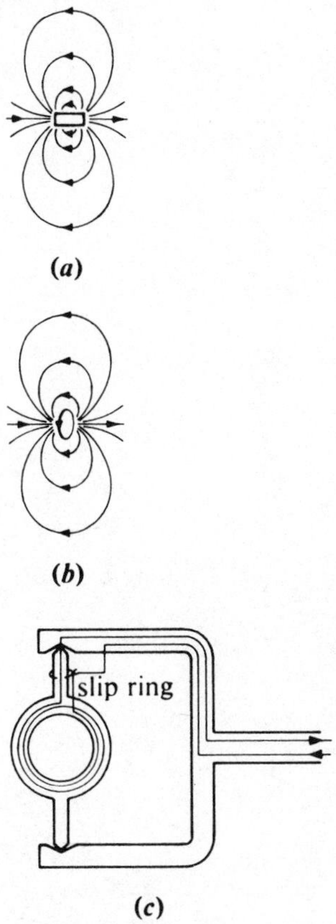

Figure 7.2 (a) Magnetic field of a small permanent magnet; (b) magnetic field of a small current loop; (c) a pivoted current-carrying coil which could be used for field plotting instead of a compass needle.

Current–current interactions It cannot be argued that, because currents exert forces on magnets, they will exert forces on each other: as Ampère pointed out, pieces of soft iron attract magnets but do not affect each other. However, experiments like that of Sec. 1.2, as well as those of Ampère himself, amply demonstrate that current–current forces exist. Since this type of force would be the one used to plot the field of the solenoid in Fig. 7.1e by using the loop of Fig. 7.2c, we are justified in describing it also as a magnetic force, as indeed we did in Sec. 1.2.

We pursue the investigation of current–current interactions first of all by using the current element of Sec. 1.7 as a testing device. To justify the laws we quote, we appeal to experiments involving current-carrying circuits in which the leads are often an unavoidable complication. It turns out, however, that a wire doubled back on itself and carrying a steady current produces no magnetic effect even if the return wire has small sinuosities in it (Fig. 7.3). This was originally established by Ampère and means that leads play no part in magnetic effects provided they follow closely the same paths to and from the circuits.

We now turn from qualitative results to the establishment of quantitative laws.

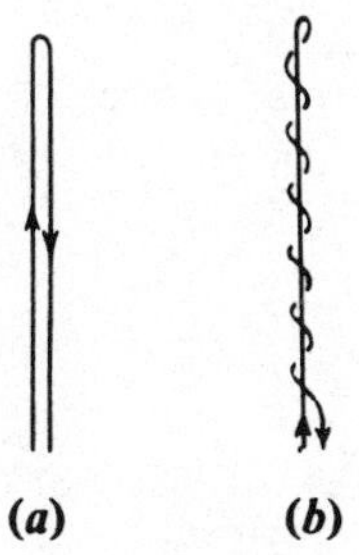

Figure 7.3 (a) A current-carrying wire doubled back on itself; (b) the same with sinuosities. Neither have external magnetic effects if the wires are close.

7.2 The magnetic B-field: force on current element

B-field: concept It became clear in Chapter 3 that there were great advantages in introducing the electric field **E** to solve electrostatic problems, instead of working directly from the fundamental Coulomb's law. There is even greater motivation for using a magnetic field quantity in a similar way, because magnetic interactions are much more complicated than the simple Coulomb force of Chapter 2. In terms of current elements, the fundamental law turns out to have the form given in Eq. (7.20) later in this chapter, but to see how that is obtained we introduce from the start the concept of a magnetic field, denoted by the symbol **B**. Our plan will be to find an acceptable method of defining a **B**-field in this section in terms of the force on a current element. We then proceed to ask what evidence there is for an expression giving the **B**-field *due to* an element (Sec. 7.3) and combine the two to give the full law (Sec. 7.4).

B-field: direction A current element can be used to define a **B**-field in the same way that a test charge was used to define **E**, and in principle to measure it as well. Unlike a test charge, however, a current element has a direction which we indicate by writing it in vector form as $I\,\mathbf{dl}$. The first thing to observe is that there are certainly magnetic forces *perpendicular* to $I\,\mathbf{dl}$ when it is placed in a magnetic field (experiment of Sec. 1.2). Ampère went even further and showed that there is no force *parallel* to the length of the element. He mounted a section of circuit on mercury cups as shown in Fig. 7.4a in such a way that it was allowed only one degree of freedom: along its own length. No system of magnets or currents brought up to it caused any movement. We conclude that the force on a current element is *perpendicular to the element.*

It is found in fact that there is a unique direction at any point in which a current element experiences no force and this we define as the **direction** of **B** at the point. Such a definition gives the same results in plotting lines of **B** as the dipole method of Sec. 7.1, although it does not give a *sense*, i.e. we cannot yet put arrows on the lines.

B-field: magnitude The **magnitude** of **B** is now defined as the force per unit current element placed perpendicular to the direction of **B**. We have to be sure that this leads to a consistent result for **B** in practice, for there is no guarantee that the definition will do so. To check this, we can use a balance such as that of Fig. 7.4b in which only the force on a straight portion XY produces any moment about the fulcrum. The upper end is in a region of negligible field and the sides experience only equal and opposite horizontal forces. Experiments with the balance show that when the **B**-field is vertical there is no vertical force component

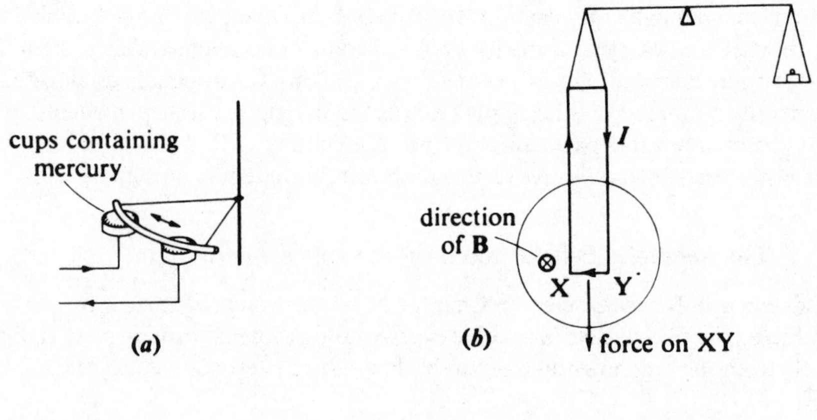

Figure 7.4 Forces on current elements: (a) Ampère's experiment; (b) the balance of Thomas, Driscoll and Hipple.

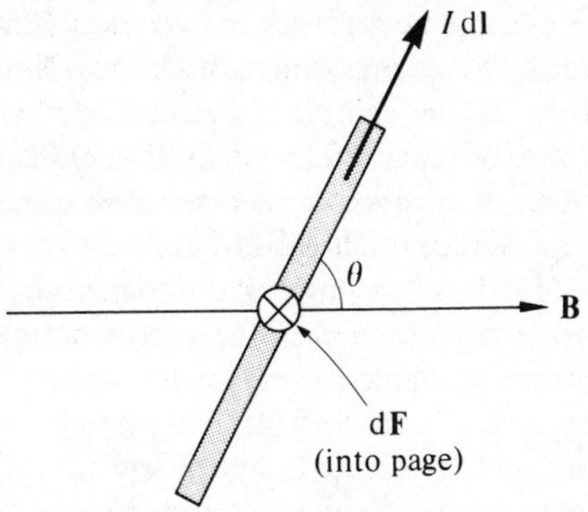

Figure 7.5 The force $\mathbf{dF} = I\,\mathbf{dl} \times \mathbf{B}$ on a current element in a **B**-field.

on XY. In other words, *the force is perpendicular not only to* **dl**, *as has already been seen, but also to* **B** *itself.* Next, when the field is horizontal as in the figure, the force on XY for a constant **B** is found to be *proportional to the current I and the length* XY. Finally, if XY is oriented at various angles θ to **B** other than $\pi/2$, the force is *proportional to* $\sin\theta$, and still perpendicular to both **dl** and **B**. It is as if $I\,\mathbf{dl}$ can be treated as a true vector whose component perpendicular to the field, $I\,\mathrm{d}l \sin\theta$, is the only one experiencing the force.

Full definition of B-field Combining all the results and definitions, we can write the force on a current element as

force on $I\,\mathbf{dl}$ in **B**: $$\boxed{\mathrm{d}F = BI\,\mathrm{d}l \sin\theta} \tag{7.1}$$

in a direction perpendicular to the plane containing **dl** and **B**. In order to give a sense to the direction of **B** which is the same as that of Sec. 7.1, the various directions must be as shown in Fig. 7.5.

Now although (7.1) coupled with the subsequent statement provides a complete definition of **B**, there is a ready-made notation which is much more concise. A vector product of two vectors was first encountered in Eq. (3.41), where it was used to express a torque: in general, if **X** and **Y** are two vectors with an angle θ between them, then $\mathbf{X} \times \mathbf{Y}$ is a vector with a magnitude $XY \sin\theta$ and a direction which is that of the advance of a right-handed screw rotated from **X** to **Y**. This is precisely how **dF** is specified if **X** is $I\,\mathbf{dl}$ and **Y** is **B**. The definition of **B** is therefore incorporated in

force on $I\,\mathbf{dl}$ in **B**: $$\boxed{\mathbf{dF} = I\,\mathbf{dl} \times \mathbf{B} \qquad \text{(definition of } \mathbf{B}\text{)}} \tag{7.2}$$

and that this indeed gives the correct direction can be verified from Fig. 7.5. In elementary work, the directions of the various quantities are sometimes remembered by what is called *Fleming's left-hand rule,* but this only works for $\theta = \pi/2$ and can be confusing. The vector product is a much more general and useful form and it is worth making an initial effort to learn it.

From the definition (7.2), the **SI unit** of **B** is the $\mathrm{N\,A^{-1}\,m^{-1}}$, known as the **tesla**, symbol T. Some use is still made of the *gauss*, equal to 10^{-4} T. For magnitudes of **B**-fields encountered in various situations, see Comment C12.1.

Comments Equation (7.2), like $\mathbf{F}=Q\mathbf{E}$, can be regarded in two ways. First, it can be a prescription for finding the force on a current element in a known **B**-field, and thus obtaining the forces and torques on complete circuits. We deal with examples below. The equation can also be regarded as a definition of **B**, although it cannot be ranked as a pure definition since it incorporates some of the experimental results, such as the dependence on directions and on **dl** (we should expect I to occur as it does because of the way we have agreed to measure current with a current balance). Note also that (7.2) implies the vector nature of the current element and the superposition principle for magnetic forces.

In its aspect as an equation defining **B**, (7.2) shows that $I\,\mathbf{dl}$ plays the part of a 'test element' similar to the test charge used to define **E**. The latter was, in Chapter 3, to be made vanishingly small in magnitude if there was any danger that it could affect the **E** to be defined or measured. There is not quite the same need to worry over the presence of a test element here, because **B**-fields which arise from currents elsewhere can be maintained constant by keeping the currents steady. However, if the **B**-field were due to a so-called 'permanent' magnet, it would be necessary to watch this effect.

As a final comment on (7.2), we note that it is not common practice to measure a **B**-field by using it directly, although the balance of Fig. 7.4b was used by Thomas, Driscoll and Hipple (1950) to measure **B** in their determination of the gyromagnetic ratio of the proton (Sec. 8.7). We deal with methods of measuring **B** in Sec. 9.4 and Comment C12.2.

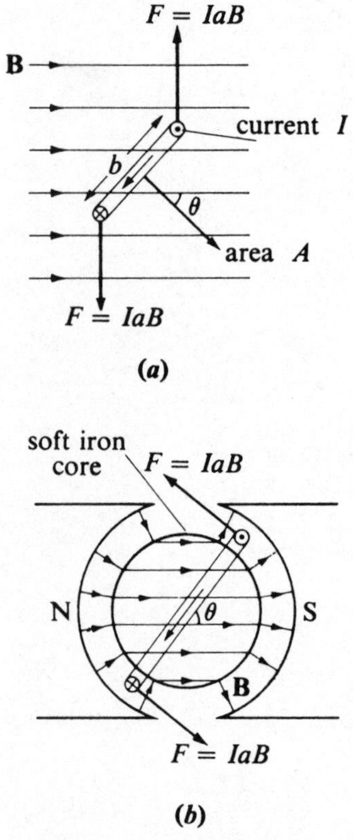

Figure 7.6 A plane current-carrying coil in (a) a uniform **B**-field and (b) a radial **B**-field.

Complete current-carrying circuits in B-fields The total force on a complete circuit L carrying a steady current I is obtained by summing the elementary $\mathbf{dF}$'s given by (7.2). Formally, therefore, we can write this force as

circuit in **B**:
$$F=I\oint_L \mathbf{dl}\times\mathbf{B} \tag{7.3}$$

If the currents are distributed throughout a volume then the appropriate current element has the form $\mathbf{J}\,d\tau$ rather than $I\,\mathbf{dl}$ (Sec. 1.7) and the force becomes a volume integral:

volume current in **B**:
$$\mathbf{F}=\int_{\mathrm{vol}}(\mathbf{J}\times\mathbf{B})\,d\tau \tag{7.4}$$

Equations (7.3) and (7.4) are 'formal' expressions because the integrals cannot be evaluated without changing the vector form of the integrands into scalars by using components. In practice, as we shall see from examples in this chapter, it is often possible to calculate forces without recourse to integration at all.

Worked example *plane rectangular circuit in uniform* ***B****-field* Consider first a rectangular coil in a uniform **B**-field as in Fig. 7.6a. The coil has sides a (perpendicular to the plane of the page) and b, while the area $A=ab$ has a normal making an angle θ with **B**. From (7.2), the forces on the sides a are each IaB as shown, and they form a couple whose moment is $IaB\times b\sin\theta$ or $IAB\sin\theta$. The forces on the b sides are equal, opposite and in the same line, and so have no resultant effect.

The total force and torque on the coil in a uniform **B** are thus:

plane circuit in uniform **B**:
$$F=0;\qquad T_\theta=IAB\sin\theta \tag{7.5}$$

the torque being in such a sense as to decrease θ. For N turns the torque becomes $NIAB \sin\theta$. Once again we can use the vector product notation to express the torque more concisely. The area A is represented by an axial vector **A** normal to the plane of the coil and with a sense related to the current by the right-hand screw rule as in Fig. 7.6a. The expressions in (7.5) can then be written as

plane circuit in uniform **B**: $\mathbf{F}=0 \qquad \mathbf{T}=I\mathbf{A}\times\mathbf{B}$ (7.6)

in which **T** is an axial vector pointing out of the page.

In a radial **B**-field (Fig. 7.6b), the torque is easily shown to be

circuit in radial **B**: $T_\theta = NABI$ (7.7)

for a coil of N turns.

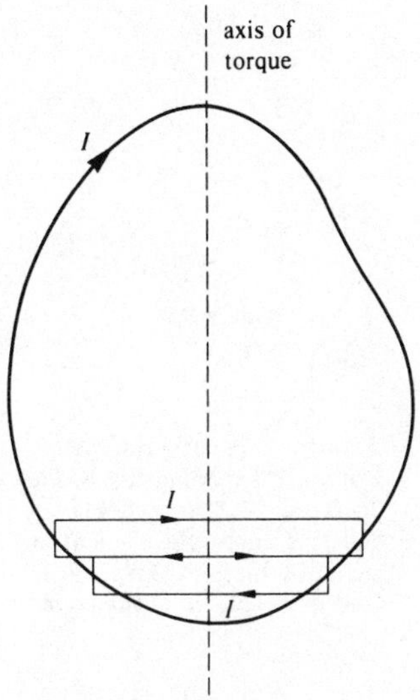

Figure 7.7 The torque on a plane coil of any shape is the same as that on a set of elementary rectangular circuits fitted into the area covered by the coil.

Generalization of example to plane circuit of any shape Equations (7.5) and (7.6) were derived for a rectangular coil but can easily be shown to apply to a plane coil of any shape, like the one shown in Fig. 7.7. The plane of this coil is to be imagined identical with that of the one in Fig. 7.6a. We also imagine the coil divided into elementary rectangles, two of which are shown, each carrying the same current I as the coil itself. The internal currents from adjacent rectangles are equal and opposite and will have no resultant forces on them. Since the uncancelled currents can be made to approximate as closely as we wish to the original circuit, the sum of the torques on the rectangles is the same as the total torque on the circuit. However, each rectangle experiences a torque given by (7.5) with the same I, B and θ. The total torque is thus obtained by summing over the areas and multiplying by $IB \sin\theta$, which yields (7.5) or (7.6), with A now standing for the area of a plane coil of any shape. This example is discussed further in Sec. 7.9.

Practical applications *moving-coil meters and electric motors* Moving-coil meters use the arrangement shown in Fig. 7.6b, with a pointer attached to the coil and with the torque of Eq. (7.7) opposed by another due to a spring. The latter torque is proportional to the deflection θ, say $c\theta$, so that in equilibrium $NABI = c\theta$ and I is proportional to θ. This gives the meter a linear scale.

In principle, most d.c. and small a.c. electric motors use the torque given by Eq. (7.5) to produce rotation of an armature around which the rotor coils are wound. However, complications arise from the need to use windings with complex geometry fed with current through commutators to ensure that direction of flow always produces a torque in the same sense and of approximately constant magnitude. The **B**-field is produced by currents in field coils fed from the same supply as the armature current, and this further complicates the connections. Finally, the very rotation of the coil in a **B**-field causes an induced electromotance and current (Chapter 9) in such a direction as to oppose the original torque. Further discussion of these matters is taken up in Sec. 9.2.

7.3 B-fields due to steady currents: the Biot–Savart law

B-field due to a current element We now seek an expression giving the value of **B** at a point due to a current element so as to complete the fundamental law of magnetic interaction. This time, however, it is not possible to appeal directly to experiment because an element cannot produce a field without the rest of the

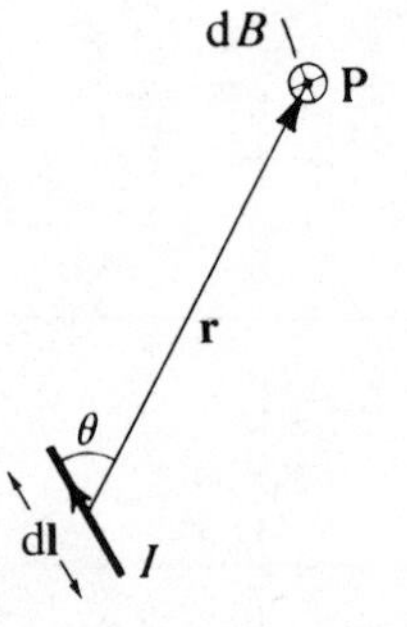

Figure 7.8 A current element $I\,\mathbf{dl}$ and its relationship to the field point P. The magnetic field $\mathbf{dB}$ due to the element is given by the Biot–Savart law and is directed into the page.

circuit doing so as well—at any rate when the current is steady. The formula will therefore only have meaning when integrated round a complete circuit. This implies that the validity of the law will be tested by the results it predicts when applied to real examples, a point we return to later.

The geometry to be used is illustrated in Fig. 7.8, where P is the field point at which dB is required in terms of I, dl, r and θ. The earliest experiments and analyses of the effect of a current element were carried out by Ampère and by Jean-Baptiste Biot and Félix Savart, all around 1820. The most important experiments from our point of view were those conducted with long straight currents, since these could be regarded as a large number of elements placed end to end. Biot and Savart were able to show that the **B**-field at a distance r from such a long current was proportional to r^{-1}. They argued that if this was to be obtained by integration over a large number of elements, each element should have a field proportional to r^{-2}. Further, by measuring the change in the field produced by the insertion of a V-shaped portion into the wire, they showed that dB was also proportional to $\sin\theta$. Finally, Ampère's experiment of Sec. 1.2 shows that the direction of the force is given if dB has the direction shown in Fig. 7.8.

These experiments lead us to conclude that the current element formula required is very likely of the form $dB = kI\,dl(\sin\theta)/r^2$ in magnitude. We write the universal constant k as $\mu_0/4\pi$, so that:

B due to current element: $$dB = \frac{\mu_0 I\,dl\sin\theta}{4\pi r^2} \tag{7.8}$$

in a direction at right angles to dl and r given by the right-hand screw rule (a screw rotated from the direction of **dl** to that of **r** will advance into the paper, i.e. in the direction of **dB**). Equation (7.8) can be expressed more concisely by using the vector product again, so that the directions are included:

B due to current element: $$\mathbf{dB} = \frac{\mu_0 I\mathbf{dl}\times\hat{\mathbf{r}}}{4\pi r^2} = \frac{\mu_0 I\,\mathbf{dl}\times\mathbf{r}}{4\pi r^3} \tag{7.9}$$

in which $\hat{\mathbf{r}}$ is a unit vector along r. This expression is known as the **Biot–Savart law**.

Value of μ_0 The constant k is expressed in the form $\mu_0/4\pi$ for reasons similar to those given in choosing $4\pi\varepsilon_0$ in Sec. 2.4. The quantity μ_0 is called the **permeability of free space**, the **magnetic constant** or simply **mu nought**. It has a value determined by the unit chosen for current and we shall see in the next section that the choice of the ampere and SI units generally means that $\mu_0/4\pi$ is *exactly* 10^{-7}. Its **SI unit** will be shown to be equivalent to the $\mathrm{H\,m^{-1}}$, so that

$$\mu_0 = 4\pi\times10^{-7}\,\mathrm{H\,m^{-1}} \tag{7.10}$$

Accuracy of the Biot–Savart law We have already pointed out that (7.9) cannot be tested directly for steady currents since the field of a current element cannot be isolated, but only through its predictions for complete circuits. One consequence of this is that any term added to (7.9) which integrated to zero round a circuit would be an equally valid law. Nevertheless, when the Biot–Savart law and the force on a current element are combined to calculate forces between circuits, it is known that the results are accurate to about 1 part in 10^6 (see Appendix C and the remarks about current balances of different designs). Moreover, it can be shown that (7.9) is a consequence of special relativity applied to Coulomb's law and, if this is accepted, the inverse square dependence is known to the same precision as that of electrostatics (Sec. 13.10). Comment C7.1

B-fields due to complete circuits The calculation of the **B**-field due to a circuit L is carried out by integrating (7.9) round the circuit. Formally:

B due to circuit: $$\mathbf{B}=\oint_L \mu_0 I\,\mathrm{d}\mathbf{l}\times\hat{\mathbf{r}}/(4\pi r^2) \qquad (7.11)$$

or for currents distributed throughout volumes:

B due to volume current: $$\mathbf{B}=\int_{\text{vol}}\frac{\mu_0\mathbf{J}\times\hat{\mathbf{r}}\,\mathrm{d}\tau}{4\pi r^2} \qquad (7.12)$$

In practice, because the integrand is a vector, the calculation first involves the resolution of d**B** into components with subsequent integration, as examples will show.

Worked example **B**-*field on the axis of a circular coil* In Fig. 7.9a, the fields due to two elements at opposite ends of a diameter are shown, each having a magnitude $\mu_0 I\,\mathrm{d}l/4\pi(x^2+a^2)$ using the Biot–Savart law. The components of the d**B**'s resolved perpendicular to the axis will cancel, while those resolved along the axis will add. Since this applies to every such pair of elements, the total field at P is simply the sum of the resolved parts along x. This is easily shown to give

on coil axis: $$B_x=\mu_0 Ia^2/[2(a^2+x^2)^{3/2}];\qquad B_y=0;\qquad B_z=0 \qquad (7.13)$$

B_x is plotted against x in Fig. 7.9b. When $x=0$ at the centre:

centre of coil: $$B_x=\mu_0 I/2a \qquad (7.14)$$

while as $x\to\infty$, $B\to\mu_0 Ia^2/2x^3$ or

at infinity along axis: $$B_x=\frac{\mu_0 2IA}{4\pi x^3} \qquad (7.15)$$

where A is the area of the coil. We look at two important applications of (7.13).

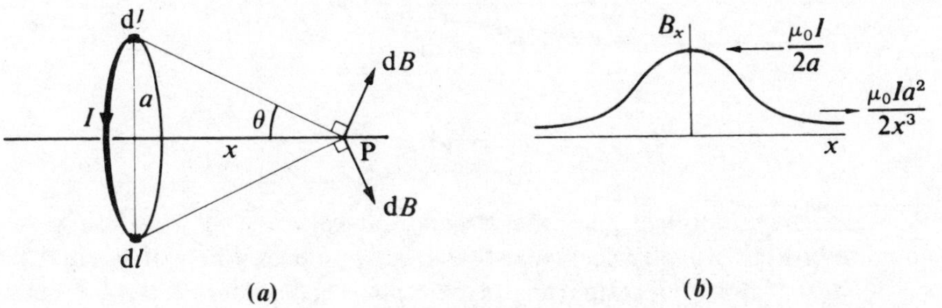

Figure 7.9 The **B**-field due to a circular current-carrying coil, at a point on its axis: (a) geometry, (b) variation of the magnitude of **B** with distance along the axis.

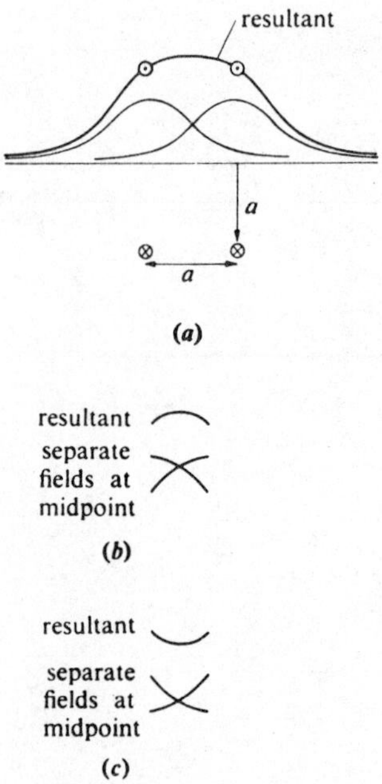

Figure 7.10 (a) Helmholtz coils; (b) fields with coils closer than (a); (c) fields with coils further apart than (a).

Application *Helmholtz coils* If two circular coils of the same radius and carrying identical currents in the same sense are arranged coaxially as in Fig. 7.10a, their fields along the axis will add. A highly uniform field will be produced at the axial point midway between the coils if the distance apart is properly adjusted. The separation should be such that the decrease in one field as we move away from the midpoint is exactly compensated by the increase in the other. As Fig. 7.10b and c shows, this will only be achieved if the variation in B_x at that point is linear, i.e. we must have $\mathrm{d}^2B_x/\mathrm{d}x^2=0$. When applied to Eq. (7.13), we obtain $x=\frac{1}{2}a$ and the coils should therefore be placed a distance apart equal to their radius. This arrangement, known as a pair of *Helmholtz coils*, produces a uniform magnetic field over a relatively large volume. It is particularly useful for experiments in which specimens being subjected to such a field need to be easily accessible.

Application **B**-*field on the axis of a solenoid* If a solenoid of n turns per unit length is closely wound, it may be regarded as a collection of n coaxial coils per unit length. An element of length $\mathrm{d}x$ as in Fig. 7.11 becomes a coil of $n\,\mathrm{d}x$ turns. At a field point P on the axis, let the angles subtended by the ends of the solenoid be α and β, and that by the typical element be θ. If the current is I, the field at P due to the element $\mathrm{d}x$ is, using (7.13),

$$\mathrm{d}B_\mathrm{P}=\mu_0 n\,\mathrm{d}x I a^2/[2(x^2+a^2)^{3/2}]$$

However, $x=a\cos\theta$, and therefore $\mathrm{d}x=a\operatorname{cosec}^2\theta\,\mathrm{d}\theta$, so that $\mathrm{d}B_\mathrm{P}=-\mu_0 nI\sin\theta\,\mathrm{d}\theta/2$. When integrated from α to β, this gives

finite solenoid: $$B_\mathrm{P}=\tfrac{1}{2}\mu_0 nI(\cos\beta-\cos\alpha) \qquad (7.16)$$

For a solenoid whose length becomes much greater than its radius, $\alpha\to\pi$ and $\beta\to0$. In the limit for an infinite solenoid, (7.16) gives

infinite solenoid: $$\boxed{B_\mathrm{P}=\mu_0 nI=\mu_0 J_s} \qquad (7.17)$$

where J_s is the current per unit length and is thus akin to the surface current density of Sec. 1.7. Equation (7.17) is found to be a good approximation to B over a fair region in the centre of long but finite solenoids. This is a further method of obtaining a nearly uniform magnetic field, and one capable of producing much stronger fields over larger volumes than Helmholtz coils.

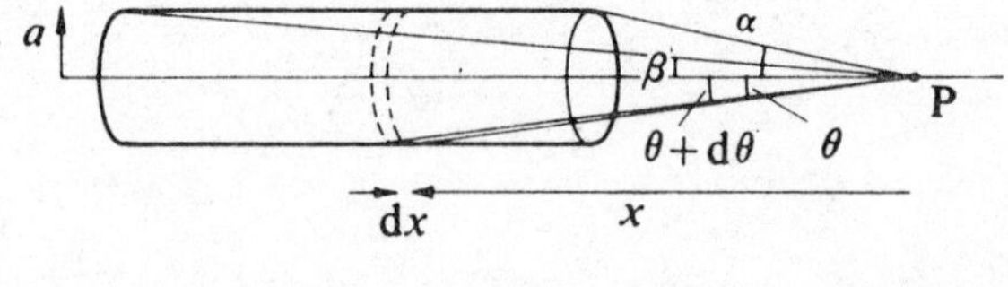

Figure 7.11 Calculation of the magnetic **B**-field at P due to a solenoid.

Worked example **B**-*field due to a straight current-carrying wire* Using the notation of Fig. 7.12, a typical current element produces a field at P equal to $\mu_0 I\,\mathrm{d}l(\sin\theta)/[4\pi(r^2+l^2)]$ at right angles to the plane of the diagram. But $l=r\cot\theta$

and therefore $dl = -r\,\text{cosec}^2\,\theta\,d\theta$, so that

$$dB = -\mu_0 I \sin\theta\, d\theta/4\pi r$$

Integration from α to β yields

finite wire: $$B = \frac{\mu_0 I}{4\pi r}(\cos\beta - \cos\alpha) \quad (7.18)$$

For a very long wire, $\alpha \to 0$ and $\beta \to \pi$ and in the limit we have

infinite wire: $$\boxed{B = \mu_0 I/2\pi r} \quad (7.19)$$

The **B**-field is thus proportional to r^{-1} as Biot and Savart originally discovered. Equation (7.18) can be used to determine fields due to circuits which are combinations of straight wires, such as rectangular coils.

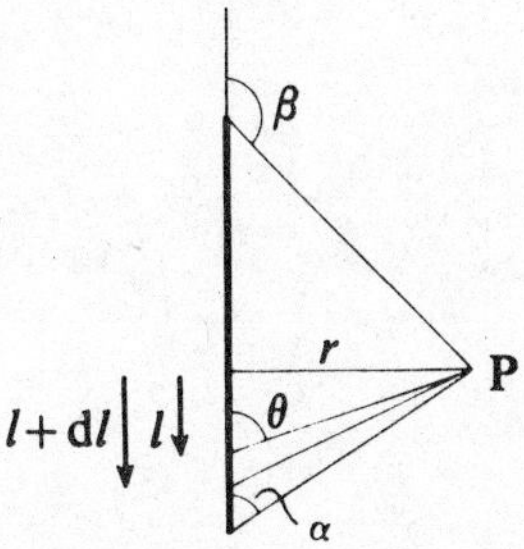

Figure 7.12 Calculation of the magnetic **B**-field at P due to a straight finite length of current-carrying wire.

7.4 Forces between currents: the fundamental law

The law of magnetic interaction We can now write down an expression for the interaction between two current elements with some understanding of its origin and significance. Let the elements be $I_1\,d\mathbf{l}_1$ and $I_2\,d\mathbf{l}_2$ separated by a vector displacement $\mathbf{r}_{12}$ directed from 1 to 2. The Biot–Savart law (7.9) will give the **B**-field due to one element at the other, which then experiences a force given by (7.2). Thus, the force on $I_2\,d\mathbf{l}_2$ *due to* $I_1\,d\mathbf{l}_1$ is

two current elements: $$\boxed{d^2\mathbf{F}_{12} = \frac{\mu_0 I_1 I_2\, d\mathbf{l}_2 \times (d\mathbf{l}_1 \times \mathbf{r}_{12})}{4\pi r_{12}^3}} \quad (7.20)$$

This can be regarded as the fundamental magnetic law of force corresponding to Coulomb's law in electrostatics.

Equation (7.20) has an odd feature. The force $d^2\mathbf{F}_{21}$ *on* $I_1\,d\mathbf{l}_1$ *due to* $I_2\,d\mathbf{l}_2$ is clearly obtained by interchanging the subscripts 1 and 2, noting that $\mathbf{r}_{12} = -\mathbf{r}_{21}$. Newton's law of the equality of action and reaction would lead us to expect that $d^2\mathbf{F}_{12} = -d^2\mathbf{F}_{21}$. *In general,* this is not so because $d\mathbf{l}_2 \times (d\mathbf{l}_1 \times \mathbf{r}_{12})$ and $d\mathbf{l}_1 \times (d\mathbf{l}_2 \times \mathbf{r}_{21})$ are not equal and opposite, and do not even have the same direction: the first is in the plane of $d\mathbf{l}_1$ and $\mathbf{r}_{12}$, the second is in the plane of $d\mathbf{l}_2$ and $\mathbf{r}_{12}$ (see Appendix B.6). The apparent paradox is resolved by realizing that *for steady currents* (7.20) has no meaning unless integrated round the complete circuits L_1 and L_2 of which the elements are a part. The force is then formally:

two circuits: $$\mathbf{F}_{12} = \oint_{L_1}\oint_{L_2} \frac{\mu_0 I_1 I_2\, d\mathbf{l}_2 \times (d\mathbf{l}_1 \times \mathbf{r}_{12})}{4\pi r_{12}^3} \quad (7.21)$$

with $\mathbf{F}_{21}$ obtained by the interchange of subscripts. It is now possible to show that $\mathbf{F}_{12} = -\mathbf{F}_{21}$. We shall see that it is possible for current elements to exist independently when the currents are *not steady* (e.g. the Hertzian dipole of Sec. 14.6) but in that case there are additional fields produced which can carry momentum away from the system. Under those circumstances, the system is not an isolated one of the two elements alone so that action and reaction would not necessarily be equal and opposite.

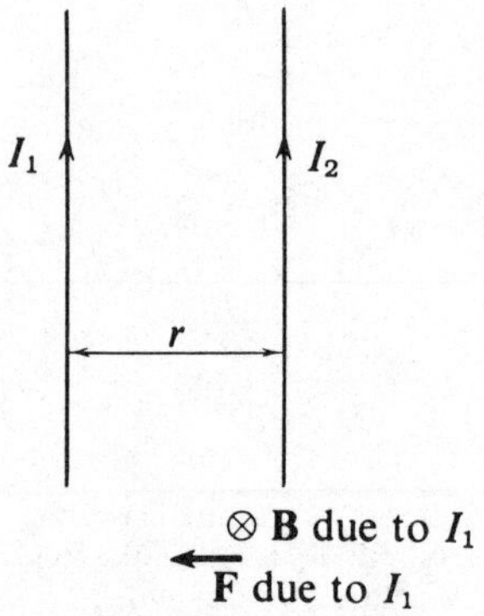

Figure 7.13 The force between two infinite straight parallel wires carrying currents.

Worked example *two infinite straight parallel currents* Two infinite wires of negligible section carrying currents I_1 and I_2 are parallel and a constant distance r apart (Fig. 7.13). The **B**-field at every point on I_2 due to I_1 is $\mu_0 I_1/(2\pi r)$ into the diagram, by (7.19). The force on a length l of I_2 is therefore, by (7.1),

two parallel currents: $$F = \frac{\mu_0 I_1 I_2 l}{2\pi r} \tag{7.22}$$

towards I_1. The reversal of I_2 would reverse the force, as the experiment of Sec. 1.2 showed. The expression (7.22) will also give an approximate value for the force between any two curved parallel currents provided their distance apart is much smaller than their radius of curvature.

The situation in Fig. 7.13 is precisely that described in the **definition of the ampere** in Sec. 1.4, so that (7.22) shows the value of μ_0 to be exactly $4\pi \times 10^{-7}$. This justifies (7.10) in the previous section.

In an older system of units known as the CGS electromagnetic units, e.m.u., the unit of current is defined so that $\mu_0 = 4\pi$ when **F** and **r** are in dyn and cm. All formulae may thus be converted to e.m.u. by the substitution $\mu_0/4\pi = 1$. See Appendix C.

7.5 Moving charges and B-fields

Equivalence of current element and moving charges A current element consists of charges moving over a small length of a circuit. In Sec. 1.7 we showed that $I\,\mathbf{dl}$ could be written as the product of the total charge *moving* in the element and the velocity:

$$I\,\mathbf{dl} = NQ\mathbf{v} \tag{7.23}$$

where N is the total number of moving charges each of magnitude Q, and where N may be large even if **dl** is small compared with other distances. It is now very tempting simply to replace $I\,\mathbf{dl}$ by $NQ\mathbf{v}$ in the formulae developed so far and assume that we then obtain equally valid expressions. There are two reasons why we should be careful about this. One is that we should be using current element formulae with no intention of integrating round a complete circuit, and we saw in the previous section that that could be suspect.

The second reason for being careful is in the interpretation of **v**. All our experimental results so far have applied to *neutral* currents as in a piece of wire which contains positive and negative charges in equal amounts. The current I is normally thought of as the movement of electrons relative to the positive metallic ions *and to the observer*. But even if the observer moved so that the electrons seemed on the average to be stationary, there would still be the same I though it would now appear as a movement of positive charges. So, as Problem 1.16 showed, the magnitude of such a neutral current is independent of any velocity an observer may have. The same is not true of a *charged* current, say of a beam of electrons *in vacuo*. Here, the value of I depends on **v** as in (7.23), it is true, but **v** is the velocity of the charges *relative to the observer*. This interpretation of **v** must be kept in mind throughout this section. It is connected with a relativistic aspect of magnetism discussed later (Sec. 13.10).

Force on a moving charge in a B-field The force given by (7.1) or (7.2), i.e. $d\mathbf{F} = I\,d\mathbf{l} \times \mathbf{B}$, coupled with (7.23) above, suggests that the force on N charges, each Q moving with a velocity $\mathbf{v}$ in $\mathbf{B}$, would be $NQvB\sin\theta$. On a single charge, therefore, we expect to find

force on Q in **B**: $$F = QvB\sin\theta \tag{7.24}$$

in a direction at right angles to both **v** and **B** such that **v**, **B** and **F** form a right-handed system as in Fig. 7.14a. Vectorially:

Lorentz force: $$\mathbf{F} = Q\mathbf{v} \times \mathbf{B} \tag{7.25}$$

The current element formula from which this is obtained was backed up by direct experiments and was used as a definition of **B**. In the same way, we shall find that the motion of charges calculated from (7.25) is always accurately confirmed by observation. Our confidence in the expression is such that it is often used as an alternative definition of **B**, although both **v** and **B** are being defined *relative to an observer* and not in any absolute sense. Consequences of (7.25) are explored below.

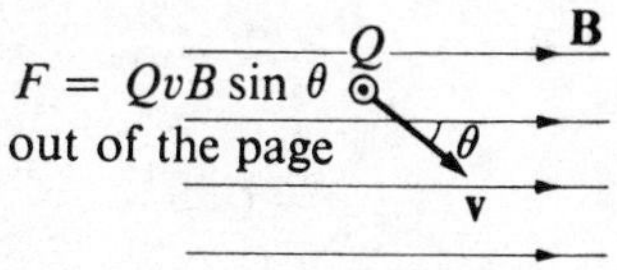

(*a*)

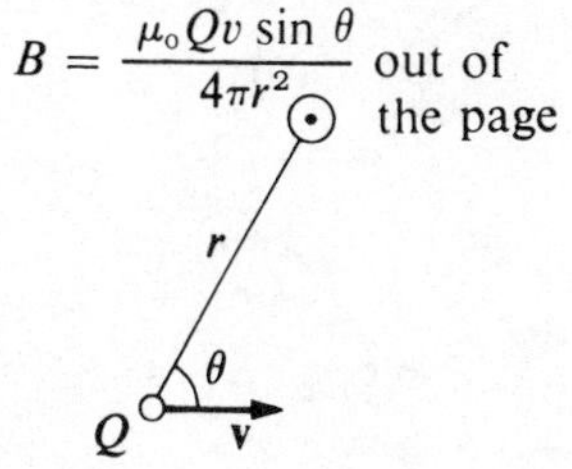

(*b*)

Figure 7.14 (a) Force $\mathbf{F} = Q\mathbf{v} \times \mathbf{B}$ on a charge moving with a velocity **v** in **B**; (b) the **B**-field at P due to a moving charge.

B-field due to a moving charge The use of (7.23) in the Biot–Savart law leads us to expect N moving charges to produce a **B**-field given by $(\mu_0 NQv\sin\theta)/(4\pi r^2)$ with the geometry of Fig. 7.14b. One charge Q would then give

$$B_P = (\mu_0 Qv\sin\theta)/(4\pi r^2) \tag{7.26}$$

in a direction given by the right-hand screw rule as in the figure. Vectorially:

B due to moving charge: $$\mathbf{B}_P = \frac{\mu_0 Q\mathbf{v} \times \hat{\mathbf{r}}}{4\pi r^2} \tag{7.27}$$

Unlike (7.25), this formula cannot be checked directly in practice and must be used with caution, particularly since the original current element formula was derived for steady currents. In fact it is found to apply at velocities much less than that of light, but it is not a happy choice as part of a fundamental magnetic law of force.

Motion of charged particles in B-fields A charged particle of mass m moving with a velocity **v** in a field **B** will experience a force given by (7.25). In an evacuated region where the mean free path is greater than the size of the overall trajectory, no other force exists and the acceleration of the particle is then $Q(\mathbf{v} \times \mathbf{B})/m$. The resultant motion depends on the initial velocity and the spatial variation of **B**. We first deal with motion in uniform fields and then look briefly at certain examples of non-uniform fields.

If the **B**-field is uniform and the initial velocity is parallel to **B**, then the vector product in (7.25) is zero: there is no force, no acceleration and the particle continues to move with a uniform velocity along the initial direction. However, the case in which **v** is initially perpendicular to **B** needs more detailed examination.

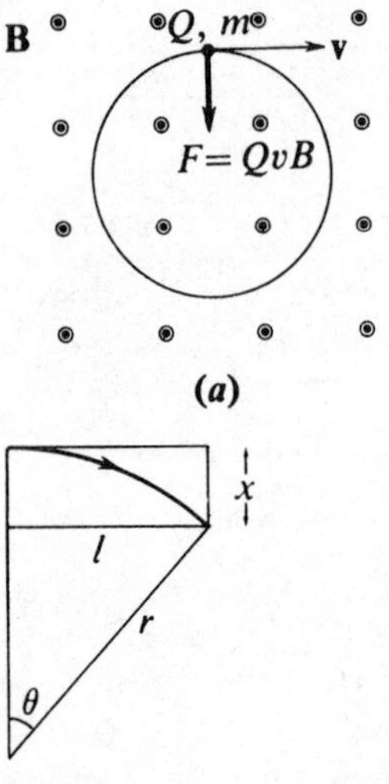

(b)

Figure 7.15 (a) The motion of a positive charged particle projected with a velocity perpendicular to a uniform **B**-field; (b) section of the trajectory in (a) when the deflection x is small.

Worked example *velocity perpendicular to uniform* **B**, *circular path* The particle is shown in its initial state in Fig. 7.15a, where **B** is towards the observer and the force given by (7.25) is simply QvB at right angles to both **v** and **B**. Because the force is perpendicular to **v**, it cannot change the *magnitude* of **v** but only its direction: the particle will move in a curve. Let the radius of curvature of the path at any instant be r, so that the acceleration towards the centre of curvature is v^2/r (a general result from kinematics of particles). Applying Newton's law $F=ma$, we obtain $QvB=mv^2/r$ or

$$r=mv/QB=p/QB \qquad \text{and} \qquad p=QBr \tag{7.28}$$

where p is the linear momentum. We have already argued that **v** is constant in magnitude, so that r must be constant as well: the trajectory is a circle of radius given by (7.28).

General applications The circular trajectories with radii given by (7.28) are a well-known feature in the physics of elementary particles. Among the applications are:

1. Proportionality between r and p for a given Q allows particles with various momenta to be distinguished in, say, a bubble chamber and enables the momenta to be calculated.
2. Proportionality between r and m for a certain v (obtained using a velocity filter described below) and a given Q can be used to compare isotopic masses in a mass spectrometer or to separate isotopes.
3. A small portion of the circular path can be regarded as a lateral deflection of a particle beam (Fig. 7.15b). The deflection x after travelling a distance l in the original directions is $r(1-\cos\theta)$ where $\sin\theta=l/r$. For *small* deflections this is

$$x\approx l^2/2r=l^2QB/2mv \tag{7.29}$$

   This is used to deflect electron beams in the cathode-ray tubes used in television and radar. The current-carrying deflection coils, unlike electrostatic deflecting plates, are located outside the tube and are therefore accessible. The deflection, if small enough, is proportional to B and hence to the current in the coils. A linear sweep current produced by auxiliary circuits achieves the required scanning of the screen.
4. A historic experiment by Bucherer (1909) also used (7.29), but as a means of determining e/m for β-particles of known velocity fixed by a velocity selector (next section). He showed that as the velocity increased from about $0.32c$ to $0.69c$, the value of e/m fell from $1.66\times10^{11}\,\mathrm{C\,kg^{-1}}$ to $1.28\times10^{11}\,\mathrm{C\,kg^{-1}}$ compared with its value of $1.76\times10^{11}\,\mathrm{C\,kg^{-1}}$ for very low velocities. These measurements were among the earliest to confirm that when the velocity of a particle approaches that of light (c), its mass m varies according to the relativistic formula

$$m=\frac{m_0}{(1-v^2/c^2)^{1/2}} \tag{7.30}$$

   in which m_0 is the rest mass.

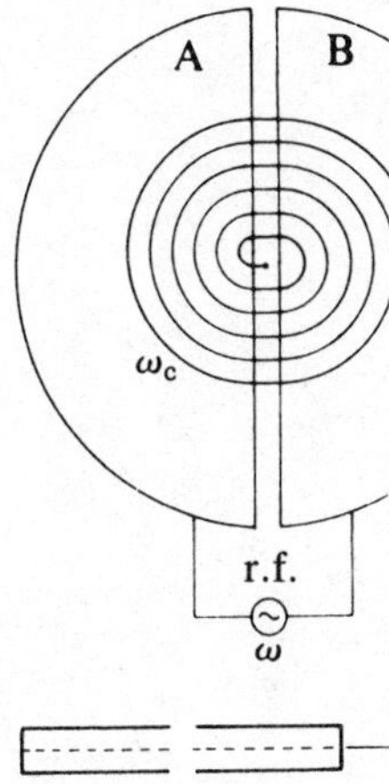

Figure 7.16 The fixed-frequency cyclotron.

Particular application *cyclotrons and the cyclotron frequency* Another important feature of (7.28) is revealed by using the angular frequency in the path $\omega = v/r$, so that

$$\omega_c = QB/m \tag{7.31}$$

The true frequency $\nu_c = QB/2\pi m$ is known as the **cyclotron frequency**. It depends only on the Q/m of the particles and on the **B**-field and is *independent of radius and velocity provided that* $v \ll c$ *and* **B** *is uniform.*

Suppose now that an alternating but uniform E-field is applied in the plane of Fig. 7.15a and across the page. If the angular frequency ω of the E-field is varied, **cyclotron resonance** is said to occur when $\omega = \omega_c$. At that frequency, and no other, those charges at the top of the orbit when the **E**-field has its maximum accelerating value will continue to experience maximum forces at exactly the right times, particularly at the top and bottom of the path. The orbits will then expand and energy is extracted from the source of the **E**-field.

Cyclotron resonance has been used in several important investigations. Sanders, Tittel and Ward (1963) determined e/m_e very precisely for free electrons while Petley and Morris (1967) obtained e/m_p for protons, detecting resonance through the expansion of orbits. It is also possible to use cyclotron resonance to obtain e/m for the carriers of current in metals and semiconductors provided the mean free paths are long enough (Kip, 1960).

The **cyclotron** itself is a particle accelerator using the same resonance (Fig. 7.16). Here, the **E**-field is applied only across the gap between the dees. When its ω is equal to ω_c, those particles crossing the gap at the instants of maximum accelerating field will remain in phase with the field. The radius, the speed, and hence the kinetic energy, will increase. However, at high speeds the relativistic increase in the mass of the particles (7.30) causes the value of ω_c to decrease, and they fall out of phase with the fixed applied frequency. Eventually they reach a stable orbit where no further increase in energy is possible. These effects become appreciable when the kinetic energy of the particles is not negligible compared with m_0c^2, i.e. around 0.51 MeV for electrons and 938 MeV for protons. For energies higher than these, modifications to the simple cyclotron are made. In the **synchrocyclotron**, the applied frequency is varied to keep particles in phase with the field variations. In the **electron synchrotron**, the same effect is achieved by slowly increasing the magnetic field, but the radius of the trajectory is kept constant so that the magnetic field has to be produced over a much smaller volume. **Proton synchrotrons** have to decrease the frequency as well as increase the **B**-field to maintain the correct phase relationship.

Worked example *general velocity in uniform* **B**, *helical path* Any velocity **v** making an angle α with **B** as in Fig. 7.17a can be resolved into a component $v \cos \alpha$ parallel to **B** and $v \sin \alpha$ perpendicular to **B**. The parallel component, as we have seen, is unaffected by the field, while the perpendicular component will produce the circular trajectory already considered. The combination of uniform motion along **B** and circular motion perpendicular to **B** produces a **helix**, as indicated in the figure. The time taken for one traverse of the circular path of length $2\pi r$ is $2\pi r/v$ or $2\pi m/(QB)$ from (7.28). In that time the particle moves a distance $2\pi m v \cos \alpha/(QB)$ or $2\pi p \cos \alpha/(QB)$, the **pitch** of the helix.

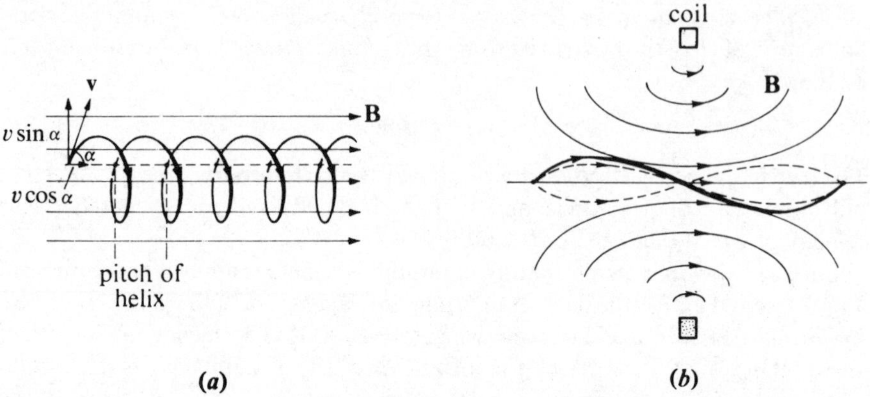

Figure 7.17 (a) The helical path of an electron in a uniform **B**-field; (b) first-order focusing by the **B**-field of a current-carrying coil.

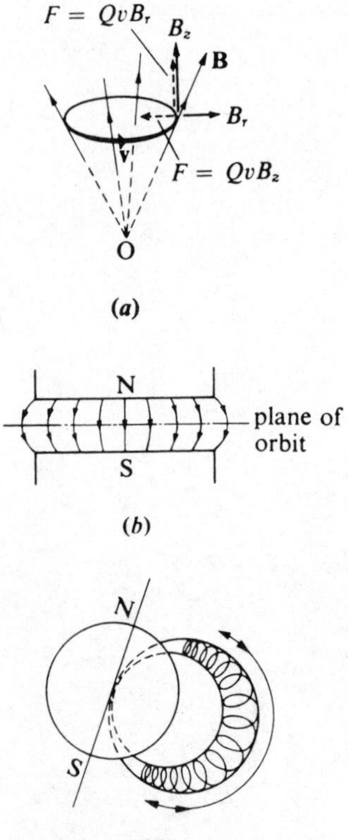

Figure 7.18 (a) The forces on an electron orbiting in a divergent **B**-field; (b) type of non-uniformity retaining particles in plane circular orbits; (c) trapping of particles in a van Allen belt.

If a beam consisting of identical particles with the same linear momenta p but with a small divergence angle α is subjected to a uniform **B** along the mean direction of motion, the pitch of all the helices is very nearly the same and a first-order focus is obtained.

Charges moving in non-uniform ***B****-fields applications* We choose for brief discussions two examples in which the magnetic fields, while non-uniform, have an axial or cylindrical symmetry:

1. The focusing property of a uniform **B**-field mentioned at the end of the previous paragraph is also possessed by the field along the axis of a circular coil (Fig. 7.17b). The coil thus forms a type of electron lens. More elaborate designs of the same type are used in **electron microscopes** and **β-ray spectrographs**.
2. If the magnetic field is divergent as in Fig. 7.18a, the forces experienced by an orbiting **electron** are as shown. The field has two components: B_z, responsible for the circular orbit, and B_r, producing a force on the electron away from O, the point from which the **B**-lines appear to originate. The electron thus follows a helical path away from O. There are two applications of this. First, a beam of particles moving in an orbit in a field like that of Fig. 7.18b will be kept in the plane of the orbit if they are inclined to stray out of it. In a uniform field they would be lost. The second application concerns the motion of particles in the earth's magnetic field, a dipolar field converging towards the poles as in Fig. 7.18c. A charge entering the field begins to describe a helical path and the non-uniformity carries it away from the nearest pole. Eventually it reaches a converging field and the pitch of the helix shortens until it becomes zero and then reverses. The particle thus oscillates continually to and fro and is trapped in the field. In certain regions of space near the earth a large number of particles are trapped in this way and form the **Van Allen belts**. The reversal of path in a converging field is called the **mirror effect** and is also used in the **containment of plasmas** (highly ionized gases).

7.6 Moving charges in E- and B-fields

The Lorentz force law We saw in Chapter 3 that the force on a stationary charge Q in a steady E-field is given by $\mathbf{F}=Q\mathbf{E}$, while in this chapter the force on Q moving with $\mathbf{v}$ in a steady **B**-field is $Q\mathbf{v}\times\mathbf{B}$. When both **E** and **B** exist at a point, the total force on a charge Q moving with a velocity **v** should be

force on Q in **E** and **B**: $$\boxed{\mathbf{F}=Q(\mathbf{E}+\mathbf{v}\times\mathbf{B})} \tag{7.32}$$

This is known as the **Lorentz force law**. It might seem obvious that it arises simply from the superposition of the two separate forces, but an assumption has been made that is often overlooked. The force $\mathbf{F}=Q\mathbf{E}$ follows from the definition of **E** only if Q is stationary, whereas (7.32) is applied to moving charges. We have assumed that $\mathbf{F}=Q\mathbf{E}$ continues to give the force on Q in **E** even when Q moves. A discussion of this point involves relativistic arguments and will be taken up in Sec. 13.10 where we shall see that (7.32) is valid.

The Lorentz force law enables us to calculate the trajectories of charged particles free to move under the action of **E** and **B** when both fields are specified over a region of space. The validity of the law can then be tested by measurement of the trajectories *in vacuo* (i.e. at a sufficiently low pressure).

In many applications of this type of motion, the **E** and **B** fields are perpendicular to each other and are made as uniform as possible. We therefore restrict our analysis to this type of geometry.

Worked example *Q moving in uniform* **E** *and* **B** *at right angles to each other* It is convenient to choose Cartesian coordinates in such a way that **E** and **B** lie along principal axes. Thus, in Fig. 7.19a, **E** is along the y axis and **B** is along the z axis (out of the page). One general deduction can be made immediately: there will be no force in the z direction on any charged particle because **E** cannot produce a

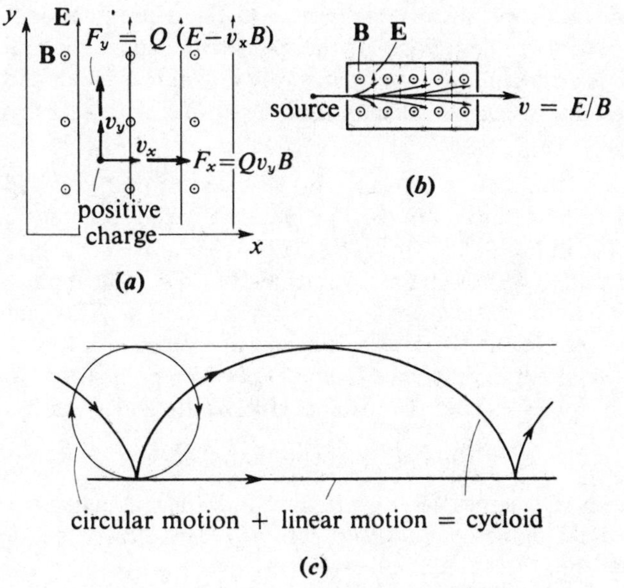

Figure 7.19 Motion of a positive charged particle in perpendicular **E**- and **B**-fields: (a) forces in the xy plane (no force is produced in the z direction); (b) the velocity filter; (c) a cycloidal trajectory is a combination of linear and circular motion in the same plane.

force at right angles to itself and **B** cannot produce a force parallel to itself. It follows that the z component of the velocity must remain unchanged throughout any motion. In particular, if v_z is initially zero, it remains zero, so that *motion which begins in the xy plane stays in that plane.* We shall therefore neglect any v_z in the analysis, and the forces in the general case are F_x and F_y as shown in Fig. 7.19a.

Application *velocity filter* An important special case arises if the particle is initially projected along x, i.e. perpendicular to both **E** and **B**, so that $v_y=0$. The only force is then $F_y=Q(E-v_xB)$, and if $v_x=E/B$ then even this is zero and the velocity remains constant. *A particle projected with* $v_x=E/B$ *and* $v_y=0$ *is undeflected.* This is the basis of the **velocity filter** or **velocity selector** illustrated in Fig. 7.19b. Of all the particles entering the fields with a velocity perpendicular to both **E** and **B**, only those with $v=E/B$ are undeflected. (In passing, note that **E**-fields alone determine *kinetic energies*, **B**-fields alone determine *momenta*, while **E** and **B** together determine *velocities*.)

Application *cycloidal paths* The general trajectory for any initial velocity can be obtained by solving the equations of motion $F_y=m\ddot{y}=Q(E-v_xB)$, $F_x=m\ddot{x}=Qv_yB$, but another method is more instructive. It is apparent from the special case just mentioned that a charge moving with $v_x=E/B$ experiences no force in the y direction and hence is effectively in a zero y field. A non-zero y force will only come into effect when v_x differs from E/B. Let us, then, put $v_x=u_x+E/B$ when the equations of motion become $F_y=Qu_xB$, $F_x=Qv_yB$. It is also apparent that if we viewed the whole system from an origin moving with an x velocity of E/B, those would be the equations of motion. The path would then be a circle of radius mv/QB with v having components u_x and v_y. The resultant motion from a *fixed* origin is thus a combination of a uniform circular motion and a uniform linear velocity in the same plane—a cycloid (Fig. 7.19c). The exact type of cycloid depends on the initial velocity.

This type of trajectory was originally used in the determination of e/m for photoelectrons (see Problem 7.13). If the **E**-field is made radial rather than uniform, the trajectories follow curved paths very similar to cycloids and these are used in **magnetrons**, which are vacuum tubes used to generate microwaves.

Application *the Hall effect* We saw in Chapter 6 that a steady **E**-field maintained in a metallic conductor produces a steady current flow. The equipotentials, as in Fig. 7.20a, are planes perpendicular to the lines of **E** and to those of the current. If we now maintain a steady **B**-field at right angles to the flow of current, the equipotentials are tilted as shown in Fig. 7.20b and c. This phenomenon is known as the **Hall effect** and its presence is revealed by the appearance of a potential difference (voltage) between points such as A and B where none previously existed. The **Hall coefficient** R_H is defined by

$$V=R_HIB/d \qquad \text{(definition of } R_H\text{)} \qquad (7.33)$$

where V is the voltage across AB, I is the current, B the magnetic field and d the thickness of the conductor parallel to **B**. R_H can clearly be determined experimentally from (7.33).

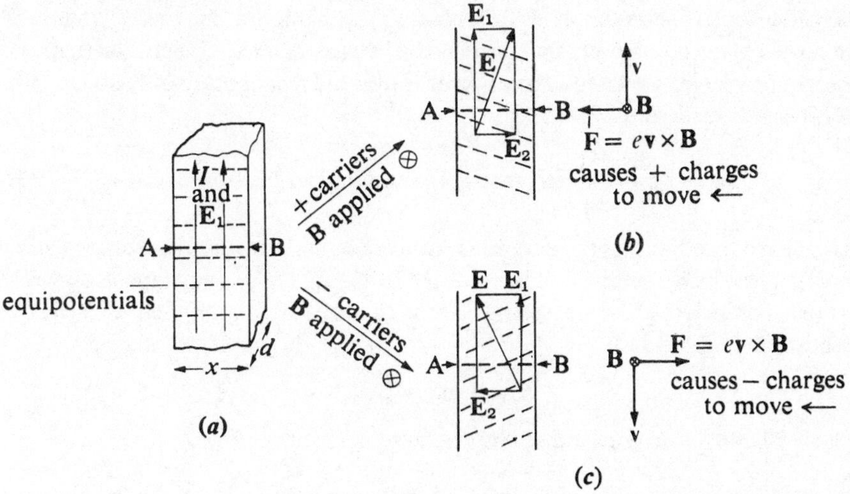

Figure 7.20 The Hall effect: (a) situation before application of **B**; (b) and (c) situation after applying a **B**-field into the page.

The way in which the effect arises is shown in Fig. 7.20b and c for positive and negative carriers, respectively. In Fig. 7.20b the velocity **v** of the carriers gives rise to the Lorentz force $e(\mathbf{v} \times \mathbf{B})$ on them in the direction shown, and this force produces a drift of positive charge to the left, leaving the right-hand side negative. These charges accumulate and set up an electrostatic field $\mathbf{E}_2$ which combines with the original $\mathbf{E}_1$ to give a resultant as shown. The equipotentials are at an angle to their original direction. The accumulation of the charge at the sides stops when evB and eE_2 are equal and opposite, for then the charge flows in its original direction with its original drift velocity **v**. (Note that for the transient lateral current we could consider (6.31) as applying with $\mathbf{E}_M = 0$ and $\mathbf{e} = \mathbf{v} \times \mathbf{B}$, a type of motional electromotance acting on the moving carriers (Chapter 9). *At equilibrium*, $\mathbf{E}_2 + \mathbf{v} \times \mathbf{B} = 0$ and the lateral current ceases.)

If x is the width of the conductor, $E_2 = vB$ and thus $V = vBx$. Since $I = nevxd$ by (1.12), $V = BI/(ned)$. Hence, from (7.33):

single carrier: $$R_H = 1/(ne) \qquad (7.34)$$

so that the Hall coefficient gives direct information about the density n of carriers. Figure 7.20c shows that for negative carriers the effect is negative so that the sign of R_H gives that of the carriers. The mobility $\mu = \sigma R_H$ from (6.34). The simple theory is only applicable when one type of carrier is present in a much higher concentration than any other. Measurements show that while many metals have negative Hall coefficients, as expected, some have positive values even when the electromechanical experiments mentioned in Sec. 1.5 show the carriers to be electrons (e.g. Mo, Zn, Cd). This is interpreted in the band theory of conduction (Sec. 6.12) as providing evidence that holes are acting as positive carriers in some metals and semiconductors.

7.7 Magnetic scalar potential, magnetic flux

Before examining other approaches to magnetic interactions, it will be an advantage to look at two quantities analogous to those we defined for **E** in Chapters 3 and 4.

Magnetic scalar potential It is possible to introduce a magnetic potential difference between two points A and B as the line integral of $-\mathbf{B}$ between them. It is more convenient for later usage, however, to use $\mathbf{B}/\mu_0$ in place of $\mathbf{B}$, so that we have, using A as a zero,

$$V_m = \int_A^B -\frac{\mathbf{B}\cdot d\mathbf{L}}{\mu_0} \text{(zero at A)} \qquad \text{(definition of } V_m \textit{ in vacuo}) \qquad (7.35)$$

It is usual to specify V_m as the magnetic *scalar* potential because there also exists a vector potential (Chapter 8). The **SI unit** for V_m is simply the ampere, A.

It follows, as in the corresponding electric case, that the component of **B** in any direction $\hat{\mathbf{L}}$ is given by

$$B_L = -\mu_0 \partial V_m / \partial L \qquad (7.36)$$

In general, **B** is clearly μ_0 times the negative gradient of V_m or

$$\boxed{\mathbf{B} = -\mu_0 \nabla V_m} \qquad (7.37)$$

while in Cartesian and plane polar coordinates respectively:

$$\left.\begin{matrix} B_x = -\mu_0 \partial V_m/\partial x, & B_y = -\mu_0 \partial V_m/\partial y, & B_z = -\mu_0 \partial V_m/\partial z \\ B_r = -\mu_0 \partial V_m/\partial r, & B_\theta = -\mu_0 (1/r) \partial V_m/\partial \theta & \end{matrix}\right\} \qquad (7.38)$$

It is not as easy to place a physical interpretation on V_m as it is with electric potential V because (7.35) does not represent the work done in conveying any real entity along the path AB. The concept of the monopole in Sec. 7.10 may help those who feel the need of such a representation.

It soon becomes clear, however, that V_m is not as useful a quantity as V. The great utility of V stems from its path-independence in an electrostatic field which is the result of $\oint \mathbf{E}_Q \cdot d\mathbf{L}$ for *any* closed path being zero. It is obvious from the shapes of **B**-lines in Fig. 7.1 that the **B**-fields of currents do *not* have $\oint \mathbf{B}\cdot d\mathbf{L} = 0$ for every closed path. V_m is therefore not in general path-independent and must be used with care.

Magnetic flux We have already encountered the concept of flux in connection with the electric field in Sec. 4.1. The reader is reminded once again that the flux of any vector **A** over a surface element d**S** regarded as a vector is given by $\mathbf{A}\cdot d\mathbf{S}$, the product of the component of **A** normal to d**S** and the area of d**S**. For a finite surface area S, the flux of **A** is the sum or integral of $\mathbf{A}\cdot d\mathbf{S}$ over S, or $\int \mathbf{A}\cdot d\mathbf{S}$.

Normally, the vector whose flux is in question should be specified exactly, e.g. flux of **E** (in Gauss's law for **E**), flux of **J** (=electric current). The flux of **B**, however, is a particularly important quantity and is usually referred to simply as the **magnetic flux** with a special symbol Φ and a special name for its unit. Thus, the magnetic flux over an area S is

$$\Phi_S = \int_S \mathbf{B}\cdot d\mathbf{S} \qquad \text{(definition of } \Phi) \qquad (7.39)$$

The **SI unit**, the T m^2, is called the **weber**, symbol Wb. Older texts will refer to the tesla as the Wb m^{-2} and the **B**-field as the *magnetic flux density*.

A sign is allocated to Φ using the following rules. The self-flux of a current-carrying circuit across the area bounded by the circuit is always counted as positive; any other magnetic flux over the same area is considered positive when in the same direction as the self-flux (see Figs 9.6a and 9.11a). Over a closed surface, the outward flux is considered positive.

7.8 Magnetic dipoles and B-fields

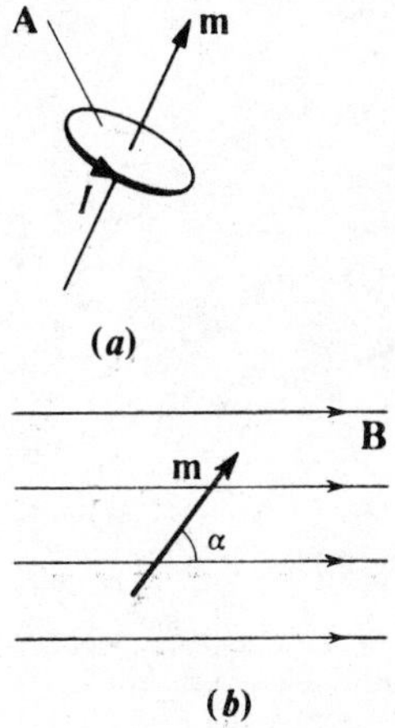

Figure 7.21 (a) A current-loop dipole: note the direction of **m** in relation to the current (right-hand screw rule); (b) a magnetic dipole in a **B**-field.

Magnetic dipole moment: concept and definition Our initial survey in Sec. 7.1 revealed that small permanent magnets and small current loops both interacted in a way strikingly similar to electric dipoles. We proceed now to develop the idea of a *magnetic dipole*, guided by the results of Sec. 3.9 and using the magnetic laws of this chapter.

We deal first with the current-loop dipole as illustrated in Fig. 7.21a, which, as we saw in Sec. 7.1, can be used instead of a compass needle to plot **B**-fields. The reason why this can be done is that the loop experiences a torque when placed in a magnetic field, a torque already calculated in Eq. (7.6) to be

$$\mathbf{T} = I\mathbf{A} \times \mathbf{B} \tag{7.40}$$

In this expression, **A** is a vector representing the area: it is directed along the axis in the sense indicated by **m** in Fig. 7.21a, so that it is related to the current flow by the right-hand screw rule. In Fig. 7.21b we represent the general position of the dipole axis at an angle α to **B** and so the torque (7.40) is $IAB \sin \alpha$ in such a sense as to turn **m** into the direction of **B**. Thus when the loop is in stable equilibrium with $\mathbf{T} = 0$, its positive axis points along **B** and enables field lines to be plotted.

Now (7.40) is precisely analogous to the torque on an electric dipole of moment **p** placed in an electric field and written as $\mathbf{T} = \mathbf{p} \times \mathbf{E}$. This suggests that we define the magnetic dipole moment **m** of a current loop as

$$\boxed{\mathbf{m} = I\mathbf{A} \qquad \text{(definition of } \mathbf{m}\text{)}} \tag{7.41}$$

Comment C7.3

and rewrite (7.40) as

magnetic dipole in **B**-field: $$\boxed{\mathbf{T} = \mathbf{m} \times \mathbf{B}} \tag{7.42}$$

It would be quite possible to use (7.41) and (7.42) as starting points for a dipole treatment of magnetism, using (7.42) to define a **B**-field. We have preferred to start from current element formulae so that (7.42) is a deduction.

If we have a small permanent magnet as our dipole, then the definition (7.41) does not apply, and we now use (7.42) as a definition of **m**. Thus the magnetic dipole moment could be defined as the torque on the dipole when its axis is perpendicular to unit field. (Note that the magnetic axis of a system which is permanently magnetized can only be found by allowing it to come to rest in a **B**-field whose direction is already known. Magnetized systems do not always come in the shape of small needles with **m** along their length!)

The **SI unit** for **m** is the $\mathrm{A\,m^2}$ and the quantity itself is sometimes called the *electromagnetic moment*.

Magnetic dipoles in B-fields The potential energy of a magnetic dipole of moment **m** making an angle θ with **B** can be calculated by a method similar to that for the electric dipole. We assume the dipole is of fixed moment and is already established in position with $\alpha=90°$ as the zero of energy (Fig. 7.21b). The work done by an external torque in rotating it to $\alpha=\theta$ against $mB\sin\alpha$ is $-mB\cos\theta$. The potential energy is therefore similar to (3.44):

magnetic dipole in **B**: $$U=-mB\cos\theta=-\mathbf{m}\cdot\mathbf{B} \tag{7.43}$$

We can now use $F_L=-\partial U/\partial L$ and $T_\theta=-\partial U/\partial\theta$ to find the force in any direction and the torque about any axis. For a uniform **B**-field, we naturally recover our starting formulae: $F=0$, $T=mB\sin\theta$. In a non-uniform field, although the torque is still the same, the force is given by an expression similar to (3.43): the x component is $m_x\partial B_x/\partial x+m_y\partial B_x/\partial y+m_z\partial B_x/\partial z$. This is discussed in Sec. 8.6.

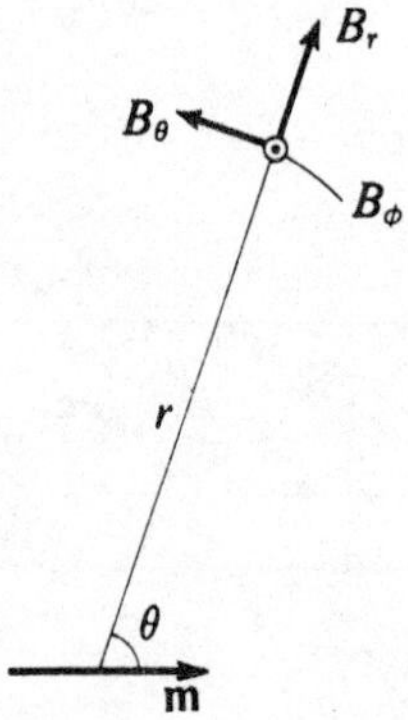

Figure 7.22 Components of the **B**-field due to a magnetic dipole.

B-fields due to magnetic dipoles In order to specify the **B**-field at a field point P produced by a dipole of moment **m**, we use the geometry of Fig. 7.22. The similarity with the electric dipole would lead us to expect expressions for the components of **B** like E_r and E_θ due to **p**, but we need to prove this.

Suppose **m** is a small current-loop dipole. If we resolve it into two components, one along $r(m\cos\theta)$ and one perpendicular to $r(m\sin\theta)$, can we find the separate **B**-fields of the components and add them? The $m\cos\theta$ component is simple, because we have already obtained the **B**-field on the axis of a circular coil at a great distance in Eq. (7.15) and found it to be $\mu_0 2IA/(4\pi r^3)$ along the axis. Hence, $m\cos\theta$ produces $B_r=\mu_0 2m\cos\theta/(4\pi r^3)$, $B_\theta=0$, $B_\phi=0$. The $m\sin\theta$ component is more difficult, but by taking a small rectangular loop as in Problem 7.20 it is possible to show that it produces $B_r=0$, $B_\theta=\mu_0 m\sin\theta/(4\pi r^3)$, $B_\phi=0$. Hence, in general,

field due to magnetic dipole: $$B_r=\mu_0\frac{2m\cos\theta}{4\pi r^3};\qquad B_\theta=\mu_0\frac{m\sin\theta}{4\pi r^3};\qquad B_\phi=0 \tag{7.44}$$

in exact analogy with the **E**-field of an electric dipole. It follows that the forces and torques between magnetic dipoles are given by the formulae of Fig. 3.23 with the substitutions $\varepsilon_0\to 1/\mu_0$, $p\to m$.

The use of $\mathbf{B}=-\mu_0\nabla V_m$ coupled with (7.44) means that the magnetic scalar potential due to a dipole is

magnetic dipole: $$V_m=\frac{m\cos\theta}{4\pi r^2}=\frac{\mathbf{m}\cdot\hat{\mathbf{r}}}{4\pi r^2} \tag{7.45}$$

or $IA\cos\theta/(4\pi r^2)$ for a current loop.

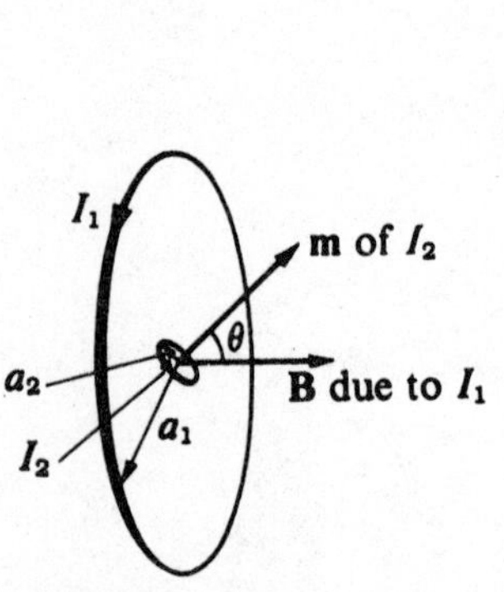

Figure 7.23 A small circular current-carrying coil at the centre of a large one can be treated as a dipole for calculating forces and torques.

Worked example *dynamometer principle* When one or both of two current-carrying circuits are small compared with their distance apart, the dipole formulae may be used. For example, in Fig. 7.23, the small coil at the centre can be treated as a dipole of moment $I_2\pi a_2^2$ at θ to a field $\mu_0 I_1/2a_1$, producing a torque

$$T_\theta=\frac{\mu_0 I_1 I_2\pi a_2^2\sin\theta}{2a_1} \tag{7.46}$$

7.9 Large circuits and their equivalent magnetic shells

To adapt the dipole formulae so that they are suitable for use with large circuits, we use the device of an **equivalent magnetic shell**. Consider the simple circuit of Fig. 7.24a, ignoring the field point P for the moment. We know from Ampère's experiments that the leads play no part in magnetic interactions so we can ignore them if they are closely wound together. The same experiments allow us to introduce a fine mesh of internal currents as in Fig. 7.24b, where each small rectangular loop carries a current I in the same sense. All the internal wires carry equal currents in opposite directions from adjacent loops, so they can neither experience any resultant force nor produce any field. The only unbalanced currents are those in the original outer circuit. It follows that the forces on, and the field due to, the original circuit are the same as those on and due to the mesh.

Because each loop of area dS is a small dipole of moment I dS, the whole mesh is like a magnetized sheet covered with N poles on one side and S poles on the other: hence the term 'magnetic shell'. Note particularly that only the periphery of the shell is important, and apart from that it may be of any shape.

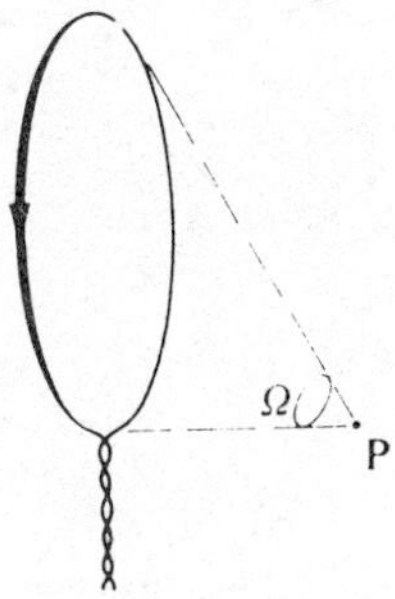

(a)

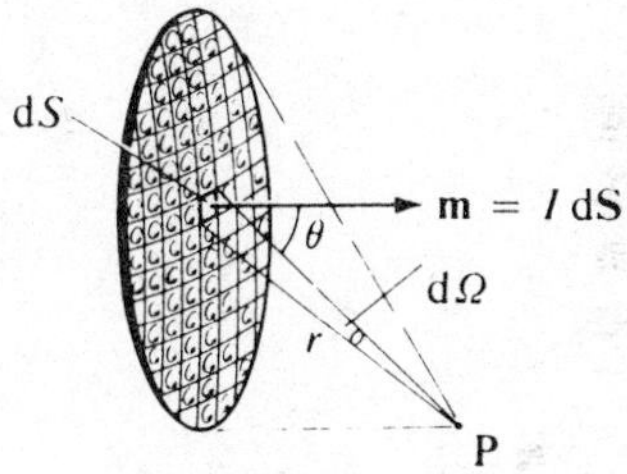

(b)

Figure 7.24 (a) A circuit carrying a current I: the leads have no magnetic effects; (b) a net in which each mesh carries the same current I in the same sense: its magnetic field at P is the same as that of (a) and it forms an equivalent magnetic shell. As long as the periphery is the same in the two cases, any bulging out of the mesh is immaterial.

Large circuits in B-fields The potential energy of a dipole **m** in a **B**-field has been shown to be $-\mathbf{m}\cdot\mathbf{B}$ (7.43). For each small loop of a shell, the moment is I dS and so its potential energy is $\mathrm{d}U=-I\mathbf{B}\cdot\mathrm{d}\mathbf{S}$. However, $\mathbf{B}\cdot\mathrm{d}\mathbf{S}$ is the external magnetic flux $\mathrm{d}\Phi$ linking the small loop, so $\mathrm{d}U=-I\,\mathrm{d}\Phi$. The potential energy of the whole shell and thus of the circuit is therefore

circuit in **B**-field
$$\boxed{U=-I\Phi} \tag{7.47}$$

From this, we can calculate forces and torques assuming, as does the derivation of (7.43), that the currents are invariant. Thus

circuit in **B**-field:
$$\boxed{F_L=-\left(\frac{\partial U}{\partial L}\right)_I=I\left(\frac{\partial \Phi}{\partial L}\right)_I;\quad T_\theta=-\left(\frac{\partial U}{\partial \theta}\right)_I=I\left(\frac{\partial \Phi}{\partial \theta}\right)_I} \tag{7.48}$$

These equations show that a circuit will be in equilibrium ($\mathbf{F}=\mathbf{T}=0$) when Φ has a stationary value. The equilibrium will be stable if a small displacement brings into play restoring forces or torques, i.e. if the flux decreases away from the point of equilibrium. Thus *for stable equilibrium a circuit embraces maximum flux*, and in any other position the forces or torques will act in such ways as to cause an increase in flux linkage. The reader should have little difficulty in applying these results to the coils of Fig. 7.6 and showing that the results are the same as those yielded by current element formulae in Eqs (7.5)–(7.7).

Although U is the mechanical potential energy of a circuit in a **B**-field **we cannot identify it with the magnetic energy**. We shall see in Chapter 9 that to maintain a current constant when it moves in a **B**-field requires an input of energy from sources of electromotance and the complete energy balance must include this.

B-fields due to large circuits We now turn to the problem of calculating the **B**-field at a field point P (Fig. 7.24a) due to a large circuit. By our previous arguments, this field will be the same as that of the equivalent shell in Fig. 7.24b.

We use the magnetic scalar potential due to a dipole given by (7.45) as $m\cos\theta/(4\pi r^2)$. The value of V_m at P due to a typical loop of moment $I\,\mathrm{d}\mathbf{S}$ is $I\,\mathrm{d}S\cos\theta/(4\pi r^2)$, but $\mathrm{d}S\cos\theta/r^2$ is the solid angle subtended at P by the loop denoted by $\mathrm{d}\Omega$ (Appendix A.1) so $\mathrm{d}V_m = I\,\mathrm{d}\Omega/4\pi$. Summing over all the loops gives

$$\text{potential due to circuit:} \quad \boxed{V_m = I\Omega/4\pi} \tag{7.49}$$

where Ω is the solid angle subtended at P by the circuit. The **B**-field is then obtained from $B_x = -\mu_0 \partial V_m/\partial x$, etc. To obtain the correct sign for V_m in (7.49)—namely positive on the N pole side—we must allocate a sign to Ω. The convention is adopted that Ω is positive at P if the current viewed from P appears to flow in a counterclockwise sense. Notice that the potential depends only on the periphery of the shell and the current, but not on the shape.

The importance of (7.49) lies in the use we make of it in the next chapter. As a means of calculating **B**-fields it suffers from the difficulty of finding solid angles. One example that is tractable is the circular coil of Fig. 7.9a, where the solid angle subtended at P by the coil is $2\pi(1-\cos\theta)$ as in Appendix A.1. V_m at P is thus $\frac{1}{2}I[1-x(x^2+a^2)^{-1/2}]$ and $B_x = -\mu_0\partial V_m/\partial x$ yields (7.13) as before.

7.10 Monopoles

Concept of a magnetic pole Great emphasis has been placed in this chapter on the analogy between magnetic and electric dipoles. An obvious extension of this suggests that a magnet could possess two equal and opposite monopoles, or simply poles, just as the electric dipole consists of two equal and opposite charges. The idea is reinforced by the shape of the field lines in Fig. 7.1a, which seems to locate separate N and S poles near the ends of the magnet. In older treatments of magnetism, the pole was the starting point from which the rest of the theory was developed. Here we merely show how the laws governing interactions in terms of poles are derived from previous results and we end with a discussion of the merits and demerits of the approach.

Pole strength and the inverse square law Assuming that a permanent dipole may be replaced by two poles separated by a displacement **l**, we define pole strength P so that the dipole moment is

$$\mathbf{m} = P\mathbf{l} \qquad \text{(definition of } P) \tag{7.50}$$

The **SI unit** of the pole is the A m.

We know that N or positive poles tend to move in the direction of **B** and S or negative poles in the opposite direction, so that when a dipole is placed in a field as in Fig. 7.25 the forces on the poles form a couple. Because we know that the torque is $mB\sin\alpha$ or $PlB\sin\alpha$, each force of the couple must be

$$\mathbf{F} = P\mathbf{B} \tag{7.51}$$

and **B** is thus the *force per unit pole*.

The **B**-field due to a dipole is given by (7.44) and, if it originates from a central field due to two unlike equal poles, the field due to a single pole must be

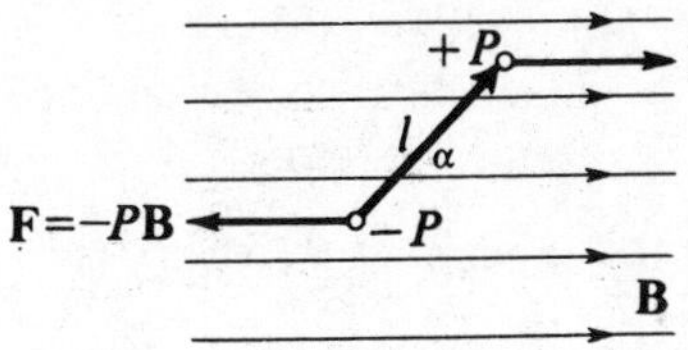

Figure 7.25 A magnetic dipole treated as two equal and opposite monopoles.

$$\mathbf{B}_{\text{pole}} = \mu_0 P\hat{\mathbf{r}}/(4\pi r^2) \tag{7.52}$$

Equations (7.51) and (7.52) combine to give a law of force between poles

$$\mathbf{F} = \mu_0 P_1 P_2 \hat{\mathbf{r}}/(4\pi r^2) \tag{7.53}$$

a law which can be approximately verified directly, using long ball-ended magnets so that the poles are more accurately located.

Merits and limitations of the pole concept Clearly the law (7.51) is so exactly similar to Coulomb's law for the force between charges that all appropriate electrostatic results can be taken over. In particular, it enables a physical interpretation to be given to V_m: the work done per unit pole in taking a pole along a path is μ_0 times V_m. It also reveals that the **B**-field due to poles obeys the circuital law $\oint \mathbf{B} \cdot d\mathbf{L} = 0$. However, there is no analogy to the conductor of charge so no results involving conducting media can be used. Moreover, in the real world only dipoles exist, not single poles, so that a law of conservation of pole strength would simply state that $\Sigma P = 0$, while Gauss's law would be just $\oint \mathbf{B} \cdot d\mathbf{S} = 0$. For the purposes of calculation we can, of course, talk of a distribution of, say, N poles as long as we remember that actual systems consist entirely of dipoles. It will become clear in Chapter 12 that a magnet is more akin to a polarized dielectric than to a charged body.

Calculations using poles can be a useful method provided the limitations are kept in mind. These are, firstly, that practical *permanent* dipoles do not exist: the moment of a magnet is affected by other fields including the earth's (although isolated elementary particles such as electrons or neutrons are permanent dipoles). Secondly, the location of the pole and thus, by (7.50), its strength, are indefinite and only the magnetic dipole moment is an accurately measurable quantity. Thirdly, evidence which we shall consider later all points to the association of angular momentum with magnetic moment and the current loop is a more appropriate model.

Comment C7.2

7.11 Conclusion

Just as with the **E**-field, we now go on to look more thoroughly at the field properties of **B** in order both to solve more complex problems and to advance the theory. Although we have been using the experience gained in handling the **E**-field in our treatment of magnetism, we can already see contrasts as well as similarities. Note the following, for example:

Like charges repel	Like currents attract
Force on Q is in the same direction as **E**	Force on I is perpendicular to **B**

Conductors exist in which **E** is zero when charges are static	No medium exists in magnetism corresponding to a conductor

We shall discover further contrasts in the next chapter when the magnetic circuital law and Gauss's law are examined.

7.12 Revision summary

Magnetic forces between currents, magnetic B-field The laws of magnetic interaction between currents or magnets can be established in terms of one of four different basic elements: the current element, the moving charge, the magnetic dipole, the monopole. Whichever approach is adopted, the magnetic field **B** is introduced at an early stage to simplify the expression of the laws. The procedure is then

- to define the element,
- to define the **B**-field in terms of a force or torque on a unit element, leading to formulae giving the action of **B**-fields *on* systems,
- to establish the **B**-field produced *by* the element, leading to calculations of fields due to large systems by integration or summation over elements.

We have chosen the current element as basic. The fundamental law of force between $I_1\,\mathrm{d}\mathbf{l}_1$ and $I_2\,\mathrm{d}\mathbf{l}_2$ separated by $\mathbf{r}_{12}$ is then as follows, and is split into two parts by using **B** just as Coulomb's law was, using **E**:

$$\mathrm{d}^2\mathbf{F}_{12}=\frac{\mu_0 I_1 I_2\,\mathrm{d}\mathbf{l}_2\times(\mathrm{d}\mathbf{l}_1\times\hat{\mathbf{r}}_{12})}{4\pi r_{12}^2} \tag{7.20}$$

$$\mathrm{d}\mathbf{F}=I\,\mathrm{d}\mathbf{l}\times\mathbf{B} \quad (7.2) \qquad \mathrm{d}\mathbf{B}_{\text{element}}=\frac{\mu_0 I\,\mathrm{d}\mathbf{l}\times\hat{\mathbf{r}}}{4\pi r^2} \tag{7.9}$$

where μ_0 takes the value $4\pi\times10^{-7}\,\mathrm{H\,m^{-1}}$ because of the definition of the ampere and where the **principle of superposition** applies.

The **B**-field is thus defined as follows: its direction is that in which a current element experiences zero force; its magnitude is the force per unit element perpendicular to that direction. Its **SI unit** is the *tesla*, T. It is also sometimes called the *magnetic induction* or *magnetic flux density*. Like **E**, it is a vector field and is capable of being represented by lines of force. It is an inverse square field but is *not central.*

Basic element	Element in B-field (from definition of **B**)		**B**-field due to element	
Current element, $I\,\mathrm{d}\mathbf{l}$	$\mathrm{d}\mathbf{F}=I\,\mathrm{d}\mathbf{l}\times\mathbf{B}$ (Lorentz force)		$\mathrm{d}\mathbf{B}=\mu_0 I\,\mathrm{d}\mathbf{l}\times\hat{\mathbf{r}}/(4\pi r^2)$ (Biot–Savart law)	
Moving charge, $Q\mathbf{v}$	$\mathbf{F}=Q\mathbf{v}\times\mathbf{B}$	(7.25)	$\mathbf{B}=\mu_0 Q\mathbf{v}\times\hat{\mathbf{r}}/(4\pi r^2)$	(7.27)
Magnetic dipole, **m** (=$I\mathbf{A}$ for current loop)	$\mathbf{T}=\mathbf{m}\times\mathbf{B}$	(7.42)	$\mathbf{B}=\mu_0\dfrac{2m\cos\theta}{4\pi r^3}\hat{\mathbf{r}}+\mu_0\dfrac{m\sin\theta}{4\pi r^3}\hat{\boldsymbol{\theta}}$	(7.44)
Monopole, P	$\mathbf{F}=P\mathbf{B}$	(7.51)	$\mathbf{B}=\mu_0 P\hat{\mathbf{r}}/(4\pi r^2)$	(7.52)

The table opposite summarizes the magnetic interaction laws for each of the possible approaches, using the **B**-field.

Magnetic scalar potential V_m between two points is defined in an analogous way to electrostatic potential as the line integral of $-\mathbf{B}/\mu_0$. If A is chosen as the zero, V_m at a point B is thus

$$V_m = \int_A^B \frac{-\mathbf{B}\cdot d\mathbf{L}}{\mu_0} \tag{7.35}$$

so that the **SI unit** of V_m is the A. The magnetic field **B** is given by $B_x = -\mu_0\,\partial V_m/\partial x$, etc., or in general by

$$\mathbf{B} = -\mu_0 \nabla V_m \tag{7.37}$$

This potential plays a much smaller role in magnetism than V in electrostatics for reasons that become clear in the next chapter.

Magnetic flux Φ is defined over an area S as the surface integral of **B** over the area, i.e.

$$\Phi_S = \int_S \mathbf{B}\cdot d\mathbf{S} \tag{7.39}$$

The **SI unit** of Φ is the *weber*, Wb.

Solution of problems: systems in B-fields Depending on the system, there is a choice of the following methods:

- For any filamentary current: the use of $d\mathbf{F} = I\,d\mathbf{l}\times\mathbf{B}$ integrated over all elements. If the current is continuously distributed throughout a volume, replace $I\,d\mathbf{l}$ by $\mathbf{J}\,d\tau$. This method yields, for example,

  outside an infinite straight current: $B = \mu_0 I/(2\pi r)$
  on the axis of an infinite solenoid: $B = \mu_0 n I$

- For moving charges: use of $\mathbf{F} = Q\mathbf{v}\times\mathbf{B}$ or $Q(\mathbf{E}+\mathbf{v}\times\mathbf{B})$ if **E** is present as well. A particularly important result is the path of a charge Q in a uniform **B** with linear momentum p perpendicular to the field: a circle of radius $r = p/QB$, covered at an angular *cyclotron frequency* $\omega_c = QB/m$.
- For small circuits: use of dipole formulae, including potential energy of dipole in **B**: $U = -\mathbf{m}\cdot\mathbf{B}$, giving forces and torques. Large magnets by integration over dipole elements (Chapter 12).
- For any circuit: use of equivalent shell formulae. Potential energy of circuit $U = -I\Phi$, where Φ is external flux. Forces and torques from $F_L = I(\partial\Phi/\partial_L)$, etc.

Solution of problems: B-fields due to systems The following methods are available using the third column of the table above:

- For any filamentary circuit: the use of Biot–Savart integrated over all elements. Replace $I\,d\mathbf{l}$ by $\mathbf{J}\,d\tau$ for volume distributions.
- For moving charges: the use of $\mathbf{B} = \mu_0 Q\mathbf{v}\times\hat{\mathbf{r}}/(4\pi r^2)$.
- For small circuits: use of dipole formulae, including scalar potential due to **m**, $V_m = m\cos\theta/(4\pi r^2)$.

- For magnets: use of dipole formulae, integrated over the volume of the magnet if necessary. Or use monopoles. (See Chapter 12.)
- For any circuit: use of equivalent shell formulae, including scalar potential at field point P: $V_m = I\Omega/4\pi$, where Ω is the solid angle subtended at P by the circuit.

Commentary

C7.1 On the Dipole Inverse Cube Law The largest handy magnetic dipole at our disposal is the earth itself whose magnetic field provides a sensitive test of the dipole inverse cube law and thus of the inverse square part of the Biot–Savart law. Deviations from the law, if any, are expected to show in a potential V_m that falls off with distance not as $1/r^2$ but as $e^{-\lambda r}/r^2$ (see Comment C4.2 for a similar expression in connection with Coulomb's law). The constant λ is such that deviations from the $1/r^2$ law would be very small within a range $1/\lambda$. If the potential took this form, the normal dipole field of the earth should be supplemented by a uniform field over its surface with a direction along the S–N axis. Sufficient information is available about the earth's field to put an upper limit to this additional field and thus a limit on λ. Satellite measurements have allowed the fields of ionospheric and solar wind currents to be allowed for and no evidence of the additional field has been found. The upper limit to λ is about $10^{-8}\,m^{-1}$.

C7.2 On Monopoles A theory of Dirac predicts that the existence of charge quantization with an elementary charge e means that a magnetic monopole should exist, which later work showed should have a mass of about 10^{-11} kg, much greater than that of the electron. Many searches for such poles have been made in experiments with high energy particles, in cosmic ray observations and in ancient rocks and sea-bed slimes, all with no positive outcome. The monopole strength is predicted to be $h/(4\pi e)$ and its detection would involve measurements of magnetic flux changes of the order of the flux quantum $h/2e$, which can be carried out using SQUIDs (see Comment C12.2). Cabrera (1982) reported an event that could be interpreted as the passage of a monopole through his system, but concluded two years later that the lack of any further such observation made it unlikely that the observation was that of a monopole.

In spite of this lack of observational evidence, some still feel that the symmetry in nature between the sources of electric and magnetic fields is *un*natural and continue to construct theories involving poles in case the symmetry should one day be restored. We should not forget, perhaps, that ideas of symmetry are man-made concepts used to impose a structure on natural laws and that the universe, whatever else it is, is not man-made.

Information on searches for monopoles up to the dates of their papers is given in Goldhaber and Smith (1975), L. W. Jones (1977), Carrigan and Trower (1982) and Schwarzchild (1984).

C7.3 On Sommerfeld and Kennelly SI Units Some authors using SI units adopt a different definition of the moment of a current loop from ours—a situation that can cause confusion. The two possible choices are designated by the names Sommerfeld and Kennelly as suggested by the Coulomb's Law Committee of the American Association of Physics Teachers (1950). That adopted in this book is the Sommerfeld (S), while the Kennelly (K) definition of $\mathbf{m}$ is $\mu_0 I\mathbf{A}$ instead of just

IA. Of course, both systems agree on fundamentals so that the following formulae for current loop dipoles are not in dispute:

$$\mathbf{T} = I\mathbf{A} \times \mathbf{B} \qquad \text{and} \qquad B_r = \frac{\mu_0 2IA \cos\theta}{4\pi r^3}$$

However, whereas our definition of **m** makes the torque $\mathbf{m} \times \mathbf{B}$, the K definition makes it $\mathbf{m} \times \mathbf{B}/\mu_0$ which is written $\mathbf{m} \times \mathbf{H}$, using the **H**-field introduced in Chapter 12 as $\mathbf{B}/\mu_0$ *in vacuo*. Moreover, B_r becomes $2m\cos\theta/(4\pi r^3)$ and $H_r = 2m\cos\theta/(4\pi\mu_0 r^3)$. Thus we have the characteristic feature of Sommerfeld that the μ_0 is in the numerator of **B**, while in Kennelly it occurs in the denominator of **H**. It is purely a matter of definition which does not affect formulae relating to currents *in vacuo* but has an effect when magnetic media are present (see Comment C12.5).

Problems

Section 7.1

7.1 A circular steel disc is magnetized uniformly in a direction parallel to one of its diameters. How could this diameter be identified using no other electric or magnetic apparatus?

Section 7.2

7.2 A uniform straight rod of mass m per unit length is hung vertically so that it can turn about its upper end, and a current I is sent through it. What is its angular deflection from the vertical at a place where the earth's horizontal field is B_0?

7.3 Find the force on a current-carrying wire in the shape of an arc of a circle placed in a plane perpendicular to a uniform **B**-field. Show that the same force is exerted if the same current flows in the corresponding chord of the circle and generalize the result.

7.4 A light circular flexible loop of radius a carries a current I and is placed in a plane perpendicular to a uniform **B**-field. Find the tension in the loop.

Section 7.3

7.5 Find the **B**-field (a) at the centre of a square coil of side a carrying a current I, (b) a distance x along the perpendicular from the centre of the coil in (a), and (c) at the centre of a regular polygon of n sides inscribed in a circle of radius r and carrying a current I.

7.6 A long thin metal strip of width $2L$ carries a current along its length. The current is distributed uniformly across the width with a surface density J_s (Sec. 1.7). Find the **B**-field (a) at a point in the plane of the strip at a distance $R(>L)$ from its centre and (b) at a point a *perpendicular* distance R from its centre. Check that (a) yields (7.19) as $L/R \to 0$. What does (b) become if the sheet is of infinite width?

Section 7.4

7.7 Two flat coils each of 10 turns have mean radii of 20 cm and are placed coaxially 1 cm apart. Find the approximate force between them if they both carry 5 A in the same sense.

Section 7.5

7.8 A wire is bent into a circle of radius a and a total charge Q is distributed uniformly around it. Find the **B**-field at the centre if the wire rotates with an angular velocity ω about the axis of the circle.

7.9 An insulating circular disc of radius a is sprayed with charge to a uniform surface density σ. If the disc rotates about its axis with an angular velocity ω, what is the **B**-field at the centre?

7.10 What is the radius of the track of a 2 MeV proton in a field of 1 T?

7.11 Calculate the cyclotron frequency for electrons in a **B**-field of 0.2 T.

Section 7.6

7.12 A stream of charged particles with various velocities is projected in a direction at right angles both to an **E**-field of 4 V cm^{-1} and a **B**-field of 10^{-3} T. What is the speed of undeflected particles? What energy have they if they are protons?

7.13 Photoelectrons are liberated from a plate by ultraviolet radiation, their initial velocity being negligible. A magnetic field **B** is maintained parallel to the plate and an **E**-field perpendicular to it. The **E**-field is produced by a second plate parallel to the first, a distance d from it and at a positive potential V with respect to it. Show that the value of d for which the current just fails to pass between the plates is $(2m_e V/eB^2)^{1/2}$, where e and m_e are the charge and mass of the electron.

7.14 The Hall angle θ_H is the angle between the direction of current flow and the resultant **E** of Fig. 7.20b. Show that $\tan\theta_H = \mu B$, where μ is the mobility of the carriers, and that θ_H for copper is always very small.

Section 7.7

7.15 A long straight wire carrying a current I lies in the same plane as a rectangular loop with sides of length a and b. The sides a are parallel to the wire and distances d and $d+b$ from it. What is the total magnetic flux linking the loop?

Section 7.8

7.16 Assuming that the earth's magnetic field is the same as that of a magnetic dipole at the centre with its axis along that of the earth, show that the angle of dip δ at a point whose latitude is λ is given by $\tan\delta = 2\tan\lambda$.

7.17 Modern magnetometers for measuring **B**-fields are described in Comment C12.2. In the older vibration magnetometer, a small permanent dipole was allowed to perform small angular oscillations in a horizontal plane first in a standard field $\mathbf{B}_0$ and then in the unknown **B**. If the periods of oscillation were, respectively, T_0 and T, show that $B = B_0 T_0^2/T^2$.

7.18 Find the magnetic moment of a circular coil of radius 0.5 cm carrying a current of 1 A in 10 turns. What is the force between two such coils arranged coaxially 10 cm apart?

7.19 Two flat coils each of 10 turns have mean radii of 20 and 2 cm. Find an approximate value for the force between them if they are coaxial, 10 cm apart and if each carries 5 A in the same sense.

7.20 Take a small rectangular current loop of sides dx, dy situated at the origin in the xy plane and carrying a current I round its four sides. Find the **B**-field at a field point P a great distance from the loop along the x axis. Show that the field has the form $\mu_0 m/(4\pi x^3)$, where m is its dipole moment. (Note that all four sides produce a field.)

7.21 A magnetic dipole of moment **m** is situated at the origin of polar coordinates and lies along the $\theta=0$ axis. Find the magnetic flux across the base of a cone of side r and semi-vertical angle θ which has its apex at the origin and its axis along the direction of **m**.

7.22 A particle with charge e and mass m is projected with linear momentum p at right angles to a uniform magnetic field B. What are the magnetic dipole moment and angular momentum possessed by the particle as a result of its subsequent motion?

Section 7.9

7.23 Expose the fallacy in the following. The magnetic scalar potential at a point in the same plane as a small current-carrying loop, but outside it, is zero because the solid angle is zero. Hence the magnetic **B**-field is zero and a current-loop dipole produces no broadside field.

Section 7.10

7.24* A sphere of radius a is magnetized uniformly parallel to a diameter so that its magnetic moment per unit volume is M. Find the **B**-field at a distance r from the centre for $r>a$ and $r<a$. (The least laborious method is to treat the sphere of dipoles as two spheres of poles of opposite sign slightly displaced from each other. The **B**-fields due to each sphere are given by expressions similar to those derived for a sphere of charge in Chapter 4.)

8

Further properties of steady magnetic fields

In Chapter 7 we concentrated on establishing the basic laws and concepts in magnetism, and discovered at the same time how to solve many simple problems. We now proceed to a further study of steady **B**-fields on similar lines to that of **E**-fields in Chapter 4 but with significant differences.

We remember that Coulomb's law led immediately to the expression $Q\hat{\mathbf{r}}/(4\pi\varepsilon_0 r^2)$ for the **E**-field due to a charge, and that Gauss's law and the circuital law for **E** followed from it. In contrast to that, there have been several expressions for **B** due to currents in Chapter 7, any of which may be taken as embodying the fundamental magnetic laws. As long as the expressions are all equivalent, we are entitled to use whichever is convenient for our purpose. We therefore begin in Sec. 8.1 with a derivation of the circuital law for **B**, dealing first with the special case of an infinite straight filament of current to illuminate what is happening and then proceeding to a more general proof. The latter could be based on the formula for the field due to a current element, which we took as the fundamental law in Chapter 7, but such a proof is mathematically complex and we prefer one based on the equivalent magnetic shell of a large circuit. However, the current element formula *is* used in Sec. 8.2 to establish a Gauss's law. In both laws, we are careful to include all sources of **B** such as permanent magnet dipoles as well as currents.

In Secs 8.3 and 8.4 the differential forms of the two laws are obtained in terms of the curl and divergence and we note the contrast between the properties of steady

E- and **B**-fields. This contrast is very marked when it comes to the question of potentials and in Sec. 8.5 we look at a new one that can be used universally for **B**-fields: the vector potential. The section could be omitted in a first-level course.

While the first five sections concern the properties of **B**-fields arising from various sources, the next two give further consideration to what happens when a **B**-field is applied to a magnetic dipole. The fact that such a dipole is associated with angular momentum means that the resultant torque produces an effect that is completely different from the corresponding electric case, with important consequences at the atomic level.

8.1 Ampère's circuital law (integral form)

Circulation and magnetomotance We first encountered the concept of the circulation of a field when examining the properties of the **E**-field in Chapter 4. We saw there that, when **E** arises only from charges, the line integral round any closed path is zero. This line integral, $\oint \mathbf{E}\cdot d\mathbf{L}$, we called the *circulation* of **E**. We also saw that there must exist other **E**-fields for which the circulation is not zero, and this led to the definition of electromotance or e.m.f. in general as the value of $\oint \mathbf{E}\cdot d\mathbf{L}$ round any closed path.

It is clear from the geometry of some of the **B**-fields in Chapter 7 that the circulation of **B**, defined as $\oint \mathbf{B}\cdot d\mathbf{L}$, is unlikely to be zero when steady currents are the source of the field. We therefore begin the further development of **B**-field properties by looking at the circuital law for **B**, rather than starting with a Gauss's law as we did for **E**.

By analogy with the electrical case, a **magnetomotance** or **magnetomotive force** (m.m.f.) can be defined, but here we must be careful to specify that the closed path L must be *in vacuo*, since the definition will be generalized in Chapter 12. Thus:

$$\mathscr{H} = \oint_L \frac{\mathbf{B}\cdot d\mathbf{L}}{\mu_0} \qquad \text{(definition of } \mathscr{H} \textit{ in vacuo}) \qquad (8.1)$$

The **SI unit** for $\mathscr{H}$, like V_m, is the A.

Derivation of Ampère's circuital law: special case Consider the **B**-field due to a current I flowing in an infinitely long straight wire. In Chapter 7, we showed that the field lines were concentric circles about the centre of the wire (Fig. 7.1c) and that the **B**-field in magnitude was given by $\mu_0 I/(2\pi r)$ [Eq. (7.19)]. Figure 8.1a illustrates the situation. If a closed path L is taken so that it coincides with the **B**-line at r from the wire, then the circulation of **B** round L is simply the length of the path $2\pi r$ multiplied by B, since d**L** and **B** are always in the same direction. Thus the circulation of **B** round L is $\mu_0 I$ (the magnetomotance is simply I). We now give a more general proof of this result by using the dipole approach of Chapter 7 in the form of the equivalent magnetic shell.

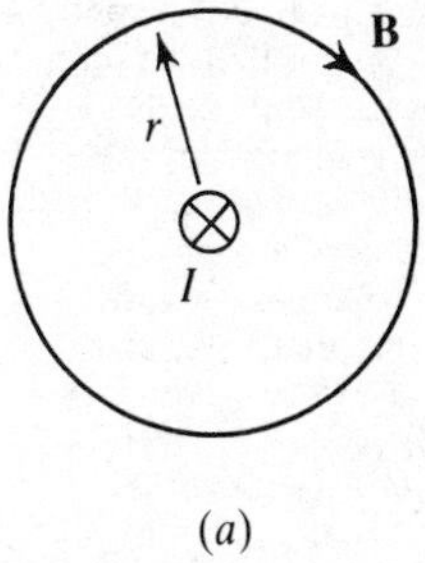

(*a*)

Figure 8.1 (*a*) Derivation of Ampère's circuital law. (a) A special case: the **B**-field due to an infinite current.

Derivation of Ampère's circuital law for steady currents: general proof Figures 8.1b and c show a cross-section through a filamentary current-carrying circuit whose plane cuts the page. In order to evaluate the line integral of **B** round a closed path, we shall first calculate it along two different paths between A and B:

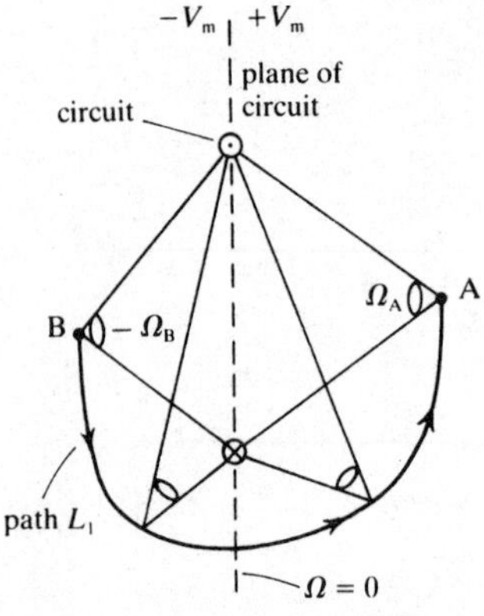

(b)

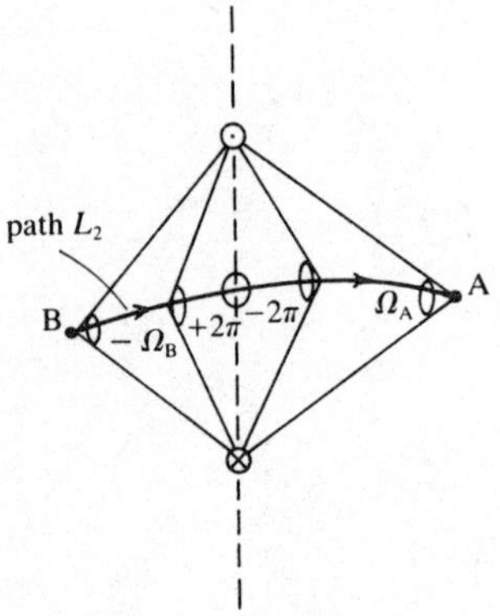

(c)

Figure 8.1 (continued) (b) general method: evaluation of magnetic potential difference between A and B along path L_1 using the equivalent magnetic shell; (c) the same along path L_2. The marked solid angles show how Ω_A and $-\Omega_B$ change along the two paths. Note that solid angles viewed from the negative side are positive (e.g. Ω_A and $+2\pi$) and vice versa.

L_1 in Fig. 8.1b and L_2 in Fig. 8.1c. We then reverse one path and add the contributions, thus completing a closed path.

From (7.35), we know that the line integral of **B** from one point to another is given by μ_0 times the difference in magnetic potential V_m between the two points. We therefore need to evaluate this difference, say ΔV_m, along L_1 and L_2. To do this, we use the expression (7.49), which tells us that the value of V_m at any point is $I\Omega/4\pi$, where Ω is the solid angle subtended by the circuit at the point. Let the solid angles at A and B be Ω_A and $-\Omega_B$, respectively, the one at B being negative by the sign convention of Sec. 7.9.

Consider first the path L_1, for which ΔV_m will be the *change* in Ω along the path from B to A multiplied by $I/4\pi$. The solid angle *increases* from $-\Omega_B$ to zero at some point, and then further *increases* to $+\Omega_A$. This gives a total *increase* of $\Omega_A+\Omega_B$. Hence

$$\Delta V_m = I(\Omega_A+\Omega_B)/4\pi \qquad \text{along } L_1 \tag{8.2a}$$

Now consider path L_2 (Fig. 8.1c). The solid angle first *decreases* from $-\Omega_B$ to -2π at some point and then, because we must change sides, further *decreases* from $+2\pi$ to Ω_A. This gives a total *decrease* of $4\pi-(\Omega_A+\Omega_B)$. Hence

$$\Delta V_m = I(\Omega_A+\Omega_B-4\pi)/4\pi \qquad \text{along } L_2 \tag{8.2b}$$

which differs from (8.2a) by I. Indeed, by choosing further paths that form more loops round the current, an infinite set of values for the magnetic potential difference can be found, all differing from (8.1a) by a multiple of I. *The magnetic scalar potential due to a current is not single-valued nor path-independent*, a fact we might have suspected from (7.49) since Ω cannot be distinguished from $\Omega+4\pi$.

Equation (8.1a) gives us the line integral of $\mathbf{B}/\mu_0$ from B to A along L_1, while the negative of (8.2b) gives us the line integral from A to B along L_2, thus completing a closed path encircling the current once. Adding (8.2a) to the negative of (8.2b) gives I, so that, in general,

Ampère's law:
$$\oint_L \mathbf{B}\cdot d\mathbf{L} = \mu_0 I \tag{8.3}$$

for a closed path L encircling a current I once only. This is known as **Ampère's circuital law** or the **circuital law for B**. We shall call the path L an **Amperian path** (just as the closed surface in Gauss's law is called a Gaussian surface). If some physical picture of (8.3) is required, the best we can do is to interpret the left-hand side as the work done per unit pole in taking a pole round L.

Derivation: dipole sources If **B** should arise from permanent dipoles rather than conduction currents, the laws of Chapter 7 show that $\oint \mathbf{B}\cdot d\mathbf{L}$ is zero exactly as it is for the **E**-field of electric dipoles, provided that path L does not intersect a dipole. Since permanent dipoles exist only as pieces of magnetic material, a path intersecting a dipole would not be entirely *in vacuo* and would then be the subject of Chapter 12. It follows that (8.3) is unaffected if **B** arises from both currents *and* dipoles, since the latter contribute nothing to the right-hand side when L is *in vacuo*.

Application of the law The use of (8.3) is similar to that of Gauss's law in electrostatics: its main value lies in summarizing the properties of magnetic fields and in the development of theory. As a method of calculating **B** it is restricted to highly symmetrical cases where the Amperian path can usually be chosen so that **B** and d**L** are parallel and B is constant. The left-hand side is then simply the product of B and the path length.

Although not single-valued, magnetic potential difference does not become useless because the 4π's which can be added to it do not enter B after differentiation. Moreover, if we restrict any paths by imagining a barrier across the circuit carrying the current, the magnetomotance always equals 0, and V_m becomes single-valued. This is clearly impossible inside a mass of material carrying a volume distribution of current and here V_m can never be used.

Note finally that (8.3) is consistent with the fact that lines of **B** *may* close on themselves, unlike those of **E** produced by static charges.

Worked example *B-field due to a long straight current-carrying wire* A straight wire of infinite length and circular section of radius a carrying a steady current I (Fig. 8.2) has not only cylindrical symmetry about the axis but constancy along the axis as well. Thus to obtain B at P a distance r from the axis outside the wire, denote by B_θ the component along the Amperian path L in the figure. B_θ is the same at all points on L by the symmetry, and the circuital theorem therefore gives

$$2\pi r B_\theta = \mu_0 I$$

or

infinite current, $r \geq a$:
$$B_\theta = \frac{\mu_0 I}{2\pi r} \tag{8.4}$$

This of course is identical with (7.19), but we can go further now.

If P is *inside* the wire and the current is distributed uniformly over the cross-section, the path L encircles a current Ir^2/a^2 and

$$2\pi r B_\theta = \mu_0 I r^2/a^2$$

or

infinite current, $r \leq a$:
$$B_\theta = \frac{\mu_0 I r}{2\pi a^2} \tag{8.5}$$

The variation of B_θ with r is plotted in Fig. 8.2. Gauss's law (Sec. 8.2) shows that B_r and B_z must be zero.

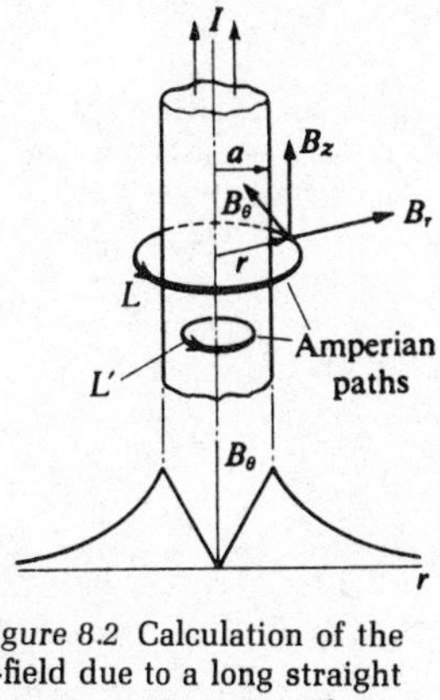

Figure 8.2 Calculation of the B-field due to a long straight current-carrying wire. The lower diagram shows how B varies with distance from the axis of the wire.

Worked example *B-field due to a toroidal coil* Applying the circuital law to the Amperian path L within a closely wound toroidal coil with n turns per unit length (Fig. 8.3) gives

$$2\pi r B = \mu_0 2\pi a n I$$

where B is the magnetic field at a distance r from the centre of the toroid. (Arguments from symmetry show that the direction of B must be tangential to the circle of radius r.) It then follows that

$$B = \mu_0 n I a / r \tag{8.6}$$

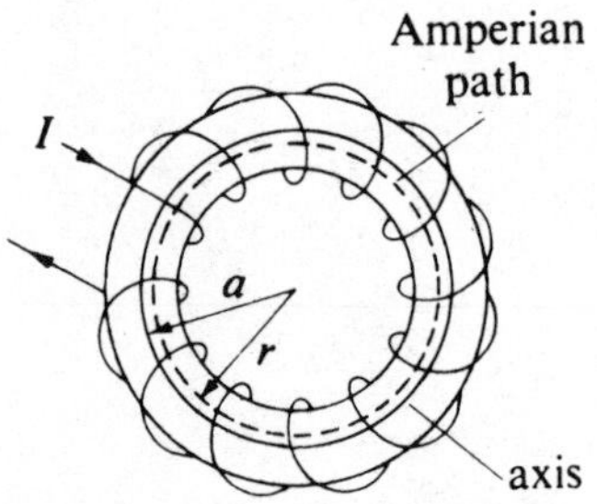

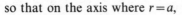

Figure 8.3 Calculation of the **B**-field due to a toroidal coil.

so that on the axis where $r=a$,

axis of toroidal coil: $$B=\mu_0 nI \qquad (8.7)$$

Since an infinite solenoid can be regarded as a toroid of infinite radius, Eq. (8.7) also gives the **B**-field on the axis of such a solenoid, which is consistent with (7.17). However, we can now go further than (7.17) and use the circuital theorem to show that $B=\mu_0 nI$ gives the field at *any* point in the interior of an infinite solenoid and not just on its axis (Problem 8.6) and finally that $B=0$ outside.

8.2 Gauss's law for B (integral form)

Derivation: steady currents Figure 8.4 shows a current element within a closed surface S. For a given r and θ, B is constant round a path such as L and a small tube with L as axis cuts S in two places so that at one intersection B is inwards and at the other it is outwards. Because B is the same, the flux across the two intersections is equal whatever the angle at which S is cut by the tube and the resultant outward flux is therefore zero. Any system of currents can be split into similar elements. The flux of B over any closed surface due to any system of steady currents is therefore zero.

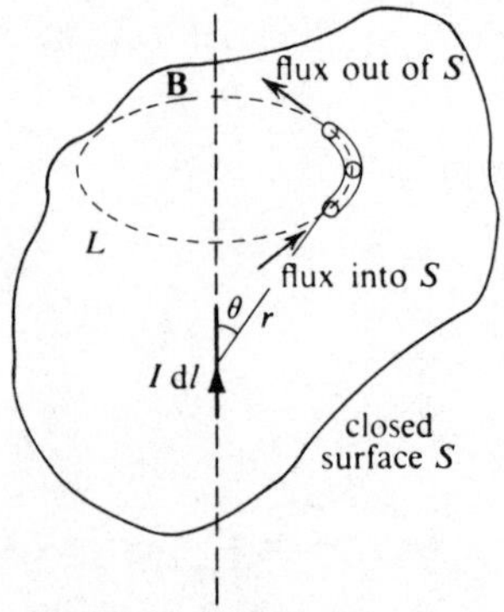

Figure 8.4 Derivation of Gauss's law for **B** due to a steady current.

Derivation: dipole sources Any closed surface S enclosing *complete* dipoles has a value for **B** at every point on it given by the same expression (7.44) as that for **E** due to electric dipoles (apart from trivial substitutions). In the electric case, the resultant charge inside S due to any number of dipoles is zero and hence the flux of **E** over S is also zero by Gauss's electrostatic law. It follows that a calculation of the flux of **B** over S would yield the same result, that is $\oint \mathbf{B}\cdot d\mathbf{S}=0$. Should S pass through any permanent dipoles, they could only form part of a piece of magnetic material which we deal with by the methods of Chapter 12.

We can thus say that for *any* magnetic sources, the flux of **B** over any closed surface entirely *in vacuo* is zero or

Gauss's law for **B**: $$\boxed{\oint_S \mathbf{B}\cdot d\mathbf{S}=0} \qquad (8.8)$$

which is the integral form of Gauss's law for **B**. The corresponding law for **E**-fields showed that charges acted as sources or sinks, i.e. flux lines of **E** began and ended on charges. Equation (8.8) shows that there are no similar sources or sinks of magnetic flux and is thus consistent with the non-existence of isolated magnetic monopoles.

Although Gauss's law has few applications in solving specific problems, the zero resultant flux has as much general significance as the zero circulation of **E** in Chapter 4. We shall examine this aspect in Sec. 8.5.

8.3 Differential form of the circuital law: curl B

Section 4.6 showed how the circuital law for **E** ($\oint \mathbf{E}\cdot d\mathbf{L}=0$ applying round a closed *path*) could be expressed in a differential form applying at a *point*. The latter form was very concisely written as $\nabla\times\mathbf{E}=0$ or curl $\mathbf{E}=0$, using the ∇ operator (Appendices B.8 and B.9).

The corresponding law for **B** is (8.3), i.e. $\oint \mathbf{B} \cdot \mathrm{d}\mathbf{L} = \mu_0 I$. Let us apply this to a small rectangular path in the xy plane as we did with **E** in Sec. 4.6 and Fig. 4.14. Then, as we saw, the left-hand side would have the value $(\partial B_y/\partial x - \partial B_x/\partial y)\,\mathrm{d}x\,\mathrm{d}y$, while the right-hand side would be simply $\mu_0 I_z$, the current through the rectangle perpendicular to the xy plane. Thus

$$\partial B_y/\partial x - \partial B_x/\partial y = \mu_0 I_z/\mathrm{d}x\,\mathrm{d}y = \mu_0 J_z \tag{8.9}$$

where J_z is the current density in the z-direction. Similarly,

$$\partial B_z/\partial y - \partial B_y/\partial z = \mu_0 J_x \tag{8.10}$$

$$\partial B_x/\partial z - \partial B_z/\partial x = \mu_0 J_y \tag{8.11}$$

The three expressions on the left of (8.9), (8.10) and (8.11) are the Cartesian components of $\nabla \times \mathbf{B}$ or **curl B**, while on the right we have the corresponding components of $\mu_0 \mathbf{J}$. Thus

$$\boxed{\nabla \times \mathbf{B} = \mu_0 \mathbf{J}} \tag{8.12}$$

The difference between this and the corresponding electric law $\nabla \times \mathbf{E} = 0$ is highly significant. It is only for vector fields with *zero* curl that a single-valued path-independent scalar potential V can be defined. Since $\nabla \times \mathbf{B}$ is zero only when $\mathbf{J} = 0$, V_m cannot be used at points inside current-bearing regions and even outside these it must be used with care.

An alternative method of transforming the integral into a differential form will be used increasingly from now on. This depends on Stokes's theorem in vector analysis which states that for any vector field **A**, $\oint_L \mathbf{A} \cdot \mathrm{d}\mathbf{L} \equiv \int (\nabla \times \mathbf{A}) \cdot \mathrm{d}\mathbf{S}$, where L is a closed path enclosing the surface S. Applied to Ampère's law, this means that $\oint \mathbf{B} \cdot \mathrm{d}\mathbf{L} \equiv \int_S (\nabla \times \mathbf{B}) \cdot \mathrm{d}\mathbf{S}$ is equal to $\mu_0 I$ or $\int_S \mu_0 \mathbf{J} \cdot \mathrm{d}\mathbf{S}$ using (1.25). Since this applies to any S, the two integrands are equal and $\nabla \times \mathbf{B} = \mu_0 \mathbf{J}$.

8.4 Differential form of Gauss's law: divergence of B

Section 4.5 showed how the integral form of Gauss's law for **E** ($\oint \mathbf{E} \cdot \mathrm{d}\mathbf{S} = Q/\varepsilon_0$) led to a differential form applying at a point: $\nabla \cdot \mathbf{E} = \rho/\varepsilon_0$. Here $\nabla \cdot \mathbf{E}$ is the divergence of **E** as described in Appendix B.9. The corresponding law for **B** is simpler since $\oint \mathbf{B} \cdot \mathrm{d}\mathbf{S} = 0$. If we apply this to the small rectangular parallelpiped of Fig. 4.13, then as in Sec. 4.5 we should obtain

$$\frac{\partial B_x}{\partial x} + \frac{\partial B_y}{\partial y} + \frac{\partial B_z}{\partial z} = 0 \tag{8.13}$$

The left-hand side is the Cartesian form of the divergence of **B** so that

$$\boxed{\nabla \cdot \mathbf{B} = 0} \tag{8.14}$$

In regions where V_m can be defined, $\mathbf{B} = -\mu_0 \nabla V_\mathrm{m}$ and hence $\nabla \cdot \nabla V_\mathrm{m} = 0$ or

$$\nabla^2 V_\mathrm{m} = 0 \tag{8.15}$$

This is Laplace's equation, identical with that obeyed by the electrostatic potential V in regions where there is no charge.

Once again, an alternative method of deriving (8.14) is to use Gauss's divergence theorem in vector analysis: for any vector field **A**, $\oint_S \mathbf{A}\cdot d\mathbf{S} \equiv \int_\tau (\nabla\cdot\mathbf{A})d\tau$, where S is a closed surface enclosing τ. Applied to **B**, this means that $\oint_S \mathbf{B}\cdot d\mathbf{S}=0$ implies that $\int_\tau(\nabla\cdot\mathbf{B})\,d\tau=0$ for any τ. Hence the integrand must be zero, or $\nabla\cdot\mathbf{B}=0$.

8.5 Vector potential[†]

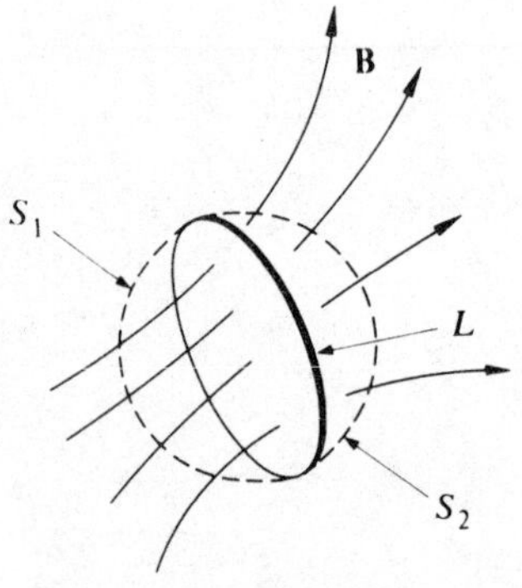

Figure 8.5 A closed path L forming the boundary of two surfaces S_1 and S_2 in a **B**-field.

Concept and definition We have seen that the magnetic scalar potential V_m has only a limited use because it is in general multivalued, even when a zero has been chosen. The reason for this has been discussed in Sec. 8.3 and is a result of curl **B** *not* being zero in regions containing currents. However, in contrast to the **E**-field from charges, the flux of **B** over a closed surface is *always* zero and so, therefore, is $\nabla\cdot\mathbf{B}$. We shall now see that this universal property of **B** enables us to define a different type of potential, this time a vector, which is valid even when currents are present.

Take any closed path L which bounds two arbitrary open surfaces S_1 and S_2 as in Fig. 8.5. Since S_1 and S_2 taken together form a closed surface, the total outward flux of **B** over them is zero. It follows that the flux Φ_S over S_1 (left to right) is the same as Φ_S over S_2 (left to right). In general, Φ_S is the same for *any* S bounded by L: it depends only on L. We now define a vector **A** at any point in the **B**-field such that its line integral round L is equal to the Φ_S across any S bounded by L; i.e.

$$\oint_L \mathbf{A}\cdot d\mathbf{L}=\Phi_S \qquad \text{(definition of } \mathbf{A}) \tag{8.16}$$

We already have experience in turning relations of this sort into differential forms applying at a point, e.g. the treatment of $\oint \mathbf{B}\cdot d\mathbf{L}=\mu_0 I$ by the method of Sec. 8.3. The right-hand sides of such integral equations are always the fluxes over an area of some vector quantity and just as I is the flux of **J**, so Φ_S in (8.16) is the flux of **B**. It follows that the differential form of (8.16) will be

$$\nabla\times\mathbf{A}=\mathbf{B} \tag{8.17}$$

which would serve as an alternative definition of **A**. In fact **B** must always be expressible in the form of (8.17) because of its zero divergence: div curl of *any* vector field is zero (see Appendix B.11).

It should be clear that **B** is obtained from **A** by a set of differentiations just as **E** is from V. This means that the same **B**-field may be given by a number of different vector potentials. In other words, *specifying the* **B**-*field does not completely determine the vector potential.* Any two vector potentials differing by the gradient of a scalar field still give the same **B** since **curl grad** is always zero (i.e. if **A** satisfies $\mathbf{B}=\nabla\times\mathbf{A}$, then so does $\mathbf{A}+\nabla\phi$). This means that we have a certain freedom in our choice of **A** and one such choice is that suggested by using the known sources of the **B**-field, as we now see.

[†] This section contains a derivation involving manipulation of vector field quantities and may be omitted at a first-level course.

A due to a current element If (8.17) is to be used at all for the determination of **B** we need an expression for **A** due to a current element. We already have that $d\mathbf{B}=\mu_0 I\, d\mathbf{l}\times\hat{\mathbf{r}}/(4\pi r^2)$. The quantity $-\hat{\mathbf{r}}/r^2$ can be written as $\nabla(1/r)$ from Eq. (B.34) of Appendix B.12, so that

$$\mathbf{B}=\oint_{\text{circuit}} -\mu_0\frac{I}{4\pi} d\mathbf{l}\times\nabla(1/r) \tag{8.18}$$

Using Eq. (B.21) in Appendix B.11, $-d\mathbf{l}\times\nabla(1/r)$ is identically the same as $\nabla\times(d\mathbf{l}/r)-(1/r)\nabla\times(d\mathbf{l})$. However, $\nabla\times(d\mathbf{l})$ is zero because $\nabla\times$ represents differentiation with respect to coordinates like x, y, z and $d\mathbf{l}$ is independent of these. Thus

$$\mathbf{B}=\oint_{\text{circuit}}\frac{\mu_0 I}{4\pi}\nabla\times\left(\frac{d\mathbf{l}}{r}\right)$$

Reversing the order of differentiation ($\nabla\times$) and integration ($\oint$), we have

$$\mathbf{B}=\nabla\times\left(\oint\frac{\mu_0 I\, d\mathbf{l}}{4\pi r}\right)$$

Comparing this with (8.17) we see that a suitable **A** for a current element is

current element: $$\boxed{d\mathbf{A}=\frac{\mu_0 I\, d\mathbf{l}}{4\pi r}} \tag{8.19}$$

If the element is part of a volume distribution or a surface current, then as usual the alternative forms

current elements: $$\boxed{d\mathbf{A}=\mu_0\mathbf{J}\, d\tau/(4\pi r); \qquad d\mathbf{A}=\mu_0 J_s\, d\mathbf{S}/(4\pi r)} \tag{8.20}$$

can be used. These correspond to the expressions (3.17) for the electric potential V due to distributions of charge.

Since $\nabla\cdot\mathbf{J}=0$ for steady currents, Eq. (8.20) shows that

steady current: $$\boxed{\nabla\cdot\mathbf{A}=0} \tag{8.21}$$

Worked example *A-lines due to currents* The vector potential is not a particularly useful addition to our repertoire of methods for solving elementary magnetic problems. Evaluation of **A** due to a complete circuit involves integration over a number of vector sources (the current elements) and then **B** has to be obtained by the complicated differentiation indicated by the operator $\nabla\times$. The virtues of vector potential become more evident in advanced work connected with radiation fields and with relativistic and quantum aspects of electromagnetism. Here we look only at some examples of **A**-fields in highly symmetrical situations.

Note that Eq. (8.19) means that the direction of **A** due to a current element is the same as that of the element itself. Thus, lines of **A** due to an infinite straight filamentary current must all be parallel to the current. It also follows that, for a

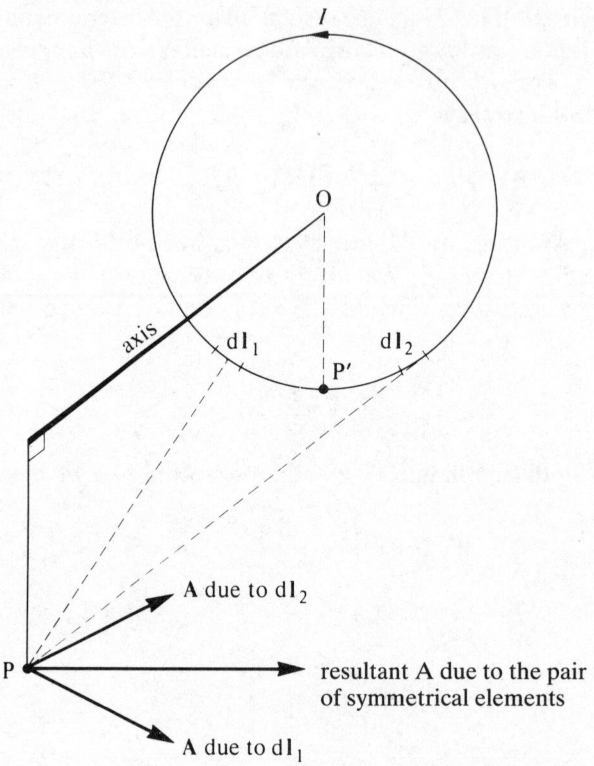

Figure 8.6 Geometry for finding the direction of **A** due to a circular current.

circular current, the lines of **A** must be coaxial circles. If in Fig. 8.6 we divide the current into pairs of elements as shown, each pair produces a *resultant* **A** at a point like P in the same direction: perpendicular to the plane containing the axis of the current and the radius OP′. Since **A** will take this direction for all points P as P′ moves round the circle, the lines of **A** must also be circles about the axis.

If we now look at the case of a long closely wound solenoid, we can calculate **A** by using the symmetry of the arrangement. The total **A**-field is the superposition of those due to all the elementary sections of the solenoid, each of which forms a circular current. The **A**-lines are thus circles about the axis as shown in Fig. 8.7 and the magnitude of **A** at any distance r from the axis will be the same at all points on a typical line as drawn in the figure. The line integral of **A** round the circular path of radius r is thus simply $2\pi rA$. By Eq. (8.16) this is equal to the flux of **B** across the circle, i.e., $\pi r^2 B$ or $\pi r^2 \mu_0 nI$, using (7.17). Thus

$$\text{infinite solenoid, } r \leq a: \qquad A = \tfrac{1}{2}\mu_0 nIr \qquad (8.22)$$

A similar calculation for points outside shows that there

$$\text{infinite solenoid, } r \geq a: \qquad A = \tfrac{1}{2}\mu_0 nIa^2(1/r) \qquad (8.23)$$

where a is the radius of the solenoid. Note that circular lines of **A** exist outside an infinite solenoid even though $\mathbf{B}=0$ over the whole exterior region.

Poisson's and Laplace's equations If we substitute $\mathbf{B}=\nabla\times\mathbf{A}$ into the law $\nabla\times\mathbf{B}=\mu_0\mathbf{J}$ (8.12), we obtain $\nabla\times(\nabla\times\mathbf{A})=\mu_0\mathbf{J}$. Using a vector identity (B.26),

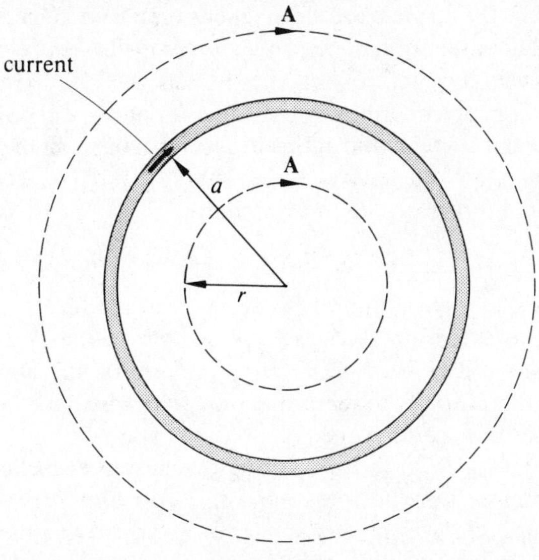

Figure 8.7 End-on view of a solenoid and its **A** lines.

the left-hand side of that equation can be written as $\nabla\nabla \cdot \mathbf{A} - \nabla^2\mathbf{A}$. Since $\nabla \cdot \mathbf{A} = 0$, we have

Poisson's equation:
$$\nabla^2\mathbf{A} = -\mu_0\mathbf{J} \tag{8.24}$$

which means that each Cartesian component of **A** obeys Poisson's equation similar to that for V in Sec. 4.7. Where no current exists we have Laplace's equation

Laplace's equation:
$$\nabla^2\mathbf{A} = 0 \tag{8.25}$$

8.6 Magnetic dipoles in B-fields: force in non-uniform B

The magnetic dipoles introduced in Chapter 7 were small enough for their structure to be immaterial. Most of the formulae, both for the **B**-fields due to dipoles and for the forces and torques on them, required a knowledge only of the dipole moment **m**. However, there are two cases needing further consideration because of the nature of the dipoles themselves. One is the force in a non-uniform **B**-field (this section) and the other concerns the motion arising from the torque (next section).

An *ideal* dipole, either electric or magnetic, has dimensions small compared with other distances. Nevertheless, in deriving formulae involving *electric* dipoles we treated the dipole as two equal and opposite charges, a model which is not valid for the *magnetic* cases (see the discussion of poles in Chapter 7). Fortunately, this does not affect the expressions for the **B**-field due to a dipole or for the torque on a dipole in a uniform **B**-field because these were the direct result of an experimentally demonstrated equivalence with the electric formulae.

However, the force on a dipole in a non-uniform field is a different matter because the corresponding electric formula was deduced using a two-charge

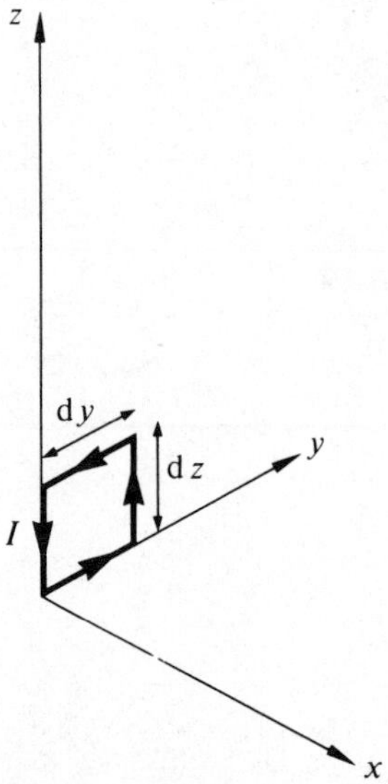

Figure 8.8 Elementary rectangular loop for calculating forces in a non-uniform **B**-field.

model. If we are to justify the formula for the magnetic dipole we must do so for a current loop. We shall not attempt here to prove the complete set of expressions which are given below in Eqs (8.27), but rather shall take a special case to illustrate why there is a problem and why the same formulae *do* apply.

Take a small rectangular current loop at the origin and in the yz plane as in Fig. 8.8. The dipole moment **m** has only an x component $m_x = I\,dy\,dz$, so according to the general formulae the forces on the loop should be

$$F_x = m_x \partial B_x/\partial x, \qquad F_y = m_x \partial B_y/\partial x, \qquad F_z = m_x \partial B_z/\partial x \tag{8.26}$$

Consider F_x. How is it that a variation of B_x along x can produce such a force when the dipole has no extension along x? The answer lies in the universal property of **B** (8.13) from which it is clear that a non-uniformity in B_x along x *must* be accompanied by non-uniformity in some direction perpendicular to x: and the loop *does* have extension in the yz plane.

It is not difficult to justify F_x in (8.26) by using the current element formula, $\mathbf{F} = I\,d\mathbf{l} \times \mathbf{B}$, to find the x forces on the sides dy and dz. F_x can only be produced by B_y acting on the sides dz and B_z acting on the sides dy. Consider first the sides dz. The forces on the nearer side is $I\,dz B_y$ to the *right*, that on the farther side is $I\,dz[B_y + (\partial B_y/\partial y)\,dy]$ to the *left*. The resultant of these two is $-I\,dy\,dz\,\partial B_y/\partial y$ along the positive x-direction. Similarly, the resultant of B_z acting on the dy's is $-I\,dz\,dy\,\partial B_2/\partial z$. Thus, using (8.13),

$$F_x = -I\,dy\,dz(\partial B_y/\partial y + \partial B_z/\partial z) = m_x \partial B_x/\partial x$$

The complete set of expressions for the force on a dipole in a non-uniform **B**-field is thus as in the electric case:

dipole in non-uniform **B**:
$$\begin{aligned} F_x &= m_x\,\partial B_x/\partial x + m_y\,\partial B_x/\partial y + m_z\,\partial B_x/\partial z \\ F_y &= m_x\,\partial B_y/\partial x + m_y\,\partial B_y/\partial y + m_z\,\partial B_y/\partial z \\ F_z &= m_x\,\partial B_z/\partial x + m_y\,\partial B_z/\partial y + m_z \partial B_z/\partial z \end{aligned} \tag{8.27}$$

The reader may like to work out that this set of equations can be very concisely written in the form

dipole in non-uniform **B**: $$\mathbf{F} = (\mathbf{m} \cdot \nabla)\mathbf{B} \tag{8.28}$$

8.7 Magnetic dipoles in B-fields: torque, angular momentum, and precession

Angular momentum of magnetic dipoles We make the sweeping statement that all magnetic dipoles possess an associated angular momentum. For an ordinary current loop this is no surprise: an electron of mass m_e, for instance, moving round a circular wire of radius r with angular velocity ω has an angular momentum $L = m_e r^2 \omega$. Since the current I is given by $e\omega/2\pi$ (Problem 1.17), the dipole moment $m = IA$ is $\frac{1}{2}er^2\omega$. Both m and L are vector quantities along the axis of the loop and so the relation between them is

current loop: $$\mathbf{m} = \frac{e}{2m_e}\mathbf{L} \tag{8.29}$$

However, there is also plenty of experimental evidence that the **m**'s of atoms and of all the fundamental particles have a similar association with **L**. In general, the ratio of **m** to **L** is known as the **gyromagnetic ratio** or, more correctly, but less euphoniously, the magnetogyric ratio, γ. Thus

$$\mathbf{m}=\gamma\mathbf{L} \qquad \text{(definition of } \gamma\text{)} \tag{8.30}$$

For a current loop, and in a similar way for the orbital motion of an electron in an atom, γ is clearly $e/2m_e$ from (8.29). We also know (Sec. 12.11) that electrons possess an intrinsic moment, the spin moment, for which γ is almost exactly e/m_e. A dimensionless quantity, the Landé g-factor, is often defined by

$$g=\frac{2m_e}{e}\gamma \qquad \text{(definition of } g\text{)} \tag{8.31}$$

so that $g=1$ for the orbital motion of a single electron and $g\approx 2$ for its spin. In a complete atom we might expect g to lie between 1 and 2. For nucleons, g is defined as $2m_p\gamma/e$.

It is the invariable association of **L** with **m** which gives the magnetic dipole quite different characteristics from those of the electric dipole. One major difference concerns quantization and we take this up later. Another difference is revealed when atomic or molecular dipoles are subjected to torques $\mathbf{T}=\mathbf{p}\times\mathbf{E}$ or $\mathbf{m}\times\mathbf{B}$ when placed in a field. An electric dipole simply turns until its moment **p** is along **E**, but a magnetic dipole precesses about the direction of **B**, as we now see.

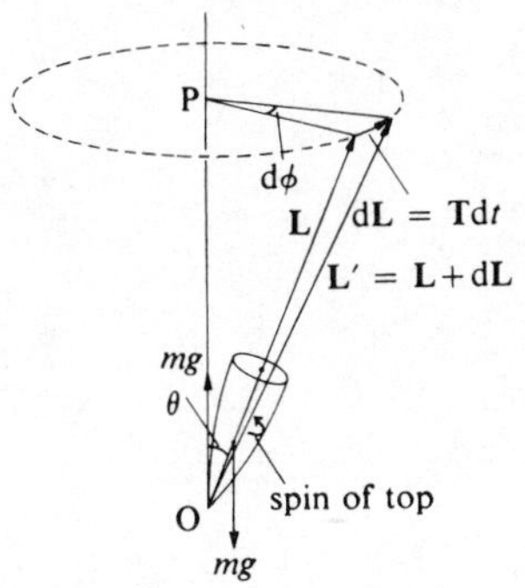

Figure 8.9 Precession of a spinning top.

Larmor precession A spinning top precesses because (a) it has an angular momentum **L** about its own axis, (b) acting on it is a torque **T** formed by the weight mg and the normal reaction at the ground and (c) **T** is an axial vector at right angles to the direction of **L** (Fig. 8.9). All these conditions are present for a magnetic dipole placed in a field **B** except that the torque will be given by $mB\sin\theta$ if **B** is along OP. So let us use the top to see the reason for the precession and to obtain a value for its frequency.

In a time dt, **T** causes a change $d\mathbf{L}=\mathbf{T}\,dt$ in angular momentum (just as a force **F** would produce a change **F** dt in linear momentum). This change, like **T**, has a direction at right angles to the plane containing the vertical OP and **L**, as shown in the figure. The addition of d**L** to **L** causes the latter to move into a new position **L**′. The new plane containing OP and **L**′ is now the one to which the next increment d**L** is perpendicular. The vector **L** thus traces out a cone as shown and the top precesses.

The angle $d\phi=dL/(L\sin\theta)$. The angular frequency of precession is therefore

$$\omega=\frac{d\phi}{dt}=\frac{dL}{dt}\frac{1}{L\sin\theta}=\frac{T}{L\sin\theta} \tag{8.32}$$

If the system is a magnetic dipole of moment **m** in a magnetic field **B** along OP then the torque is $-mB\sin\theta$, the negative sign occurring because, for a negative electron, **m** and **L** are in opposite directions. Substituting this in (8.32) gives

$$\omega_L=-mB/L=-\gamma B \tag{8.33}$$

The frequency $\nu_L=\omega_L/2\pi$ is often known as the **Larmor frequency**, a name originally restricted to the precession of orbital electrons.

Quantization It is known from quantum mechanics that the angular momentum of any system can only take certain discrete values, multiples of a fundamental unit $h/2\pi$, where h is Planck's constant. It follows that a similar discreteness will show itself in the magnetic moment **m**. From (8.29) a natural unit for atomic magnetic moments would be $eh/(4\pi m_e)$. This is a quantity known as the **Bohr magneton** and denoted by μ_B. Nucleons (protons and neutrons) also have intrinsic magnetic moments but here the unit is the **nuclear magneton** $eh/(4\pi m_p)$ (where m_p is the mass of the proton), a quantity much smaller than μ_B.

A further quantum effect also reveals itself. Stern and Gerlach in 1922 passed a beam of atoms through a non-uniform **B**-field. The forces on the atomic magnetic moments, given by (8.27), deflected the beam with unexpected results: the deflections showed that the atomic dipoles can only take up certain angles with the direction of **B**, a phenomenon known as **spatial quantization**. The angles are such that the component of angular momentum along **B** may only be $m_J h/2\pi$, where the quantum number m_J has a limited number of values differing by unity (for example 0, ± 1, etc., or $\pm\frac{1}{2}$, $\pm\frac{3}{2}$, etc.). It follows that the magnetic moment precesses about **B** at such angles that its component along **B**, by (8.30), is $\gamma m_J h/2\pi$. It follows from (7.43) that an atom with energy W placed in an external field **B** has a total energy $W - \gamma m_J hB/2\pi$: i.e. the original energy level is split into a set of levels separated by $\gamma hB/2\pi$ or by $g\mu_B B$. The separation depends on the value of g for the atom which is, as a consequence, sometimes known as the **splitting factor**.

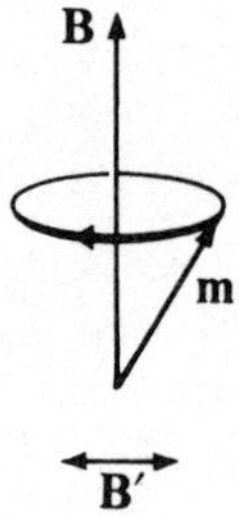

(*a*)

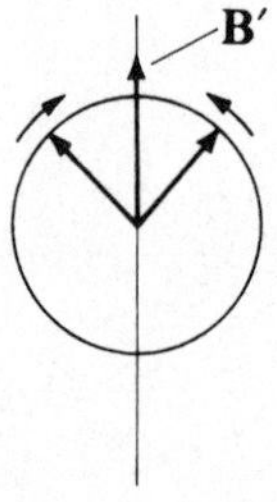

(*b*)

Figure 8.10 Paramagnetic resonance. **B**′ in (a) is a sinusoidally alternating field at right angles to a steady **B**. As shown in (b), **B**′ can be regarded as a combination of two fields rotating in opposite senses. One of these components will be rotating in the same sense as the precessing dipole and is responsible for resonance at the Larmor frequency.

Determination of γ_p The gyromagnetic ratio of the proton, γ_p, is important both in relation to the value of fundamental constants generally and to the measurement of **B**-fields, using (8.33). The determination of γ_p using Larmor precession can be carried out as follows. The protons in a sample of water are aligned at right angles to a measurable field **B** by a strong auxiliary magnetic field. The auxiliary field is switched off and Larmor precession then starts at 90° to **B** in all the protons. Although the precession becomes random in a few seconds, that is long enough for it to induce an electromotance in a tuned detector coil around the sample. The coil feeds a frequency meter so that ω_L can be determined. This is the *weak field* or *induction* method for γ_p (Vigoureux, 1971) as opposed to the *strong field* or *resonance* method described below.

Practical applications *paramagnetic resonance* (ESR *and* NMR) A magnetic dipole precesses in a steady magnetic field with an angular frequency $\omega_L = \gamma B$ so that a determination of ω_L and B yields the gyromagnetic ratio. A small alternating magnetic flux density **B**′ of angular frequency ω applied perpendicular to **B** can be considered as two rotating fields with opposite senses (Fig. 8.10a and b). When $\omega = \omega_L$, one of the components is rotating in the same sense and with the same frequency as the precessing dipole. If we look at the whole system from a set of coordinates rotating with an angular velocity ω_L, both the moment **m** and the in-phase component of **B**′ will appear stationary and **m** will thus now precess about **B**′, changing its angle with **B**. In doing so it will absorb energy from the oscillating source since the potential energy is $-mB\cos\theta$. In practice ω is often kept constant and B is swept over a small range, thus varying ω_L.

In quantum-mechanical terms, the same result is obtained if we regard the alternating field **B**′ as supplying energy in quanta of $h\nu$ or $h\omega/2\pi$. If these match the separation of atomic energy states ($\gamma hB/2\pi$), the atom can be raised to the next higher state with the resonance absorption of the quantum. This will occur when $\omega = \gamma B$ as above.

higher state with the resonance absorption of the quantum. This will occur when $\omega = \gamma B$ as above.

Electron spin resonance (ESR) occurs in substances containing electrons with unpaired spins and occurs in the microwave region of the frequency spectrum (see Problem 8.15). It is a valuable analytical tool in investigating chemical reactions where free radicals occur as an intermediate product and in structural analysis of transition metal complexes.

Nuclear magnetic resonance (NMR) provides an even more important method of analysis, particularly since protons give a strong resonance, e.g. at 13 MHz in a **B**-field of 0.3 T. The method is being used to provide a non-invasive imaging of the human body for medical diagnosis, which is all the more powerful in that it can be adapted to trace the distribution of specific nuclei belonging to a variety of chemical elements.

One measurement using NMR is of particular importance: the determination of the gyromagnetic ratio of the proton γ_p. Thomas, Driscoll and Hipple (1950) measured γ_p absolutely with great precision in terms of a magnetic field determined with the balance of Fig. 7.4b. This value provides a rapid and accurate method of measuring a magnetic field using proton resonance.

8.8 Conclusion

The general equations governing the behaviour of electric and magnetic fields that have been developed so far can be expressed in differential form as follows:

Maxwell's equations for steady fields *in vacuo*:

$$\nabla \cdot \mathbf{E} = \rho/\varepsilon_0 \qquad \nabla \times \mathbf{E} = 0$$
$$\nabla \cdot \mathbf{B} = 0 \qquad \nabla \times \mathbf{B} = \mu_0 \mathbf{J} \tag{8.34}$$

It is clear that the **E**-field equations are completely independent of those governing the **B**-field, and that is why we have been able to treat electric and magnetic problems separately. This simple state of affairs has been brought about by deliberately excluding time variations in the fields and we must now proceed to include these, examining first the effect of changing **B**-fields.

It might be thought that a chapter on magnetic energy, similar to Chapter 5 on electric energy, could be included first. It will become evident in the next chapter, however, that magnetic energy, the motion of conductors and time variations are so intimately connected that they must all be dealt with together.

8.9 Revision summary

The **B**-field introduced in Chapter 7 to deal with interactions between currents has been used in this chapter to express magnetic laws in very general ways. *Ampère's circuital law* and *Gauss's law* for **B** have both been developed in integral forms, applying to closed paths and surfaces, respectively, and differential forms applying at a point. A magnetic **scalar potential**, V_m, can be defined by analogy with the electric potential but with validity limited to regions where there is no current and even then without the path-independence that makes V so useful. A magnetic **vector potential**, **A**, can also be defined because of the zero divergence of

B, while *magnetomotance* or magnetomotive force (m.m.f.) $\mathscr{H}$ is the line integral of $-\mathbf{B}/\mu_0$ evaluated round a closed path:

$$\mathscr{H} = \oint_L \frac{\mathbf{B}\cdot d\mathbf{L}}{\mu_0} \qquad (8.1)$$

All the results obtained are summarized in the following table:

	Integral forms		Differential forms	
D Ampère's circuital law	$\oint_L \mathbf{B}\cdot d\mathbf{L} = \mu_0 I$	(8.3)	Curl: $\nabla\times\mathbf{B} = \mu_0\mathbf{J}$	(8.12)
E Gauss's law	$\oint_S \mathbf{B}\cdot d\mathbf{S} = 0$	(8.8)	Divergence: $\nabla\cdot\mathbf{B} = 0$	(8.14)
F Scalar potential, V_m (only valid if I and **J** are zero)	$\int_A^B -\frac{\mathbf{B}}{\mu_0}\cdot d\mathbf{L} = V_m$	(7.35)	Gradient: $\mathbf{B} = -\mu_0\nabla V_m$	(7.37)
G Vector potential, **A**	$\oint_L \mathbf{A}\cdot d\mathbf{L} = \Phi_S$	(8.16)	Curl: $\nabla\times\mathbf{A} = \mathbf{B}$	(8.17)

The reader should contrast this with the table in Sec. 4.10 which exhibits the corresponding properties of **E**-fields due to charges.

Properties of B-fields

From **D** above: **B**-fields arise from currents and lines of **B** *may* close on themselves. We say that $\nabla\times\mathbf{B} = \mu_0\mathbf{J}$ indicates that electric currents are *vortices* of **B**.

From **E** above: **B**-flux does not *originate* at any point, nor at any surface or volume. There are no *sources* of **B** equivalent to electric charges as sources of **E**.

Potentials

The scalar potential V_m is only single-valued and path-independent in singly connected regions where I and **J** are zero. It has properties similar to those of V, e.g. it obeys Laplace's equation

$$\nabla^2 V_m = 0 \qquad (8.15)$$

The vector potential **A** is not unique if only **B** is given: there are an infinite number of **A**'s that yield the same **B**. However, if the currents giving rise to the field are given, then d**A** due to a current element is $\mu_0 I\, d\mathbf{l}/(4\pi r)$ and **A** can in principle be found by integration. It obeys Poisson's equation

$$\nabla^2\mathbf{A} = -\mu_0\mathbf{J} \qquad (8.24)$$

Solution of problems: B-fields due to systems

We can add two further methods to those of Chapter 7:

- Ampère's circuital law in integral form for highly symmetrical distributions of current,
- calculation of **A** using $\mu_0 I\, d\mathbf{l}/(4\pi r)$ and then **B** from $\nabla\times\mathbf{A}$. This is not a method recommended for use in elementary problems.

Solution of problems: systems in B-fields

- Magnetic dipoles in **B**-fields are subject to a torque **T** and a force **F** given by

$$\mathbf{T} = \mathbf{m} \times \mathbf{B}; \qquad \mathbf{F} = (\mathbf{m} \cdot \nabla)\mathbf{B} \qquad (7.42)\ (8.28)$$

Magnetic dipoles possess angular momentum due to the orbital motion or spin of their elementary particles. The relation between magnet dipole moment **m** and angular momentum **L** is $\mathbf{m} = \gamma \mathbf{L}$, where γ is the *gyromagnetic ratio.* The torque on a dipole in a **B**-field causes precession, because of the angular momentum, at the Larmor frequency

$$\omega_L = -\gamma \mathbf{B} \qquad (8.33)$$

The *g-factor* for a magnetic dipole is defined as $2m_e\gamma/e$ for systems of electrons and $2m_p\gamma/e$ for systems of nucleons. The *g*-factor for orbital electrons is 1 and for spin is almost exactly 2.

Since angular momentum **L** is quantized in units of $h/2\pi$, a natural unit for magnetic moment is the Bohr magneton $\mu_B = eh/(4\pi m_e)$ for electron systems and the nuclear magneton $\mu_N = eh/(4\pi m_p)$ for systems of nucleons.

Commentary

C8.1 On Lines of Force We have frequently illustrated **E**- and **B**-fields with lines of force, but the time has come to give a warning about their use beyond that as a pictorial aid. There are two ways in which it is easy to give the lines properties that are quite inadmissible.

One way concerns *steady fields* of the type so far encountered. We have seen that **B** has no sources so that lines of **B** do not begin or end anywhere (strictly speaking there is no reason why one set of **B** lines should not end on one side of an area provided an exactly similar set started from the other side since it is the *flux* of **B** which must not change). Even accepting the continuity of **B** lines, it is only in highly symmetrical situations that a **B** line traced out by small test dipole would come back exactly to its starting point and close on itself. In real situations this closure is unlikely to occur because of deviations from strict symmetry arising from distortions in the currents and from stray external fields. For instance, a slightly kinked solenoid will not produce closed lines of **B**, and nor will a straight current-carrying wire in an external uniform **B**-field along the wire (the **B**-lines here form an infinite helix). See Slepian (1951) and McDonald (1954).

The second inadmissible property is much more important and concerns *motion and changes in time* which are the concern of the next chapter. Motion of an object or particle through a field has a meaning because the velocity can be measured with respect to the basic laboratory reference system. However, to talk of the motion of lines of force is quite meaningless because there is no way, even in principle, of identifying a single line and measuring its change in position. *The only safe concepts are those of the magnitude and direction of* **B** *or* **E** *at any point and their fluxes over areas*, and all of these may change in time to give effects we explore later. None of these should be taken to imply any motion of individual lines. For example, consider either a *perfectly* cylindrical bar magnet rotating about its long axis of symmetry or an infinite parallel plate capacitor moving so that its plates stay in the same plane. In neither case does the field at any point change in magnitude or direction and no meaning could be attached to any motion of field lines.

C8.2 On Alternating-Gradient Focusing The motion of charged particles in electromagnetic fields has long been a lively branch of classical electromagnetism. One aspect of importance in recent decades has been the stability of orbits pursued in circular accelerators. Our discussion will be in terms of magnetic fields broadly of the type shown in Fig. 7.18b, where there is axial symmetry about the vertical line NS so that only the radial and axial components B_r and B_z need be specified (z being *downwards* to give a positive B_z). We have already argued that the type of curvature shown in the figure is necessary if particles straying from the orbital plane are to be refocused. This curvature means that $\partial B_r/\partial z$ must be everywhere negative. Because (8.12) applies with $\mathbf{J}=0$, we have from the cylindrical polar form that $\partial B_r/\partial z = \partial B_z/\partial r$, which is thus also negative: B_z must decrease as r increases. However, this decrease must not be too great or particles deviating from the orbit in the horizontal plane will not be refocused. The limiting rate of decrease allowed is in fact when $B_z \propto r^{-n}$ with $n<1$, because $n=1$ means that (7.28) is satisfied for any r and the orbits are neutral, i.e. any orbit is stable. The factor n is known as the *field index* and is defined more precisely by $-\partial(\ln B)/\partial(\ln r)$ because n may vary with r. If $n=0$ everywhere, the field is uniform and if n is negative the curvature is opposite to that of Fig. 7.18b, a configuration that leads to stability against radial wandering but none against vertical displacement. Machines with n lying in the desirable range between 0 and 1 are said to use *weak focusing* (sometimes the field index is defined as the negative of ours so that n then lies between 0 and -1).

Synchrotrons are circular accelerators using an orbit of fixed radius, so that the magnetic field need only be produced over a limited region round the orbit itself. If only weak focusing is used, the particles may wander some way from the orbit so that large evacuated chambers and extensive fields are needed. With higher energies the need for stronger focusing to limit this wandering has led to the use of *alternating-gradient* (AG) fields. The magnetic field is produced in segments round the orbit, each having a **B**-field with a very high gradient (n as much as 300). The segments are arranged to have n alternately positive and negative, so that one is convergent in the vertical plane and divergent in the horizontal plane while the next is vertically divergent and horizontally convergent. The overall effect is analogous to that with optical lenses of equal power but opposite sign: when separated by a finite distance, the combination is always converging. The saving in the size of magnets and vacuum chambers is enormous.

Problems

Section 8.1

8.1 A steady current I flows in one direction in the solid inner conductor, radius a, of a coaxial cable, and in the opposite direction in the outer conductor whose inner radius is b and outer radius c. Find the magnetic field at various distances from the axis.

8.2 An infinite plane conducting sheet of negligible thickness carries a uniform surface current density J_s (see Sec. 1.7). Use the circuital theorem and symmetry to find the **B**-field outside the sheet.

8.3 The current in a wire consists of n electrons per unit volume moving with a velocity v. What is the effect on these electrons of the magnetic field set up by the current itself within the wire? What is the force on one electron at a distance r from the centre of the wire?

8.4 A cylindrical beam of electrons, density n, velocity v, can be treated as a continuous distribution of charge. What is the resultant force due to **E**- and **B**-fields on an electron a distance r from the axis?

Section 8.2

8.5 A **B**-field is specified by the conditions $B_x = kx$, $B_y = ky$, $B_z = 0$, k being constant. Sketch the shape that **B**-lines would take and explain why such a field is impossible.

8.6 An infinite solenoid has its axis along z. Using Ampère's circuital law, Gauss's law and symmetry, show that for the **B**-field due to a current I in the solenoid (a) B_θ is everywhere zero, (b) B_r is everywhere zero, (c) B_z is zero outside the solenoid and equal to $\mu_0 nI$ everywhere inside, n being the number of turns per unit length.

8.7 A straight current-carrying wire of infinite length and negligible section lies along z. Show that the cylindrical components of B denoted by B_r and B_z are everywhere zero.

Section 8.3

8.8 If a **B**-field is specified everywhere by $B_x = ky$, $B_y = -kx$, $B_z = 0$, k being constant, find an expression for the current density **J** which would give rise to it.

Section 8.4

8.9 Are the following possible as specifications of **B**-fields?

(a) $B_x = kx$, $B_y = ky$, $B_z = kz$.

(b) $B_x = kx$, $B_y = 0$, $B_z = -kz$.

(c) $B_x = k(x^2 + y^2)$, $B_y = -k(x^2 + y^2)$, $B_z = 0$.

8.10 Suppose the magnetic field of Fig. 7.18b is a cross-section through that of a pair of very long pole pieces perpendicular to the page. Argue that if the field becomes weaker towards the edges, the curvature of the field must be as shown, and that stronger fields towards the edges would demand the opposite curvature.

Section 8.5

8.11* Show that the vector potential outside an infinite solenoid is given by Eq. (8.23).

8.12* A uniform magnetic field $\mathbf{B}_0$ has only a z-component. Show that a vector potential given by $A_x = -\frac{1}{2}B_0 y$, $A_y = \frac{1}{2}B_0 x$, $A_z = 0$ is suitable, but that so is $A_x = -B_0 y$, $A_y = A_z = 0$. Explain.

8.13* Show that the vector potential at a point a distance R from the centre of a finite straight wire of length $2L$ and carrying a current I is $(\mu_0 I/2\pi)\log_e[p + (1 + p^2)^{1/2}]$, where $p = L/R$. Hence show that B tends to $\mu_0 I/(2\pi R)$ as L tends to infinity, even though A itself tends to infinity.

Section 8.6

8.14 A magnetic dipole of fixed moment **m** is moved slowly along the axis of a circular coil of radius a carrying a fixed current. How does the force on the dipole vary with its distance x from the centre of the coil if **m** always lies along the axis? At what distance, if any, has the force a maximum value?

Section 8.7

8.15 Calculate the Larmor frequencies for a single orbital electron and for a single spinning electron in a **B**-field of 0.2 T.

8.16 Show that the ratio of the Larmor and cyclotron frequencies of a proton in the same **B**-field is equal to its magnetic dipole moment in nuclear magnetons.

Section 8.8

8.17 Compare and contrast

(a) the properties of **E**-fields due to charges and **B**-fields due to steady currents,

(b) the properties of electric charges and magnetic poles,

(c) the properties of electric and magnetic dipoles.

9

Electromagnetic induction, inductance, and magnetic energy

Faraday's discovery of the earliest[†] induced current on 29 August 1831 began six months of brilliant experiment in which he established most of the qualitative laws of electromagnetic induction.

In Sec. 9.1 we summarize his results and those of his contemporaries using modern concepts, remembering that these were only then being painfully developed. Not the least of Faraday's achievements was the expression of his results in terms of flux and lines of force, ideas which later proved so fruitful.

However, we also wish to relate this new set of laws to our study of electromagnetism so far in this book. Up to this point we have apparently only considered steady conditions: our **E**-fields and **B**-fields have not varied with time, our current-carrying conductors have not moved. Yet an electric current is, after all, a *moving* charge upon which a **B**-field exerts a force $Q\mathbf{v} \times \mathbf{B}$. Is it not possible, if we move a conductor through a **B**-field, that $Q\mathbf{v} \times \mathbf{B}$ acting on the free charges will cause a current to flow? It is by examining the consequences of this in Secs 9.2 and

[†] Strictly this was the first induced current *recognized as such.* Arago in 1824 had observed the motion of a suspended magnet induced by a rotating copper disc beneath it (Arago's disc) but could not account for it.

9.3 that we arrive at quantitative laws of electromagnetic induction and see the effects of time variations in **B**-fields.

In later sections we examine further consequences of the laws, introducing the concepts of self-inductance and mutual inductance, deriving expressions for magnetic energy and, finally, seeing the effect on the general properties of **B**-fields and **E**-fields in our further development of Maxwell's equations.

9.1 Electromagnetic induction: experimental basis

In this section we describe briefly experiments that originally led to the laws of electromagnetic induction and at the same time we establish the meaning of some new terms that will be needed. It will be noticed that the early results were largely qualitative and it is only in Secs 9.2 and 9.3 that we develop quantitative laws.

Faraday's experiments These experiments of 1831–32 included the following:

- *Mutual induction*: two coils, denoted A and B, are arranged so that, if a steady current flows in A, some of its magnetic flux links (i.e. crosses) B. If the current in A then *changes*, a current is induced in B. This phenomenon is known as *mutual* induction. (Faraday used two helical wires wound on a common iron core and later dispensed with the core.)
- *Relative motion causing change in flux linkage*: a coil is arranged so that it is linked by some of the magnetic flux from a source M of magnetic field, which may be either a magnet or a current (Fig. 9.1a and b). If *relative motion* occurs between the coil and M such that the flux linkage with the coil is *changed*, a current is induced in the coil. (Faraday plunged a bar magnet into a solenoid; moved a coil near a magnet and between its poles; made one current-carrying coil approach another.)
- *Cutting of flux by a moving conductor*: part of a conducting circuit which *moves* and thereby cuts magnetic flux (Fig. 9.1c) has a current induced in it.

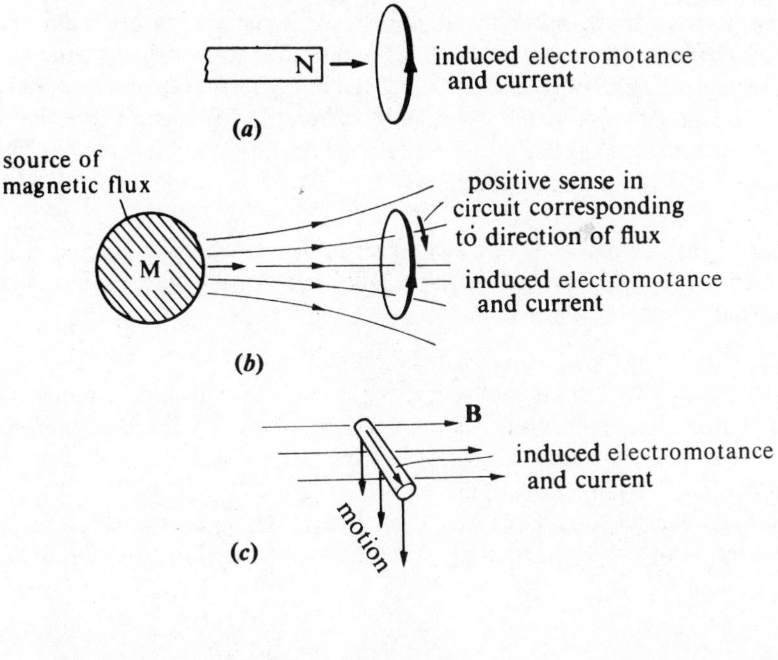

Figure 9.1 Examples of Faraday's and Lenz's laws. (a) Induced current flows in such a direction as to produce a **B**-field repelling the north pole; (b) generalization of (a): the induced current produces flux in opposition to increase of flux from moving source; (c) induced current flows in such a direction that the force on it from the **B**-field ($I\,d\mathbf{l} \times \mathbf{B}$) is upwards and thus opposes the motion.

(Faraday's disc, illustrated in Fig. 9.3a, was the first continuous generator. He also passed a straight wire between the poles of a U-magnet.)

- *Electromotance or current?* The induced currents produced under the same conditions are proportional to the conductance of the circuit. This means that given changes produce a definite *electromotance* rather than a definite current.
- *Magnitude of electromotance*: greater rates of change produce larger electromotances, but there is no explicit reference to the exact dependence in these early researches.
- *Self-induction* is the electromotance induced in a circuit because of a change in its *own* current. This was first discovered by Joseph Henry (1797–1878) in 1832 and was independently investigated by Faraday in 1834.

Faraday's laws of electromagnetic induction may thus be summarized as follows. An electromotance is induced in

1. A rigid stationary circuit across which there is a time-varying magnetic flux.
2. A rigid circuit moving in a steady **B**-field in such a way that the magnetic flux across it changes.
3. *Part* of a circuit which moves and, in doing so, *cuts* magnetic flux.

Electromotances of type 1 we shall call **transformer electromotances** and those of types 2 and 3 **motional electromotances**. Phenomena encountered in practice can often be placed in one category or the other, although it is clearly possible both for a field to be changing in time and for a circuit to be simultaneously moving through it. As far as the **magnitude** of these electromotances is concerned, it was not until 1845 that Neumann expressly assumed that they were proportional to the rate of change of flux linkage or to the rate of flux cutting. Faraday finally showed in 1851–1852 that motional electromotances were proportional to the rate of flux cut but the law for stationary circuits seems merely to have been assumed.

Lenz's law Faraday's original descriptions of the **direction** or **sense** of the induced electromotances and currents were confused and Emil Lenz in 1834 gave the first clear statement. Lenz's law is now usually given in the form: **whenever a change produces an induced current, the direction of current flow is such as to produce effects opposing the change**. Some examples of its operation are included in Fig. 9.1, which should be carefully studied.

Allocation of signs to magnetic flux and induced electromotance A sign convention is often needed in connection with Faraday's and Lenz's laws, both for flux and electromotance.

(a) *Magnetic flux linking a complete circuit*: as at the end of Sec. 7.7, the direction of any self-flux that exists is allocated a positive sign, which determines the sign of other flux. If there is no self-flux, then we allow any external magnetic flux to establish the positive direction of flux linkage. Thus, in Fig. 9.1a and b, positive flux is from left to right.

(b) *Electromotance induced round a complete circuit*: the positive sense is that of the rotation of a right-handed screw advancing in the direction of positive flux allocated as in (a). Thus in Fig. 9.1b, a right-handed screw would advance from left to right (the positive flux direction) if it were rotated in the sense shown by the small separate arrow.

(c) *Electromotance in a moving part of a circuit*: this is more difficult. If the velocity of the movement is **v** and the field is **B**, then the positive induced electromotance is in the direction of the vector product $\mathbf{v} \times \mathbf{B}$ (for reasons appearing in the next section).

(d) *Flux cut by a moving part of a circuit*: this is allocated a positive sign if it produces an induced electromotance in the direction of $\mathbf{v} \times \mathbf{B}$ (see Fig. 9.4a).

9.2 Motional electromotance

Guided by the above experiments, we now see how the laws established in previous chapters for steady conditions can be used to derive new laws which apply when circuits move and when **B**-fields vary in time. In this section, various expressions for *motional* electromotance are derived and applied. These will then be used in Sec. 9.3 to deduce the laws for *transformer* electromotances.

E-field in a conductor moving in a B-field Any charge Q moving with a velocity **v** in a magnetic field **B** experiences a force

$$\mathbf{F} = Q\mathbf{v} \times \mathbf{B} \qquad (7.25) = (9.1)$$

When a straight conducting wire moves in a magnetic field this force is exerted on all the charges in it, so that by definition (3.1) they will experience an electric field given by

moving conductor: $$\mathbf{E} = \mathbf{F}/Q = \mathbf{v} \times \mathbf{B} \qquad (9.2)$$

It will be recalled that (3.1) is restricted to cases where Q is stationary. Now, although an external observer describes the charges as moving with the conductor, an observer attached to the conductor would describe them as stationary. It follows that the **E**-field of (9.2) exists *from the point of view of the conductor.*

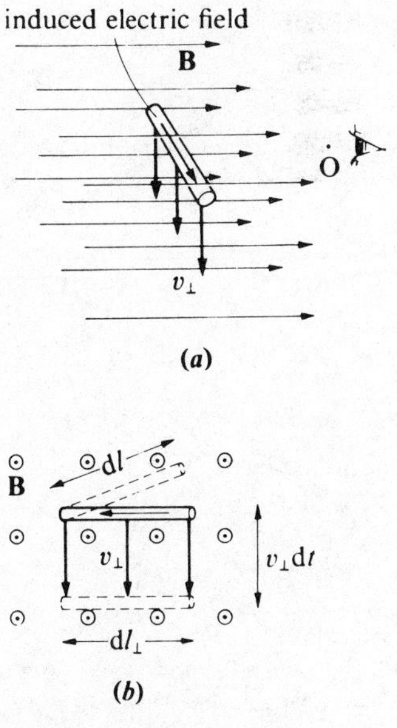

Figure 9.2 (a) Induced electromotance in a moving conductor: any component of the velocity parallel to **B** produces no effect, so only $v_\perp$ is drawn; (b) view of (a) from O. Any component of **dl** parallel to **v** has no electromotance along its length, so that only $dl_\perp$ need be considered.

Figure 9.2 shows the relation between directions for the special case of a wire at right angles to **B** moving with a velocity perpendicular both to the wire and to **B**. The induced electric field will cause a current if the wire is connected to a conducting circuit and the direction of the current is remembered in elementary physics by Fleming's right-hand rule. Once the vector product has been mastered, however, (9.2) is more concise and more general and shows that only the component of velocity perpendicular to $\mathbf{B}(v_\perp)$ contributes to **E**. Thus, (9.2) could be expressed as

moving conductor: $$E = v_\perp B \quad \text{in magnitude} \qquad (9.3)$$

This **E**-field, unlike that of Chapter 3, is not produced by static charges and falls into the category we have denoted in Sec. 6.1 by the symbol $\mathbf{E_M}$.

Induced motional $\mathscr{E}$: current element formula If an **E**-field given by (9.2) or (9.3) exists along an element of conductor **dl**, the contribution to the electromotance is given by (6.4) as $d\mathscr{E} = \mathbf{E} \cdot \mathbf{dl}$ or

moving circuit element: $$\mathrm{d}\mathscr{E} = (\mathbf{v} \times \mathbf{B}) \cdot \mathbf{dl} \tag{9.4}$$

However, this can be simplified by realizing that the scalar product means that **dl** should be resolved in the direction of **E**; i.e. only the component of **dl** perpendicular both to **B** and the velocity $v_\perp$ is significant. Let us call this component $\mathrm{d}l_\perp$ as in Fig. 9.2b. We then have

moving circuit element: $$\mathrm{d}\mathscr{E} = v_\perp B\, \mathrm{d}l_\perp \quad \text{in magnitude} \tag{9.5}$$

as a current element formula for motional electromotance. This can be summed or integrated to obtain the total electromotance round a complete circuit, so that formally we should have

moving circuit: $$\mathscr{E} = \oint (\mathbf{v} \times \mathbf{B}) \cdot \mathbf{dl} = \oint v_\perp B \mathrm{d}l_\perp \tag{9.6}$$

A positive sign should be allocated to one sense round the circuit using, if possible, the rule given under Lenz's law in the previous section. The contribution to $\mathrm{d}\mathscr{E}$ is any element is positive if $\mathbf{v} \times \mathbf{B}$ has a component along the positive direction.

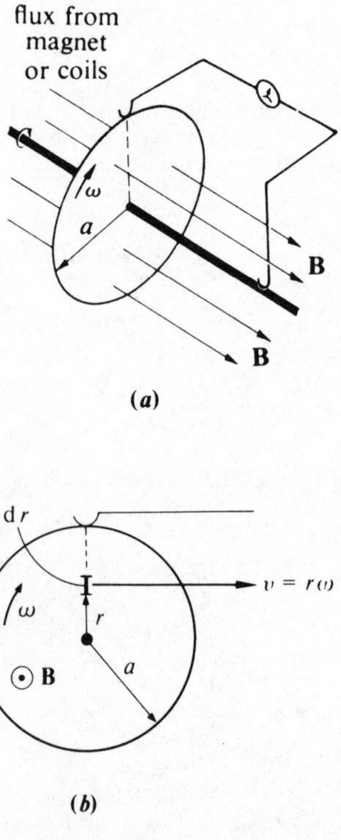

Figure 9.3 Faraday's disc. (a) General view; (b) geometry for calculation of motional electromotance. At the element dr, $(\mathbf{v} \times \mathbf{B})$ is downwards so that this is the positive direction of the induced electromotance and E-field.

Induced motional $\mathscr{E}$: flux-cutting law An alternative way of looking at the above current element formula is obtained by seeing that $(\mathbf{v} \times \mathbf{B}) \cdot \mathbf{dl}$ is the rate at which magnetic flux is cut by the moving element. This is most easily proved using the form $v_\perp B\, \mathrm{d}l_\perp$. Figure 9.2b shows that $v_\perp \mathrm{d}t\, \mathrm{d}l_\perp$ is the area traced out by the element in time dt. Since B is the flux density normal to this area, the flux cut in time dt is $v_\perp B\, \mathrm{d}t\, \mathrm{d}l_\perp$, or just $\mathrm{d}\mathscr{E}\, \mathrm{d}t$. It follows that $\mathrm{d}\mathscr{E}$ is given by the flux cut per unit time. Summing over all elements we get

moving circuit: $$\mathscr{E} = \frac{\mathrm{d}\Phi}{\mathrm{d}t} \quad (\Phi \text{ is flux cut}) \tag{9.7}$$

The flux-cutting law is most useful when only a part of a circuit is moving. The direction of $\mathscr{E}$ is then most easily obtained from that of $(\mathbf{v} \times \mathbf{B})$. Note that the sign of $\mathscr{E}$ is the same as that of $\mathrm{d}\Phi_{\mathrm{cut}}$: if $\mathscr{E}$ is positive in relation to a sign convention, then so is $\mathrm{d}\Phi_{\mathrm{cut}}$ (see Fig. 9.4a).

Worked example *Faraday's disc, a homopolar generator* Suppose a conducting disc is rotating in a uniform **B**-field as shown in Fig. 9.3a. We can calculate the electromotance generated between the rim and the axis by using either the current element formula or the flux-cutting law.

First, consider an element dr of the radius situated at a distance r from the axis as in Fig. 9.3b. The linear velocity of the element and all the charges in it is $r\omega$. The force on one charge Q is given by QvB or $Qr\omega B$ along the radius, and this constitutes an **E**-field F/Q or $r\omega B$. The contribution of the element to the total electromotance is $E\,\mathrm{d}r$ or $r\omega B\,\mathrm{d}r$ [as could be obtained directly from (9.5)]. The total $\mathscr{E}$ is the integral of this from $r=0$ to $r=a$. Hence $\mathscr{E} = a^2\omega B/2$.

Alternatively, using the flux-cutting rule, we see that the angular velocity of the radius of length a is ω. This radius sweeps out area at a rate $\pi a^2 \times \omega/2\pi$, i.e. $\omega a^2/2$. Thus the flux cut per unit time is $a^2\omega B/2$, and this is the total electromotance as before.

Induced motional $\mathscr{E}$: flux-linking law for rigid circuits Provided a circuit does not change its shape during motion, the law (9.7) can be reexpressed in terms of flux **linked.**

Consider a rigid circuit moving as in Fig. 9.4b. The flux **cut** by it in time dt is that flux which crosses the curved surface of the cylinder traced out by it. Call this $-\mathrm{d}\Phi_{\mathrm{cut}}$ (it is negative in the case illustrated, by the rules already discussed, but note that it is *inwards* over the cylinder). The fluxes **linked** by the same circuit in the initial and final positions are those across the *ends* of the cylinder, Φ_{i} and Φ_{f}.

Gauss's law for **B**-fields states that the outward flux over any closed surface is zero. The outward flux over the ends of the cylinder in Fig. 9.4b is $\Phi_{\mathrm{f}}-\Phi_{\mathrm{i}}$ or $\mathrm{d}\Phi_{\mathrm{link}}$. The outward flux across the curved surface of the same cylinder is $-(-\mathrm{d}\Phi_{\mathrm{cut}})$. Hence

$$\mathrm{d}\Phi_{\mathrm{cut}}+\mathrm{d}\Phi_{\mathrm{link}}=0$$

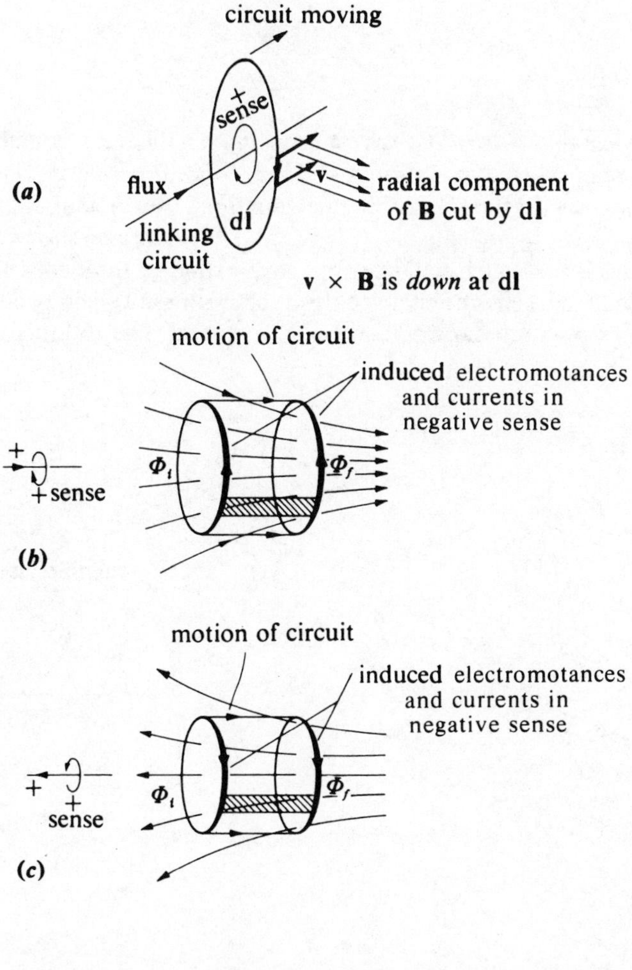

Figure 9.4 Flux cutting and flux linking in the motion of a rigid circuit. (a) If the **B**-field has a component perpendicular to **dl** in the direction shown, the flux cut by **dl** is positive because **v** × **B** is in the positive sense of the circuit as determined by the original flux; (b) and (c) total outward flux over the closed cylindrical surfaces must be zero by Gauss's law for **B**-fields. It follows that the change in flux linkage (over the ends of the cylinder) is the negative of the flux cut (i.e. flux over the curved surface of the cylinder).

Similar arguments apply to other cases such as that of Fig. 9.4c, where the total outward flux is $-d\Phi_{cut} - d\Phi_{link}$. It is always true that $d\Phi_{cut} = -d\Phi_{link}$. Hence for a moving rigid circuit, (9.7) becomes

$$\text{rigid moving circuit:} \qquad \boxed{\mathscr{E} = -\frac{d\Phi}{dt} \qquad (\Phi \text{ is flux linked})} \qquad (9.8)$$

which is usually known as **Faraday's law of electromagnetic induction**. The negative sign means that if the motion is one that increases the flux linkage, the electromotance is in a negative sense round the circuit. It therefore produces currents which oppose the motion.

Note that Eq. (9.8) cannot be used in general for circuits that are not rigid. Many paradoxes arise from just such misuse. Further comment on (9.8) is made at the end of Sec. 9.3.

Worked example *rotating coil in a uniform* **B**-*field* A coil of N turns each of mean area A rotates in a uniform magnetic **B**-field at an angular velocity ω (Fig. 9.5a). At any instant when the normal to the plane of the coil makes an angle θ with **B**, the flux linkage is $NAB\cos\theta$. If the rotation is uniform then $\theta = \omega t$ and Φ_{link} is $NAB\cos\omega t$. By (9.8) the induced electromotance is thus

$$\text{rotating coil:} \qquad \mathscr{E} = \mathscr{E}_0 \sin\omega t \qquad (9.9)$$

where $\mathscr{E}_0 = NAB\omega$.

Practical application *alternating current generators* If the leads from the above rotating coil are connected to slip rings and the induced current is taken off by brushes in contact with the rings, the time variation is sinusoidal and forms a single-phase alternating current (a.c.) (Fig. 9.5b). The figure also shows the form of a three-phase output which could be obtained by winding three coils at 120° to each other. Small a.c. generators are of this type, with the **B**-field produced by further windings on stationary pole-pieces (the stator). The rotating coils are

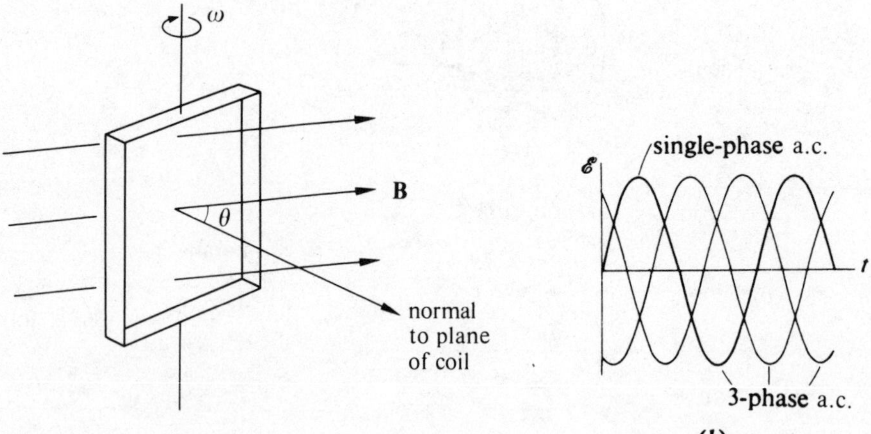

Figure 9.5 (a) Simple alternating current generator. The plane coil has N turns each of area A. (b) The form of single-phase and three-phase a.c.

wound on a steel drum armature which completes the magnetic circuit (Sec. 12.7). Direct current can be generated by using commutators instead of slip rings and a large number of windings each of which can be arranged to be connected to the output only at the point of maximum induced electromotance.

In large a.c. generators, the field coils and their pole-pieces form the rotor, while the armature coils are static because of the heavy duty wiring needed to carry the much greater currents.

Practical application *standardization of the watt* An ingenious method devised at the UK National Physical Laboratory for relating the watt and hence the ampere to the mechanical SI units involves two stages of experimentation. In the first, a current-carrying coil of many turns is suspended in a **B**-field and experiences a force in the vertical x direction given by (7.48) as $I(\partial\Phi/\partial x)$ where I is the steady current flowing in it and Φ is the magnetic flux linked by it. This force is balanced by a weight mg, so that $mg = I(\partial\Phi/\partial x)$. The same coil is then moved in the x direction in the same **B**-field with a velocity $u = \partial x/\partial t$, so that an induced electromotance $\mathscr{E}$ of magnitude $\partial\Phi/\partial t$ is induced in it. Since $\partial\Phi/\partial t$ can be written as $(\partial\Phi/\partial x)(\partial x/\partial t)$ or $u(\partial\Phi/\partial x)$, it follows that the product $\mathscr{E}I$ is simply mgu, so that a wattage is expressed solely in terms of mechanical quantities. By balancing the electromotance $\mathscr{E}$ against the voltage drop across a standard resistor R carrying the current I, the current is then given by $(mgu/R)^{1/2}$ (Kibble, Smith and Robinson, 1983; see also Appendix C).

9.3 Transformer or changing-field electromotance

Flux-linking law for $\mathscr{E}$ Equation (9.8) was derived for a rigid circuit moving in a magnetic field that did not change in time. However, Faraday's experiment and common experience tells us that what matters is the *relative* motion between the rigid circuit and the sources of the **B**-field. We shall get the same induced electromotances and currents whether the circuit moves with **v** relative to the source of **B** or the source of **B** moves with $-\mathbf{v}$ relative to the circuit. This means that (9.8) applies also to a stationary circuit across which the Φ_{link} changes in time because the source of the flux is moving.

However, in Chapter 7 we saw that there is no distinction to be drawn between **B**-fields from various sources: they all produce the same effects. In that case, it should not matter *why* the flux of **B** across a stationary circuit changes in time, so we expect (9.8) to apply if Φ_{link} alters for any reason whatsoever. There need be no motion of the source of flux, but perhaps just a varying current in another stationary coil (i.e. mutual induction).

We make just one alteration to (9.8) before we apply it to transformer electromotances. We make the derivative a partial one because only changes in Φ with *time* are relevant, not changes with position. Hence transformer electromotances are given by

$$\text{transformer electromotance:} \qquad \boxed{\mathscr{E} = -\frac{\partial\Phi}{\partial t}} \qquad (9.10)$$

where Φ is the flux linked by the stationary circuit. This is **Neumann's law**.

Applications of the law to self-induction and mutual induction are to be found in later sections of this chapter. The assumptions made in the above derivation are amply justified by direct consequences, particularly in a.c. circuits (Chapter 10).

Equation (9.10) applies when Φ_{link} changes only with time, and Eq. (9.8) when it changes only with position. If both types of change occur, the total electromotance will be the sum of the two and (9.8) could be taken to imply that result. This would be the case if a rigid circuit moved through a **B**-field that was non-uniform and varied in time.

9.4 Induced currents and charges

If the electromotances produced by electromagnetic induction occur in conducting material, they will cause currents to flow. We consider first the currents and charges in circuits formed by wires, and later in the section deal briefly with currents in larger volumes of conductor.

Circuit of conducting wire According to Eq. (6.32), a total electromotance $\mathscr{E}$ in a conducting circuit of total resistance R produces a current $I = \mathscr{E}/R$. If the electromotance is entirely due to electromagnetic induction, then the current is given by

circuit with flux change:
$$\boxed{I = -\frac{1}{R}\frac{\mathrm{d}\Phi}{\mathrm{d}t}} \tag{9.11}$$

while the charge flowing in a circuit in which the flux linkage changes from Φ_{i} to Φ_{f} in a time from t_{i} to t_{f} will be

$$Q = \int_{t_{\text{i}}}^{t_{\text{f}}} I\,\mathrm{d}t = -\frac{1}{R}\int_{\Phi_{\text{i}}}^{\Phi_{\text{f}}} \mathrm{d}\Phi$$

i.e.

circuit with flux change:
$$\boxed{Q = \frac{\Delta\Phi}{R}} \tag{9.12}$$

where $\Delta\Phi$ is the change in flux linkage.

Practical application *search coil or exploring coil* An important method for the measurement of Φ or **B** depends upon the electromotance induced in a small coil of N turns each of mean area A placed in the field. The flux linkage is changed in some way (see next paragraph) by an amount $\Delta\Phi$ over a time Δt and it is then given by $R\int I\,\mathrm{d}t$ or $\int \mathscr{E}\,\mathrm{d}t$ using (9.7) or (9.8) and (9.11), both integrals being over the time interval Δt. The coil therefore needs to be connected to an instrument which will carry out the integration of current or voltage. The classical instruments used for this are the moving-coil fluxmeter and ballistic galvanometer which effectively integrate the current, but electronic integrators have become available giving $\int \mathscr{E}\,\mathrm{d}t$ directly. The change in **B**-field will be $\Delta\Phi/NA$, yielding a mean value for B over the area of the coil (see also Sec. 12.10). For rough measurements it may be sufficient to calculate NA from the geometry of the coil. For greater accuracy the arrangement is calibrated with a standard B obtained from a current-carrying solenoid or with a standard $\Delta\Phi$ from a mutual inductance [Eq. (9.29)].

If the **B**-field is steady and invariable, then changes in Φ can only be made by moving the coil, often by rotating it. If, however, **B** can be varied (e.g. by switching off a current) or if time variations in **B** are required anyway, the coil can be stationary and the induced transformer electromotance used. See also Comment C12.2.

Induced volume currents When a conductor *moves* with a velocity **v** through a magnetic field **B**, there is an **E**-field of $\mathbf{v} \times \mathbf{B}$ generated at every point (as we saw in Sec. 9.2). From Eq. (6.31), the resultant current density at any point where the conductivity is σ will be $\sigma(\mathbf{v} \times \mathbf{B})$. In general, there may also be other electric fields not dependent on the motion, which can be included in the $\mathbf{E}_Q$ term of (6.31). The current density at any point in a moving conductor is therefore given by

conductor moving in **E** and **B**:
$$\boxed{\mathbf{J} = \sigma(\mathbf{E}_Q + \mathbf{v} \times \mathbf{B})} \qquad (9.13)$$

In the notation of Sec. 6.1, the $(\mathbf{v} \times \mathbf{B})$ field can be classified either as an $\mathbf{E}_M$ (from the point of view of the conductor) or as an **e** given by $\mathbf{v} \times \mathbf{B}$ (from the laboratory point of view). It seems that whether the field experienced is classified as electric or magnetic depends on the observer. This is a relativistic aspect which we discuss further in Chapter 13.

Eddy currents The currents induced in massive metallic components (e.g. in the armatures of generators and motors and in transformer cores) are known as **eddy currents** and are generally undesirable because of the resultant RI^2 losses. In addition, if the metal is moving, the currents flow in such directions as to produce forces opposing the motion and cause what is known as *eddy current damping.* Eddy currents are minimized by laminating the metal, using high-resistance materials or dispersing the metal in a non-conducting matrix.

On the other hand, eddy current damping can be put to good use (e.g. in electricity supply watt-hour meters, where the Arago's disc principle is used to control the speed of the motor). In addition, test instruments using induced eddy currents are available for measuring the resistivity, magnetic permeability or surface roughness of components used in large structures such as nuclear reactors and also for detecting the presence of defects such as cracks or voids (Holt, 1988).

9.5 Self-inductance

Concept and definition In the absence of magnetic materials, the **B**-field at any point due to a single circuit of given shape is proportional to the current in it, since the current is measured by its magnetic effect. It follows that the total flux Φ linking a circuit due to its own current I is proportional to I. We define the self-inductance L of a circuit as the ratio of Φ to I so that

$$\Phi = LI \qquad \text{(definition of } L\text{)} \tag{9.14}$$

and L is thus a constant for a circuit of given shape and size. The self-flux of a circuit defines the positive direction for Φ so that L is essentially positive. Its **SI unit**, the $\mathrm{Wb\,A^{-1}}$, is called the **henry**, symbol H.

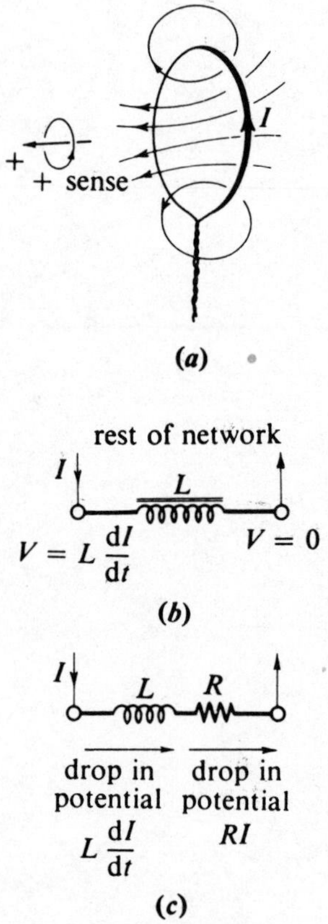

Figure 9.6 (a) Self-flux of a circuit: increase of flux due to an increase in I causes an electromotance in the negative sense and this opposes the growth of I; (b) self-inductance as a circuit element; (c) the equivalent circuit of a self-inductor with resistance.

Self-induction When the current in a rigid circuit changes, the induced electromotance, by (9.10), is

$$\mathscr{E} = -L\,\mathrm{d}I/\mathrm{d}t \tag{9.15}$$

the negative sign indicating that when I increases, $\mathscr{E}$ is negative and is thus in a direction opposing the growth of I (Fig. 9.6a). When a device is specially constructed so that (9.14) or (9.15) applies, it is known as an **inductor** and has the symbol in Fig. 9.6b in a circuit diagram, crossed by an arrow if variable and with parallel lines adjacent to it if it has a ferromagnetic core.

In a network with varying currents, an inductor acts as an electromotance $-L\,\mathrm{d}I/\mathrm{d}t$ with an internal resistance which cannot be avoided. Like a cell, therefore, it can be represented by an equivalent network consisting of a pure inductance L in series with its resistance R as in Fig. 9.6c. The pure inductance has no resistance and thus has a potential difference across its ends equal to the electromotance: once the direction of I has been chosen arbitrarily in any problem, the potential drop across an inductor L is $L\,\mathrm{d}I/\mathrm{d}t$ just as that across R is RI (Fig. 9.6c).

Growth and decay of current in an inductor In Fig. 9.7 the resistance R includes the internal resistance of the cell and of the inductor as well as any other resistance in the circuit. If I is the current flowing at any time t after the switch K_1 is closed then we again assume that, in spite of varying currents, the steady current laws apply at any instant (cf. the treatment in Sec. 6.8). Equating the total potential drop round the circuit to zero gives

$$\mathscr{E} - L\,\mathrm{d}I/\mathrm{d}t - RI = 0 \tag{9.16}$$

or

$$L\,\mathrm{d}I/\mathrm{d}t + RI = \mathscr{E} \tag{9.17}$$

which has the same form as (6.22) and is solved in the same way. The variable is now I and the initial condition is $I=0$ when $t=0$ because any instantaneous increase in I would mean that $\mathrm{d}I/\mathrm{d}t$ was infinite and this would produce an infinite back-electromotance in L. Thus

growth of I in L:
$$I = I_0(1 - \mathrm{e}^{-Rt/L}) \tag{9.18}$$

where $I_0 = \mathscr{E}/R$ and is the current in the circuit when $r \to \infty$.

When the switch is connected to K_2 the current is given by

$$L\,\mathrm{d}I/\mathrm{d}t + RI = 0 \tag{9.19}$$

which, with the initial condition $I = I_0$ when $t=0$, gives

decay of I in L:
$$I = I_0\mathrm{e}^{-Rt/L} \tag{9.20}$$

The growth (9.18) and the decay (9.20) of the current are plotted in Fig. 9.8. The time L/R is the time constant or relaxation time of the circuit just as CR was in Sec. 6.8.

Because *any* circuit has self-inductance, a contact which is broken while a current is flowing is equivalent to the introduction of a very high resistance. The time constant L/R thus suddenly decreases to a very small value and the *rate of collapse* of current is very high even if the current itself is small. The resultant induced electromotance is often sufficient to cause a spark to jump across the gap. A quantitative analysis is more complex than at first sight appears, because of the capacitance also introduced.

If the time constant L/R is large compared with the time of growth in (9.18), then the approximation $e^{-x} \approx 1-x$ can be used and the growth is nearly linear: Eq. (9.18) becomes $I=(I_0R/L)t$. If some form of rapid decay is provided in such a way that linear growth and a sharp decay are repeated, a saw-tooth current waveform is produced. In this way, the LR combination can form the basis of the **linear current sweep circuits** required for television and radar receivers, where the beam deflections for line or frame scanning are produced magnetically (Sec. 7.5).

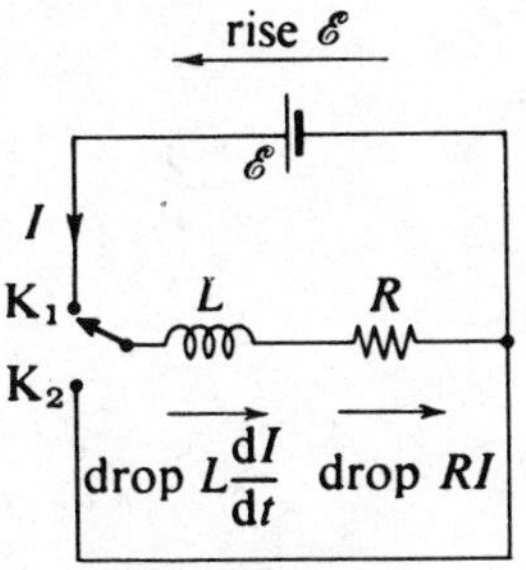

Figure 9.7 Potential differences at any instant when current grows in a self-inductance with resistance. When the switch is thrown to K_2, the decaying current causes the same potential differences, so that $L\,dI/dt + RI = 0$ ($L\,dI/dt$ is negative).

Inductor applications An inductor is not such a simple component to manufacture as a capacitor and for that reason is not so common. However, in combination with capacitors, inductors form an essential part of filters and smoothing circuits where their opposition to variations in current strength comes into play. They are also used in tuned circuits (Sec. 10.6) and current sweep circuits (above). We shall see in Chapter 10 how their properties are consistently to be contrasted with those of capacitors.

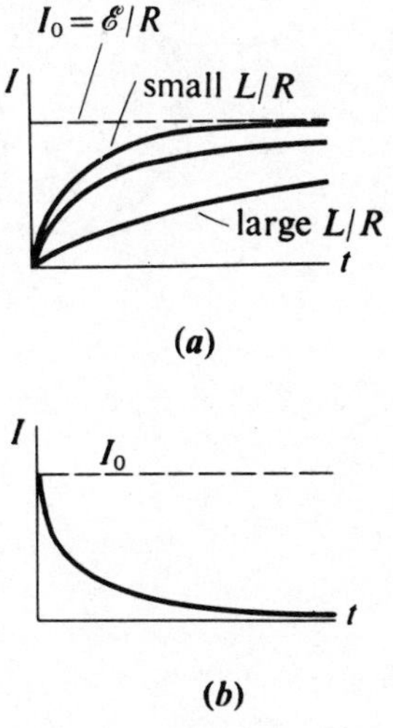

Figure 9.8 (a) Growth and (b) decay of current in the network of Fig. 9.7.

9.6 Calculation of self-inductance

Like capacitance, self-inductance can either be localized between two terminals in a circuit (a *lumped* component) or *distributed* throughout the length of a circuit. In simple cases, both of these can be calculated by using the definition (9.14). The first example below illustrates the calculation for the most common form taken by an inductor, i.e. a solenoid. We then deal with several examples of distributed self-inductance, and finish with the problem of combinations and the effect of media other than air or vacuum on the value of L.

Worked example *self-inductance of a solenoid* If the turns of a helically wound solenoid are close enough and if the length is much greater than the diameter, the **B**-field to a good degree of accuracy is $\mu_0 nI$ over the whole cross-section from (7.17), n being the number of turns per unit length. The flux across each turn of area A is thus $\mu_0 nIA$ and, because there are nl turns, the total self-flux is $\mu_0 n^2 IlA$. Hence

ideal solenoid: $$L_\infty = \mu_0 n^2 lA = \mu_0 N^2 A/l \qquad (9.21)$$

N being the total number of turns. This gives the self-inductance of an ideal solenoid considered as part of an infinite solenoid. Finite solenoids have $L=bL_\infty$ where b is a function of the ratio of diameter to length which can only be obtained by more advanced methods. The following table gives values of b for various coils from which it is seen that (9.21) is excellent for yielding an order of magnitude.

Diameter/length	0	0.5	1.0	10.0
b	1	0.8	0.7	0.2

Worked examples *distributed inductance of transmission lines* Transmission lines were found in Sec. 5.2 to possess distributed capacitance and it is clear from Fig. 9.9 that either type of line will also possess self-inductance. We first obtain a value for L by neglecting the flux within the wires or conductors themselves. For the twin cable of Fig. 9.9a, the magnetic field B at r from one wire is $\mu_0 I/2\pi r + \mu_0 I/2\pi(d-r)$. The flux across an elementary strip is B multiplied by the area $l\,dr$ and the total flux is

$$\Phi = \int_{a}^{d-a} \mu_0 I[dr/r + dr/(d-r)]/2\pi \qquad \text{per unit length}$$

and hence

$$\text{twin cable:} \qquad L = \frac{\mu_0}{\pi}\log_e\left(\frac{d-a}{a}\right) \qquad \text{per unit length} \qquad (9.22)$$

$$\approx \frac{\mu_0}{\pi}\log_e\frac{d}{a} \qquad \text{per unit length if } d \gg a$$

In a similar way

$$\text{coaxial cable:} \qquad L = \frac{\mu_0}{2\pi}\log_e\frac{b}{a} \qquad \text{per unit length} \qquad (9.23)$$

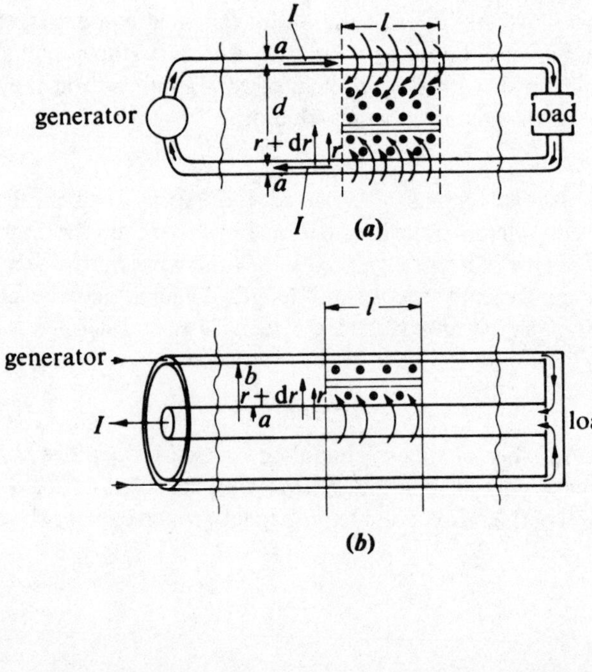

Figure 9.9 Magnetic flux in (a) a twin cable, (b) a coaxial cable.

for a coaxial cable (Fig. 9.9b), for here the current in the outer conductor produces no flux inside it.

Skin effect The neglect of flux within the conductors in the above calculations would be valid if all the current flowed in the outer surfaces of the twin wires and in the adjacent surfaces of the coaxial cable. When the current alternates at high frequency this condition in fact arises as follows and is known as the **skin effect**. Figure 9.10 shows a current-carrying loop of wire drawn with an exaggerated cross-section for clarity. If the current is steady, the elementary filament in the centre of the wire (midway between numbers 1 and 2) will link most flux, numbers 1 and 2 will link less, while 3 and 4 will link only the flux external to the wire. If the current changes in time, the induced electromotance opposing the change is greatest in the centre filament and least in filaments 3 and 4. For alternating electromotances, the effect is to create a greater impedance (Sec. 10.4) to current flow along the centre than along the surface, with the result that more current tends to flow in the surface. The effect is very marked at high frequencies and makes (9.22) and (9.23) highly accurate. (See Sec. 14.3 for another aspect of this effect.)

Worked example *self-inductance of a wire* At lower frequencies the current will be more evenly distributed and the flux in a wire will contribute to self-inductance. We cannot take the easy way out and assume that if the diameter of the wire is negligible the flux in the wire may be neglected, for the flux density B near the surface given by $\mu_0 I/(2\pi r)$ (7.19) tends to infinity as r tends to zero and the flux linked by the circuit, and hence the self-inductance, also become infinite [see also (9.22) and (9.23)].

However, assuming that the cross-section remains finite and the current density across it uniform, we can divide the current into infinitesimal filaments for

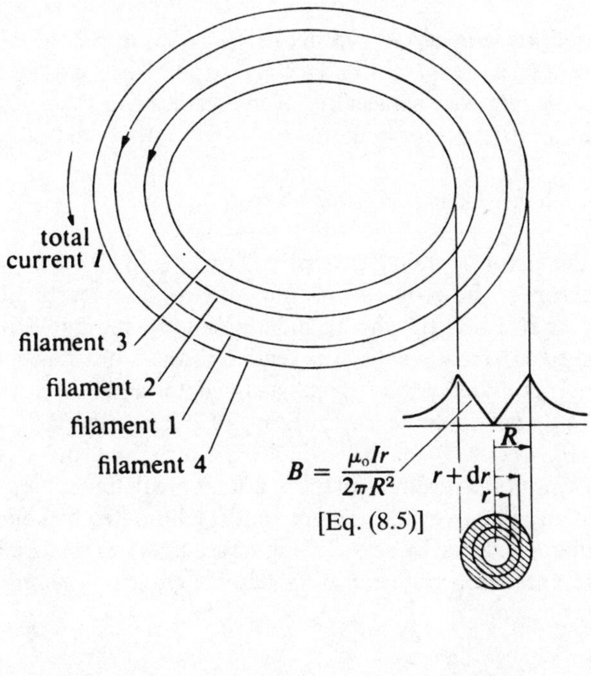

Figure 9.10 A wire circuit carrying a current I.

each of which the self-flux tends to zero. We then calculate the flux linked by every filament due to the current in all the others and average the L's of the filaments so obtained over the cross-section.

In Fig. 9.10, all the filaments at a distance r from the central axis of the wire link the flux in the shaded area. By a method similar to that for the transmission lines, this is

$$\int_r^R \mu_0 I x l \, \mathrm{d}x / 2\pi R^2$$

or $\mu_0 I l(R^2 - r^2)/(4\pi R^2)$ for a length l of wire. The number of filaments linking this particular flux is a fraction $2\pi r \, \mathrm{d}r/(\pi R^2)$ of the total filaments in the wire and hence the contribution of this bundle to the self-inductance is

$$\frac{\mu_0 l(R^2 - r^2) r \, \mathrm{d}r}{2\pi R^4}$$

L is therefore the integral of this from 0 to R or

length of wire: $$L = \mu_0 l / 8\pi$$

Thus even a straight wire has a self-inductance of about $1/20\,\mu\mathrm{H\,m^{-1}}$ due solely to the flux in the wire. This will only apply when the skin effect is negligible and will even then be small compared with the contribution from the flux outside. (We have neglected the permeability of the material of the wire, but this is only of importance when it is ferromagnetic—iron or nickel, for instance.)

Combinations of self-inductors It is easy to show from the definition that inductors in series have an equivalent inductance equal to the sum of the separate inductances provided there is no mutual flux linkage between one and another. Parallel and more complex combinations will be left until we deal with the general theory of circuits containing varying currents in Chapter 10.

Inductors in practice: permeability As we shall see in Chapter 12 the effect of a medium filling the whole of the space round a coil is to change its inductance by a factor μ, known as the **relative permeability** of the medium so that, if L_0 is the inductance *in vacuo* and L_m in the medium,

$$L_m/L_0 = \mu_r \qquad \text{(definition of } \mu_r\text{)} \tag{9.24}$$

which means that the factor μ_0 should be replaced by $\mu_r\mu_0$ in all the expressions for inductance in terms of the geometry of the inductor. For nearly all media, however, μ_r is so close to 1 that the change in inductance is negligible, but for a limited class, described as ferromagnetic, μ_r is very large and is not constant as the magnetic field changes: thus although ferromagnetic materials allow us to obtain large inductances in a small volume, the linearity of L is sacrificed.

Standard inductors, whether fixed or variable, are air-cored and wound on formers in such a way that the positions of the windings are definite; temperature variations of inductance due to expansion are almost eliminated by compensating the increase in diameter by an increase in length [see Eq. (9.21) in which A and l can be made to increase in proportion by a suitable choice of materials].

Inductors for use in electronic **power circuits** will be iron-cored at low frequencies and air-cored at high frequencies because the reactance ωL (Sec. 10.4) rather than L itself is the quantity required. In **microwave circuits**, inductance can be provided by sections of transmission line (Sec. 10.10), while in miniaturized or **hybrid** circuits it can be produced by metallic strips in the shape of loops or spirals mounted on a non-conducting substrate.

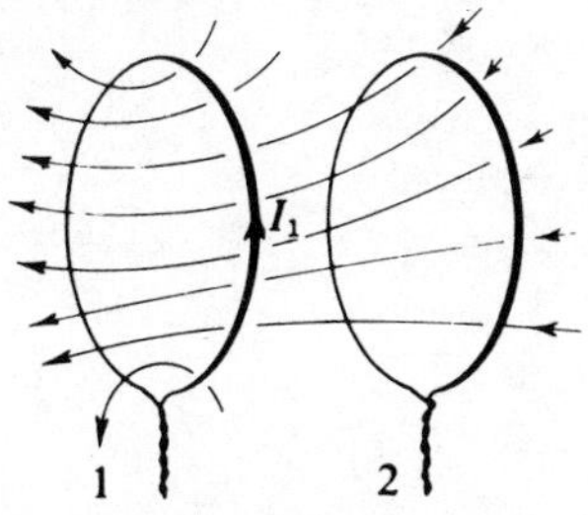

(*a*)

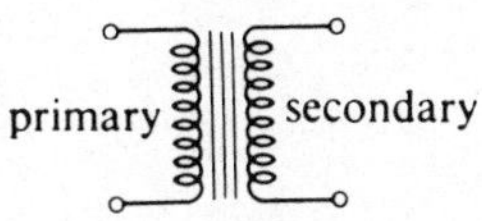

(*b*)

Figure 9.11 (a) Mutual inductance: magnetic flux links circuit 2 due to current I_1 in circuit 1; (b) network symbol for mutual inductance.

9.7 Mutual inductance

Concept and definition When a current I_1 flows in one of two circuits as in Fig. 9.11 the flux Φ_2 linking the other, for a given geometry, is proportional to I_1 in the absence of magnetic materials so that

$$\Phi_2 = M_{12}I_1 \qquad \text{(definition of } M_{12}) \tag{9.25}$$

where M_{12} is a constant for the pair of circuits in their specified positions known as their **mutual inductance** and measured, like L, in henries. Similarly, a current I_2 in 2 produces a flux linkage

$$\Phi_1 = M_{21}I_2 \qquad \text{(definition of } M_{21}) \tag{9.26}$$

and we shall later (Sec. 9.9) show that $M_{12} = M_{21} = M$.

The sign given to the flux linking a circuit has been taken as positive when in the same direction as any self-flux. If there is no self-flux, then only the external flux can be used to allocate a sign, taken conventionally as positive. It follows that, if currents flow in both of two circuits, M may be positive or negative (see Fig. 9.13 and Sec. 10.1).

Mutual induction If the current I_1 changes, the induced electromotance in circuit 2 is given by

$$\mathscr{E}_2 = -M\frac{dI_1}{dt} \tag{9.27}$$

the significance of the negative sign being indicated in Fig. 9.1b. As a circuit element, a **mutual inductor** is represented as in Fig. 9.11b, the parallel lines indicating an iron core if present. With such a core the inductor is usually a transformer (Sec. 10.8), one coil of which is used for the input of power (the primary) and the other for the output (the secondary).

In circumstances where the circuits are in relative motion and I_1 is simultaneously varying with time,

$$\mathscr{E}_2 = -M\frac{dI_1}{dt} - I_1\frac{dM}{dt} \tag{9.28}$$

a combination of transformer and motional electromotances.

A known current I reversed in the primary of a standard mutual inductor gives a **standard flux change**

$$\Delta\Phi = 2MI \tag{9.29}$$

in the secondary, and this can be used to calibrate a fluxmeter. The charge flowing in the secondary circuit is, from (9.12), $Q=2MI/R$.

Neumann's formula for M By using expressions involving vector potential from Sec. 8.5, we can derive a general formula for mutual inductance. Consider a circuit C_1 carrying a current I_1 which produces a flux Φ_2 across a second circuit C_2. M is then given by Φ_2/I_1. However, Φ_2 can be written as $\oint \mathbf{A}\cdot d\mathbf{l}_2$ round C_2 according to (8.16), where $\mathbf{A}$ is the vector potential due to C_1. Now using (8.19), $\mathbf{A}$ can be written as $\oint \mu_0 I_1\, d\mathbf{l}_1/(4\pi r_{12})$ taken round C_1. Altogether this means that

$$M=\oint_{C_2}\oint_{C_1}(\mu_0\, d\mathbf{l}_1\cdot d\mathbf{l}_2)/(4\pi r_{12}) \qquad (9.30)$$

which is known as *Neumann's formula* for mutual inductance. It is clearly symmetrical between the two circuits and provides a proof that $M_{12}=M_{21}$. It also reveals the purely geometrical character of M.

9.8 Calculation of mutual inductance

In simple cases, mutual inductance may be calculated by finding the magnetic flux linked by *either* circuit due to unit current in the other.

Worked example *coaxial solenoids* A convenient form of fixed mutual inductance is a pair of coaxial solenoids as in Fig. 9.12a. Using the notation of the figure, a current I in the longer solenoid produces a B-field of $\mu_0 n_1 I$ inside itself. This field will be approximately constant over the whole length of the short solenoid, each turn of which links a flux $\mu_0 n_1 IA$. Hence

coaxial solenoids: $$M=\mu_0 n_1 n_2 lA \qquad (9.31)$$

Worked example *small coil on the axis of a large coil* In Fig. 9.12b, the small coil is situated in a flux density $B=\mu_0 N_1 I a_1^2/[2(a_1^2+x^2)^{3/2}]$ when I flows in the large one. The flux linked is $N_2\pi a_2^2 B\cos\theta$ and so

two coils: $$M=(\mu_0 N_1 N_2 \pi a_1^2 a_2^2\cos\theta)/[2(a_1^2+x^2)^{3/2}] \qquad (9.32)$$

Two coupled coils The coils of Fig. 9.13 have a mutual inductance of magnitude M. If Q and R are connected and a current I passed through both coils in series

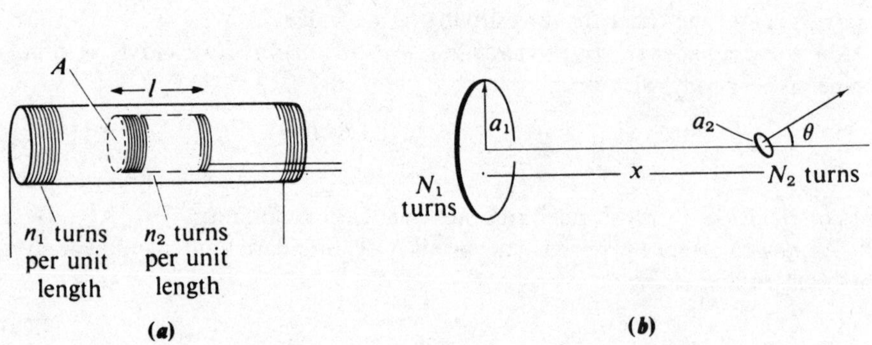

Figure 9.12 Geometry for calculation of mutual inductance between (a) coaxial solenoids, (b) two plane coils.

then, considered as one circuit, the total flux linkage is L_AI+L_BI+2MI and the self-inductance is

coupled coils, $+M$: $$L_1=L_A+L_B+2M \qquad (9.33)$$

If, on the other hand, Q and S are connected and a current passed from P to R the total flux linkage is now L_AI+L_BI-2MI (the self-fluxes must be positive and the mutual fluxes oppose them) so that

coupled coils, $-M$: $$L_2=L_A+L_B-2M \qquad (9.34)$$

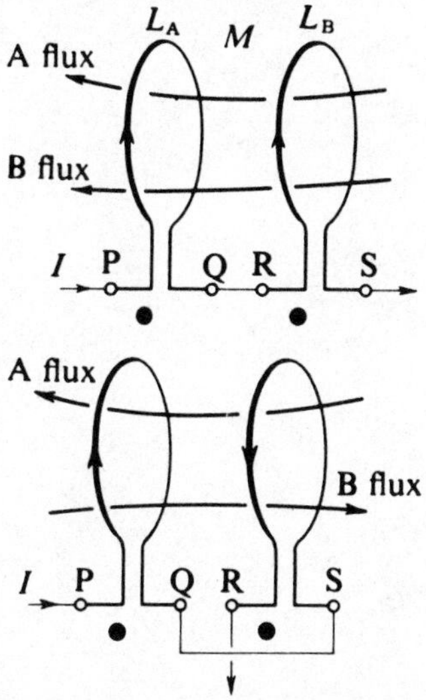

Figure 9.13 Two coupled coils. The dot convention is used even when the coils are separate windings of a mutual inductor and has the following meaning: if the current enters both coils (or leaves both coils) at the dot-marked terminals, the mutual and self-fluxes reinforce each other; otherwise the fluxes are in opposition. See also Sec. 10.1.

The same effect could be obtained by retaining the connections but rotating one coil through 180°—a method which can therefore be used to obtain a self-inductance varying smoothly from L_1 to L_2. Alternatively, because $M=\frac{1}{4}(L_1-L_2)$ from (9.33) and (9.34), a mutual inductance may be obtained from measurements only of self-inductance, which are often simpler.

A transformation which is sometimes useful is illustrated in Fig. 9.14. Two coupled coils with self-inductances L_A and L_B and a mutual indutance M connected between three points A, B, and C are equivalent to the star or Y of L's if the effective inductances between the three points taken in pairs are equal.

AB: $L_A+L_B\pm 2M=L_1+L_2$;
BC: $L_B=L_2+L_3$;
CA: $L_A=L_3+L_1$

Solved for L_1, L_2 and L_3, these yield

M–L transformation: $$L_1=L_A\pm M;\; L_2=L_B\pm M;\; L_3=\mp M \qquad (9.35)$$

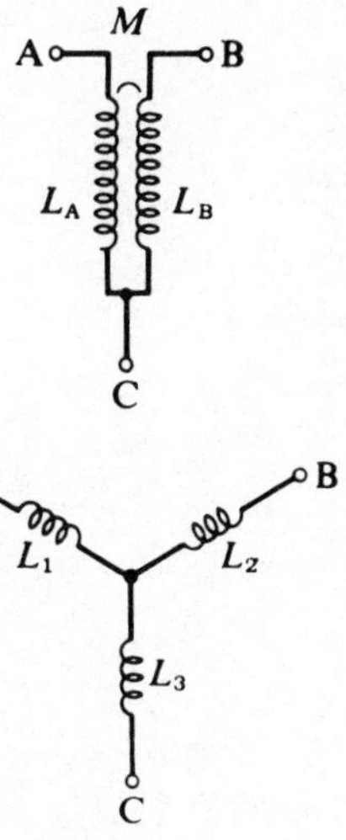

Figure 9.14 These two networks have equivalent effects between A, B and C if $L_1=L_A\pm M$, $L_2=L_B\pm M$, $L_3=\pm M$.

Coefficient of coupling Some care is needed when considering the fluxes linking circuits of many turns. Suppose a coil A has N_A turns and the flux *through each turn* is Φ_A when I_A flows in it. Then its self-inductance L_A is given by $N_A\Phi_A=L_AI_A$. Similarly for a second coil B, $N_B\Phi_B=L_BI_B$. If these two coils are placed so that a mutual inductance M_{AB} exists between them, then the flux linking *each turn* of B due to A cannot be greater than Φ_A itself, i.e. it would be $k_1\Phi_A$ where $k_1\leq 1$. Hence the total mutual flux linking B is $N_Bk_1\Phi_A$, so that

$$M_{AB}=N_Bk_1\Phi_A/I_A=N_Bk_1L_A/N_A$$

by the relation for L_A above. It similarly follows that

$$M_{AB}=N_Ak_2L_B/N_B$$

where $k_2\leq 1$. Multiplying the two expressions for M_{AB} gives $M_{AB}^2=k_1k_2L_AL_B$, so that

$$\boxed{M_{AB}/\sqrt{(L_AL_B)}=k \qquad \text{(definition of } k\text{)}} \qquad (9.36)$$

where k is a constant for a given pair of circuits in given positions known as the **coefficient of coupling**, with a value which cannot exceed 1. Coils with k above about 0.5 are said to be tightly coupled.

Effect of medium As with self-inductance, the presence of materials other than vacuum in the region of two circuits affects the mutual inductance. If such a medium fills the whole of the space in which **B**-fields exist, then the mutual inductance is increased by the relative permeability μ_r. In practice, the effect is only significant for ferromagnetic and similar substances where μ_r can be large and non-linear. The non-linearity, as we shall see in Sec. 12.5, means that M_{12} is no longer equal to M_{21}. In this chapter we assume that any materials are linear.

9.9 Magnetic energy 1: in terms of L, M and I

A set of wire circuits carrying currents possesses more energy than the same set with no currents because of the work done in establishing them against induced electromotances. We shall calculate this energy in stages, first in a single circuit, next in two circuits, and finally in a set.

Magnetic energy due to self-inductance Consider the growth of current I in a circuit containing self-inductance L and resistance R as in Sec. 9.5. The rate at which energy is being transferred from the source of electromotance to the rest of the circuit at any instant is $I\mathscr{E}$ and from (9.17) we see that this is equal to $LI\,\mathrm{d}I/\mathrm{d}t+RI^2$. The second term is recognizable as the rate of consumption of energy in the resistance and it follows that the first represents the rate of storage in the inductance. In a time $\mathrm{d}t$, therefore, the energy stored is $LI\,\mathrm{d}I$ and, integrating from $I=0$ to $I=I_0$, the final current, gives

single circuit:
$$\boxed{U_M=\tfrac{1}{2}LI_0^2} \tag{9.37}$$

The justification for calling the energy *stored* is that, when the current decays with no input of energy from $\mathscr{E}$, a further amount of energy $\frac{1}{2}LI_0^2$ is dissipated in R, as the reader can verify by integrating $RI^2\,\mathrm{d}t$ from 0 to ∞ using (9.20).

Magnetic energy due to mutual inductance Consider now two circuits with self-inductances L_1 and L_2 and a mutual inductance M_{21} as defined in (9.26). We do *not* now assume that $M_{21}=M_{12}$.

First insert an electromotance (e.g. a cell) in circuit 1 so that the current in it grows as above to its final value I_1, keeping any induced currents in 2 at zero by extra electromotances (these do no work since $I_2=0$). The energy stored, as above, is $\frac{1}{2}L_1I_1^2$.

Now insert an electromotance in circuit 2 to make the current in it grow to its final value I_2. When it has done so, naturally an energy $\frac{1}{2}L_2I_2^2$ is stored in 2, but while I_2 is changing an extra electromotance $-M_{21}\,\mathrm{d}i_2/\mathrm{d}t$ is induced in circuit 1, i_2 being the instantaneous value. We insert in 1 an adjustable electromotance of value $+M_{21}\,di_2/\mathrm{d}t$, thus keeping I_1 constant (and not upsetting the $\frac{1}{2}L_1I_1^2$ already stored). This time, however, the extra electromotance passes the current I_1 and thus puts energy into the system at the rate $I_1M_{21}\,\mathrm{d}i_2/\mathrm{d}t$. The total energy put in by this extra electromotance is therefore $I_1M_{21}\,\mathrm{d}i_2$ integrated from 0 to I_2, or $M_{21}I_1I_2$.

Hence for a pair of circuits

$$U_M=\tfrac{1}{2}L_1I_1^2+M_{21}I_1I_2+\tfrac{1}{2}L_2I_2^2$$

and because the same situation can be reached by allowing I_2 to grow first and finding that U_M is then $\frac{1}{2}L_1I_1^2+M_{12}I_1I_1+\frac{1}{2}L_2I_2^2$, it follows that $M_{12}=M_{21}$:

there is only one mutual inductance M between two circuits and the magnetic energy is

two coupled circuits: $$U_M = \tfrac{1}{2}L_1 I_1^2 + M I_1 I_2 + \tfrac{1}{2}L_2 I_2^2 \tag{9.38}$$

This can also be written as

$$U_M = \tfrac{1}{2}I_1(L_1 I_1 + M I_2) + \tfrac{1}{2}I_2(L_2 I_2 + M I_1) \tag{9.39}$$
$$= \tfrac{1}{2}I_1 \Phi_1 + \tfrac{1}{2}I_2 \Phi_2 \tag{9.40}$$

where Φ_1 and Φ_2 are *total* magnetic fluxes linking each circuit.

Magnetic energy in a set of circuits For any number of circuits with self-inductances $L_1, L_2, \ldots, L_i, \ldots$, and mutual inductances $M_{12}, M_{13}, M_{23}, \ldots, M_{ij}, \ldots$, Eq. (9.38) can be extended to

N coupled circuits: $$U_M = \sum_i \tfrac{1}{2}L_i I_i^2 + \sum_i \sum_j M_{ij} I_i I_j \tag{9.41}$$

and by splitting each M term into two halves and grouping as in (9.39)

N coupled circuits: $$U_M = \sum_i \tfrac{1}{2} I_i \Phi_i \tag{9.42}$$

It is easy to show that for small increments of all the currents by dI_i, the increment in U_M is

$$dU_M = \sum I_i \, d\Phi_i = \sum \Phi_i \, dI_i \tag{9.43}$$

because I and Φ are linearly related (but see Sec. 12.5).

9.10 Magnetic energy 2: in terms of B

We now express U_M in terms of B in a way similar to that in which we expressed U_E in terms of E in Sec. 5.6. To do this, consider a single circuit carrying a current I linked by its own flux of total amount Φ (Fig. 9.15). Take a closed tube containing a constant element $d\Phi$ of the total flux (the area dS and the value of B may change around the tube but the product $B\,dS = d\Phi$ remains the same). The total energy U_M is $\tfrac{1}{2}I\Phi$ by (9.42) and an amount $\tfrac{1}{2}I\,d\Phi$ could be associated with the tube. By the circuital law (8.3), $I = \int B\,ds/\mu_0$ round any closed path (where we use ds instead of dL for an increment of path to avoid confusion with self-inductance). It follows that

$$dU_M = \tfrac{1}{2}\,d\Phi \oint B \; ds/\mu_0 = \tfrac{1}{2} \oint B^2 \; d\tau/\mu_0$$

is the energy associated with the tube, where $d\tau$ is an element of volume. If we integrate over the two variables represented by dS so as to include all the tubes, we finally obtain

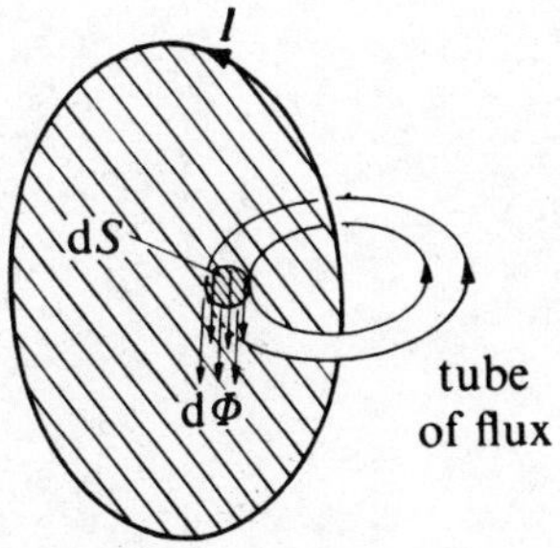

Figure 9.15 Calculation of the magnetic energy of a current-carrying circuit in terms of **B**.

magnetic energy: $$U_M = \int_\tau \frac{B^2}{2\mu_0} d\tau \tag{9.44}$$

The **energy per unit volume** associated with a magnetic field is thus given by $B^2/2\mu_0$. The same qualifications apply here concerning the location of energy as in the electric case of Sec. 5.6. Note too that (9.42) and (9.44) are alternative expressions for magnetic energy as Problem 9.18 shows.

9.11 Forces, torques and changes in energy

We use a method similar to that in Sec. 5.7 to find the forces or torques between two current-carrying circuits, the difference being that in the electric case the sources of the field (the charges) remained constant without extra energy being needed. Here, the sources of field are *currents* and if these are to be constant during any changes, extra working electromotances are needed.

Changes at constant current In Fig. 9.16, the internal forces of attraction are shown dotted and the external forces, keeping the circuits apart and in equilibrium, in full. Circuit B is subject to an internal force **F**, which we are trying to find, and an external force $\mathbf{G}=-\mathbf{F}$. For any displacement $\mathrm{d}\mathbf{L}$ of B let extra electromotances be introduced into the two circuits to keep I_{A} and I_{B} constant, and let the change in mutual inductance be $\mathrm{d}M$. The increment in magnetic energy is

$$\mathrm{d}U_{\mathrm{M}}=I_{\mathrm{A}}I_{\mathrm{B}}\,\mathrm{d}M \tag{9.45}$$

from (9.38). The extra electromotances will be of magnitude $I_{\mathrm{B}}\,\mathrm{d}M/\mathrm{d}t$ in A and $I_{\mathrm{A}}\,\mathrm{d}M/\mathrm{d}t$ in B so that both will be supplying energy to the whole system at a rate $I_{\mathrm{A}}I_{\mathrm{B}}\,\mathrm{d}M/\mathrm{d}t$. Thus *both* will lose an amount of energy in time $\mathrm{d}t$ equal to $I_{\mathrm{A}}I_{\mathrm{B}}\,\mathrm{d}M$ so that the increment in the energy of the system (circuits + batteries) is $-I_{\mathrm{A}}I_{\mathrm{B}}\,\mathrm{d}M$ or $-\mathrm{d}U_{\mathrm{M}}$. The work done *on* the circuit B by the external force is $\mathbf{G}\cdot\mathrm{d}\mathbf{L}$ or $-\mathbf{F}\cdot\mathrm{d}\mathbf{L}$ and therefore

$$-\mathbf{F}\cdot\mathrm{d}\mathbf{L}=-I_{\mathrm{A}}I_{\mathrm{B}}\,\mathrm{d}M=-\mathrm{d}U_{\mathrm{M}}$$

or

$$\text{force on circuit:}\quad \boxed{F_x=I_{\mathrm{A}}I_{\mathrm{B}}\frac{\partial M}{\partial x}=\left(\frac{\partial U_{\mathrm{M}}}{\partial x}\right)_I,\ \text{etc.}} \tag{9.46}$$

Similarly the torque about any axis θ is given by

$$\text{torque on circuit:}\quad \boxed{T_\theta=I_{\mathrm{A}}I_{\mathrm{B}}\frac{\partial M}{\partial \theta}=\left(\frac{\partial U_{\mathrm{M}}}{\partial \theta}\right)_I} \tag{9.47}$$

Changes at constant current thus mean that, of the energy supplied by the batteries, *half goes to increasing the internal magnetic energy and half to doing external work.* These changes correspond to those at constant potential in electrostatics. Changes at constant flux linkage (corresponding to constant charge) yield formulae like $F_x=-(\partial U_{\mathrm{M}}/\partial x)_\Phi$ but they are in practice of little use, and give the same forces and torques in a static situation as do (9.46) and (9.47), as they must.

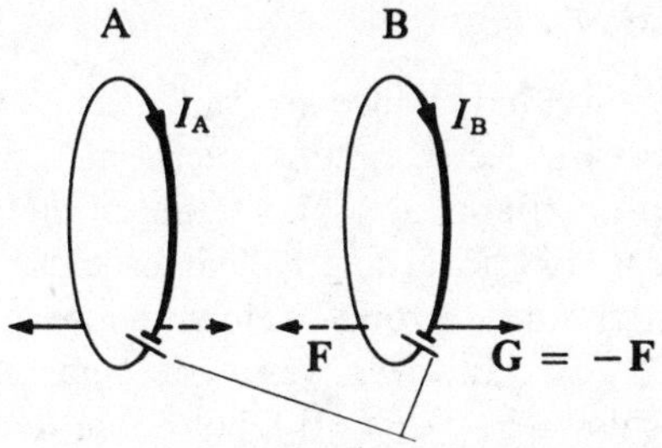

Figure 9.16 Forces between current-carrying circuits. The required internal forces **F** are drawn with broken lines, the external forces maintaining equilibrium **G** with full lines.

Relationship between magnetic energy and potential energy The potential energy U of a current I linked by an external magnetic flux Φ was given in Eq. (7.47) by $U=-I\Phi$, and forces and torques can be obtained from this by using $Fx=-(\partial U/\partial x)_I$, etc. The above example of a pair of circuits with a mutual inductance M is a special case for which $\Phi_B=MI_A$, so that the potential energy of the pair is

potential energy: $$\boxed{U=-MI_AI_B} \tag{9.48}$$

which will clearly yield (9.46) and (9.47) above.

However, confusion can arise between the potential energy given by $-MI_AI_B$ and the term in the magnetic energy U_M due to mutual inductance which has the opposite sign $+MI_AI_B$. The same question is not asked in electrostatics because U_E and U are equal: why the difference? The answer lies in that electric energy depends solely on the *positions* of the charges and is thus identical with potential energy, whereas the magnetic energy of currents depends on both the positions and motion of charges and cannot be completely characterized as either potential or kinetic. When we calculate potential energy we find the work done by external mechanical forces in moving the circuits from infinity to their final positions *with constant currents*: this is bound to be different from the total energy needed to establish the currents in their final positions because, as we have seen, batteries must put in $+2MI_AI_B$ to keep the currents constant. The total energy, therefore, increases by $2MI_AI_B+U$ or $+MI_AI_B$, thus accounting for the + sign in U_M.

Worked example The small coil of Fig. 9.12b will experience a force F_x and a torque T_θ given by

$$F_x=\frac{3\mu_0I_1I_2N_1N_2\pi a_1^2a_2^2x\cos\theta}{2(a_1^2+x^2)^{5/2}}$$

$$T_\theta=\frac{\mu_0I_1I_2N_1N_2\pi a_1^2a_2^2\sin\theta}{2(a_1^2+x^2)^{3/2}} \tag{9.49}$$

using (9.32), (9.46) and (9.47).

9.12 Differential form of Faraday's law

The electromotive E-field In dealing with inductance and magnetic energy, the treatment has been concerned mainly with electric circuits and the more practical aspects. We now return to field properties and look at the effects of electromagnetic induction on the general laws governing **E**- and **B**-fields.

The existence of an induced electromotance $\mathscr{E}$ round a closed path means that an E-field also exists because $\mathscr{E} = \oint \mathbf{E} \cdot d\mathbf{L}$. However, we saw in Chapter 4 that electrostatic fields, denoted by $\mathbf{E}_Q$, satisfy $\oint \mathbf{E} \cdot d\mathbf{L} = 0$. It follows that the electric fields in this chapter are different in character from the $\mathbf{E}_Q$'s and we denote them generally by $\mathbf{E}_M$ as in Sec. 6.1.

Can we express the properties of $\mathbf{E}_M$ in forms similar to those of $\mathbf{E}_Q$ in Chapter 4? First we must recognize that the fields are assumed to exist even when there is no conducting material. In all of Faraday's experiments and in the subsequent arguments there was always an induced *current* because the electromotance and the $\mathbf{E}_M$ acted on free charges in a conductor. However, just as $\mathbf{E}_Q$'s were given an existence independent of any charges, so we take the $\mathbf{E}_M$'s as present in the absence of conductors. The betatron below is evidence that this is a valid viewpoint. For that reason we change the symbol for an element of path from $d\mathbf{l}$, used at the beginning of the chapter, to $d\mathbf{L}$, in order to indicate that it need not be a conducting element.

If we are to examine the properties of $\mathbf{E}_M$ in the absence of conductors, then motional electromotances are irrelevant because the motion concerned is that of a conductor. Only transformer electromotances are relevant, as given by (9.10): $\mathscr{E} = -\partial\Phi/\partial t$, but it is generalized to apply to *any* closed path L across which the **B**-field varies in time.

Since $\mathscr{E}$ can be written as $\oint \mathbf{E}_M \cdot d\mathbf{L}$ and Φ as $\int \mathbf{B} \cdot d\mathbf{S}$ over the surface S enclosed by L, Eq. (9.10) can be expressed as

$$\oint_L \mathbf{E}_M \cdot d\mathbf{L} = -\frac{\partial}{\partial t}\int_S \mathbf{B} \cdot d\mathbf{S}$$

Reversing the order of integration and differentiation gives

$$\boxed{\oint_L \mathbf{E}_M \cdot d\mathbf{L} = \int_S \left(-\frac{\partial \mathbf{B}}{\partial t}\right) \cdot d\mathbf{S}} \tag{9.50}$$

If a conducting path coincides with L then a current will flow. The electromotance may be non-localized in the sense of Sec. 6.1 and may exist round the whole of L.

We now use the same method as with the circuital laws for $\mathbf{E}_Q$ and $\mathbf{B}$ (Secs 4.6 and 8.3) to put (9.50) into a differential form applying at a *point*. The line integral of $\mathbf{E}_M$ on the left can be transformed into the surface integral of **curl** $\mathbf{E}_M$ over S, using Stokes's theorem. Since (9.50) now equates two surface integrals over any surface S, the integrands must be equal and hence:

$$\boxed{\nabla \times \mathbf{E}_M \equiv \text{curl}\, \mathbf{E}_M = -\partial \mathbf{B}/\partial t} \tag{9.51}$$

Note that (9.50) implies that lines of $\mathbf{E}_M$ may close on themselves, unlike those of $\mathbf{E}_Q$: $\mathbf{E}_M$ is a *non-conservative* field.

Application *the betatron* In the betatron, charged particles move in a circular path in a **B**-field as in Fig. 9.17. If what we have been saying is correct, then a time variation in the **B**-field should be accompanied by an $\mathbf{E}_M$-field round the orbit, thus accelerating the particles.

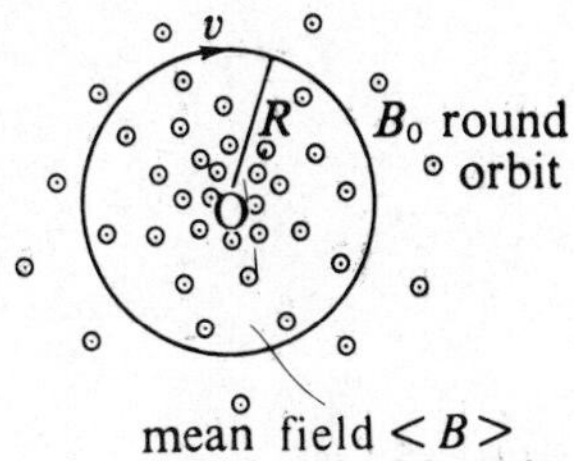

Figure 9.17 Orbit of a charged particle in the betatron.

Let us make the **B**-field symmetrical about the axis through O so that it varies with r but not with θ. Suppose the magnitude of **B** at the orbit radius R is B_0. Then (7.28) shows that the particles travel in a circular path with linear momentum

$$p = QRB_0 \tag{9.52}$$

Now let **B** vary in time. If $\langle B \rangle$ is the mean value of the magnetic field over the area inside the orbit, the total flux linkage with the path is $\Phi = \pi R^2 \langle B \rangle$. The electromotance is $\partial\Phi/\partial t$ and because of the symmetry we can equate it to $2\pi R E_M$ (i.e., $\oint \mathbf{E} \cdot d\mathbf{L}$ round the path):

$$2\pi R E_M = \partial\Phi/\partial t = \pi R^2\, \partial\langle B \rangle/\partial t$$

so that $E_M = \frac{1}{2} R \partial \langle B \rangle/\partial t$. If this E-field exists, the particle should experience a tangential force QE, causing p to change according to Newton's law:

$$\mathrm{d}p/\mathrm{d}t = QE = \tfrac{1}{2} QR\, \partial\langle B \rangle/\partial t \tag{9.53}$$

For the betatron to work, the orbit should remain the same size as p increases. Hence, in (9.52), R is constant and we have also that

$$\mathrm{d}p/\mathrm{d}t = QR\partial\, B_0/\partial t \tag{9.54}$$

Comparison of (9.53) and (9.54) shows that we need $B_0 = \frac{1}{2}\langle B \rangle$, i.e. that the magnetic field *at* the orbit should be half the mean value over the area of the orbit. This is a **B**-field that decreases with increasing distance from O.

In this instrument, the final momentum for all particles is fixed independently of their mass. Only the lightest particles, electrons, achieve reasonably high *energies*. Values of 300 MeV have been produced.

Vector potential and the divergence of $\mathbf{E}_M$ Our attention has been concentrated on the circulation of $\mathbf{E}_M$ and its circulation per unit area or curl. The other important aspect of field quantities is the flux over a closed surface and the flux per unit volume or divergence. To investigate this, we shall need the general properties of vector potential introduced in Sec. 8.5, but not the detail. Readers willing to accept a qualitative argument that this chapter has provided no evidence that lines of flux of $\mathbf{E}_M$ can originate or end in any region other than infinity can omit this sub-section.

Certainly, none of the results of this chapter alter the properties of **B**-fields, so that there are still no sources of **B**-flux and hence $\nabla \cdot \mathbf{B} = 0$ (Sec. 8.2). It is therefore still possible to define a vector potential **A** through (8.17): $\mathbf{B} = \nabla \times \mathbf{A}$. If this is substituted into (9.51), the **curl** of $\mathbf{E}_M$ is equated to the **curl** of $-\partial\mathbf{A}/\partial t$, after interchanging the time and space differentiations. It follows that

$$\boxed{\mathbf{E}_M = -\frac{\partial \mathbf{A}}{\partial t}} \tag{9.55}$$

with the possible addition of a term having zero curl. However, $\mathbf{E}_M$ does not exist in the absence of time variations so this added term is zero.

Equation (9.55) is consistent with the results of Sec. 8.5, where we saw that a **B**-field due to a long solenoid gave **A**-lines which were concentric circles round the solenoid. When such a **B**-field varies in time, the lines of $\mathbf{E}_M$ must follow the same circles because of (9.55). This results in the electromotance round the path which is predicted by Faraday's law.

Now under steady or quasi-steady conditions, (8.21) showed that $\nabla \cdot \mathbf{A} = 0$. Taking the divergence of both sides of (9.55) then shows that

$$\boxed{\nabla \cdot \mathbf{E}_M = 0} \tag{9.56}$$

and hence that the flux of $\mathbf{E}_M$ does not start or finish anywhere except possibly at infinity.

9.13 Conclusion

General equations for E-fields We have the following contrasting properties of electrostatic and electromotive **E**-fields:

$$\nabla \cdot \mathbf{E}_Q = \rho/\varepsilon_0 \qquad \nabla \times \mathbf{E}_Q = 0$$
$$\nabla \cdot \mathbf{E}_M = 0 \qquad \nabla \times \mathbf{E}_M = -\partial \mathbf{B}/\partial t$$

We call the total **E**-field at any point, $\mathbf{E} = \mathbf{E}_M + \mathbf{E}_Q$, the **Maxwell E**-field. The equations just quoted show that this field obeys the general laws:

$$\nabla \cdot \mathbf{E} = \rho/\varepsilon_0 \qquad \nabla \times \mathbf{E} = -\partial \mathbf{B}/\partial t \tag{9.57}$$

and that, since $\mathbf{E}_Q$ can be written as $-\nabla V$ and $\mathbf{E}_M$ as $-\partial \mathbf{A}/\partial t$,

$$\boxed{\mathbf{E} = -\partial \mathbf{A}/\partial t - \nabla V} \tag{9.58}$$

where the first term is the contribution of $\mathbf{E}_M$ and the second that of $\mathbf{E}_Q$.

The **B**-fields are apparently unaffected by any of the above, so that we still have $\nabla \cdot \mathbf{B} = 0$ and $\nabla \times \mathbf{B} = \mu_0 \mathbf{J}$ from Chapter 8. We therefore have

Maxwell's equations for slowly varying fields *in vacuo*:

$$\boxed{\begin{array}{ll} \nabla \cdot \mathbf{E} = \rho/\varepsilon_0 & \nabla \times \mathbf{E} = -\partial \mathbf{B}/\partial t \\ \nabla \cdot \mathbf{B} = 0 & \nabla \times \mathbf{B} = \mu_0 \mathbf{J} \end{array}} \tag{9.59}$$

The next steps In the next chapter we take up network theory once more, now that we have further elements L and M as well as R and C, but this time we deal with varying currents. Following that, in Chapters 11 and 12, media other than vacuum are introduced and their effect on the laws established so far are examined. It is not until Chapter 13 that we return to the consideration of time variations more generally, and ask whether a time-varying **E**-field will produce any new effects as have time-varying **B**-fields.

References Faraday's work is very well documented, largely as a result of his own notebooks. Good introductions are to be found in Martin (1949) and Williams (1963), but his own *Experimental Researches in Electricity* (1839) is a masterly account of his methods.

9.14 Revision summary

Induced electromotances can be classified into two types:

- Motional electromotance, given by
  either the current element formula

$$d\mathscr{E} = (\mathbf{v} \times \mathbf{B}) \cdot d\mathbf{l} = v_{\perp} B \, dl_{\perp} \qquad (9.4)\ (9.5)$$

  or the flux-cutting law

$$\mathscr{E} = d\Phi_{cut}/dt \qquad (9.7)$$

  or the flux-linking law

$$\mathscr{E} = -d\Phi_{link}/dt \quad \text{for rigid circuits only} \qquad (9.8)$$

- Transformer electromotance, given by

$$\mathscr{E} = -\partial\Phi_{link}/\partial t \quad \text{in stationary rigid circuits} \qquad (9.10)$$

Electromotive E-fields In a motional electromotance, the **E**-field is $\mathbf{v} \times \mathbf{B}$ and exists relative to the moving conductor. It will set up a current density at any point given by

$$\mathbf{J} = \sigma(\mathbf{v} \times \mathbf{B}) \qquad \text{from } (9.13)$$

In a transformer electromotance, the **E**-field is denoted by $\mathbf{E}_M$ and is given by

$$\oint_L \mathbf{E}_M \cdot d\mathbf{L} = \int_S \left(-\frac{\partial \mathbf{B}}{\partial t} \right) \cdot d\mathbf{S} \qquad (9.50)$$

This can be generalized by assuming that it exists independently of the stationary conducting path. In that case we have the corresponding differential form applying at any point:

$$\nabla \times \mathbf{E}_M = -\partial \mathbf{B}/\partial t \qquad (9.51)$$

Since a vector potential can still be defined through $\mathbf{B} = \nabla \times \mathbf{A}$, $\mathbf{E}_M$ will equal $-\partial\mathbf{A}/\partial t$. $\mathbf{E}_M$ is a non-conservative field, unlike $\mathbf{E}_Q$.

Inductance *Self-inductance* L of a circuit is the flux linkage per unit current in itself so that

$$\Phi = LI \qquad (9.14)$$

Mutual inductance M between two circuits is the flux linkage through one per unit current in the other

$$\Phi_1 = MI_2; \qquad \Phi_2 = MI_1 \qquad (9.25)\ (9.26)$$

Calculation of L and M in simple cases proceeds directly from these definitions. L and M are strictly geometrical properties in the absence of media and all expressions for them have the form $\mu_0 \times$ a factor with the dimensions of length. The **SI unit** for both L and M is the *henry*, H.

As network elements, L is a two-terminal element with $V = L\,dI/dt$
M is a four-terminal element with $V_1 = M\,dI_2/dt$

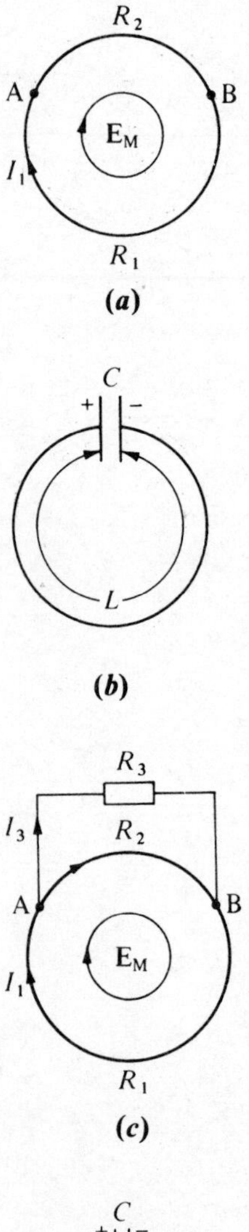

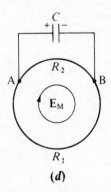

Figure 9.18 Electromagnetic induction in a symmetrical ring.

Magnetic energy U_M In the absence of ferromagnetic materials, U_M can be expressed for a number of circuits as

$$U_M=\sum_i \tfrac{1}{2}L_iI_i^2+\sum_i\sum_j M_{ij}I_iI_j=\sum_i \tfrac{1}{2}I_i\Phi_i \qquad (9.41)\ (9.42)$$

where Φ_i is the total magnetic flux linking the ith circuit.

In terms of the **B**-field, a more general expression arises:

$$U_M=\int_\tau \frac{B^2}{2\mu_0}\,\mathrm{d}\tau \qquad (9.44)$$

Forces and torques between two circuits are given by

$$F_x=I_AI_B\frac{\partial M}{\partial x}; \qquad T_\theta=I_AI_B\frac{\partial M}{\partial \theta} \qquad (9.46)\ (9.47)$$

General field equations If the total **E**-field at any point is now taken to be the sum of electrostatic and electromotive contributions, the general differential equations are

$$\nabla\times\mathbf{E}=-\partial\mathbf{B}/\partial t \qquad \nabla\cdot\mathbf{E}=\rho/\varepsilon_0 \qquad (9.57)$$

while the two equations for **B** of Chapter 8 are so far unaffected:

$$\nabla\times\mathbf{B}=\mu_0\mathbf{J} \qquad \nabla\cdot\mathbf{B}=0 \qquad (9.59)$$

Commentary

C9.1 On Electromotance Again Some intriguing questions arise when only non-localized electromagnetically induced electromotances are acting. Suppose, for instance, that a cylindrical bar magnet is pushed along the axis of the conducting ring in Fig. 9.18a. The changing flux produces an $\mathbf{E}_M$-field whose lines will be concentric circles around the magnet. Some of the field will coincide with the ring and cause the current $I=\mathscr{E}/(R_1+R_2)$ to flow, $\mathscr{E}$ being the total electromotance round the ring. However, no charge accumulates at any point so that no $\mathbf{E}_Q$-field exists (notation of Sec. 6.1). This means that no potential difference can be meaningfully said to exist between any two points such as A and B in the ring. However, if a capacitor is inserted as in Fig. 9.18b, charge will build up on its plates until the V across them is equal to $\int \mathbf{E}_M\cdot\mathbf{dl}$ round the path L and the flow stops. V is less than $\mathscr{E}$ because L is less than the complete circuit.

Now return to Fig. 9.18a. If no potential difference can be said to exist between A and B, does this mean that a voltage-measuring device connected between A and B would read zero? If so, this would contradict our experience that electromotances from rotary generators do produce measurable voltages in external wires. If not, what does the device read?

The answer lies in realizing that the measuring instrument itself has now to be included in the arrangement. Let us assume for the sake of argument that the flux changes are confined to the region encircled by the ring: so for any path entirely outside the ring $\oint \mathbf{E}_M\cdot\mathbf{dl}=0$. Figure 9.18c illustrates the situation when a voltmeter of resistance R_3 is connected across AB. The current I_3 is, using (6.32), $R_2\mathscr{E}/(R_1R_2+R_2R_3+R_3R_1)$. The voltmeter will therefore read a voltage on its scale given by R_3I_3 or $R_2\mathscr{E}/(R_1+R_2+R_1R_2/R_3)$ because the meter is essentially a current-measuring device calibrated to read volts. If instead we use an electrometer acting as a capacitor (Fig. 9.18d), then charge will build up on its plates and flow will stop when there is no resultant field acting on the electrons in the wires connecting C to A and B. In that case V will equal the $\int \mathbf{E}_M \cdot d\mathbf{l}$ along R_2, that is $R_2\mathscr{E}/(R_1+R_2)$. This is the same as would be recorded by a voltmeter whose $R_3 \rightarrow \infty$. Note that if R_1 and R_2 are unequal, the readings depend on which side of the ring the instrument is connected (see Moorcroft, 1969, 1970). Laithwaite (1969) poses some other electromagnetic problems connected with induced electromotances.

Problems

Section 9.1

9.1 Show how Lenz's law can be used to find the direction of induced current flow in (a) a circular coil rotating about a diameter in a uniform magnetic field and (b) a stationary circular coil towards which a coaxial current-carrying coil is moving.

Section 9.2

9.2 Estimate the magnitude of the potential differences produced by the motion of a conductor in the earth's magnetic field for which B is 4.3×10^{-5} T and the angle of dip is 64° 9′. Take as an example a car bumper of length $1\frac{1}{2}$ m travelling at 100 km h^{-1}.

9.3 The Faraday disc of Fig. 9.3 is to be used as a motor by including a battery in the circuit shown. If the current flowing is I and if the magnetic field $\mathbf{B}$ is uniform, show that the torque exerted on the disc is $\Phi I/2\pi$, where Φ is the flux crossing the whole disc. Find I if the electromotance of the battery is $\mathscr{E}$ and the resistance of the complete circuit is R. Account for the power consumed from the battery.

9.4 Two long horizontal parallel bars are separated by a distance a, and are connected at one end by a resistance R. A uniform magnetic field $\mathbf{B}$ is maintained vertically. A straight rod of mass m is laid across the bars at right angles so as to complete a conducting circuit. If the rod is given an impulse that causes it to move with an initial velocity v_0 parallel to the bars, find an expression for the velocity at any subsequent time t. Neglect the resistance of the bars and rod, and assume no friction.

9.5 In the simple a.c. generator in which the electromotance is given by (9.9), show that the torque opposing the motion is $\Phi_m I \sin \omega t$, where I is the armature current and Φ_m is the maximum flux linkage with the coil.

9.6 If the a.c. generator of Problem 9.5 is converted to a d.c. motor by including a battery of electromotance $\mathscr{E}$ in series with the coil and using a simple commutator, find the current in the coil when rotating at an angular velocity ω and hence find the mechanical power output. Assume that the flux density is constant.

9.7 The three-phase output of an a.c. generator is used to produce three magnetic **B**-fields given in magnitude by $B_0 \sin \omega t$, $B_0 \sin(\omega t + 2\pi/3)$, $B_0 \sin(\omega t + 4\pi/3)$. If the directions of the three fields are at 120° to each other and they are superposed in the same region of space, show that the resultant field is one of magnitude $3B_0/2$ rotating at an angular velocity ω.

Section 9.3

9.8 A circular brass hoop of radius a and resistance R is placed with its plane perpendicular to a uniform magnetic field whose magnitude fluctuates. If $B = B_0 \sin \omega t$, what is the induced current in the loop?

9.9* A brass disc of radius a, thickness b and conductivity σ has its plane perpendicular to a uniform magnetic **B**-field which varies according to

$B = B_0 \sin \omega t$. Assuming that eddy currents flow in concentric circles about the centre of the disc, find the total current flowing at any instant and the mean power dissipated as heat. Comment on the result as an indication of the factors affecting eddy current losses in iron.

Section 9.4

9.10 A metal hoop of radius a and resistance R is held with its plane in the magnetic meridian of the earth, and falls over through 90° to the east. Find the total quantity of charge which flows while the hoop is falling, if the earth's field has a magnitude B and an angle of dip δ. Evaluate the charge for $a = 0.5$ m, $R = 0.2\,\Omega$ and the earth's field as in Problem 9.2.

Section 9.5

9.11 In the circuit of Fig. 9.7 find expressions for the potential differences across L and R subsequent to the closing of K_1, and plot the variation against time.

9.12 Show that when a current I_0 flowing in an inductor decays according to Eq. (9.20), the energy dissipated as heat is always $\frac{1}{2}LI_0^2$ whatever the value of R.

Section 9.6

9.13 Estimate the order of magnitude of the self-inductance of an air-cored solenoid of length 20 cm with one layer of 10 turns per cm each turn forming a circle of radius 2 cm.

Section 9.7

9.14 The self-inductance of a coil is 20 mH. When a current flows in it, one-fifth of its magnetic flux links a second coil. What electromotance is induced in the second coil if a current in the first collapses at a uniform rate of $0.5\,\mathrm{A\,s^{-1}}$?

Section 9.8

9.15 Two coils A and B have self-inductances L_1 and L_2. When a steady current flows in A, a quarter of its magnetic flux links B. Find the proportion of the flux from a current in B which links A. Evaluate the coefficient of coupling.

9.16 Find the mutual inductance between an infinitely long straight wire and a one-turn rectangular coil whose plane passes through the wire and two of whose sides are parallel to the wire. The sides parallel to the wire are of length a, the other sides are of length b and the side nearest the wire is a distance d from it.

Section 9.9

9.17 A stationary circuit carries a current I_1 while a second circuit carrying I_2 moves towards it, the mutual inductance at any instant being M. What power must be provided to maintain I_1 and I_2 constant if the two circuits are rigid?

Section 9.10

9.18 Show that both Eqs (9.37) and (9.44) give the same value for the magnetic energy stored in a long solenoid of length l, area A, with n turns per unit length each carrying a current I.

Section 9.11

9.19 Find the force on the coil in Problem 9.16 when a current I flows in both circuits. Check the result by using Eq. (7.2).

Section 9.12

9.20 A magnetic dipole has a moment $\mathbf{m}$ lying along the axis of a circular ring of radius a and it moves with a velocity v towards the ring. Calculate the E-field in the ring when the dipole is at a distance x from its centre.

10

Varying currents in linear networks

When a *steady* current flows in a network of conductors, resistance is the only property affecting the distribution of currents and voltages, as we saw in Chapter 6. As soon as the currents change in time, however, both capacitance and inductance become important factors as well. All three properties must be possessed by any real network, both concentrated in circuit components represented by ideal elements R, L, C and M (Sec. 10.1) and distributed along connecting wires, cables and transmission lines (Sec. 10.10).

We first consider the sudden application or removal of a steady source of voltage to some networks (Secs 10.2 and 10.3) and see that only transient changes are produced.

We then consider the application of sinusoidally varying voltages and currents to networks of gradually increasing complexity (Secs 10.4 onwards). The general solution in all cases is found to be the sum of (a) the same transient variations as those of Secs 10.2 and 10.3 and (b) a steady-state solution which can be obtained by algebraic methods rather than by solving differential equations. These algebraic methods form the basis of what is conventionally known as alternating current (a.c.) theory, and they involve the use of complex numbers to represent sinusoidally varying quantities (phasors). Section 10.12 deals with the question of non-sinusoidal waveforms.

The whole chapter is based on already established principles together with the assumption of quasi-steady conditions. Rather than attempt a revision summary of the large amount of network theory, we end by taking stock of the theory generally (Sec. 10.13) and discuss its limitations.

10.1 Network elements and network equations

V–I relationships for ideal elements As in the solution of steady current networks, we replace an actual set of components by a set of ideal elements which match the real network as closely as possible. Sometimes a real component may be adequately represented by a single ideal element, sometimes by several. The ideal elements we can use are the *passive* ones of resistance R, capacitance C, self-inductance L and mutual inductance M, together with the *active* ones of voltage and current generators $\mathscr{E}$ and $\mathscr{I}$. We assume to begin with that the elements are localized between terminals or are *lumped*, in contrast to systems we meet later where R, L and C are *distributed*.

The relationships between the potential difference V and the current I for the **passive** elements, R, L, C and M are shown in Fig. 10.1 which is a summary of results from previous chapters. Note particularly the assumed directions for current in relation to V. For a capacitor, if I has been chosen so that it flows *away* from the assumed positive plate, then $I = -\mathrm{d}Q/\mathrm{d}t$. For a mutual inductor, details of the directions of windings are avoided by using the dot notation of Fig. 9.13. A current flowing in at the dotted end of one coil produces a voltage $V = M\,\mathrm{d}I/\mathrm{d}t$ across the other which is higher at its dotted end. The *sign* attached to this voltage will depend on the assumed direction of current in the second coil. In any given circuit with directions for all currents decided, taking one dot to the other end of its coil will reverse the sign of M.

R, C, L and M will be assumed independent of the magnitude of charge or current. Many components in practice may be non-linear—resistors because of temperature variation, inductors because of ferromagnetic cores, capacitors because of non-linear dielectrics—but the conditions of use are frequently such that the effect of non-linearity is small enough to be neglected. We shall not be concerned here with grossly non-linear components, such as rectifiers, where the non-linearity can be utilized.

The **active** elements, the voltage and current generators, will be assumed to generate electromotance $\mathscr{E}$ or current $\mathscr{I}$ independent of the load placed on them.

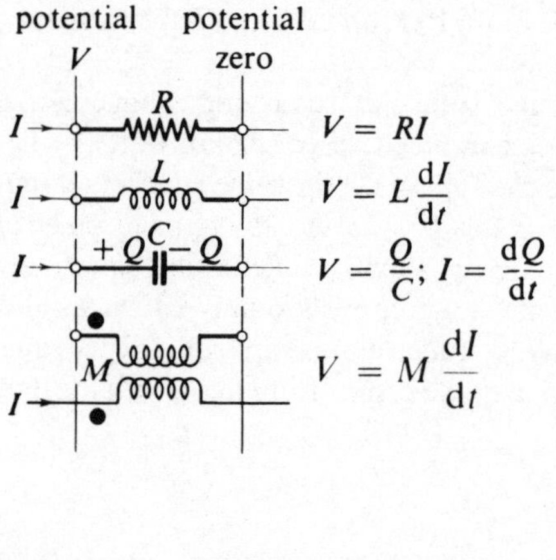

Figure 10.1 Ideal network elements and their V–I relationships.

Any internal R, L or C will if necessary be included explicitly in the network. We shall mostly use voltage generators as in Chapter 6.

Network laws and equations We assume that time variations are slow enough for steady current laws to apply at any instant (quasi-steady conditions). The basic laws we need are Kirchhoff's laws K1 and K2, summarized by

K1: $\Sigma I=0$ at a function or node
K2: $\Sigma V=0$ round a mesh or loop

The result of applying these laws and the relationships of Fig. 10.1 to any network is to yield sets of differential equations. These equations are *linear* because of the linearity of the V–I relationships and *have constant coefficients* because R, L, C and M are constant. The general form of the equations is

$$A_n\frac{\mathrm{d}^n y}{\mathrm{d}t^n}+A_{n-1}\frac{\mathrm{d}^{n-1}y}{\mathrm{d}t^{n-1}}+\cdots+A_1\frac{\mathrm{d}y}{\mathrm{d}t}+A_0 y=f(t) \tag{10.1}$$

where the left-hand side represents the passive elements and the right-hand side the effect of applied voltages or currents from generators. The quantity y may be charge, current or voltage. Similar equations appear in the theory of vibrations of other linear systems.

Direct solution of the equations in every problem is a cumbersome method of proceeding and one to be avoided if possible. However, if we are to understand why other methods work, we cannot shirk the initial effort of going through such solutions. Eventually, the differential equations will be replaced by algebraic ones that are much easier to solve.

It is essential to grasp that solutions to equations like that of (10.1) have the following properties:

1. *Superposition applies.* Firstly, the sum of any two solutions is itself a solution. Secondly, if the right-hand side consists of $f_1(t)+f_2(t)$, the solution is the sum of the two solutions with $f_1(t)$ alone and $f_2(t)$ alone.
2. *The general solution* is *any* solution containing n constants which are determined by the conditions specified at a particular time (the initial conditions). For example, Eq. (6.22) is like (10.1) with $n=1$: it has a general solution with *one* constant determined by *one* initial condition.
3. The *homogeneous* equation with $f(t)$ put equal to zero has a solution corresponding to **free** motion of the system. This solution is called the complementary function (CF) and contains n constants determined by initial conditions.
4. It follows from 1, 2 and 3 that the general solution of (10.1) is the sum of the CF and any particular solution (known as a particular integral PI). This is because (a) by 1, the sum CF + PI is a solution and (b) by 3, the sum CF + PI contains n arbitrary constants. The PI is the extra part of the solution determined by $f(t)$ and thus represents the **forced** motion of the system.
5. Property of the CF: it is generally *transient* in that its amplitude tends to zero as $t\to\infty$.
6. Property of the PI: If $f(t)$ contains any exponential or sinusoidal terms, $e^{\lambda t}$ or $\sin\omega t$, the PI can only contain exponential or sinusoidal terms with the same λ or ω. It is not otherwise possible for the solution to satisfy (10.1) *for all t*.

The classical method of analysing networks separate the transient behaviour from the steady state and this procedure is adopted here. Nevertheless, both types of behaviour are determined entirely by the elements in the network and must be related. This is a matter taken up in Sec. 10.12.

An understanding of transients is not necessary for a.c. theory and readers wishing to deal immediately with the latter should proceed straight to Sec. 10.4.

10.2 Transients in a series *LCR* circuit

The *decay* of charge and current in CR and LR circuits (Secs 6.8 and 9.5) yields homogeneous equations for Q and I, respectively. The solutions are examples of transient effects as we should expect because there is no input of energy after $t=0$ and R is dissipative. The *growth* of Q and I in the same circuits obeys the same equations with a constant $\mathscr{E}$ on the right-hand side. The solutions were found by changing the variables and getting a homogeneous equation again. We now see that they could be obtained by adding the decay solution (CF) to a PI (i.e. $Q_0=\mathscr{E}C$ or $I_0=\mathscr{E}/R$) provided the signs are carefully watched.

We now turn to a series combination of three components L, C and R, and examine the variations with time of charge on the capacitor and current in the circuit.

Setting up the equation Consider first the capacitor of Fig. 10.2a, which has an initial charge Q_0 at $t=0$ and, by the closing of a switch, then begins to discharge through L and R. The situation illustrated is that at a subsequent time t. Application of K2 ($\Sigma V=0$) gives $Q/C-L\dot{I}-RI=0$, where the dot indicates differentiation with respect to time. Because $I=-\mathrm{d}Q/\mathrm{d}t$, we have

$$L\ddot{Q}+R\dot{Q}+Q/C=0 \qquad \text{at } t=0\text{: } I=0,\ Q=Q_0 \tag{10.2}$$

The initial condition $I=0$ is for the same reason as in Eq. (9.17).

In Fig. 10.2b, the capacitor is initially uncharged but is charged by the closing of a switch at $t=0$. The situation illustrated is that at a time t, and application of K2 together with $I=\mathrm{d}Q/\mathrm{d}t$ gives

$$L\ddot{Q}+R\dot{Q}+Q/C=\mathscr{E} \qquad \text{at } t=0\text{: } I=0,\ Q=0 \tag{10.3}$$

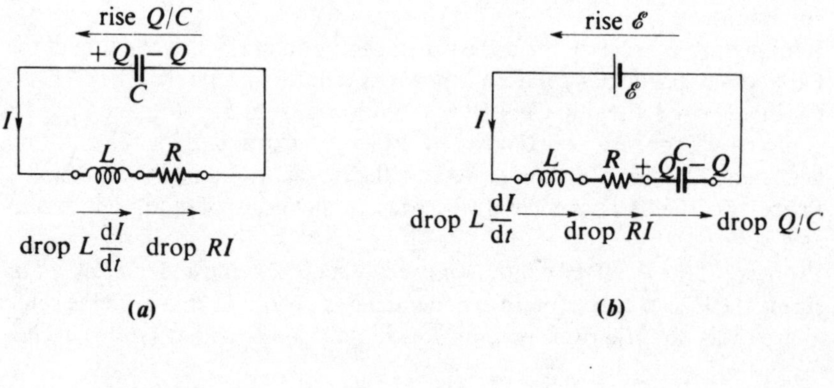

Figure 10.2 Transients in a series *LCR* circuit, showing the voltage steps round the circuit during (a) the discharge of C and (b) the charging of C, both through L and R.

The CF of this equation is the same as (10.2), although with different initial conditions and therefore different constants. A particular integral is clearly $Q_0 = \mathscr{E}C$, so the solution of (10.3) is easily obtained from that of (10.2). We concentrate on (10.2).

Solving the equation The form of Eq. (10.2) is simplified by dividing throughout by L and making the substitutions

$$\boxed{\alpha = R/2L \qquad \omega_0^2 = 1/LC} \tag{10.4}$$

The coefficient α thus includes the dissipative effects in the circuit. The motivation behind the use of ω_0 is seen by considering the case when $R=0$. Equation (10.2) then becomes that of an undamped harmonic oscillator whose angular frequency† is $1/(LC)^{1/2}$ or ω_0.

The equation to be solved is therefore

$$\ddot{Q} + 2\alpha\dot{Q} + \omega_0^2 Q = 0 \tag{10.5}$$

with initial conditions $\dot{Q}=0$ and $Q=Q_0$ at $t=0$. The conventional method of solution involves the assumption that Q has the form $Q_0 e^{\lambda t}$ and then obtaining an equation for λ by substitution in (10.5). This method is to be found in many texts on vibrations, but does not deal naturally with the case when $\alpha = \omega_0$. We therefore adopt a different approach (Lenz, 1979).

It is known that Q is a function of time, so we lose no generality if we make the substitution

$$Q = q\mathrm{e}^{-\alpha t} \tag{10.6}$$

where q is also a function of time. The reason for this choice is that when (10.6) is used in (10.5) the term in $\mathrm{d}q/\mathrm{d}t$ is eliminated and the equation for q found to be simply

$$\ddot{q} + (\omega_0^2 - \alpha^2)q = 0 \tag{10.7}$$

Now this is an equation which can be solved by elementary methods, although the form of the solution depends on the value of $(\omega_0^2 - \alpha^2)$. When this quantity is positive, (10.7) is the equation of an undamped harmonic oscillator with a natural angular frequency ω_N given by

$$\boxed{\omega_\mathrm{N}^2 = \omega_0^2 - \alpha^2} \tag{10.8}$$

so that q has the form $A\sin(\omega_\mathrm{N}t + \phi)$ and Q is given by (10.6) as a damped harmonic motion. The system is said to be *underdamped*.

When $\alpha = \omega_0$, the system is *critically damped*, and (10.7) becomes simply $\mathrm{d}^2q/\mathrm{d}t^2 = 0$. So q has the form $At + B$. Finally, if $\alpha > \omega_0$, the system is *over-damped* and q has the form $A\mathrm{e}^{+\omega' t} + B\mathrm{e}^{-\omega' t}$, where $\omega'^2 = \alpha^2 - \omega_0^2$.

† Note that although quantities denoted generally by ω are strictly angular frequencies, it is common practice to refer to them loosely simply as frequencies.

To summarize these results:

Case 1 $\alpha=0$ or $R=0$ (**undamped**): $\ddot{q}+\omega_0^2 q=0$ and hence

$$Q=A\sin(\omega_0 t+\phi) \tag{10.9}$$

Case 2 $\alpha^2<\omega_0^2$ or $R^2<4L/C$ (**underdamped**): $\ddot{q}+\omega_N^2 q=0$ and hence

$$Q=Ae^{-\alpha t}\sin(\omega_N t+\phi) \tag{10.10}$$

Case 3 $\alpha^2=\omega_0^2$ or $R^2=4L/C$ (**critically damped**): $\ddot{q}=0$ and hence

$$Q=e^{-\alpha t}(At+B) \tag{10.11}$$

Case 4 $\alpha>\omega_0^2$ or $R^2>4L/C$ (**overdamped**): $\ddot{q}-\omega'^2 q=0$ and hence

$$Q=e^{-\alpha t}(Ae^{\omega' t}+Be^{-\omega' t})$$

or

$$Q=Ae^{-\alpha t}\sinh(\omega' t+\phi) \tag{10.12}$$

The constants A and ϕ, or A and B, are obtained by applying the initial conditions $\dot{Q}=0$, $Q=Q_0$ at $t=0$, and the current from $I=-\mathrm{d}Q/\mathrm{d}t$. For instance, the complete solution for the underdamped case is

$$Q=Q_0(\omega_0/\omega_N)e^{-\alpha t}\sin(\omega_N t+\phi); \qquad I=Q_0(\omega_0^2/\omega_N)e^{-\alpha t}\sin\omega_N t \tag{10.13}$$

where $\tan\phi=\omega_N/\alpha$.

For completeness, we give the solutions also for critical damping:

$$Q=Q_0(1+\alpha t)e^{-\alpha t}; \qquad I=Q_0\alpha^2 te^{-\alpha t} \tag{10.14}$$

and for overdamping:

$$Q=Q_0(\omega_0/\omega')e^{-\alpha t}\sinh(\omega' t+\phi); \qquad I=Q_0(\omega_0^2/\omega')e^{-\alpha t}\sinh\omega' t \tag{10.15}$$

Equations (10.13), (10.14) and (10.15) are plotted in Fig. 10.3a and the corresponding curves for the charging of the capacitor in Fig. 10.3b.

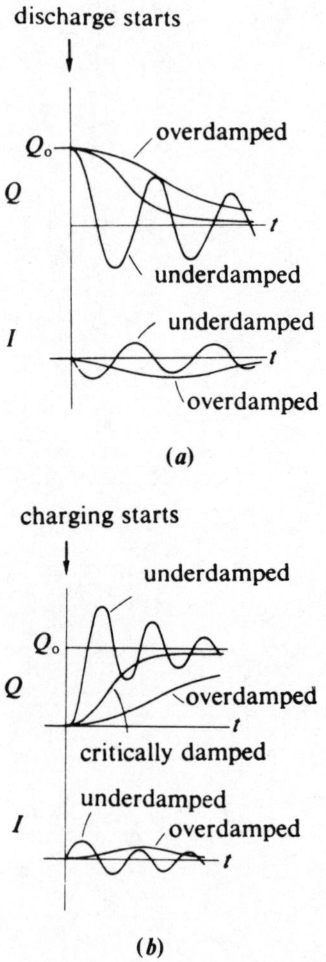

Figure 10.3 The variations of Q and I with time in the circuits of Fig. 10.2 (a) during the discharge of C, (b) during the charging of C, both through L and R.

Natural frequency and logarithmic decrement The natural angular frequency of damped oscillation is, from (10.8),

$$\text{series } LCR \text{ circuit:} \qquad \boxed{\omega_N=(\omega_0^2-\alpha^2)^{1/2}=\left(\frac{1}{LC}-\frac{R^2}{4L^2}\right)^{1/2}} \tag{10.16}$$

The *decrement* of decaying oscillations, Δ, is defined as the ratio of successive maxima. Hence $\Delta=e^{-\alpha t}/e^{-\alpha(t+T)}$, where T is the period $2\pi/\omega_N$. Thus $\Delta=e^{\alpha T}$. The logarithmic decrement, Λ, is defined as $\log_e\Delta$ and so

$$\Lambda=\alpha T=\pi R/\omega_N L \tag{10.17}$$

$$=2\pi/(4L/CR^2-1)^{1/2} \tag{10.18}$$

Note that some prefer to define Δ as the ratio of a maximum to the next minimum, so that their Λ will be only half of the one used here.

Analogy with mechanical systems The motion of a particle of mass m attracted to an origin by a force μx and subjected to a resistance $k\,dx/dt$ is governed by the differential equation

$$m\frac{d^2x}{dt^2}+k\frac{dx}{dt}+\mu x=0 \qquad (10.19)$$

which is of exactly the same form as (10.2) and has similar solutions. The analogy is expressed by the equivalents

$$x\leftrightarrow Q; \qquad \frac{dx}{dt}\leftrightarrow I; \qquad m\leftrightarrow L; \qquad k\leftrightarrow R; \qquad \mu\leftrightarrow\frac{1}{C} \qquad (10.20)$$

Just as mass is a measure of the resistance to changes in motion of the particle, so L is a measure of the inertia of the circuit opposing changes in current: without the middle terms, the oscillations are harmonic with energy conserved but continuously changing from kinetic, $\frac{1}{2}mv^2$ or $\frac{1}{2}LI^2$, to potential, $\frac{1}{2}\mu x^2$ or $\frac{1}{2}Q^2/C$, and back again. The middle terms contain the factors responsible for the dissipation of energy from the system occurring at a rate $k(dx/dt)^2$ or RI^2, while damped harmonic oscillations take place if $k^2<4\mu/m$ or $R^2<4L/C$.

10.3 Transients in coupled circuits

Networks more complex than that of the last section can be solved in a similar way, but when more than one mesh occurs the currents are governed by a set of simultaneous differential equations whose solutions are usually quite complicated. An important class of network is one with two meshes each with its own natural frequency and possessing a common element which couples them together. Our main interest lies in oscillations with small or negligible damping and, because the presence of resistance complicates formulae without greatly affecting the main results, we shall ignore it.

Worked example *mutually coupled circuits* A common and important coupling element is the mutual inductance. Consider the network of Fig. 10.4 in which C_1 with an initial charge Q_0 is discharging, the situation illustrated being that at time t. The mesh currents are $I_1=-dx/dt$, $I_2=+dy/dt$, and by applying $\Sigma V=0$ to each mesh:

$$\frac{x}{C_1}-L_1\frac{dI_1}{dt}+M\frac{dI_2}{dt}=-\frac{y}{C_2}-L_2\frac{dI_2}{dt}+M\frac{dI_1}{dt}=0 \qquad (10.21)$$

so that

$$x/C_1+L_1\ddot{x}+M\ddot{y}=0 \qquad (10.22)$$

$$y/C_2+L_2\ddot{y}+M\ddot{x}=0 \qquad (10.23)$$

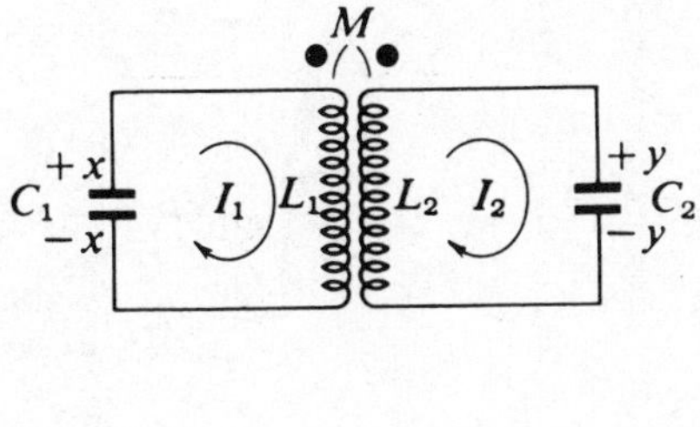

Figure 10.4 Mutually coupled circuits.

We look at the special case when the two circuits have the same *uncoupled* natural frequency ω_0 so that $\omega_0^2 = 1/L_1C_1 = 1/L_2C_2$. We also use the coefficient of coupling $k^2 = M^2/L_1L_2$. Now we expect the solution to be oscillatory when there is negligible damping and so we try solutions of the form $x = A\sin(\omega_N t + \phi)$, $y = B\sin(\omega_N t + \psi)$, where ω_N is to be found by substituting in (10.22) and (10.23). When this is done, the following equation is obtained for ω_N:

$$\omega_N^4(1-k^2) - \omega_N^2(1/L_1C_1 + 1/L_2C_2) + 1/L_1L_2C_1C_2 = 0 \qquad (10.24)$$

or

$$\omega_N^4(1-k^2) - 2\omega_0^2\omega_N^2 + \omega_0^4 = 0 \qquad (10.25)$$

from which

coupled circuits:
$$\boxed{\omega_N = \frac{\omega_0}{\sqrt{(1 \pm k)}}} \qquad (10.26)$$

giving *two* natural frequencies separated by an amount depending on the tightness of coupling. The charges are given by

$$x = A\sin\left(\frac{\omega_0}{\sqrt{(1+k)}}t + \phi_A\right) + B\sin\left(\frac{\omega_0}{\sqrt{(1-k)}}t + \phi_B\right) \qquad (10.27)$$

with four constants as demanded by the fourth-order equation (10.24). If the solution for y is assumed to be of the form $C\sin(\omega_N t + \phi)$, then substitution in (10.22) gives $C = \mp A(L_1/L_2)^{1/2}$, the upper sign corresponding to that of (10.26). Thus

$$y = -(L_1/L_2)^{1/2}\left[A\sin\left(\frac{\omega_0}{\sqrt{(1+k)}}t + \phi_A\right) - B\sin\left(\frac{\omega_0}{\sqrt{(1-k)}}t + \phi_B\right)\right] \qquad (10.28)$$

In general, both x and y contain two terms which, for small k, have nearly equal frequencies so that beats are produced as in Fig. 10.5: the currents will be given by similar curves. It should be remembered that the presence of resistance will produce decaying oscillations and will slightly affect the frequencies.

The result is analogous to that for two coupled mechanical systems, such as pendulums coupled through their suspensions, and these are treated in detail in textbooks on mechanics (see, for example, Becker, 1954). The two components of x and y are known as the *normal modes of oscillation* of the system. In giving only one oscillator an initial displacement (corresponding to the charge Q_0 on C_1) both modes are excited and the motion is such that the energy initially associated

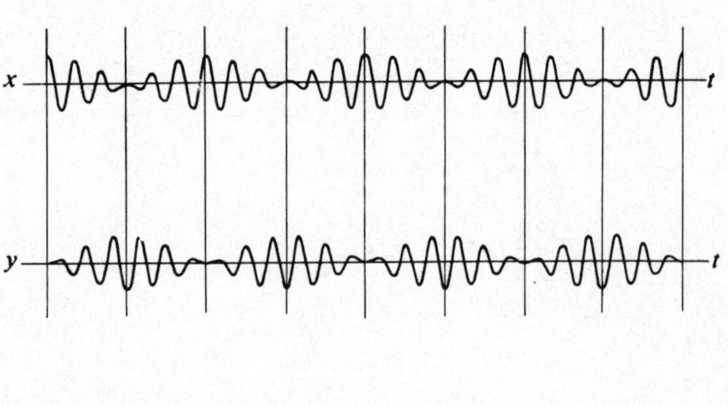

Figure 10.5 The time variations of charge on the capacitors of the mutually coupled circuits in Fig. 10.4.

completely with the first oscillator is transferred completely to the second, then back to the first and so on, as is evident from Fig. 10.5.

The natural frequencies of other two-mesh networks can be obtained by similar methods (see Problem 10.5).

10.4 Basic a.c. theory: impedance

We now insert into networks sinusoidally varying sources generating, for example, a voltage like $\mathscr{E}=\mathscr{E}_0 \sin \omega t$. These sources provide a continuous input of energy and the network is now analogous to a mechanical system undergoing forced vibrations. When Kirchhoff's laws are applied to such networks at any instant, the differential equations obtained differ from those of the last two sections only by the addition of the applied voltages or currents. Thus if the circuit of Fig. 10.2a incorporated a voltage generator $\mathscr{E}_0 \sin \omega t$ in series, the equation governing Q would be

$$L\frac{\mathrm{d}^2Q}{\mathrm{d}t^2}+R\frac{\mathrm{d}Q}{\mathrm{d}t}+\frac{Q}{C}=\mathscr{E}_0 \sin \omega t \tag{10.29}$$

and the current is given by $I=-\mathrm{d}Q/\mathrm{d}t$.

In Sec. 10.1 we saw that the general solution of (10.29) would be the sum of a CF and a PI. However, the CF is identical with the transient solutions of Sec. 10.2 so that it always tends to zero as $t\to\infty$ because of the damping present. The PI constitutes the *steady-state* part of the solution and is the important one for a.c. theory. We shall therefore ignore the transient CF's, bearing in mind only that they will be present when alternating sources are first switched on: they will affect the total voltages and currents over the first few cycles until their amplitudes become negligible.

Although the calculation of PI's for simple circuits can be obtained by standard methods from the theory of differential equations, we prefer methods in which these equations are replaced by algebraic ones. Such methods lead to expressions so closely analogous to those for d.c. networks that all the results of Chapter 6 can be taken over virtually unchanged. The remainder of this section shows two equivalent routes by which the solution of a.c. problems can be reduced to algebra.

Replacement of differential equations: route 1 The *linearity* of the network differential equations has an important consequence when the applied voltages and currents have a single frequency. As noted in Sec. 10.1, the solutions (PI's) for Q, I and V everywhere in the network must all vary sinusoidally with the same angular frequency. This follows from the fact that differentiation of a sine or cosine term generates terms with the same value of ω. Inspection of (10.29) shows that Q must be of the form $Q_0 \sin(\omega t+\delta)$ or the two sides could not be equal for *all* times t.

It follows that when the applied electromotance is $\mathscr{E}_0 \sin \omega t$, say, all currents have the form $I_0 \sin(\omega t+\alpha)$ and all voltages the form $V_0 \sin(\omega t+\beta)$. To solve a network we still apply Kirchhoff's laws in the form $\Sigma I=0$ at a junction and $\Sigma V=0$ round a mesh. Our basic problem is therefore one of adding a number of sinusoidally varying quantities of the same frequencies but with differing

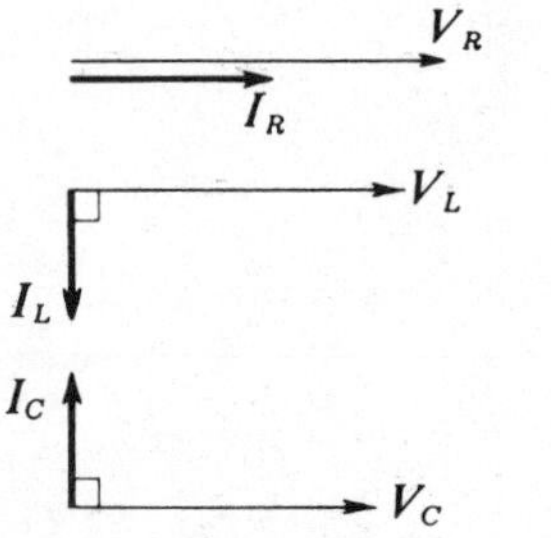

Figure 10.6 The phase relationships between V and I in ideally pure circuit elements. The three figures are phasor diagrams for R, L and C respectively.

amplitudes and phases. Such quantities are called *phasors* and their summation can be carried out either by using a phasor diagram or by using complex numbers. The reader should refer continually to Appendix A.7 for the rest of this section if unfamiliar with these methods.

Pure resistance, capacitance and inductance If we can find the relationships between V and I for single circuit elements, Kirchhoff's laws can be used, together with the methods for the addition of phasors, to cope with more complicated networks. So we pass a current $I=I_0 \sin \omega t$ through a pure R, a pure L and a pure C in turn and obtain the potential differences from the relationships of Fig. 10.1:

For R: $V=RI=RI_0 \sin \omega t$ and hence

$$V=V_0 \sin \omega t \qquad \text{where } V_0=RI_0$$

For L: $V=L\,\mathrm{d}I/\mathrm{d}t=\omega L I_0 \cos \omega t$ and hence

$$V=V_0 \sin (\omega t+\tfrac{1}{2}\pi) \qquad \text{where } V_0=\omega L I_0$$

For C: $V=Q/C=\int I\,\mathrm{d}t/C=-(I_0 \cos \omega t)/\omega C$ and hence

$$V=V_0 \sin (\omega t-\tfrac{1}{2}\pi) \qquad \text{where } V_0=I_0/\omega C$$

These three important relations can be summarized by using either the phasor diagrams of Fig. 10.6 or a complex representation as follows:

In R, $V_0=RI_0$ and V and I are in phase:	$\mathbf{V}=R\mathbf{I}$	(10.30)
In L, $V_0=\omega L I_0$ and V leads I by $\frac{1}{2}\pi$:	$\mathbf{V}=\mathrm{j}\omega L\mathbf{I}$	
In C, $V_0=I_0/\omega C$ and V lags on I by $\frac{1}{2}\pi$:	$\mathbf{V}=\mathbf{I}/\mathrm{j}\omega C$ or $-\mathrm{j}\mathbf{I}/\omega C$	

The quantities ωL and $1/\omega C$, known as **reactances**, play the same part as R in the relation between the *magnitudes* of V and I, but the phase differences must not be forgotten and the presence of the j's ensures that they will not be.

For a mutual inductance M, the same treatment shows that the current $\mathbf{I}$ in one coil gives a voltage $\mathbf{V}=\pm \mathrm{j}\omega M\mathbf{I}$ across the other. The phasor diagram for a pure M will resemble that for L or C depending on the sign.

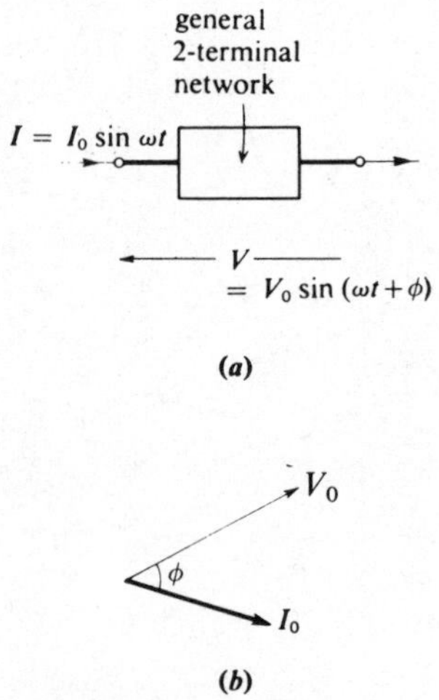

Figure 10.7 (a) Current and voltage for a general two-terminal network; (b) the phasor diagram for (a).

Impedance Any two-terminal network of R's, L's and C's will have a current of amplitude I_0, say, through it and a potential difference of amplitude V_0 across it as in Fig. 10.7a. The phase difference between the current and potential difference will in general be, say, ϕ, so that the network behaves neither as a pure resistance ($\phi=0$) nor as a pure reactance ($\phi=\frac{1}{2}\pi$). Suppose, however, that the origin of time is chosen so that the current is $I=I_0 \sin \omega t$ with a complex representation $\mathbf{I}=I_0 e^{\mathrm{j}\omega t}$. The corresponding complex voltage is $\mathbf{V}=V_0 e^{\mathrm{j}(\omega t+\phi)}$ (see Appendix A.7). By analogy with the definition of resistance for d.c. networks in Eq. (6.8), we define a *complex impedance* of the two-terminal network by

$$\mathbf{Z}=\mathbf{V}/\mathbf{I} \qquad \text{(definition of } \mathbf{Z}) \tag{10.31}$$

Note that whereas the complex $\mathbf{V}$ and $\mathbf{I}$ include the time variation $e^{\mathrm{j}\omega t}$, the complex $\mathbf{Z}$ does *not*, because the terms cancel in (10.31). Thus $\mathbf{Z}$ is not a phasor but merely a complex number which can be obtained in terms of the network elements R, L and C and the frequency ω, as we see below.

Like any complex number, $\mathbf{Z}$ can be expressed in two forms: a **polar** form $re^{j\theta}$, involving the **modulus** r and the **argument** θ, and a **Cartesian form** $a+jb$, involving the **real and imaginary parts** a and b. Each form has its own particular use and interpretation.

Polar form of impedance For the general two-terminal network of Fig. 10.7, we have that

polar form: $$\mathbf{Z}=\mathbf{V}/\mathbf{I}=(V_0/I_0)e^{j\phi}=Ze^{j\phi} \tag{10.32}$$

where the modulus of $\mathbf{Z}$, which can be written as $|\mathbf{Z}|$ or just Z, is the ratio of the magnitudes of voltage and current. For this reason, it is often called simply the **impedance** of the network: its **SI unit** is the ohm. The argument of $\mathbf{Z}$ gives ϕ, the phase difference between voltage and current.

In the polar form, $\mathbf{Z}$ can be regarded as an operator which converts $\mathbf{I}$ into $\mathbf{V}$ through $\mathbf{V}=|\mathbf{Z}|e^{j\phi}\mathbf{I}$. The modulus changes the magnitude of $\mathbf{I}$ while $e^{j\phi}$ rotates it through an angle ϕ in the complex plane (Fig. 10.7b).

Cartesian form of impedance Since $e^{j\phi}=\cos\phi+j\sin\phi$, (10.32) gives

$$\mathbf{Z}=(V_0\cos\phi)/I_0+j(V_0\sin\phi)/I_0 \tag{10.33}$$

The real and imaginary parts of $\mathbf{Z}$ define the **resistance** R and **reactance** X of the network:

$$\left.\begin{aligned} R&=\operatorname{Re}\mathbf{Z}=(V_0\cos\phi)/I_0 \quad \text{(Definition of } R) \\ X&=\operatorname{Im}\mathbf{Z}=(V_0\sin\phi)/I_0 \quad \text{(Definition of } X) \end{aligned}\right\} \tag{10.34}$$

and it follows from these and from (10.33) that

Cartesian form: $$\mathbf{Z}=R+jX \tag{10.35}$$

Note that $V_0\cos\phi$ is the component of $\mathbf{V}$ in phase with $\mathbf{I}$ and $V_0\sin\phi$ the component in quadrature. This makes the definitions (10.34) consistent with the simple examples of Eq. (10.30). It also follows from (10.35) and (10.32) that

$$|\mathbf{Z}|=(R^2+X^2)^{1/2}=V_0/I_0; \qquad \arg\mathbf{Z}=\tan^{-1}(X/R)=\phi \tag{10.36}$$

Complex impedances of network elements It is clear from the above discussion that the relationships established in (10.30) mean that circuit elements R, L, C and M have the following complex impedances:

$$\begin{aligned} \mathbf{Z}\text{ of } R &= R \\ \mathbf{Z}\text{ of } L &= j\omega L \\ \mathbf{Z}\text{ of } C &= -j/\omega C \text{ or } 1/j\omega C \\ \mathbf{Z}\text{ of } M &= \pm j\omega M \end{aligned} \tag{10.37}$$

Generalized network theorems The laws governing d.c. networks in Chapter 6 were derived from the definition of resistance $V=RI$ together with the laws $\Sigma I=0$

$\Sigma V=0$. A.C. networks are governed by $\mathbf{V}=\mathbf{ZI}$ coupled with $\Sigma\mathbf{I}=0$ and $\Sigma\mathbf{V}=0$, so that all the theorems of Chapter 6 can be taken over with the substitution of *complex* potential differences, currents and impedances for V, I and R, respectively.

Hence,

$$\left.\begin{aligned}&\text{impedances in series:} &&\boxed{\mathbf{Z}_{\text{eq}}=\mathbf{Z}_1+\mathbf{Z}_2+\mathbf{Z}_3+\cdots=\Sigma\mathbf{Z}_i}\\&\text{impedances in parallel:} &&\boxed{1/\mathbf{Z}_{\text{eq}}=1/\mathbf{Z}_1+1/\mathbf{Z}_2+1/\mathbf{Z}_3+\cdots=\Sigma 1/\mathbf{Z}_i}\end{aligned}\right\}\quad(10.38)$$

while the delta–Y transformation, Kirchhoff's laws, the superposition theorem, and Thévenin's theorem all apply provided that $\mathbf{Z}$ is linear.

If at any time the instantaneous values of an $\mathbf{I}$ or a $\mathbf{V}$ are required, then either the real or imaginary part must be taken as appropriate. This is not, however, a common requirement.

Phasor diagrams are best started with one of the potential differences or currents common to more than one component (see Sec. 10.6). Note that arrows are still used in circuit diagrams for $\mathbf{V}$ and $\mathbf{I}$ even though we are dealing with a.c., because we still wish to assign relative directions to them at any instant.

Two simple circuits are dealt with here to illustrate the application of the method. More complicated examples will be encountered later in the chapter.

Worked example *L and R in series* It immediately follows from (10.37) and (10.38) that the equivalent impedance is given by:

L and R in series:

$$\mathbf{Z}=R+\mathrm{j}\omega L,\qquad |\mathbf{Z}|=(R^2+\omega^2L^2)^{1/2},\qquad \phi=\tan^{-1}\omega L/R \qquad (10.39)$$

The same result can be obtained from a phasor diagram started by drawing the common current I and then inserting the voltage RI across R in phase with I and the voltage ωLI across L at $\pi/2$ to I. The resultant voltage is given by the diagonal of the rectangle of which RI and ωLI are the two adjacent sides. It is clear that this also yields (10.39).

Worked example *C and R in parallel* Once again, (10.37) and (10.38) give the equivalent impedance as

$$\mathbf{Z}=\mathbf{Z}_1\mathbf{Z}_2/(\mathbf{Z}_1+\mathbf{Z}_2)=-\mathrm{j}R/\omega C(R-\mathrm{j}\omega C)$$

and hence

$$C \text{ and } R \text{ in parallel:}\qquad |\mathbf{Z}|=\left[\frac{R^2}{1+\omega^2C^2R^2}\right]^{1/2},\qquad \phi=\tan^{-1}\omega CR \qquad (10.40)$$

The phasor diagram this time is best started by drawing the common voltage V and inserting the currents through C and R.

Admittance and susceptance Corresponding to the definition of d.c. conductance in (6.8) as I/V, we can define a **complex admittance** $\mathbf{Y}$ by

$$\boxed{\mathbf{Y}=1/\mathbf{V}=1/\mathbf{Z}\qquad \text{(definition of } \mathbf{Y})} \qquad (10.41)$$

Using the same general two-terminal network as in Fig. 10.7, equation (10.32) gives us that $\mathbf{Y}=I_0 e^{-j\phi}/V_0$. The real part of $\mathbf{Y}$ is called the **conductance** G and the imaginary part the **susceptance** B, so that

$$G=\text{Re}\,\mathbf{Y}=(I_0\cos\phi)/V_0 \qquad \text{(definition of } G\text{)} \qquad (10.42)$$
$$B=\text{Im}\,\mathbf{Y}=-(I_0\sin\phi)/V_0 \qquad \text{(definition of } B\text{)} \qquad (10.43)$$

and

$$\boxed{\mathbf{Y}=G+jB} \qquad (10.44)$$

Replacement of differential equations: route 2 A more analytical approach, somewhat shorter than the above, is to use the linearity of the original differential equations like (10.29) to go straight to a complex representation of all time-varying quantities like I and V.

We have already seen that an equation like (10.29) with the applied electromotance given by $\mathscr{E}_0\sin\omega t$ has solutions for Q, I, V, etc., of the form $Q_0\sin(\omega t+\phi)$, etc. Altering the time zero by a quarter period shows that if the applied electromotance were $\mathscr{E}_0\cos\omega t$, the solutions would now be $Q_0\cos(\omega t+\phi)$, etc., where all the Q_0, ϕ, etc., are the same as before. Because of the linearity, it follows that if we take as the applied electromotance the linear combination $\mathscr{E}_0\cos\omega t+j\mathscr{E}_0\sin\omega t$ (i.e., $\mathscr{E}_0 e^{j\omega t}$), the solutions would be of the form $Q_0 e^{j(\omega t+\phi)}$. We now have the enormous gain in simplicity given by the use of exponentials. For instance, by substituting the values just obtained into (10.29), the $e^{j\omega t}$ cancels throughout. Since $I=-dQ/dt$ from Fig. 10.2a with a source of electromotance in series, the solution for the current is of the form $\mathbf{I}=\mathscr{E}_0 e^{j\omega t}/\mathbf{Z}$, where $\mathbf{Z}=R+j(\omega L-1/\omega C)$ as we expect from our previous treatment. Thus the solution of a set of differential equations is once again reduced to an algebraic problem.

The use of the complex exponential function $e^{j\omega t}$ instead of real functions $\sin\omega t$ and $\cos\omega t$ is common throughout vibration and wave theory. It is permissible because the expressions are linear: it is desirable because exponential functions are easy to handle mathematically.

10.5 R.M.S. values and a.c. power

R.M.S. values What do we mean by an a.c. of 3 A or an alternating voltage of 240 V? It would be possible, with a sinusoidally varying current $I=I_0\sin\omega t$, to quote the peak current I_0, but this is not a very useful quantity when the variations are not sinusoidal. In fact, we choose to quote a value such that the mean heating effect in a pure resistor is the same whether the current is steady or not.

In a pure R the instantaneous rate of production of heat is RI^2 no matter how I varies. Over a period of time the rate of heat production is $R\langle I^2\rangle$, where the brackets indicate an average taken with respect to time. If we define a **root mean square** (r.m.s.) current by

$$\boxed{I_{\text{rms}}=\sqrt{\langle I^2\rangle} \qquad \text{(definition of } I_{\text{rms}}\text{)}} \qquad (10.45)$$

then the mean rate of heat production is RI_{rms}^2. For a steady current, I_{rms} and I are the same. It follows that, provided we already quote r.m.s. values, the heating effects of say, 3 A d.c. and 3 A a.c. are the same when passing through a resistor. This applies equally to potential differences: an alternating voltage quoted simply as 240 V is understood to be an r.m.s. value.

For a sinusoidally varying current $I_0 \sin \omega t$, we have

$$I_{\text{rms}}^2 = \langle I_0^2 \sin^2 \omega t\rangle = \langle \tfrac{1}{2}I_0^2 - \tfrac{1}{2}I_0^2 \cos 2\omega t\rangle = \tfrac{1}{2}I_0^2$$

because the mean value of $\cos 2\omega t$ over a complete number of cycles is zero. Hence, for a sinusoidally varying current,

sinusoidal current: $$I_{\text{rms}} = I_0/\sqrt{2} \tag{10.46}$$

and it follows similarly that $V_{\text{rms}} = V_0/\sqrt{2}$. In Sec. 10.4 we restricted the treatment to sinusoidal currents and voltages and it follows that *all the relations established so far involving V_0 or I_0 either as a ratio or in a scale diagram could equally well have used V_{rms} and I_{rms}.*

Equation (10.46) does not apply in general. For other types of variation the mean square value must be found and (10.45) applied. (Note that for any time-varying quantity $f(t)$, the mean value over a period T is defined as $\int f(t)\,dt$ from t to $t+T$, divided by T.)

Power in a.c. circuits The instantaneous rate of consumption of energy in a general two-terminal network such as that of Fig. 10.7a is VI, where $V = V_0 \sin(\omega t + \phi)$ and $I = I_0 \sin \omega t$. The power consumed is thus

$$\begin{aligned} P &= V_0 I_0 \sin \omega t \sin(\omega t + \phi) \\ &= V_0 I_0 \sin^2 \omega t \cos \phi + V_0 I_0 \sin \omega t \cos \omega t \sin \phi \end{aligned}$$

The mean power over a cycle is thus

$$\langle P\rangle = \tfrac{1}{2}V_0 I_0 \cos\phi = \tfrac{1}{2}RI_0^2$$

because $\langle \sin^2 \omega t\rangle = \tfrac{1}{2}$ as above, because $\langle \sin \omega t \cos \omega t\rangle = \langle \tfrac{1}{2}\sin 2\omega t\rangle = 0$, and where R is the real part of the complex impedance. Hence

$$\boxed{\langle P\rangle = V_{\text{rms}} I_{\text{rms}} \cos\phi = RI_{\text{rms}}^2} \tag{10.47}$$

Similarly, the mean power consumption from a voltage generator $\mathscr{E}_{\text{rms}}$ supplying I_{rms} at a phase difference ϕ is

$$\boxed{\langle P\rangle = \mathscr{E}_{\text{rms}} I_{\text{rms}} \cos\phi} \tag{10.48}$$

Cos ϕ is known as the **power factor** and is clearly 1 for a pure resistance and 0 for a reactance. Note that, because power is not a linear function of current or potential difference, it is not permissible to calculate it using the complex representation. A device sometimes adopted, however, is to use the complex conjugate of quantities, obtained by replacing j by $-$j everywhere and denoted by a star: thus if $\mathbf{I} = I_0 e^{j\omega t}$ then $\mathbf{I}^* = e^{-j\omega t}$. In that case the mean power can be written as $\tfrac{1}{2}\text{Re}(\mathbf{I}^*\mathbf{V})$.

Impedance matching We now find the conditions for maximum power transfer to a load by a source of alternating voltage with internal impedance (Fig. 10.8). The

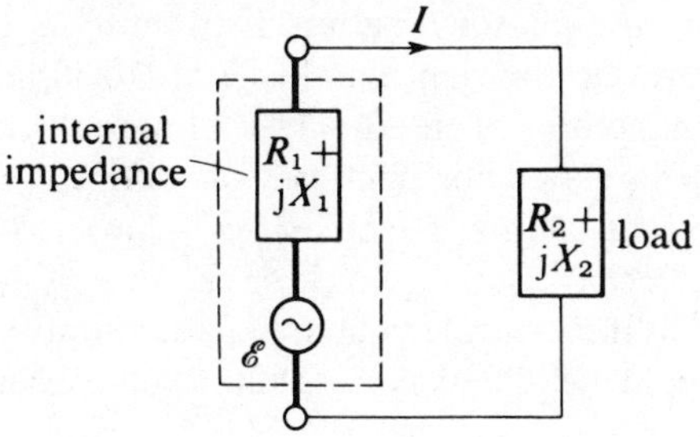

Figure 10.8 Voltage source with internal impedance and load.

current I_0 is $\mathscr{E}_0/[(R_1+R_2)^2+(X_1+X_2)^2]^{1/2}$, and hence, from (10.47),

$$\langle P\rangle = R_2 I^2_{\text{rms}} = R_2\mathscr{E}^2_{\text{rms}}/\{(R_1+R_2)^2+(X_1+X_2)^2\} \tag{10.49}$$

If the load is given, then $\langle P\rangle$ is a maximum when $R_1=0$ and $X_1=-X_2$, but if, as is more common, the internal impedance is fixed then the power transfer depends on what parameters in the load can be chosen. If R_2 and X_2 are independent, differentiation of (10.49) with respect to R_2 shows that the optimum value of R_2 is $[R_1^2+(X_1+X_2)^2]^{1/2}$. If X_2 can be chosen as well, then $X_2=-X_1$ and $R_2=R_1$ are the conditions for maximum $\langle P\rangle$. The load is then **matched** to the source.

Power ratios and the decibel, dB The ratio of two powers P_2 and P_1 is expressed in *bels* by $\log_{10}(P_2/P_1)$. More commonly, the *decibel* (dB) is used and a power gain in dB is thus $10\log_{10}(P_2/P_1)$. The half-power points defining the width of the resonance curves shown later in Fig. 10.11 represent a reduction of almost exactly 3 dB from the peak power.

10.6 Resonance in LCR networks

The differential equations governing a.c. circuits and those governing forced vibrations in linear mechanical systems are identical in form. We should therefore expect resonance phenomena to occur in the first case just as they do in the second. In this section we examine the response of simple series and parallel *LCR* networks.

Series resonance The series *LCR* circuit of Fig. 10.9a carries the same current through each of its components and this is the starting phasor in the diagrams of Fig. 10.9b and c. In Fig. 10.9b the potential differences are drawn from the same point and compounded in the usual way to form the resultant, but Fig. 10.9c illustrates a more useful method of arriving at the same result by using a polygon of phasors in which the corners can be labelled with the letters corresponding to junctions in the circuit.

The complex impedance of the combination is

LCR in series: $\mathbf{Z}=R+\mathrm{j}(\omega L-1/\omega C)$

and the magnitude of the current is thus

$$LCR \text{ in series:} \qquad I=\frac{V}{[R^2+(\omega L-1/\omega C)^2]^{1/2}} \tag{10.50}$$

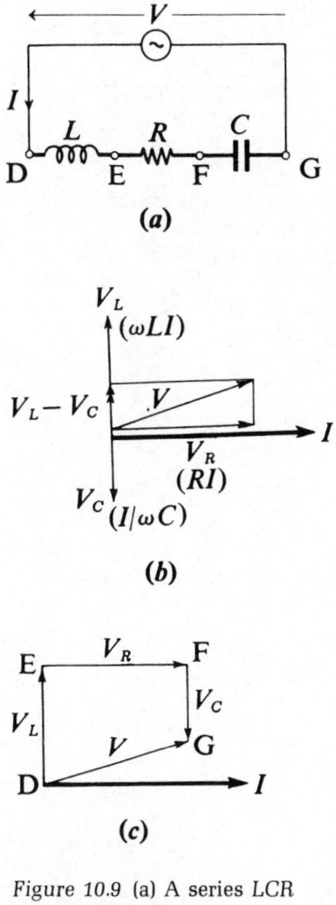

Figure 10.9 (a) A series *LCR* circuit with voltage generator; (b) the phasor diagram for (a) using a parallelogram: $V^2=V_R^2+(V_L-V_C)^2$ giving (10.50); (c) the phasor diagram for (a) using a polygon.

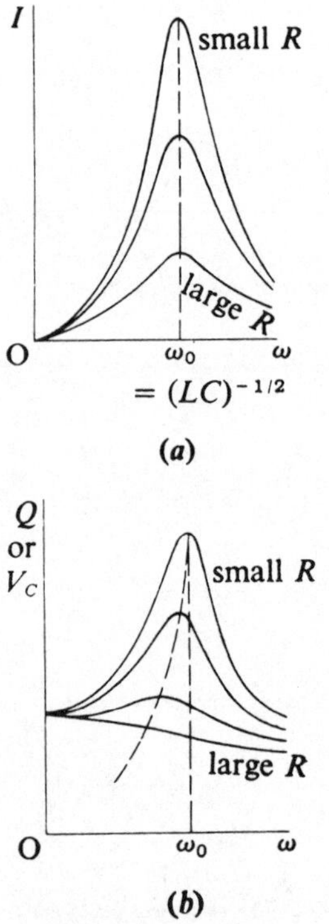

Figure 10.10 Resonance curves for a series LCR circuit. (a) Variation of current with frequency; (b) variation of charge on capacitor with frequency.

which can also be obtained from the phasor diagram. If either ω, L or C are varied, keeping all other quantities on the right-hand side of (10.50) constant, then the variation of I shows a maximum when $\omega L = 1/\omega C$. Figure 10.10 shows the form of the curves when ω is varied for circuits with different R's. The occurrence of a maximum current at $\omega_0 = 1/(LC)^{1/2}$ is known as **current resonance** and corresponds to velocity resonance in a mechanical system. The resonant value ω_0 is *not* the same as the natural frequency of free vibrations in the same circuit (ω_N of Sec. 10.2) except when $R = 0$. Note that the voltage across R is also a maximum at ω_0.

The phasor diagram shows that, also at ω_0, **V** and **I** are in phase and the impedance is purely resistive. The condition $\omega = \omega_0$ is then also known as **phase resonance**.

Corresponding to amplitude resonance in a mechanical system is **charge resonance**. The charge circulating is I/ω and is thus obtainable from (10.50), curves of Q against ω being given in Fig. 10.10b. The maximum is obtained as usual by differentiation and occurs at an angular frequency ω_Q given by

$$\omega_Q^2 = \omega_0^2 - R^2/2L^2 \qquad (10.51)$$

It should be noted that neither charge nor current resonance occur at the natural frequency of the circuit given by (10.16) but that the three approach each other as R tends to zero. (See also Problem 10.19.)

$\hat{Q}$ of a circuit and sharpness of resonance The quantity

$$\hat{Q} = 2\pi \times \frac{\text{energy stored}}{\text{energy loss in one period}} \qquad \text{(definition of } \hat{Q}) \qquad (10.52)$$

is known as the **quality factor** or $\hat{Q}$ of a periodic system and for the series LCR circuit is $2\pi \times \frac{1}{2}LI_0^2/\frac{1}{2}RI_0^2T$ or $\omega L/R$. This varies with frequency and it is quite common to use the value at current resonance, $\omega_0 L/R$ or $(L/C)^{1/2}/R$, which we shall denote by $\hat{Q}_0$ (the unconventional 'hat' on the Q is adopted here to avoid confusion with the symbol for charge). Some previous results can be expressed anew: the critical damping condition in Sec. 10.2, $L/CR^2 = \frac{1}{4}$, becomes $\hat{Q}_0 = \frac{1}{2}$; the logarithmic decrement from (10.17) and (10.18) is $2\pi(4\hat{Q}_0^2 - 1)^{-1/2}$ or simply $\pi/\hat{Q}_0$ if R is negligible; and the complex impedance of a series LCR circuit can be written as

$$LCR \text{ in series:} \qquad \mathbf{Z} = R[1 + \mathrm{j}\hat{Q}_0(\omega/\omega_0 - \omega_0/\omega)] \qquad (10.53)$$

which becomes $\mathbf{Z} = R(1 + \mathrm{j}2\hat{Q}_0\Delta\omega/\omega_0)$ near resonance, with $\Delta\omega = \omega - \omega_0$.

The sharpness of a resonance curve is defined by the width at *half-power* when the current has a value $I_{max}/\sqrt{2}$, I_{max} being the resonant current V/R (Fig. 10.11). The width is thus $(\omega_1 - \omega_2)$, where ω_1 and ω_2 are the roots of

$$I_{max}/\sqrt{2} = V/R\sqrt{2} = V/[R^2 + (\omega L - 1/\omega C)^2]^{1/2}$$

using (10.50). The equation for ω is $\omega^2 \pm (R/L)\omega - 1/LC = 0$ whose roots are $\pm R/2L \pm (R^2/4L^2 + 1/LC)^{1/2}$. The second negative sign is inadmissible because it gives negative ω's. Hence

$$\Delta\omega = (\omega_1 - \omega_2) = R/L = \omega_0/\hat{Q}_0 \qquad (10.54)$$

and a **high-$\hat{Q}$ circuit is one with low damping ($\Lambda=\pi/\hat{Q}_0$) and sharp resonance.**

If (10.50) is plotted for variation of C at a fixed frequency, a resonance curve of similar shape is obtained whose width can be shown to be $\Delta C=2C_0/\hat{Q}_0$ for high $\hat{Q}$.

Voltage magnification The maximum potential difference across R occurs at the resonant angular frequency ω_0 and is equal to the applied voltage. The maximum potential difference across C occurs at ω_Q, slightly lower than ω_0, while the maximum potential difference across L occurs at ω_L, slightly higher than ω_0 (see Problem 10.19). At ω_0, the potential differences across L and C are opposite in phase but equal in magnitude to $\omega_0 L I_{max}$ or $\hat{Q}_0$ times the applied voltage. $\hat{Q}_0$ is thus the factor by which the applied voltage is magnified across L or C at resonance.

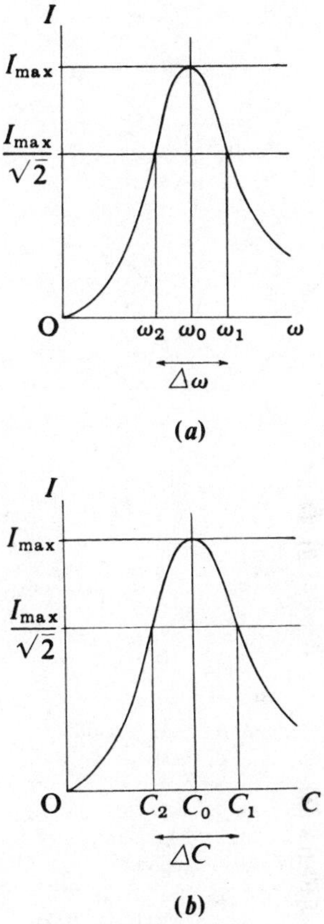

Figure 10.11 Width of resonance curves. (a) Half-width of current against frequency as in Fig. 10.10a; (b) half-width of current against capacitance at a fixed frequency.

Parallel resonance In a parallel connection of components, the voltage is common to all branches and the current divides. The differential equations describing such networks are similar to those for series circuits but the roles of I and V are interchanged. Thus for C, R and L all in parallel we should have $I=C\,dV/dt+V/R+\int V\,dt/L$. If this is compared with equations representing other forced damped harmonic systems, it is seen that in this case the *current* must be regarded as the stimulus and the *voltage* as the response. A constant *current* generator is therefore appropriate as the source for a parallel circuit, and resonance will be identified by a maximum in the *voltage*.

If the calculation is carried through with L, C and R all in parallel, as mentioned above, the algebra is the same as that for a series LCR circuit with the interchange of L and C, of I and V, and the substitution of R for $1/R$ (a network formed by such an interchange is called the *dual* of the original). The variation of voltage with frequency is the same as the variation of I shown in Fig. 10.10a and resonance occurs at $\omega_0=(LC)^{-1/2}$.

As a slightly more complicated but realistic example, consider the network of Fig. 10.12 in which the resistance of the inductor is represented by R. The admittance of such a combination is

$$\begin{aligned}\mathbf{Y}&=\mathrm{j}\omega C+1/(R+\mathrm{j}\omega L)\\&=[(1-\omega^2LC)+\mathrm{j}\omega CR]/(R+\mathrm{j}\omega L)\end{aligned}\qquad(10.55)$$

This gives

C and LR in parallel:

$$\frac{|\mathbf{I}|}{|\mathbf{V}|}=|\mathbf{Y}|=\left[\frac{\omega^2C^2R^2+(\omega^2LC-1)^2}{R^2+\omega^2L^2}\right]^{1/2}\qquad(10.56)$$

Voltage resonance occurs when $|\mathbf{Y}|$ is a minimum so that the resonant frequency would be given by $\partial|\mathbf{Y}|/\partial\omega=0$. This calculation is generally tedious and it is common practice to use **phase resonance** as the principal condition because this is achieved by finding when $\mathbf{Y}$ is real. From (10.55), the imaginary part of $\mathbf{Y}$ is $\omega C-\omega L/(R^2+\omega^2L^2)$ and this is zero for an angular frequency ω_P given by

$$\omega_P^2=\omega_0^2-R^2/L^2\qquad(10.57)$$

in which $\omega_0^2=1/LC$. At ω_P, $|\mathbf{Y}|$ from (10.56) is CR/L and $|\mathbf{Z}|=L/CR$, sometimes known as the **dynamic resistance** of the network. Voltage resonance in fact occurs

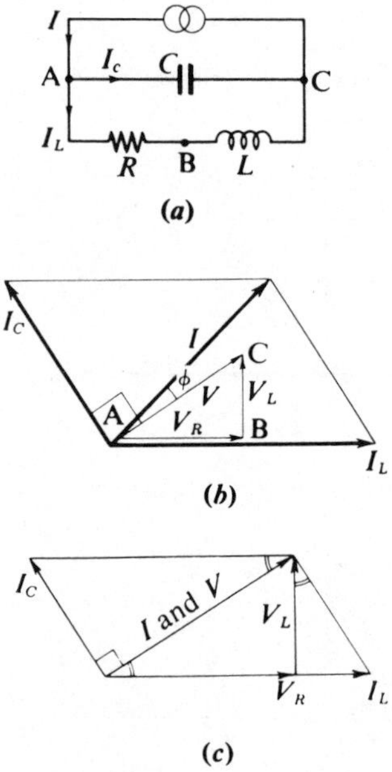

Figure 10.12 (a) A realistic parallel LCR circuit with current generator. (b) Phasor diagram for (a): start by drawing I_L. The points ABC in the phasor diagram correspond to those in (a). (c) Phasor diagram at phase resonance (I and V in phase). By similar triangles, $V_L/V_R = I_C/I$ or $\omega L/R = \omega CV/I$. Hence $V/I = L/CR$.

at a frequency slightly different from ω_P, but in practice useful circuits have very high $\hat{Q}$'s and then all the resonant frequencies are very close together.

The $\hat{Q}$ of the circuit will be $\omega L/R$ by (10.52) and $\hat{Q}_0$ will again be taken as $\omega_0 L/R$. At resonance the phasor diagram of Fig. 10.12c shows that the phase angle between I_C and I_L is $\frac{1}{2}\pi + \phi$, where $\phi = \tan^{-1}\omega_0 L/R = \tan^{-1}\hat{Q}_0$. Thus for a high-$\hat{Q}$ circuit, I_L and I_C will be nearly π out of phase, equal in magnitude and each will be in quadrature with the main current. The magnitude of both will be $\omega_0 CV$ while the main current is V/dynamic resistance or VCR/L. The ratio of I_C or I_L to the main current at resonance is thus $\omega_0 L/R$ or $\hat{Q}_0$, which is again a magnification factor but this time of current.

Application of resonant circuits *electronic amplifying and oscillating circuits* Audio-frequency amplifiers should ideally have a constant gain over a wide range of frequencies (wide-band amplification) so that resonant circuits are not normally included in their design. Radio-frequency amplifiers, on the other hand, are used for receiver pre-amplification in order to respond only to a narrow band of frequencies centred on the carrier frequency of the desired signal and not to all the other signals coming in. As a result, they incorporate one or more resonant circuits which are tuned to the required frequency by variation of their capacitance or inductance. The information carried by the incoming electromagnetic wave causes the frequencies contained in the signal to spread over a finite band around the carrier frequency (the signal bandwidth). The $\hat{Q}$ of the tuned circuit must therefore be high enough to select only the carrier frequency of the required signal and to reject others, but must not be so high that the information contained in what are known as sidebands is also lost.

Electronic circuits used as local oscillators in radio receivers or transmitters produce oscillations at frequencies determined by resonant circuits in various forms. At radio frequencies, the circuits will be ordinary LCR combinations but in the microwave region these are replaced by resonant cavities (see Sec. 14.4).

10.7 Four-terminal networks: general parameters

Input and output impedance Two junctions or nodes of complicated networks are often given special significance because we wish to feed into them a voltage or current from a source. These nodes then become *input terminals* and the impedance of the network looking into them is known as the **input impedance** $\mathbf{Z}_{in}$ defined by $\mathbf{V}_1 = \mathbf{Z}_{in}\mathbf{I}_1$ as in Fig. 10.13a. $\mathbf{Z}_{in}$ may be calculated by the methods of Sec. 10.4. An input admittance $\mathbf{Y}_{in}$ is similarly defined as $\mathbf{I}_1/\mathbf{V}_1$ and both $\mathbf{Z}_{in}$ and $\mathbf{Y}_{in}$ can be regarded as operators: if $\mathbf{V}_1$ is regarded as the stimulus applied to the network, then the response $\mathbf{I}_1$ is given by $\mathbf{Y}_{in}$ operating on $\mathbf{V}_1$. The functions $\mathbf{Y}_{in}$ or $\mathbf{Z}_{in}$ contain all the information needed about the network to calculate its response to any stimulus.

In the rest of this chapter we shall be frequently concerned also with the output from a second pair of terminals, the *output terminals*. The directions of current and voltage at the four terminals are conventionally taken as shown in Fig. 10.13b, but we shall often break the convention for various reasons. Many electric and electronic circuits can be regarded as a series of such four-terminal networks, the output of each stage feeding into the next in the series. In this section we introduce some useful parameters of such networks when all the circuit elements within them are passive.

If a network such as that of Fig. 10.13b, together with any generators across the input, is replaced by its Thévenin or Norton equivalent between the output terminals, the impedance of such an equivalent is known as the **output impedance** $\mathbf{Z}_{\text{out}}$. If a series of networks are connected so that the output of each is fed into the input of the next, then maximum power will be transferred if $\mathbf{Z}_{\text{out}}$ for each stage is the complex conjugate of $\mathbf{Z}_{\text{in}}$ for the following stage (impedance matching, Sec. 10.5).

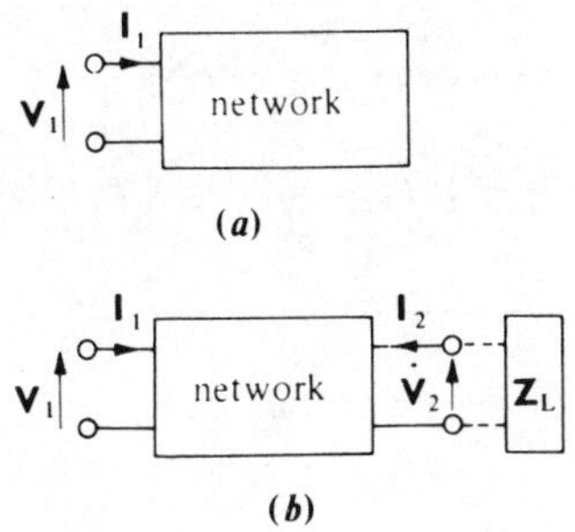

Figure 10.13 (a) A general two-terminal (or one-port) network; (b) a general four-terminal (or two-port) network.

Voltage gain, characteristic impedance, propagation constant We are often interested in the way a particular output from a network depends upon the input. This gives rise to transfer parameters or functions. For instance the ratio of output to input voltage

$$\mathbf{A}=\mathbf{V}_2/\mathbf{V}_1$$

is the **voltage transfer function** or **voltage gain**. Similarly, a current gain $\mathbf{I}_2/\mathbf{I}_1$ can be defined, as well as transfer or mutual admittances or impedances such as $\mathbf{I}_2/\mathbf{V}_1$, etc.

All the quantities so far mentioned in this section, $\mathbf{Z}_{\text{in}}$, $\mathbf{Z}_{\text{out}}$, $\mathbf{A}$, etc., will depend not only on the network itself but also upon what is actually connected across the input and output terminals. For instance, $\mathbf{Z}_{\text{in}}$ will clearly depend on the value of the load $\mathbf{Z}_{\text{L}}$ in Fig. 10.13b. However, it is not difficult to show that any passive linear four-terminal network, however complicated, can be replaced by only three equivalent impedances connected as a Y or a delta (also called T and π, as in Sec. 10.9). We shall not enter into a discussion of this aspect of network theory, but note that only three parameters are needed to define the behaviour of any such network.

If a four-terminal network is not only passive and linear, but is symmetrical (looks the same from input terminals as from output terminals) there are only two independent parameters and these may be specified in various convenient ways. With filters and lines (Secs 10.9 and 10.10), these parameters are the **characteristic impedance** $\mathbf{Z}_{\text{k}}$ and the **propagation constant** γ. $\mathbf{Z}_{\text{k}}$ is defined as the value of $\mathbf{Z}_{\text{L}}$ making $\mathbf{Z}_{\text{in}}=\mathbf{Z}_{\text{L}}$. If the load is in fact made equal to $\mathbf{Z}_{\text{k}}$, the network is said to be *correctly terminated.* The propagation constant is defined as the value of log $\mathbf{A}$ for a correctly terminated network, and because $\mathbf{V}_1/\mathbf{I}_1=\mathbf{V}_2/\mathbf{I}_2=\mathbf{Z}_{\text{k}}$ in the case of correct termination, γ also equals $\log_e \mathbf{I}_2/\mathbf{I}_1$). For n four-terminal networks with the same γ connected output to input, the overall propagation constant is $n\gamma$.

10.8 Coupled circuits, transformers

In this section we consider networks with two meshes: a *primary* mesh with input terminals across which a voltage generator $\mathbf{V}$ may be connected and a *secondary* mesh with no generator but with a load $\mathbf{Z}_{\text{L}}$. The two meshes are coupled through an impedance common to both, beginning with mutual inductance coupling.

Mutual inductance coupling (non-resonant) We first consider coupling by a mutual inductance M between circuits with negligible capacitance so that

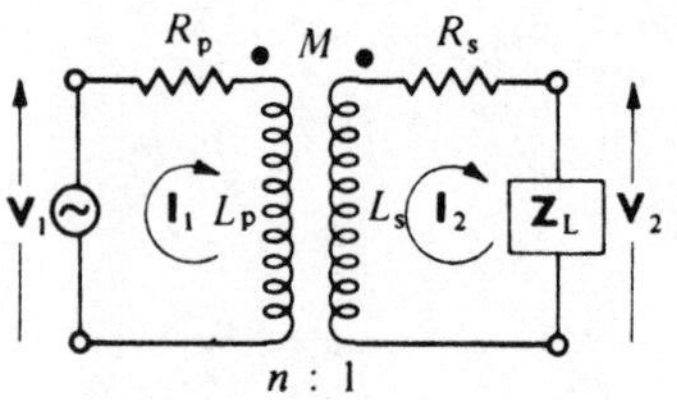

(a)

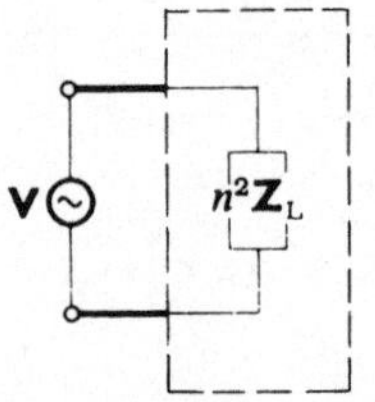

(b)

Figure 10.14 (a) Non-resonant circuits coupled by M: R_p, L_p, R_s, L_s are the resistances and inductances of the primary and secondary windings respectively. (b) The reflected impedance of the load as it appears to the generator in an ideal transformer ($R_p = R_s = 0$; $M^2 = L_pL_s$).

resonance does not occur at low frequencies. The two meshes are labelled with subscripts 1 and 2 (Fig. 10.14a) and in line with this we denote the total impedance round the meshes by $\mathbf{Z}_1$ and $\mathbf{Z}_2$. We then have that $\mathbf{Z}_1 = R_p + j\omega L_p$, $\mathbf{Z}_2 = R_s + j\omega L_s + \mathbf{Z}_L$. Applying Kirchhoff's second law to the two meshes:

$$\mathbf{V}_1 = \mathbf{Z}_1\mathbf{I}_1 - j\omega M\mathbf{I}_2 \tag{10.58}$$
$$0 = \mathbf{Z}_2\mathbf{I}_2 - j\omega M\mathbf{I}_1 \tag{10.59}$$

From (10.59) we obtain $\mathbf{I}_2/\mathbf{I}_1$, and substitution in (10.58) gives $\mathbf{Z}_{in}$:

$$\mathbf{I}_2/\mathbf{I}_1 = j\omega M/\mathbf{Z}_2; \qquad \mathbf{Z}_{in} = \mathbf{V}_1/\mathbf{I}_1 = \mathbf{Z}_1 + \omega^2 M^2/\mathbf{Z}_2 \tag{10.60}$$

In addition, the ratio of the output to input voltage $\mathbf{V}_2/\mathbf{V}_1$ is given by $-\mathbf{Z}_L\mathbf{I}_2/\mathbf{V}_1$ and (10.60) then gives

$$\frac{\mathbf{V}_2}{\mathbf{I}_1} = \frac{-j\omega M\mathbf{Z}_L}{\mathbf{Z}_1\mathbf{Z}_2 + \omega^2 M^2} \tag{10.61}$$

The effect of the secondary circuit on $\mathbf{Z}_{in}$ is, by (10.60), to increase it by $\omega^2 M^2/\mathbf{Z}_2$. However, if $\mathbf{Z}_2$ is expressed as $R_2 + jX_2$, we then have

$$\mathbf{Z}_{in} = \left(R_p + \frac{\omega^2 M^2 R_2}{|\mathbf{Z}_2|^2}\right) + j\left(\omega L_p - \frac{\omega^2 M^2 X^2}{|\mathbf{Z}_2|^2}\right) \tag{10.62}$$

so that the effective primary resistance is increased and the effective primary inductance is decreased if X_2 is inductive. Since L_s is included in X_2 we see that the effect of the secondary winding is partially to cancel the primary inductance.

Application *the ideal transformer* By an ideal transformer is meant a four-terminal network like that of Fig. 10.14a but with three conditions: complete transference of energy from generator to load, no flux leakage, and with very large L_p and L_s. The first of these three conditions means that no resistances occur except in the load ($R_p = R_s = 0$). The second means that coupling between primary and secondary is perfect so that by the results of Sec. 9.8, $M^2 = L_pL_s$ and $L_p/L_s = n^2$, where n is the ratio of the total number of turns on the primary and secondary windings, N_p and N_s. The third condition means that L_p and L_s are large enough to swamp any other reactances or resistances and that the primary and secondary currents are vanishingly small.

$$\left.\begin{aligned} \frac{\mathbf{V}_2}{\mathbf{V}_1} &= \frac{-j\omega M\mathbf{Z}_L}{j\omega L_p(j\omega L_s + \mathbf{Z}_L) + \omega^2 M^2} = -\frac{M}{L_p} \\ \text{or} \\ \frac{\mathbf{V}_2}{\mathbf{V}_1} &= -\frac{N_s}{N_p} = -\frac{1}{n} \end{aligned}\right\} \tag{10.63}$$

This could also have been seen by realizing that, if Φ is the magnetic flux per turn, $\mathbf{V}_1 = +N_p\,d\Phi/dt$ and $\mathbf{V}_2 = -N_s\,d\Phi/dt$. Equation (10.63) shows that the potential difference is stepped up or down in the inverse ratio of n and that the potential differences are π out of phase. Using only the conditions of no loss and no leakage, (10.60) gives

$$\mathbf{I}_2/\mathbf{I}_1 = j\omega M/(j\omega L_s + \mathbf{Z}_L); \qquad \mathbf{Z}_{in} = j\omega L_p\mathbf{Z}_L/(j\omega L_s + \mathbf{Z}_L)$$

after some manipulation. Taking the reciprocal of $\mathbf{Z}_{in}$ we find

$$\frac{1}{\mathbf{Z}_{in}}=\frac{1}{n^2\mathbf{Z}_L}+\frac{1}{j\omega L_p}$$

i.e., the effect on $\mathbf{Z}_{in}$ of the load is to shunt the primary inductance with an effective impedance $n^2\mathbf{Z}_L$. If we now add the condition that L_s is to be very large, we have both

$$\mathbf{I}_2/\mathbf{I}_1=n \quad \text{and} \quad \mathbf{Z}_{in}=n^2\mathbf{Z}_L \tag{10.64}$$

The current is thus stepped up in the inverse ratio of the voltage and the load as seen by the generator is simply $n^2\mathbf{Z}_L$, known as the **reflected impedance** (Fig. 10.14b). This property can be used for impedance matching (Sec. 10.5). A load $\mathbf{Z}_L$ connected directly across a generator may not match its output impedance. By inserting a transformer with a well-chosen value of n, a better match can often be achieved.

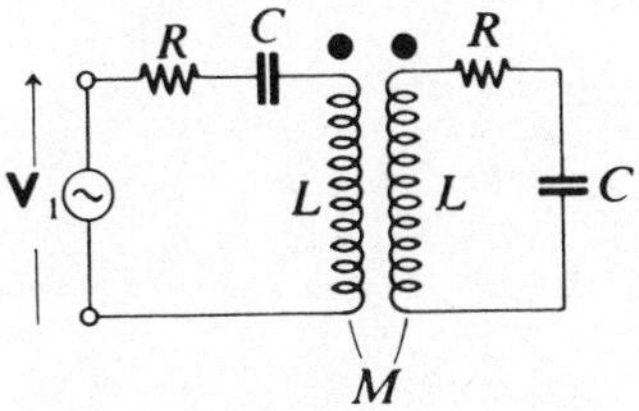

Figure 10.15 Identical resonant circuits coupled by a mutual inductance.

Mutual inductance coupling (resonant) If both primary and secondary circuits incorporate capacitance then resonance is possible. To simplify the algebra while showing the main features, we shall use identical primary and secondary circuits. Figure 10.15 shows the arrangement upon which we use the same method of analysis as we did for the network of Fig. 10.14a. We then have $\mathbf{Z}_1=\mathbf{Z}_2=R+j(\omega L-1/\omega C)$. No ideal conditions are yet assumed and we introduce at the start the coupling coefficient k so that $M=kL$. This time we are interested in the way the responses (currents $\mathbf{I}_1$ and $\mathbf{I}_2$) vary with frequency and tightness of coupling with a constant voltage generator $\mathbf{V}_1$. Equations (10.58) and (10.59) will apply and give us

$$\mathbf{I}_1=\mathbf{V}_1\mathbf{Z}_2/(\mathbf{Z}_1\mathbf{Z}_2+\omega^2M^2); \qquad \mathbf{I}_2=j\omega M\mathbf{V}_1/(\mathbf{Z}_1\mathbf{Z}_2+\omega^2M^2) \tag{10.65}$$

We consider in turn the currents **at** the angular frequency $\omega_0=1/\sqrt{(LC)}$, **near** ω_0 and finally **remote from** ω_0.

At ω_0, $\mathbf{Z}_1=\mathbf{Z}_2=R$ and (10.65) gives

$$\mathbf{I}_1=\frac{\mathbf{V}_1R}{R^2+\omega_0^2k^2L^2}; \qquad \mathbf{I}_2=\frac{-j\omega_0kL\mathbf{V}_1}{R^2+\omega_0^2k^2L^2} \tag{10.66}$$

These show that $\mathbf{I}_1$ at ω_0 increases continually as the coupling becomes looser while $\mathbf{I}_2$ first increases and then falls, its maximum occurring when $k=R/\omega_0L=1/\hat{Q}_0$.

Near ω_0, let $\omega-\omega_0=d\omega$, when $(\omega L-1/\omega C)$, or $\omega_0L(\omega/\omega_0-\omega_0/\omega)$ can be written to a high accuracy as $2L\,d\omega$. It follows that, near ω_0,

$$\mathbf{I}_1=\frac{\mathbf{V}_1(R+j2L\,d\omega)}{(R+j2L\,d\omega)^2+\omega_0^2k^2L^2}; \qquad \mathbf{I}_2=\frac{-j\omega_0kL\mathbf{V}_1}{(R+j2L\,d\omega)^2+\omega_0^2k^2L^2} \tag{10.67}$$

Treating $d\omega$ as the variable, we can obtain the maxima and minima by differentiating and equating to zero the moduli of $\mathbf{I}_1$ and $\mathbf{I}_2$. For $\mathbf{I}_2$, only the denominator contains the variable and its modulus is

$$[R_2-4L^2(d\omega)^2+\omega_0^2k^2L^2]^2+16L^2(d\omega)^2R^2$$

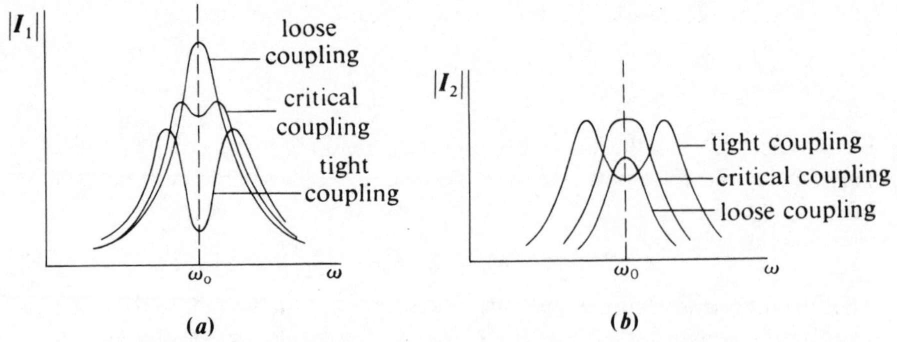

Figure 10.16 Currents in (a) the primary mesh and (b) the secondary mesh in the coupled circuits of Fig. 10.15 and their variation with frequency.

Thus for maxima and minima of $|\mathbf{I}_2|$,

$$4L^2(\mathrm{d}\omega)^3+(R^2-\omega_0^2k^2L^2)\,\mathrm{d}\omega=0$$

The three roots of this equation can be written

$$\mathrm{d}\omega=0 \qquad \text{or} \qquad \pm\frac{\omega_0}{2\hat{Q}_0}\sqrt{[(\hat{Q}_0k)^2-1]}$$

Thus if $\hat{Q}_0k>1$ there are three real roots, $\mathrm{d}\omega=0$ being a minimum and the other two being maxima, while if $\hat{Q}_0k<1$ there is only one real root at $\omega=\omega_0$, which is now a maximum. $k_c=1/\hat{Q}_0$ is critical coupling where three roots coincide, tighter coupling giving two peaks and looser coupling only one (Fig. 10.16). Since k cannot be greater than 1, $\hat{Q}_0$ must also be at least 1 if a double peak is to occur. The variations in $|\mathbf{I}_1|$ are similar.

With tight coupling the peaks are **remote from** ω_0 and the resistances can usually be neglected. This results in the denominators of $\mathbf{I}_1$ and $\mathbf{I}_2$ both being given by $\omega^2k^2L^2-(\omega L-1/\omega C)^2$ which has a minimum value of zero when

$$\omega=\frac{\omega_0}{\sqrt{(1\pm k)}} \tag{10.68}$$

the same frequencies as those of (10.26).

Application *real transformers* Energy losses in real transformers result from joule heating in the windings (copper losses) and in the core (eddy current losses) together with hysteresis loss (Sec. 12.6) also in the core. Copper losses are equivalent to resistances r_p and r_s in Fig. 10.17a while the core or iron losses are represented by a shunt r_c. The flux leakage which makes k less than 1 can be represented by splitting off $(1-k)L_p$ from the inductance of the primary winding so that only kL_p takes the current which produces the magnetic flux in the core. The effect of the secondary is that of leakage inductance, copper losses r_s and the load reflected back into the primary as if through an ideal transformer with a turns ratio n: 1. The reader should be able to show from (10.62) that this is so. There will in addition be capacitance associated with the primary and secondary windings, because of the potential difference between adjacent turns.

Power transformers operate at low frequencies, usually 50–60 Hz, and are used to step up or step down voltages or currents. They are wound on ferromagnetic

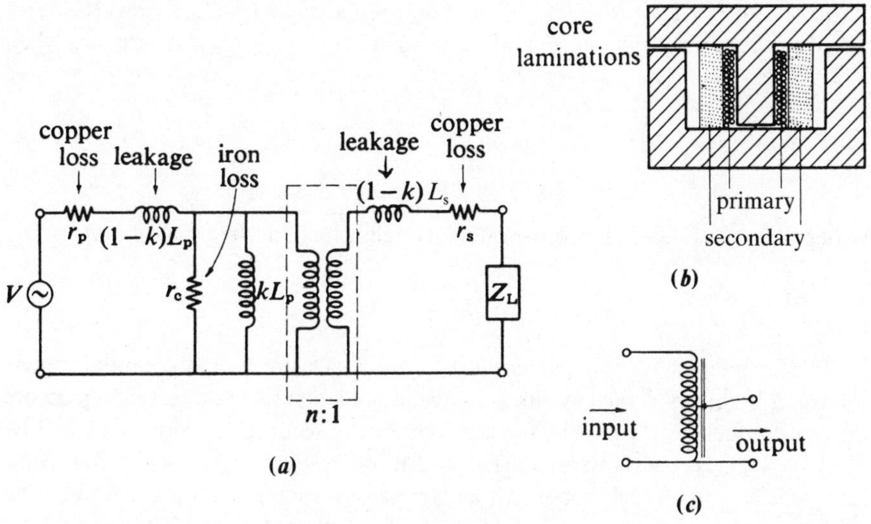

Figure 10.17 (a) Equivalent circuit of a real transformer with losses. The ideal transformer in the broken rectangle indicates that the effect of the secondary on the primary is that of $n^2 \times$ the impedance on the right, in parallel with kL_p; (b) construction of a small power transformer; (c) an autotransformer circuit.

cores which reduce leakage and make k nearly 1. The resultant iron losses are reduced by lamination (minimizing eddy currents) and by choosing a material with a narrow hysteresis loop such as silicon steel. At such low frequencies the effect of self-capacitance is also negligible and the transformer thus approximates to an ideal one. The construction is shown in Fig. 10.17b. The auto-transformer shown in Fig. 10.17c is used when isolation of primary from secondary for d.c. potentials is not important: this form reduces flux leakage still further and also reduces copper losses.

Over the range from about 200 Hz to 20 kHz, *audio-frequency transformers* are used mainly for impedance matching or isolation of d.c. At the lower end of the range the same considerations apply as to power transformers, but at the upper end the possibility of resonance occurs because of the self-capacitances. Since a uniform response over the range is required, that at low frequencies has to be improved by increasing the incremental inductance (Sec. 12.6). At *radio frequencies*, from about 10 kHz to 10 MHz, the need on the other hand is for a narrow band of frequencies to be passed on with a sharp cut-off outside it. The properties of coupled resonant circuits are useful for this and r.f. transformers are invariably tuned. *Pulse transformers* present special difficulties since a train of pulses contains a very wide range of frequencies. Transmission without distortion needs a core with a constant permeability up to high frequencies as well as high resistivity so that ferrites (Sec. 12.1) are commonly used.

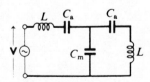

Figure 10.18 Direct-coupled resonant circuits. The total primary and secondary capacitances are both given by $C = C_aC_m/(C_a + C_m)$, and the coefficient of coupling is then C/C_m.

Direct coupling (resonant) Circuits may be directly coupled through a self-inductance, or, as in Fig. 10.18, through a capacitance. The definition of k is now generalized to $X_m/\sqrt{(X_pX_s)}$, where X_m is the mutual reactance and X_p and X_s are the total reactances *of the same kind as* X_m in the primary and secondary. In Fig. 10.18 for example, $X_m = 1/\omega C_m$ and both X_p and X_s are $(C_a + C_m)/(\omega C_aC_m)$. If we use the notation that $C = C_aC_m/(C_a + C_m)$, then $X_p = X_s = 1/\omega C$ and $k = C/C_m$.

The equations for the two mesh currents will take the same form as (10.58) and

(10.59) with $\mathbf{Z}_1 = \mathbf{Z}_2 = \mathrm{j}(\omega L - 1/\omega C)$ and ωM replaced by $1/\omega C_{\mathrm{m}}$. From (10.65) we see that the currents will be a maximum when $\mathbf{Z}_1\mathbf{Z}_2 - (1/\mathrm{j}\omega C_{\mathrm{m}})^2$ is zero, i.e., when

$$[\mathrm{j}(\omega L - 1/\omega C)]^2 - (1/\mathrm{j}\omega C)^2 = 0$$

or

$$(\omega/\omega_0)^4 - 2(\omega/\omega_0)^2 + 1 - C^2/C_{\mathrm{m}}^2 = 0$$

where $\omega_0 = 1/\surd(LC)$. Hence the currents again have double maxima at

$$\omega = \frac{\omega_0}{\surd(1 \pm k)}$$

The properties of coupled circuits of all types are broadly similar: loose coupling means that the circuits are virtually independent and resonance occurs at one frequency in the primary and one in the secondary (coincident in Fig. 10.16). Tighter coupling produces a double resonance in both circuits at frequencies more widely separated as k increases. For near-critical coupling the response is broad at the peak but drops on both sides more sharply than that for a single resonant circuit: such a network can be used to pass only a band of frequencies with nearly constant amplification.

More detailed analysis shows that when the primary and secondary resonate at the same frequency but have unequal $\hat{Q}_0$'s, $\hat{Q}_{\mathrm{p}}$ and $\hat{Q}_{\mathrm{s}}$, critical coupling is given by $k_{\mathrm{c}} = 1/\surd(\hat{Q}_{\mathrm{p}}\hat{Q}_{\mathrm{s}})$. If the primary is non-resonant and the secondary resonant, the currents show maxima which are shifted from ω_0. Finally, if the resonant frequencies of primary and secondary are unequal, the two peaks obtained at tight coupling are of unequal height.

10.9 Filters and attenuators

Ladder networks Four-terminal networks with an output-to-input voltage ratio of less than 1 and independent of frequency are known as **attenuators** – the potential divider of Fig. 6.16d is an example. If the ratio is almost 1 for a range of frequencies but falls off sharply outside this range, the network is a **filter**—the coupled resonant circuits of Sec. 10.8 are examples if the coupling is near critical.

Both filters and attenuators become more versatile if they consist of more than one stage, and a common form is the symmetrical ladder network shown in Fig. 10.19 which can be divided into a number of identical T- or π-sections as shown (these can be interrelated by the Y–Δ transformation).

Individual T-sections have a characteristic impedance (Sec. 10.7) calculated by connecting $\mathbf{Z}_{\mathrm{k}}$ across the output and equating $\mathbf{Z}_{\mathrm{in}}$ to $\mathbf{Z}_{\mathrm{k}}$. Applying this to the T-section of Fig. 10.19b gives

$$\mathbf{Z}_{\mathrm{k}} = \tfrac{1}{2}\mathbf{Z}_1 + \mathbf{Z}_2 - \frac{\mathbf{Z}_2^2}{\mathbf{Z}_2 + \tfrac{1}{2}\mathbf{Z}_1 + \mathbf{Z}_{\mathrm{k}}}$$

This simplifies to

$$\mathbf{Z}_{\mathrm{kT}} = \surd(\mathbf{Z}_1\mathbf{Z}_2 + \tfrac{1}{4}\mathbf{Z}_1^2) \qquad (10.69)$$

The π-section of Fig. 10.19c similarly has a characteristic impedance $\mathbf{Z}_{\mathrm{k}\pi}$ given by $\mathbf{Z}_{\mathrm{kT}}\mathbf{Z}_{\mathrm{k}\pi} = \mathbf{Z}_1\mathbf{Z}_2$ (Problem 10.26). We shall work wholly in terms of T-sections.

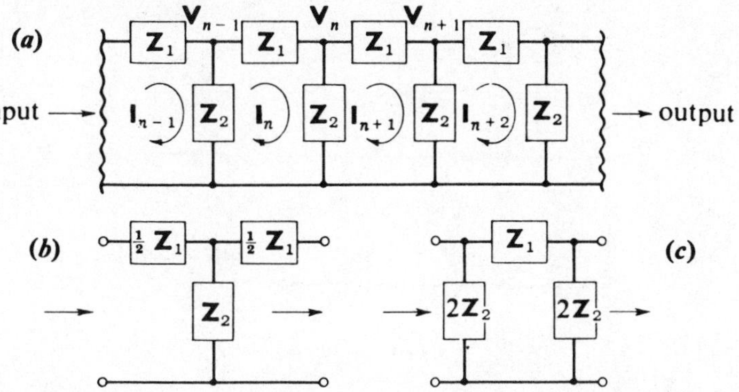

Figure 10.19 (a) A ladder network that is symmetrical as regards input and output. It can be built up either (a) from identical T-sections by splitting $\mathbf{Z}_1$ or (b) from identical π-sections by splitting $\mathbf{Z}_2$.

Each T-section of the ladder network of Fig. 10.19a is correctly terminated if the whole network is either infinite or terminated by $\mathbf{Z}_{kT}$—that is why $\mathbf{Z}_k$ is sometimes called the *iterative impedance.* Take a typical single T-section terminated by $\mathbf{Z}_{kT}$ and denote the mesh current in its secondary by $\mathbf{I}_{n+1}$ and in its primary by $\mathbf{I}_n$. The ratio of these currents is obtained by applying Kirchhoff's second law to the secondary mesh in Fig. 10.19b terminated by $\mathbf{Z}_{kT}$. This gives $\mathbf{I}_{n+1}/\mathbf{I}_n = \mathbf{Z}_2/(\mathbf{Z}_{kT}+\frac{1}{2}\mathbf{Z}_1+\mathbf{Z}_2)$. If we substitute (10.69) into this and put $\mathbf{Z}_1/2\mathbf{Z}_2 = r$, we obtain $\mathbf{I}_{n+1}/\mathbf{I}_n = 1+r-(r^2+2r)^{1/2}$.

The ratio is the same for any n in a correctly terminated network. It can then be shown (Problem 10.28) that the ratio is also equal to $\mathbf{V}_{n+1}/\mathbf{V}_n$ which we defined in Sec. 10.7 as the voltage transfer ratio **A**. It is this quantity that determines the attenuating and filtering action of the network. Remembering that the propagation constant γ was defined so that $\mathbf{A}=e^{\gamma}$, we have

$$\mathbf{A}=e^{\gamma}=1+r-(r^2+2r)^{1/2} \tag{10.70}$$

For n sections the voltage ratio will be $\mathbf{A}^n$ and the propagation constant $n\gamma$. In general γ will be complex so let

$$\gamma=\alpha+j\beta; \qquad e^{\gamma}=e^{\alpha}\cos\beta+je^{\alpha}\sin\beta \tag{10.71}$$

This means that e^{α} represents the **attenuation** and β the **phase change**.

We restrict r to real values so that $\mathbf{Z}_1$ and $\mathbf{Z}_2$ are either both resistive or both reactive. Equation (10.70) then shows that it is the value of r^2+2r that is important. If $r^2+2r \geq 0$, **A** is real, so $\beta=0$ and the attenuation is e^{α} per section where (10.70) shows that $\cosh\alpha=(e^{\alpha}+e^{-\alpha})/2=1+r=1+\mathbf{Z}_1/2\mathbf{Z}_2$. On the other hand, if $r^2+2r<0$, then $e^{\alpha}\cos\beta=1+r$ and $e^{\alpha}\sin\beta=(-r^2-2r)^{1/2}$ whence $e^{2\alpha}=1$ so that $\alpha=0$ and γ is imaginary. In that case there is no attenuation but a phase change of β per section where $\cos\beta=1+r$. Thus there is no attenuation only when $r^2+2r<0$ and this occurs when r lies between 0 and -2 (i.e. $\mathbf{Z}_1/4\mathbf{Z}_2$ lies between 0 and -1). To summarize:

$-1 \geq \mathbf{Z}_1/4\mathbf{Z}_2 \geq 0$: attenuation of e^{α} per section, where $\cosh\alpha=1+\mathbf{Z}_1/2\mathbf{Z}_2$; **I** and **V** change phase by 0 or π (10.72)

$-1 < \mathbf{Z}_1/4\mathbf{Z}_2 < 0$: no attenuation; **I** and **V** change phase by β per section, where $\cos\beta=1+\mathbf{Z}_1/2\mathbf{Z}_2$ (10.73)

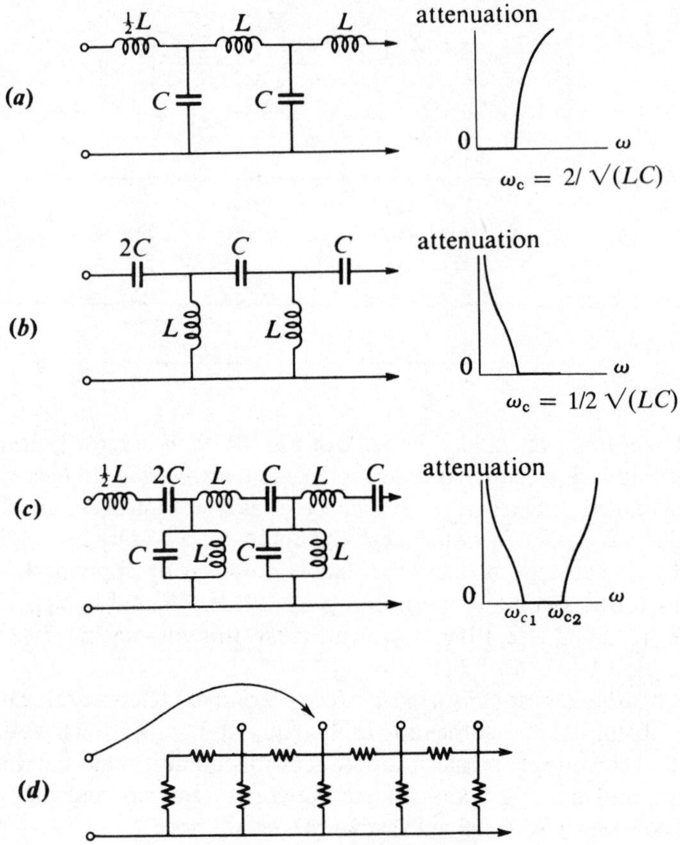

Figure 10.20 (a) A low-pass filter and its attenuation curve; (b) a high-pass filter; (c) a band-pass filter; (d) a ladder attenuator.

Filters A ladder network thus passes without attenuation frequencies for which $\mathbf{Z}_1/4\mathbf{Z}_2$ lies between 0 and -1 and attenuates the rest. Filters thus require $\mathbf{Z}_1$ and $\mathbf{Z}_2$ to be reactances of *opposite sign.* If $\mathbf{Z}_1 = \mathrm{j}\omega L$ and $\mathbf{Z}_2 = 1/\mathrm{j}\omega C$ then frequencies are passed for which $\frac{1}{4}\omega^2 LC$ lies between 0 and 1, i.e. all frequencies up to $\omega = 2/\sqrt{(LC)}$ are passed. Figure 10.20 shows such a low-pass filter together with a high- and a band-pass filter.

Sections with the same $\mathbf{Z}_k$ but with different $\mathbf{Z}$'s can be connected to obtain desirable characteristics in the pass band or in the attenuation curve and if the load is not equal to $\mathbf{Z}_k$, terminal half-sections (Problem 10.29) can be used for matching. Note that a low-pass filter is used as a smoothing circuit after rectification of a.c. in d.c. power supplies.

Delay lines When an alternating voltage is fed into a low-pass filter such as that in Fig. 10.20a, a phase lag of β occurs in $\mathbf{I}$ and $\mathbf{V}$ at each section. A total lag of 2π rad would correspond to a time delay of a complete period $2\pi/\omega$, so that a phase shift of β corresponds to a time delay of β/ω per section. We have already seen that β is given in this case by $\cos\beta = 1 + r = 1 - 2\omega^2/\omega_c^2$. Writing $\cos\beta$ as $1 - 2\sin^2(\beta/2)$, we have that $\sin(\beta/2) = \omega/\omega_c$. Provided β is small enough ($\beta^2/24 \ll 1$), $\beta = 2\omega/\omega_c$. Thus the delay per section for a low-pass filter is $(LC)^{1/2}$ provided that ω is well below ω_c. This forms an *artificial delay line* (see also next section under 'Special Cases').

Attenuators If $\mathbf{Z}_1$ and $\mathbf{Z}_2$ are both resistances, r is always real and positive, and attenuation therefore takes place without change of phase. A ladder network is superior to a simple potential divider because with proper termination the load on the generator is independent of the number of sections and therefore of the attenuation: this is not the case with a potential divider unless it feeds a system with a very high input impedance. For a correctly terminated ladder network of resistors the ratio of output to input power per section is $V_{n+1}I_{n+1}/V_nI_n$. The power gain is thus $20\log_{10}\mathrm{e}^{\alpha}$ and since $\mathrm{e}^{\alpha}<1$ this will be a loss of $20\alpha\log_{10}\mathrm{e}\,\mathrm{dB}$ per section.

10.10 Transmission lines

General theory Electric power is often carried over a distance by cables consisting of two conductors which may be thought of as conveying current to and from the generator. We have seen in Secs 5.2 and 9.6 that such cables will possess distributed capacitance and self-inductance, and we see also that they will have resistance along the conductors and conductance between them. If C, L, R and G are the magnitudes of these parameters per unit length of the cable (R for both conductors) then a small section of length $\mathrm{d}x$ can be represented diagrammatically as in Fig. 10.21a, x being the distance measured from left to right. The complete line is thus a ladder network in which the impedances are continuous instead of lumped and we might expect to find similarities between these lines and filters.

Let the complex current and potential difference at the input end of the elementary section be $\mathbf{I}$ and $\mathbf{V}$, when the change $\mathrm{d}\mathbf{V}$ is a result of a fall in potential along the impedance $(R+\mathrm{j}\omega L)\,\mathrm{d}x$. If we let $\mathbf{Z}=R+\mathrm{j}\omega L$,

$$\mathbf{Z}\,\mathrm{d}x\,\mathbf{I}=-\mathrm{d}\mathbf{V}$$

or

$$-\frac{\partial\mathbf{V}}{\partial x}=\mathbf{ZI} \tag{10.74}$$

The increment in current $\mathrm{d}\mathbf{I}$ is a result of that bypassed by the admittance $\mathbf{Y}\,\mathrm{d}x=(G+\mathrm{j}\omega C)\,\mathrm{d}x$ so that

$$-\frac{\partial\mathbf{I}}{\partial x}=\mathbf{YV} \tag{10.75}$$

the partial derivatives being used because variations with time occur as well. It follows that

$$\frac{\partial^2\mathbf{I}}{\partial x^2}=\mathbf{YZI};\qquad \frac{\partial^2\mathbf{V}}{\partial x^2}=\mathbf{YZV} \tag{10.76}$$

and hence

$$\mathbf{I}=\mathbf{I}_1\mathrm{e}^{\gamma x}+\mathbf{I}_2\mathrm{e}^{-\gamma x};\qquad \mathbf{V}=\mathbf{V}_1\mathrm{e}^{\gamma x}+\mathbf{V}_2\mathrm{e}^{-\gamma x} \tag{10.77}$$

where $\gamma=\sqrt{(\mathbf{YZ})}$,† called the propagation constant as in the last section. In general γ is complex so let

$$\gamma=\alpha+\mathrm{j}\beta \tag{10.78}$$

† Some prefer to define $\mathbf{P}=\sqrt{(-\mathbf{YZ})}$ as a propagation constant so that $\mathbf{P}=\mathrm{j}\gamma$.

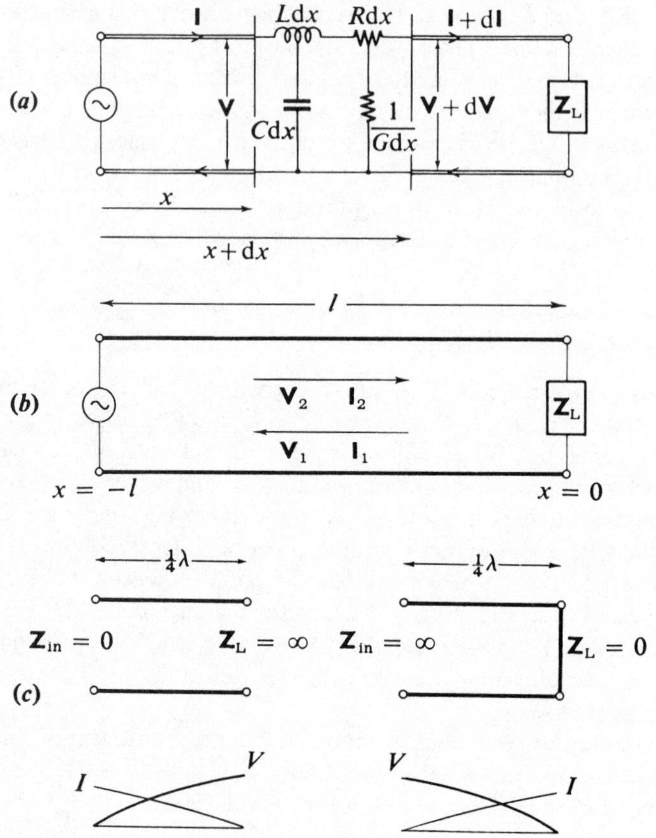

Figure 10.21 (a) Transmission line parameters; (b) reflection from the end of a mismatched line; (c) voltage and current variations in quarter-wave lines.

The physical meaning of these terms is brought out when we remember that the instantaneous values of current and potential difference are given by the real or imaginary parts of **I** and **V** and we have

$$I_0 e^{j\omega t} = I_{01} e^{\alpha x} e^{j(\omega t + \beta x)} + I_{02} e^{-\alpha x} e^{j(\omega t - \beta x)} \tag{10.79}$$

with a similar expression for **V**. The second term is an attenuated wave moving from left to right, the first a similar wave travelling from right to left, α is thus an attenuation constant and β a phase constant determining the wave velocity, which is ω/β (and therefore $\beta = 2\pi/\lambda$, λ being the wavelength). The reader should confirm that this is so by considering the variation of each term, first at constant x and varying t, and then at constant t and varying x.

Because α and β are in general frequency-dependent, any input waveform containing a range of frequencies is distorted and dispersion occurs.

Characteristic impedance Returning to the solution (10.77), we find by substitution into (10.75) that

$$-\gamma \mathbf{I}_1 e^{\gamma x} + \gamma \mathbf{I}_2 e^{-\gamma x} = \mathbf{Y}\mathbf{V}_1 e^{\gamma x} + \mathbf{Y}\mathbf{V}_2 e^{-\gamma x}$$

for *all* x and hence, equating coefficients,

$$-\gamma \mathbf{I}_1 = \mathbf{Y}\mathbf{V}_1; \qquad \gamma \mathbf{I}_2 = \mathbf{Y}\mathbf{V}_2 \qquad \text{or} \qquad \frac{\mathbf{V}_2}{\mathbf{I}_2} = -\frac{\mathbf{V}_1}{\mathbf{I}_1} = \sqrt{\frac{\mathbf{Z}}{\mathbf{Y}}} = \mathbf{Z}_k \tag{10.80}$$

$\mathbf{Z}_k$ has all the properties of a characteristic impedance, for as we proceed down the line it is the constant ratio of **V** to **I** for each travelling wave and so for a line of infinite length with only one wave generated, say from left to right, it is the input impedance at any point looking away from the generator. If, instead, the line is finite but terminated by a load $\mathbf{Z}_k$, the ratio of **V** to **I** is still $\mathbf{Z}_k$ even at the load and no reverse wave occurs: the input impedance is thus $\mathbf{Z}_k$ for a correctly terminated line.

Special cases At high frequencies (1 MHz and above) R and G become negligible compared with ωL and ωC, so that $\gamma = j\omega(LC)^{1/2}$ and $\mathbf{Z}_k = (L/C)^{1/2}$, which is purely resistive. The attenuation is therefore negligible for short lines. Moreover, the velocity (ω/β from above) is $1/(LC)^{1/2}$ and thus independent of frequency so that dispersion does not occur. These properties are consistent with those of a continuous low-pass filter with sectional inductance and capacitance $L\,dx$ and $C\,dx$. According to the results of the last section, this should result in a delay time of $(LC)^{1/2}$ per unit length and thus a velocity of $1/(LC)^{1/2}$. In addition, the cut-off frequency $\omega_c = 2/(LC)^{1/2}\,dx$ becomes infinite as dx tends to zero and all frequencies should therefore be passed without attenuation. From the expressions for L and C of twin or coaxial cables already derived [Eqs (5.10), (5.11), (9.22) and (9.23)], the velocity of waves in such cables will be $1/(\varepsilon_0\mu_0)^{1/2}$. This is almost exactly 3×10^8 m/s, the velocity of light *in vacuo*, and it can be shown to be the same for twin cables of any constant cross-section. The effect of dielectric and magnetic media is to lower this to $1/(\varepsilon_r\varepsilon_0\mu_r\mu_0)^{1/2}$, the velocity of light in such media (Sec. 14.1).

At lower frequencies, particularly those used in telegraphy and telephony, the losses cannot be ignored. If however, it were possible to arrange that

$$R + j\omega L = y^2(G + j\omega C) \tag{10.81}$$

y being any constant, γ would be $y(G + j\omega C)$. The attenuation constant would thus be yG and the velocity $1/yC$, both independent of frequency, and the line would be distortionless. Unfortunately (10.81) can only be achieved if $R = y^2G$ and $L = y^2C$, i.e. if $RC = LG$, and in practice $RC \gg LG$. To increase G would introduce unacceptable attenuation and the best which can be done is to increase L either by winding permalloy tape round the conductor (for long low-frequency telegraphy lines) or by inserting coils in series with the cables at intervals of a fraction of a wavelength (for telephony). This does not achieve (10.81) but produces some improvement.

Standing waves in loss-less or HF lines When $\gamma = j\beta$, the solution (10.77) becomes

$$\mathbf{I} = \mathbf{I}_1 e^{j\beta x} + \mathbf{I}_2 e^{-j\beta x}; \qquad \mathbf{V} = \mathbf{V}_1 e^{j\beta x} + \mathbf{V}_2 e^{-j\beta x} \tag{10.82}$$

in which $\mathbf{V}_1 = -\mathbf{Z}_k\mathbf{I}_1$ and $\mathbf{V}_2 = \mathbf{Z}_k\mathbf{I}_2$ by (10.80). If, as in Fig. 10.21b, we take the zero to be at the load $\mathbf{Z}_L$ we shall obtain a reflected wave unless $\mathbf{Z}_L = \mathbf{Z}_k$, and from (10.82)

$$\mathbf{Z}_L = \frac{\mathbf{V}_1 + \mathbf{V}_2}{\mathbf{I}_1 + \mathbf{I}_2} \tag{10.83}$$

We define a **reflection coefficient** ρ as $\mathbf{V}_1/\mathbf{V}_2$ at $x = 0$, and it is clearly also $-\mathbf{I}_1/\mathbf{I}_2$ by (10.80). Hence, from (10.83),

$$\mathbf{Z}_L = \frac{\mathbf{V}_2(1+\rho)}{\mathbf{I}_2(1-\rho)} = \mathbf{Z}_k \frac{1+\rho}{1-\rho} \tag{10.84}$$

or
$$\rho = \frac{\mathbf{Z}_L - \mathbf{Z}_k}{\mathbf{Z}_L + \mathbf{Z}_k} \tag{10.85}$$

The current and potential difference (10.82) are thus given by

$$\mathbf{I} = \mathbf{I}_2(e^{-j\beta x} - \rho e^{j\beta x}); \qquad \mathbf{V} = \mathbf{Z}_k \mathbf{I}_2(e^{-j\beta x} + \rho e^{j\beta x}) \tag{10.86}$$

where $\mathbf{I}_2$ is determined by the generator.

The input impedance at any point a distance l from the load is $\mathbf{V}/\mathbf{I}$ evaluated at $x = -l$ and so

$$\mathbf{Z}_{in} = \mathbf{Z}_k \frac{e^{j\beta l} + \rho e^{-j\beta l}}{e^{j\beta l} - \rho e^{-j\beta l}}$$

which, using (10.85) and $e^{\pm j\beta l} = \cos \beta l \pm j \sin \beta l$, gives

$$\mathbf{Z}_{in} = \mathbf{Z}_k \frac{\mathbf{Z}_L + j\mathbf{Z}_k \tan \beta l}{\mathbf{Z}_k + j\mathbf{Z}_L \tan \beta l} \tag{10.87}$$

For a quarter-wavelength line, $\beta l = \frac{1}{2}\pi$ and

$$\mathbf{Z}_{in} = \mathbf{Z}_k^2/\mathbf{Z}_L \tag{10.88}$$

a property which enables it to be used to transform one impedance into another. In particular, if open-circuited, $\mathbf{Z}_{in}$ is zero, while if short-circuited $\mathbf{Z}_{in}$ is infinite, so that a short-circuited quarter-wave line acts at its input end as an open circuit and can be used to support lines without affecting their properties. (See Fig. 10.21c.)

The potential difference given by (10.86) represents two travelling waves in opposite directions superposed and is thus a standing wave. If we write ρ as $|\rho| e^{j\theta}$, then (10.86) gives the magnitude of $\mathbf{V}$ as

$$|\mathbf{V}| = |\mathbf{Z}_k||\mathbf{I}_2|[|\rho|^2 + 1 + 2|\rho| \cos(2\beta x + \theta)]^{1/2} \tag{10.89}$$

As x varies, $|\mathbf{V}|$ passes through minima and maxima—the nodes and anti-nodes of the standing waves—as the cosine term fluctuates. The ratio of the maximum to minimum $|\mathbf{V}|$ is known as the **voltage standing wave ratio** (VSWR) and is $(1+|\rho|)/(1-|\rho|)$. The *positions* of the nodes are given by $2\beta x + \theta = (2n+1)\pi$ so that they occur at values of x given by $n + \frac{1}{2}\pi/\beta - \theta/2\beta$. Because $\beta = 2\pi/\lambda$, the nodes are separated by $\lambda/2$, while the distance of the first node from the end is $\theta/2\beta$ or $\lambda\theta/4\pi$. Observations of the positions and amplitudes of the standing waves thus enables the wavelength to be determined as well as the magnitude and phase angle of ρ. This, through (10.85), gives $\mathbf{Z}_L$ in terms of $\mathbf{Z}_k$.

10.11 Measurement of impedance

The use of an a.c. voltmeter and an a.c. ammeter to obtain the impedance between two terminals through the ratio V/I is subject to large errors and in any case yields only the magnitude $|\mathbf{Z}|$. Precise methods fall into three classes according to the frequency range. Bridge methods are used at low frequencies and with special precautions up to 50 or 100 MHz. At radio frequencies, resonance methods are more common while in the microwave region (1000 MHz and above) VSWR methods with transmission lines are used. Some of these methods also incidentally provide means of measuring frequency or wavelength.

A.C. bridges—components The general construction and properties of resistors, capacitors and inductors have already been discussed in previous chapters, but although components may be designed as one of these only, they inevitably possess the properties of all three when used with a.c. Connecting leads will also have stray capacitance and inductance.

Wirewound **resistors** possess considerable inductance when simply wound on a bobbin and a small capacitance because of the potential difference between adjacent turns. Non-inductive windings ensure that adjacent wires carry opposing currents as far as possible (as in the bifilar type in which the wire is doubled back on itself) but since this often increases the self-capacitance fairly complex windings must be used. At high frequencies, the skin effect (Sec. 9.6) causes an increase in resistance because of the reduction in the effective cross-section of the wire: this is minimized by using multistranded wire with strands of very small diameter.

Inductors possess an inherent resistance and a self-capacitance for the same reason as resistors and they are thus equivalent to the network of Fig. 10.12. Because of this they are less convenient than capacitors as standards in a.c. bridges, particularly when continuous variation is required.

Capacitors usually have dielectric losses which can be represented by a resistance r in series with the capacitance C. The **loss angle** δ is the complement of ϕ, the phase different between current and voltage, i.e. $\delta=\pi/2-\phi$. The quantity normally quoted is the **loss tangent** $\tan\delta$, given by

$$\tan\delta=\omega Cr \tag{10.90}$$

and is related to the power factor $\cos\phi$ defined in (10.48). Note that if the imperfect capacitor were represented by C in *parallel* with a resistance R, then $\tan\delta$ would be $1/\omega CR$.

A.C. bridges—methods Conventional a.c. bridges are usually similar in form to the d.c. Wheatstone bridge, as in Fig. 10.22a. Where they are not, as in some mutual inductance bridges, they can be reduced to the conventional form by the Y–Δ transformation or by the equivalent circuit of Fig. 9.14. The source is invariably an electronic oscillator and the detector is preceded by an amplifier which may be sensitive to the phase as well as to the amplitude of an input signal. Both source and detector may be coupled to the bridge network by transformers for impedance matching and isolation, and shielding of all components and leads becomes necessary at higher frequencies to minimize strays.

For complete balance of the bridge, the points B and D must be at the same potential *at all times* so that both amplitude and phase of the alternating potentials at these points must be equalized. In general, therefore, we expect two balance conditions to emerge. By the same argument as used for the d.c. Wheatstone bridge, the balance condition is

$$\frac{\mathsf{Z}_1}{\mathsf{Z}_2}=\frac{\mathsf{Z}_3}{\mathsf{Z}_4} \tag{10.91}$$

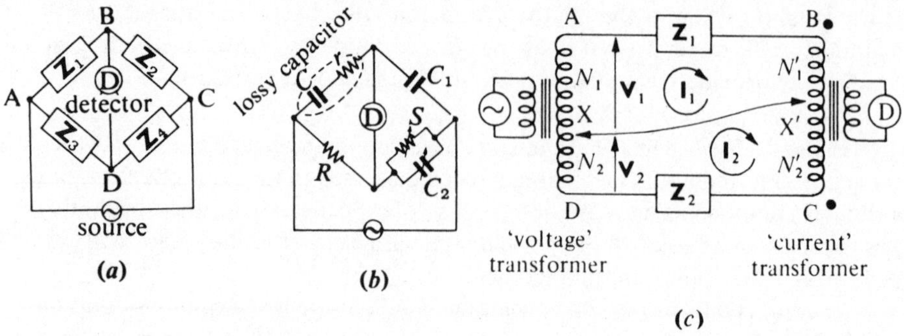

Figure 10.22 (a) A general a.c. bridge; (b) the Schering bridge; (c) the transformer ratio-arm bridge.

On cross-multiplying and equating real and imaginary parts, two conditions are obtained which are achieved by alternately varying two components in the network. As an example, the **Schering bridge** of Fig. 10.22b is used for the determination of capacitance and power factor of a capacitor, or of relative permittivity and dielectric loss for a material filling a capacitor. The condition (10.91) yields the balance conditions $C=C_1S/R$, $r=C_2R/C_1$ (i.e. $\tan\delta=\omega C_2S$) and these can be achieved by alternate adjustment of S and C_2. The balance condition is independent of frequency so that the presence of harmonics in the source is immaterial.

In recent years the **transformer ratio-arm bridge** has become widely used because of its great advantages over conventional types of bridge. One form is shown in Fig. 10.22c. The two transformers are toroidal in form with a very high permeability core (such as supermalloy) so that the flux leakage is negligible and the resistances small enough for ideal transformer conditions (10.63) and (10.64) to apply. The voltages applied to the two loops are then accurately in the ratio $\mathbf{V}_1/\mathbf{V}_2=N_1/N_2$, where N_1 and N_2 are the total numbers of turns in the two sections of the voltage transformer winding. The loop currents $\mathbf{I}_1$ and $\mathbf{I}_2$ flow through the current transformer primary in such directions that their magnetic fluxes oppose each other. The bridge is balanced by adjusting the tappings and $\mathbf{Z}_2$ until a detector across the secondary winding indicates zero flux in the primary. Then $N'_1\mathbf{I}_1=N'_2\mathbf{I}_2$. The absence of any resultant flux in the current transformer at balance and the negligible resistance of the windings means that B, X′ and C are virtually at the same potential and hence that $\mathbf{V}_1=\mathbf{Z}_1\mathbf{I}_1$, $\mathbf{V}_2=\mathbf{Z}_2\mathbf{I}_2$. Hence

$$\frac{\mathbf{Z}_1}{\mathbf{Z}_2}=\frac{N_1}{N_2}\frac{N'_1}{N'_2} \tag{10.92}$$

Although this seems to indicate that only like impedances can be compared (because $\mathbf{Z}_1/\mathbf{Z}_2$ must be real) the standard $\mathbf{Z}_2$ can consist of, say, a bank of resistors in parallel with a bank of capacitors, each bank being connected to different tappings on the voltage transformer so that their currents can be separately balanced.

The advantages of this type of bridge are (a) the use of two ratios enabling far wider ranges of impedance to be compared than is normal; (b) the high accuracy, because the flux leakage is so small that the voltage and current ratios are equal to the turns ratios to better than 1 part per million; (c) the stability and permanency of the ratios without the need for calibration; (d) the high sensitivity, because as soon as the bridge goes out of balance the large inductances of the two sections of the primary current winding come into play; (e) the elimination of strays: if XX′ is earthed, strays from B and C are short-circuited by the current transformer windings at balance, and those from A and D merely shunt the voltage input and lower the sensitivity slightly without affecting the balance condition.

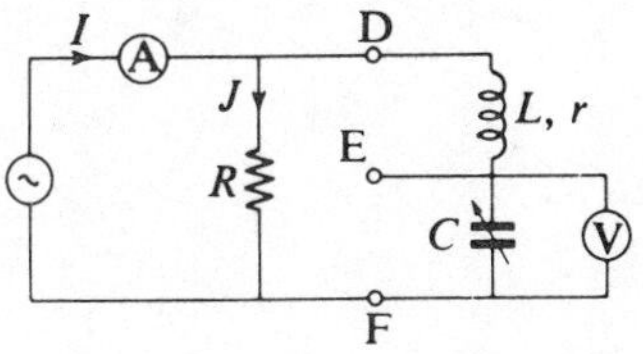

Figure 10.23 The Q-meter.

Resonance methods Inductance and capacitance can be measured very simply by using the properties of the series *LCR* circuit (Problem 10.36). Typical of these methods is the *Q*-meter (Fig. 10.23) in which a small resistance $R(\ll r)$ is in parallel with a calibrated variable capacitor together with a coil whose $\hat{Q}$ and L are required. A voltmeter V with an effectively infinite resistance is connected across C. C is adjusted so that the meter V indicates resonance and from Sec. 10.6 we know that the voltage across C is $\hat{Q}_0 V$ or $\hat{Q}_0 RI$ because most of the current passes through R, and I and J are nearly equal. For a given instrument the value of I used is adjusted to a set value and the scale of V can be calibrated directly to give $\hat{Q}_0$. In addition, the calibration of C allows L to be calculated at the known frequency. An unknown capacitance can be determined by connecting it across EF and finding the change in C needed to retune to resonance, while its power factor is obtained from the change in $\hat{Q}_0$.

VSWR methods Measurement of the standing waves in a mismatched coaxial line enables wavelength and the load impedance to be obtained from relations given at the end of Sec. 10.10. A section of the line has a lengthwise slot cut in the outer conductor and a small probe projecting into the field between the two conductors assumes a potential measured by an electronic detector.

10.12 Non-sinusoidal voltages and currents

So far, we have assumed that a.c. networks are driven by voltage or current generators whose time variations are purely sinusoidal, i.e. contain only one frequency. As long as the components are sufficiently linear, only this frequency occurs in any output. In practice, the applied voltages or currents may be periodic but not necessarily sinusoidal. In that case, however, Fourier's theorem allows us to replace such variations with sums of sinusoidal components, each component having a frequency which is a multiple of the fundamental and each with an amplitude and phase peculiar to the time variation in question. Thus a square-wave variation of frequency ω and amplitude ± 1 can be expressed as $(4/\pi)[(\sin \omega t)+(\sin \omega t)/3+(\sin 5\omega t)/+\cdots]$. For any one such component all the previous work will apply, and we can see from Eq. (10.1) that if $y=u$ is a solution for the component $f_1(t)$ and $y=v$ is a solution for $f_2(t)$ then $y=u+v$ is a solution for $f_1(t)+f_2(t)$; we need therefore only add the effects of the various components (known as harmonics).

However, most linear circuits have properties which vary with frequency so that the relative amplitudes of the components and their phases will change as we move from one point of a network to another, and any output waveform will differ from the input giving rise to it (this is **distortion**, already referred to in Sec. 10.10). Inductances with reactances ωL, for instance, will tend to accentuate the higher harmonics in the potential difference across it, while capacitors will do the opposite. Distortion sometimes needs to be minimized, for instance in audio-frequency amplifiers where good reproduction of music is required or in pulse amplifiers where we may wish to preserve the pulse shape because of the information it contains about the effect which produced it (possibly the transition of a nuclear particle through a radiation detector). At other times we may wish to alter the shape of the input so that certain functions can be performed, for instance, the shortening or delay of a pulse. We shall consider only the simple series *RC* circuit to illustrate the distorting effects which can be produced (note that the transients of Secs 10.2 and 10.3 are the response of the networks to a single step in the applied voltage).

Figure 10.24 shows the circuit, whose time constant is $\tau = CR$, with an input in the form of pulses of duration T repeated at regular intervals. The 'square wave' is equivalent to applying suddenly a steady electromotance for a time T and then removing it. An output may be taken either across R or across C and since we have seen that the latter will be deficient in the higher frequency harmonics it will form a low-pass filter for *any* input, though with no sharp cut-off. For the square wave we apply the results of Sec. 6.8 and obtain the outputs shown in the figure. If $\tau \ll T$, the output from R is a train of very short duration pulses often needed in electronic circuits as triggers, while across C the output almost faithfully reproduces the input waveform. On the other hand, if $\tau \gg T$, it is the output from R which tends to reproduce the input.

For a general input potential difference of the form $V(t)$ we know that the sum of $V_R = RI$ and $V_C = Q/C$ is at all times equal to $V(t)$. Moreover, because $I = dQ/dt$, the relation between V_R and V_C is always

$$V_R = \tau \frac{dV_C}{dt} \qquad \text{or} \qquad V_C = \frac{1}{\tau}\int V_R \, dt \tag{10.93}$$

If τ is very small then for not too rapid changes of potential difference, $V_R \ll V_C$ and the waveform across C reproduces $V(t)$, while V_R is proportional to dV_C/dt which is very nearly $dV(t)/dt$. Conversely, if τ is very large, V_R reproduces $V(t)$ and V_C is proportional to its integral. Figure 10.24 exhibits these properties. A capacitor and resistor in series can therefore be used as a simple differentiating or integrating circuit.

General network functions We have already mentioned the fact that all the responses of a linear network to stimuli—whether transient or periodic, sinusoidal or non-sinusoidal—depend purely on the network elements and the way they are connected. This suggests the existence of a *universal* function for a

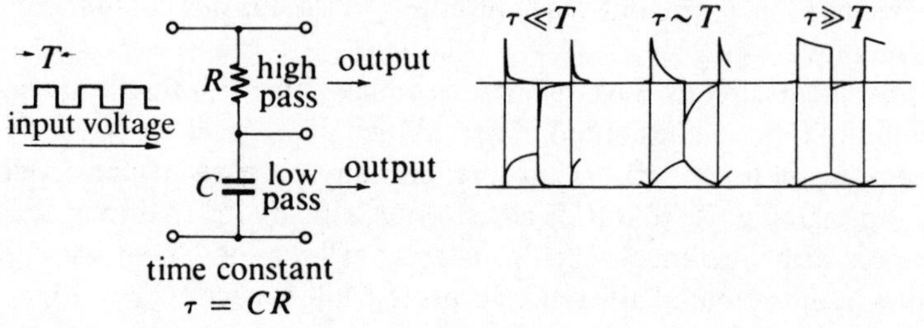

Figure 10.24 Wave-shaping by a CR circuit. For $\tau \ll T$, the output across R is the derivative of the total input, while for $\tau \gg T$, the output across C is the integral of the total input.

given network that would represent its responses to *any* stimulus. For a two-terminal network this would be an impedance or admittance function, Z or Y, depending on whether the current or the voltage were regarded as the stimulus. For a four-terminal network it could be one of the transfer functions. As an illustration we examine the determination of such a Z or Y for a two-terminal network.

The important property of all solutions we have met so far is that they are of the form $V = Ae^{st}$ or $I = Be^{st}$, or combinations of several such terms, where s may be real, imaginary or complex, according to the stimulus and network. If these forms are used, there arise functions $Z(s)$ and $Y(s)$ which relate V and I at the terminals through $V = Z(s)I$ or $I = Y(s)V$. The calculation of Z and Y is performed in much the same way as for an a.c. network and the functions, once obtained, yield solutions for the transient and steady-state cases as well as for any other.

As an example we take a series combination of L and R, where the total voltage across the pair is $V = RI + L\,\mathrm{d}I/\mathrm{d}t$. If the response current to any applied V has the form $I = Be^{st}$, then $V = (R + sL)I$ so that

$$Z(s) = R + sL \qquad \text{or} \qquad Y(s) = 1/(R + sL)$$

are the impedance and admittance functions of the network. To obtain the steady-state sinusoidal response to $V = V_0 e^{j\omega t}$ put $s = j\omega$. Then $I = Y(s)V = V_0 e^{j\omega t}/(R + j\omega L)$ as in a.c. theory—Eq. (10.39). If, on the other hand, we wish to examine the transient response, let us assume that at $t = 0$ a current I_0 is flowing and the terminals are short-circuited. The short circuit means that $V = 0$ so that $Z(s) = 0$ or $s = -R/L$. Hence $I = I_0 e^{-Rt/L}$ as in Eq. (9.20).

The general rules for finding the $Z(s)$ for any network are (a) that inductances are represented by sL, capacitances by $1/sC$, resistances by R, and (b) that the individual impedance representations are combined using the series and parallel formulae or the Δ–Y transformation. $Y(s)$ is $1/Z(s)$, but s^2 must not be put equal to -1. The free motion of the system is obtained by equating $Z(s)$ to zero and any imaginary part of s yields the natural frequency of oscillations in the system.

The response of networks to non-sinusoidal stimuli is most easily calculated using Laplace transforms.

10.13 The limits of lumped network theory

For much of the work on a.c. we have assumed that we can use lumped components (resistors, capacitors, inductors) together with wires playing little part in the operation of a network beyond their function as connectors. We have also assumed that we can represent the behaviour of the network by lumped circuit elements (R, C, L and M) joined by lines representing short circuits and that Kirchhoff's laws can be applied. At low frequencies, the representation and the laws work well because we can ignore the unpleasant facts about the processes occurring inside capacitors and inductors (the fields being well confined) and concentrate our interest on the I–V relationship at the terminals.

However, two related effects occur at high frequencies that limit the usefulness of conventional components and wires, of the representation by lumped elements and of Kirchhoff's laws. One concerns the finite time taken for the transmission of changes and the other the leakage of fields causing radiation.

Time delays We saw in Sec. 10.10 that changes in I and V propagate themselves along connecting wires and cables with a velocity of about $c = 3 \times 10^8$ m/s. For a network of overall dimensions about 0.3 m, say, the delay in propagation from one end to the other is about 10^{-9} s, and this is comparable with the period of the a.c. at frequencies ν approaching 1000 MHz (1 GHz). The variations in I and V along wires cannot then be neglected and must be treated in terms of waves as in Sec. 10.10, with a wavelength $\lambda = c/\nu$ short enough to be comparable with the size of the network. Cables can no longer be ignored as mere connecting links, but may well be the cause of mismatching between a source and a load because of the relation (10.88). There is then a consequent reduction in the power transferred which makes itself evident in standing waves set up in an incorrectly terminated line.

Flux leakage The time delay in the connecting wire is due to their L and C, i.e. to the magnetic and electric fields surrounding them. Leakage of both electric and magnetic flux will also occur even from well-designed components. We can, it is true, take account of leakage by including further elements in the circuit diagram to represent strays (as in Fig. 5.13) and this becomes increasingly necessary as the frequency rises. However, this must not conceal the important fact that such fields are unconfined, and we shall see in Chapter 14 that this leads to a loss of energy through radiation. The loss once again becomes appreciable when the wavelength becomes comparable with the size of the component or network: at such frequencies they become efficient transmitting antennae (aerials). To prevent radiation as well as to avoid strays, we are led to the shielding of all components by metallic enclosures and the screening of all connecting wires, making them into coaxial cables.

We can also expect the design of components to change radically at frequencies corresponding to wavelengths of 1 m or less (microwaves). Firstly, the resistance of a wire becomes so great because of the skin effect that conductors need to provide large surface areas for current flow. Secondly, the reactances ωL and $1/\omega C$ need very small L and C to be as effective as at low frequencies. However, a small L means a very simple geometry (less than one turn) and a small C means widely separated plates. Both of these will radiate enormously unless constructed in an entirely different way.

The next steps: waveguides and cavities In Sec. 10.10 we saw that it was still possible to retain network concepts in the case of transmission lines by regarding R, C and L as distributed rather than lumped and by confining the fields in the coaxial case. However, we shall see in Chapter 14 that such lines are a special case of a *waveguide*—a tube along which changes in **E**- and **B**-fields can be propagated. In general, the distribution of current and voltage in the walls of such guides are so complicated that it is simpler to work in terms of fields, and this we shall do. Transmission lines happen to be cases in which the distribution of I and V is simple enough for network concepts to be used.

We shall also see that the place of the lumped component is taken by sections of guide—in particular, a resonant network is replaced by an electromagnetic cavity resonator. Conventional networks as we have known them in this chapter will have disappeared completely and with them the network theory. We thus return to the field laws from which we began.

Commentary

C10.1 On Integrated Circuits and Miniaturization No attempt has been made in this book to deal with the area broadly included under the heading of *electronic circuitry*, since even the basic principles involved would need a volume of their own, while any attempt to give an account of the current state of the art could rapidly become outdated through new developments. Nevertheless, in view of the large amount of circuitry covered in this chapter, it is worth while surveying the electronics field if only to put the material into perspective.

The earliest form of electronic circuit used an *LCR* combination incorporating a spark gap which, on breakdown, caused an oscillatory discharge of the capacitor according to Eq. (10.13) and Fig. 10.3a and the transmission of damped wave trains. Together with crystal detectors, these formed the basis of telegraphic systems during the nineteenth century, particularly for use at sea. The discovery by Thomas Edison in the 1880s of the unidirectional current across a vacuum produced by thermionic emission from a filament led to J. A. Fleming's diode in 1904, which could then be used as a basis for the detection of electromagnetic waves. In 1907, de Forest patented a Fleming 'valve' incorporating a third electrode, the grid, forming a triode and opening the way to a multitude of applications such as amplification and generation of alternating voltages and currents. Except for the addition of further grids, the development of basic active components remained relatively static until after the Second World War. All the devices used over this period, from radio receivers and transmitters to radar systems and early computers, were based on thermionic vacuum tubes.

The modern revolution began with the discovery of the rectifying action of the *p-n* junction and the invention of the bipolar transistor in 1948 by Bardeen, Brittain and Shockley, followed later by the field effect transistor (FET). At first, the new solid-state components were used as discrete circuit elements simply to replace their thermionic equivalents, with the advantages of greater reliability, smaller size and lower production and running costs. In 1960, however, small-scale *integration* (SSI) began with the use of planar methods of producing circuits which incorporated logic gates, transistors, diodes, resistors and capacitors in the same basic silicon chip. The first chips were 1 or 2 mm across and contained about a dozen individual components.

Rapid development took place through medium-scale integration (MSI) in the 1960s (counting circuits, shift registers, random access memories) and large-scale integration (LSI) in the 1970s (microprocessors) to the present very large-scale integration (VLSI) where a single chip may contain 2500 components per square millimetre on a total area of 7 mm square.

Alongside this, *hybrid circuits* have been developed for many specialized applications where complete integration in a single chip is either undesirable, uneconomic or impossible. These circuits are mounted on a plane substrate, usually alumina, and the various components (which may include silicon chips) are mounted on the substrate and interconnected in various ways. The mounting and interconnection of many components is carried out by depositing a thick or thin metallic layer (metallization) in a pattern predetermined by a technique similar to photolithography. The reduction in size from circuits with discrete components can be by factors as high as 20. Hybrid circuits are used for electronic control circuits in vehicles, domestic equipment such as microwave ovens and sewing machines, electronic typewriters, power tools, etc.

Whether the circuits are integrated or hybrid, they incorporate a wide variety of active circuit elements in addition to the passive elements considered in this chapter. Nevertheless, except for logic and counting circuits that depend on the operation of switches or gates using binary arithmetic, equivalent circuits can be used in many cases and solved by methods based on the principles developed here.

A general article on materials used in information technology and communications is that of Mayo (1986).

Problems

Section 10.1

10.1 Take a two-terminal network of R's, L's and C's of your own choosing with a voltage V between the terminals and a current I flowing into one and out of the other. Find the differential equation relating V and I, and show that if V is given by $V_0 \sin \omega t$, the equation takes the form of (10.1). For example, you could take L in series with R_1 both in series with a parallel combination of C and R_2.

Section 10.2

10.2 A 1 μF capacitor is discharged through a coil of self-inductance 10 mH and resistance 200 Ω. At what time after the discharge begins does the current reach its maximum value?

10.3 Capacitors of 0.5 μF and 0.1 μF and an inductor of self-inductance 12 mH and negligible resistance are connected in series. Find the natural angular frequency for discharge of the 0.5 μF capacitor by the methods of Sec. 10.2 without assuming that the capacitors can be replaced by a single equivalent one.

10.4 Put $CR^2/4L = \delta$ in (10.16) and (10.18) and show how ω_N and Λ vary with δ when damping is small.

Section 10.3

10.5* In the network of Fig. 10.18, the voltage generator is short-circuited and all the capacitors are equal to C_1. Neglecting resistance, find the natural angular frequencies of the oscillatory discharge of one of the capacitors.

Section 10.4

10.6 Draw phasor diagrams for the currents and voltages in (a) a capacitance C and resistance R in series, (b) the same C and R in parallel, both being across a supply of constant amplitude and angular frequency ω. At what angular frequency is the impedance of (a) twice that of (b)? What is the phase difference between the supply voltage and supply current at this frequency?

10.7 A purely resistive 25 W 100 V lamp is to be supplied from an a.c. source of 200 V r.m.s. and 50 Hz frequency. What capacitor, placed in series with the lamp, will suitably adjust the current flowing, no ohmic resistance being added?

10.8 Find the conductance and susceptance of (a) L and R in parallel and (b) L and R in series.

10.9 When connected to a d.c. supply, the voltage across a coil measured by a voltmeter is 1 V when the current through it measured by an ammeter is 1 mA. The same coil across a 50 Hz a.c. supply gives a.c. meter readings of 10 V and 4 mA. What are the resistance and inductance of the coil, assuming that the meters are ideal?

10.10 Find the complex impedance of a coil with self-inductance L and resistance R in parallel with a capacitor C having an imperfect dielectric equivalent to a series resistance also R. Find (a) the value of R which makes the impedance purely resistive at all frequencies and (b) a value of ω which makes the impedance purely resistive for all values of R.

10.11 A coil of self-inductance L and resistance R is connected in series with a switch and the combination is placed in parallel with a capacitor of capacitance C. An alternating voltage is applied across C, and the current supplied by this source is measured by a milliammeter. The value of C is adjusted until the milliammeter reading remains the same whether the switch is open or closed. Show that $2\omega^2 LC=1$ (Turner's method for measuring a large L).

10.12 In a conventional a.c. bridge network (Fig. 10.22a), $\mathbf{Z}_1=\mathbf{Z}_2=R$, $\mathbf{Z}_3$ is a capacitor C and $\mathbf{Z}_4$ a variable resistance S. If the detector is removed, draw a phasor diagram of voltages and currents in the network (using the head-to-tail type of diagram as in Fig. 10.9c for $\mathbf{V}$). Show that the phase of the output across BD relative to that of the input can be varied from 0 to π, without change of amplitude, by varying S (phase-shift network).

10.13 A simple induction motor consists of a flat coil with one turn of area A able to rotate about an axis. A magnetic field of constant magnitude B_0 (produced as in Problem 9.7) has its direction rotating about the same axis with an angular velocity ω with respect to the coil. Show that the mean torque on the coil is $A^2B_0^2\omega R/[2(R^2+\omega^2L^2)]$, where R and L are the resistance and self-inductance of the coil.

Section 10.5

10.14 Find the r.m.s. current when the following are superposed (i.e. fed into the same wire):

(a) a steady current I_2 and an alternating current $I_1 \sin \omega t$,

(b) two alternating currents with the same frequency but with a phase difference α and peak values I_1 and I_2,

(c) two alternating currents as in (b) but with different frequencies ω_1 and ω_2.

10.15 What is the mean value of $I=I_0 \sin \omega t$ over *half* a cycle from $t=0$?

10.16 A certain load consumes 200 W when supplied with 100 V r.m.s. at a particular frequency. When the frequency is doubled, the power consumed is halved. What is the resistance of the load and its reactance X at the original frequency, if X is inductive?

10.17 A 500 W 200 V resistive lamp is placed in series with an inductance to enable it to work from a 250 V 50 Hz supply. What capacitor connected across the supply will make the power factor of the circuit unity?

Section 10.6

10.18 A resonant current of 50 mA flows in a circuit consisting of 2 mH, 20 Ω and 0.0003 μF all in series. What is the applied voltage? If the current is reduced to 30 mA by changing the frequency but not the voltage, find the new frequency and the phase difference between voltage and current.

10.19 In a series LCR circuit with an applied voltage of constant amplitude and variable frequency, the maximum voltages across the L, the C and the R occur at angular frequencies ω_L, ω_Q and ω_R, respectively. Show that $\omega_Q^2=\omega_0^2(1-1/\hat{Q}_0^2)$, $\omega_R=\omega_0$ and $\omega_L^2=\omega_0^2/(1-1/2\hat{Q}_0^2)$, where $\omega_0=1/(LC)^{1/2}$ and $\hat{Q}_0=\omega_0 L/R$.

10.20 In a series LCR circuit, the alternating applied voltage is of fixed amplitude and frequency while the capacitance is variable. Show that the curve of current against C has a half-power width ΔC and a peak value C_0 related by $2C_0/\Delta C=\hat{Q}_0$, provided that $\hat{Q}_0 \gg 1$.

10.21 Find the true voltage resonant frequency in the network of Fig. 10.12a.

10.22 An alternating voltage is applied across the central capacitance of the network of Problem 10.5. Show that the resonant frequencies of the network are the same as the natural frequencies obtained in that problem.

Section 10.7

10.23 Find $\mathbf{A}$ for a four-terminal network in which the input is across R and C in series and the output is taken across R. Express $\mathbf{A}$ in the form $re^{j\theta}$ and find an equation between r and θ. Plot the relation in the (r, θ) plane and show that it has the form of a semi-circle. (This is a *Nyquist plot*, useful in the analysis of active networks such as those of amplifiers and oscillators.)

Section 10.8

10.24 Show that the two peaks in the $|I_2|$ against ω curve of Fig. 10.16 are of equal height V/R for ω near ω_0.

10.25 Use the method of Sec. 10.8 to find the condition that $|I_1|$ of (10.67) shall have two maxima and one minimum. Show that this condition is already fulfilled at critical coupling $k = 1/\hat{Q}_0$.

Section 10.9

10.26 Show that the characteristic impedance for the π-section of Fig. 10.19 is $\mathbf{Z}_1\mathbf{Z}_2/\mathbf{Z}_{kT}$, where $\mathbf{Z}_{kT}$ is given by (10.69). (First convert π to T by the delta–Y transformation.)

10.27 If $\mathbf{Z}_{oc}$ and $\mathbf{Z}_{sc}$ are the input impedances of a symmetrical T-section when the output terminals are open-circuited and short-circuited, respectively, show that $\mathbf{Z}_{oc}\mathbf{Z}_{sc} = \mathbf{Z}_k^2$. Verify that the same condition applies to a symmetrical π-section and to a loss-less transmission line using (10.87).

10.28 Show that, in the ladder network of Fig. 10.19a, $\mathbf{V}_{n+1}/\mathbf{V}_n = \mathbf{I}_{n+1}/\mathbf{I}_n$.

10.29 A half T-section ($\frac{1}{2}\mathbf{Z}_1$ in series, $2\mathbf{Z}_2$ in parallel) terminates a T-section ladder network with elements $\mathbf{Z}_1$ and $\mathbf{Z}_2$. Show that the network is correctly terminated if a load with impedance $\mathbf{Z}_{k\pi}$ is connected across the half-section, where $\mathbf{Z}_{k\pi}$ is given by Problem 10.26.

10.30 Show that the cut-off frequency for the high-pass filter of Fig. 10.20b is given by $\omega_0 = 2/(LC)^{1/2}$.

10.31 Show that the pass band of the filter in Fig. 10.20c is from $\omega_1 = (\sqrt{2} - 1)/(LC)^{1/2}$ to $\omega_2 = (\sqrt{2} + 1)/(LC)^{1/2}$.

10.32 Show that the shape of the attenuation curve (α against ω) for the low-pass filter of Fig. 10.20a is given by $\cosh \alpha = \frac{1}{2}\omega^2 LC - 1$. If a rectifier output with a 100 Hz ripple is to be smoothed, what attenuation in power per T-section is achieved if $L = 20$ H and $C = 5\,\mu$F?

Section 10.10

10.33 Find the characteristic impedance of a loss-less coaxial cable in which the diameter of the inner conductor is one-third of that of the inner surface of the outer conductor. If the voltage standing wave ratio in the cable is 2.5, what is the resistance of the termination?

10.34 What is wrong with the following argument? An electric shock cannot be obtained when playing a jet of water on to a live electric cable provided that the velocity of the water exceeds that of the current carriers given by a calculation like that of Problem 1.9. (Warning: the argument *is incorrect*, so don't try it.)

Section 10.11

10.35 Estimate the order of magnitude of stray capacitance and inductance introduced by unshielded connecting leads in an a.c. bridge. At what frequency would the effects become comparable with, say, a 1000 Ω resistor in one arm of the bridge?

10.36 In a series LCR circuit, the L is the secondary of a transformer feeding in a voltage of fixed amplitude and frequency. The capacitor C is variable and calibrated, and has an electrometer of effectively infinite resistance placed across it. The electrometer reading is a maximum when $C = 5.42\,\mu$F. When an unknown capacitance is connected in parallel with C, the electrometer reading is a maximum for $C = 3.15\,\mu$F. What is the unknown capacitance? How would you proceed to measure the unknown if it was larger than the maximum value of C?

10.37 Find the balance conditions for conventional a.c. bridges with the following components in the arms 1, 2, 3, 4, respectively, of Fig. 10.22a: (a) L and r in series (unknown); R_1; R_2; R_3 and C in parallel (Maxwell's bridge); (b) L and r in series (unknown); R_1; R_2 and C_2 in series; C_1 (Owen's bridge); (c) C_1 and R_1 in parallel; C_2 and R_2 in series; R_3; R_4 (Wien's bridge for an unknown frequency).

Section 10.12

10.38 Find the impedance function $Z(s)$ for a series LCR circuit and hence establish that its natural angular frequency is given by Eq. (10.16).

10.39 Although this chapter is confined to linear circuit elements, it is worth mentioning the effects of non-linearity on frequency response through the following problems. The current in a non-linear device is given by $I = a + bV + cV^2$, say, where V is the voltage and a, b and c are constants. If the applied voltage has the form $V_0 + V_1 \sin \omega t$, show that the effect of the alternating component on the current is to increase the mean value by $\frac{1}{2}cV_1^2$ and to introduce a harmonic of angular frequency 2ω. Show further that, if the applied voltage is $V_0 + V_1 \sin \omega_1 t + V_2 \sin \omega_2 t$, terms involving sum and difference frequencies occur in the current in addition to those expected from the first example.

Section 10.13

10.40 Lengths of cable with characteristic impedance of 70 Ω are to be used as circuit elements at a frequency corresponding to a wavelength of 1 m. What length of short-circuited cable would be equivalent to an inductance of 0.1 μH? If the same length of cable were open-circuited, what circuit would it be equivalent to? (Neglect losses in the cable at this high frequency.)

11

Dielectric materials

No substance is a perfect insulator, but there are many whose electric conductivity is small enough to be neglected as a first approximation. Such substances are known as **dielectric materials** or simply as **dielectrics** and we devote this chapter to a study of their behaviour. So far, we have two pieces of experimental evidence to guide us. Firstly, we have the effect of an insulator on capacitance, described in Sec. 5.4 in terms of the **relative permittivity** ε_r. Secondly, we have the idea of **polarization** arising from phenomena described in Secs 1.1 and 3.8: that positive and negative charges move small distances in opposite directions when an **E**-field is applied.

In Sec. 11.1, we see how far the description in terms of relative permittivity will take us and it becomes immediately apparent that only the most restricted types of material can be dealt with, namely those that are *linear*, *isotropic*, *homogeneous* and *effectively infinite* in extent. One of the principal reasons for delaying a consideration of dielectrics and magnetic media until so late in this book is precisely the complexity that real materials can bring to the properties of electric and magnetic fields: the vacuum properties are already quite complex enough for a first course. However, material media are undoubtedly of great importance and must be included in our general theory at this stage.

In Sec. 11.2, we turn instead to the more fruitful concept of polarization and use it in Secs 11.2–11.7 to develop a field theory that applies to dielectrics, incidentally showing how relative permittivity can be generalized. The introduction of a new electric field, the **D**-field or electric displacement, enables us to express the general laws in a comparatively simple form once more.

The model of a dielectric used for the development of field theory is a smoothed-out or macroscopic one, which cannot explain the observed properties of dielectrics but has to take them as given. Physicists are not satisfied with such a situation and require more detailed models based on the atomic structure of the materials to provide an explanation of their behaviour. In Secs 11.8–11.10, therefore, we examine such models in order to relate the continuum approach of field theory to atomic aspects.

11.1 Relative permittivity of linear, isotropic, homogeneous media

The **relative permittivity** ε_r of a substance was defined in (5.14) by

$$\frac{C_m}{C_0}=\varepsilon_r \qquad \text{(definition of mean } \varepsilon_r\text{)} \tag{11.1}$$

where C_0 is the capacitance of a capacitor *in vacuo* and C_m is that of the same capacitor with the insulating medium filling the whole of the space in which an **E**-field exists. The ε_r's of actual materials are thus obtained from measurements of a capacitance first with and then without the dielectric medium.

The ε_r that is obtained in this way is clearly an average in some sense over the whole volume of material, and we shall eventually need a definition giving ε_r at any point (Sec. 11.3). However, there is a class of material for which such an average is a significant general property: these are linear, isotropic, homogeneous dielectrics, which we shall denote by the initials LIH, omitting one of the letters if we wish to be more general.

Linearity is recognized by the proportionality between Q and V in a capacitor containing the dielectric, i.e. ε_r does not vary with voltage or with **E**-field. By **isotropy**, we mean that ε_r is independent of the orientation of the specimen between the capacitor plates, i.e. ε_r does not vary with the direction of the **E**-field. Finally, by **homogeneity** we mean that the value of ε_r is the same for every part of the dielectric so that the size of the capacitor used is immaterial.

For LIH dielectrics, ε_r is found experimentally to be independent of the shape of the capacitor, to be always greater than 1, and to vary with frequency, temperature, etc. The value in a steady field is known as the **static relative permittivity**. Most substances show little variation in ε_r at low frequencies and it is the static value we shall be dealing with for most of the chapter.

Effect of ε_r on electrostatic formulae When the vacuum round a capacitor is completely filled with an infinite LIH medium the capacitance clearly increases by ε_r. If there are no connections to the plates, Q will remain constant and, because $Q=VC$, the voltage will fall by ε_r. Since this effect is independent of the shape of the plates, we can make the following general statement: the *potential difference between* and the *potential of* conductors carrying any charge falls by a factor ε_r when an infinite LIH dielectric is introduced. The same applies to the *mean* potential gradient and **E**-field.

It does *not* follow that **E** and V *at every point* will fall by ε_r, so that strictly speaking we can go no further, but we shall make an assumption (justified in Sec. 11.4) that this is so for LIH materials. This would then mean that all expressions for V due to specified charge distributions in Chapters 3 and 4 must be modified by the replacement of ε_0 by $\varepsilon_r\varepsilon_0$ if the charges are embedded in an infinite LIH dielectric. For example,

$$\left.\begin{aligned} &\text{point charge:} && V=\frac{Q}{4\pi\varepsilon_r\varepsilon_0 r}\\ &\text{surface charge:} && V=\int_S \frac{\sigma\,\mathrm{d}S}{4\pi\varepsilon_r\varepsilon_0 r}\end{aligned}\right\} \qquad (11.2)$$

etc. In addition, the **E**-fields *defined as negative potential gradient* must be similarly modified so that, for example,

$$\left.\begin{aligned} &\text{point charge:} && \mathbf{E}=\frac{Q}{4\pi\varepsilon_r\varepsilon_0 r^2}\hat{\mathbf{r}}\\ &\text{surface charge:} && \mathbf{E}_{\text{normal}}=\frac{\sigma}{\varepsilon_r\varepsilon_0}\end{aligned}\right\} \qquad (11.3)$$

We must not assume, however, that **E** is also the force per unit charge on any charges in the medium. Without that assumption, (11.3) cannot be extended to Coulomb's law itself: it does not necessarily become $F=Q_1Q_2/(4\pi\varepsilon_r\varepsilon_0 r^2)$ in an infinite LIH medium, much less in any other. The meaning of **E** inside a dielectric is discussed in Secs 11.2 and 11.8.

Some explanation of the reduction of V and **E** above is afforded by the polarization of the dielectric, which produces bound charges on its surfaces adjacent to the charged conductors. Because the bound charges are of opposite sign to the conduction charge, the effective magnitude of the latter is reduced from Q to Q/ε_r, σ to σ/ε_r, etc. We shall justify this interpretation at the end of Sec. 11.4. It should be clear, however, that we cannot proceed with a study of dielectric behaviour until we have developed the concept of polarization more quantitatively.

11.2 Polarization, E-field, and susceptibility

Polarization: concept and definition We now look for an explanation of dielectric behaviour in terms of the polarization which we believe takes place. There is now no restriction to LIH media. In an unpolarized element of an insulator as in Fig. 11.1a, the centroids of the positive and negative charges coincide and no external field is produced. The effect of polarization is shown in Fig. 11.1b, where small movements of the charges in opposite directions have taken place. The element has become an electric dipole and we define the polarization **P** at a point as the electric dipole moment per unit volume at the point. Thus for a volume element $\mathrm{d}\tau$, the dipole moment is $\mathbf{P}\,\mathrm{d}\tau$ and so

(a) (b) (c)

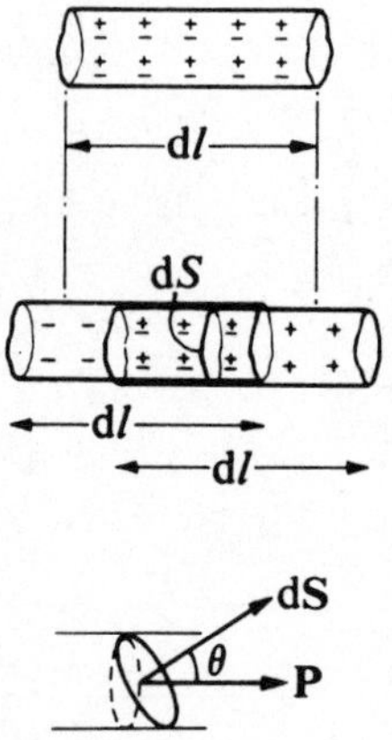

Figure 11.1 (a) An element of unpolarized material; (b) the displacement of charges in polarized material: the element becomes an electric dipole; (c) the charge crossing **dS** during the process of polarization is $P\,\mathrm{d}S\cos\theta$ or $\mathbf{P}\cdot\mathrm{d}\mathbf{S}$.

Electric dipole moment **dp** of volume $\mathrm{d}\tau=\mathbf{P}\,\mathrm{d}\tau$	(definition of **P**) (11.4)

Another interpretation of **P** is extremely important. The moment $\mathbf{P}\,\mathrm{d}\tau$ can be written as $P\,\mathrm{d}S\,\mathbf{dl}$, and this shows that the unbalanced charge at each end of the element in Fig. 11.1b is $P\,dS$ (dipole moment $\mathbf{p}=Q\mathbf{l}$). This amount of charge must therefore have crossed any area $d\mathbf{S}$ when the polarization took place. For any area $d\mathbf{S}$ not normal to $\mathbf{P}$, the charge per unit area crossing $d\mathbf{S}$ is given by the normal component of $\mathbf{P}$. We can therefore state quite generally that when polarization takes place

$$\boxed{\text{Resultant polarization charge crossing } d\mathbf{S}=\mathbf{P}\cdot d\mathbf{S}} \tag{11.5}$$

and this is sometimes regarded as an alternative definition of $\mathbf{P}$. The **SI unit** of $\mathbf{P}$ is the $\mathrm{C\,m^{-2}}$.

Surface and volume polarization charges If $\mathbf{P}$ is uniform throughout a block of dielectric, the positive charges at one end of an element such as that of Fig. 11.1b will overlap the negative ones at the end of the adjacent element and will be of the same density. The only unbalanced charges will occur at the outer surfaces and these will have a density given by (11.5). Thus if P_n is the normal component of $\mathbf{P}$ at the surface we have

$$\boxed{\text{Surface density of polarization charges, } \sigma_P=P_n} \tag{11.6}$$

If $\mathbf{P}$ is not uniform in the body of a material, there may be unbalanced volume charges in addition to σ_P. To find an expression for these, consider a closed surface S entirely within a dielectric. When the dielectric is polarized, the total charge *leaving* the volume inside S can be written $\oint \mathbf{P}\cdot d\mathbf{S}$ using (11.5), where the surface integral is taken over S. Since this is not necessarily zero, there will in general be a polarization charge Q_P left in the volume of opposite sign to that which has left it. Hence

polarization charge:
$$\boxed{Q_P=-\oint_S \mathbf{P}\cdot d\mathbf{S}} \tag{11.7}$$

We can convert this to a differential form operating at a point in various ways. One is to regard it like the integral form of Gauss's law in Chapter 4 and to use the same method as in Sec. 4.5. Or we can convert the surface integral on the right to a volume integral of $\nabla\cdot\mathbf{P}$ using (B.17) in Appendix B, and at the same time write Q as the volume integral of $\rho\,d\tau$. These methods are in fact the same and lead to

$$\boxed{\text{Volume density of polarization charge, } \rho_P=-\nabla\cdot\mathbf{P}} \tag{11.8}$$

The charges σ_P and ρ_P are known as **Poisson's equivalent distribution** because they can be used to replace the effects of a polarized insulating medium by smoothed-out charge distributions given by (11.6) and (11.8). A more formal derivation of the distribution is indicated in Problem 11.3.

E-field in the dielectric Suppose now that an electric field $\mathbf{E}_0$ is maintained over a region of free space by steady external sources. Let a piece of dielectric be introduced into this region. It will become polarized and its polarization charges produce extra fields *outside* the dielectric as if they arise from σ_P and ρ_P above.

Call this additional field $\mathbf{E}_P$. We now **define** the total **E**-field everywhere, including the interior of the dielectric, as

$$\boxed{\mathbf{E}=\mathbf{E}_0+\mathbf{E}_P} \tag{11.9}$$

where $\mathbf{E}_P$ is the field due to σ_P and ρ_P. Once again it must be emphasized that this does not necessarily give the force on any charges there may be in the interior of the dielectric through $\mathbf{F}=Q\mathbf{E}$. The question of these forces is discussed in Sec. 11.8. The point about the **E**-field defined by (11.9) is that it is a field that arises from specified distributions of charge and that it will therefore have the same properties as any other electrostatic field.

Electric susceptibility If the **E**-field at a point in a dielectric as defined by (11.9) is **E** and the polarization at the same point is **P**, we define an electric susceptibility χ_e by

$$\boxed{\mathbf{P}=\varepsilon_0\chi_e\mathbf{E} \qquad \text{(definition of } \chi_e\text{)}} \tag{11.10}$$

where the ε_0 is inserted so that χ_e shall be dimensionless. (See Comment C11.1 for the definition in other unit systems.) A dielectric is linear if χ_e is independent of the magnitude of **E**, is homogeneous if it is independent of position and is isotropic if independent of the direction of **E**.

11.3 Gauss's law, electric displacement (D-field), and permittivity

Gauss's law with dielectrics In Chapter 4 we saw that the properties of electrostatic fields could be summarized by

$$\oint_L \mathbf{E}\cdot d\mathbf{L}=0; \qquad \oint_S \mathbf{E}\cdot d\mathbf{S}=\Sigma Q/\varepsilon_0 \tag{11.11}$$

where the first is a line integral round a closed path L and the second, Gauss's law, a surface integral over a closed surface S. These will be unaffected by the presence of dielectrics if the definition (11.9) is used, except that Gauss's law must include both conduction charges Q_c and polarization charges Q_P on the right-hand side:

$$\oint_S \mathbf{E}\cdot d\mathbf{S}=\Sigma Q_c/\varepsilon_0+\Sigma Q_P/\varepsilon_0 \tag{11.12}$$

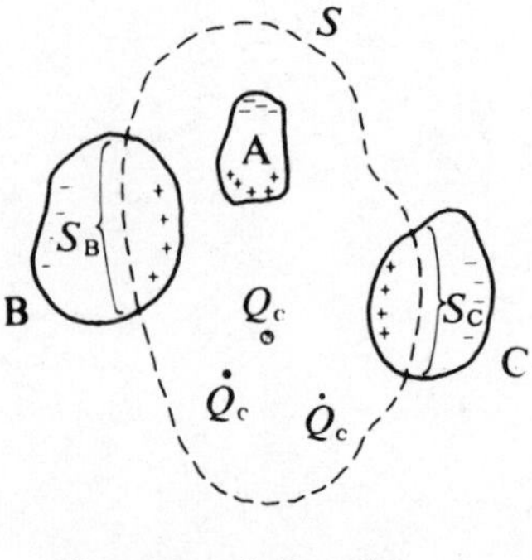

Figure 11.2 A Gaussian surface S with dielectrics present.

A typical situation is illustrated in Fig. 11.2 where the Gaussian surface S intersects some dielectrics such as B and C and completely encloses others such as A. Only B and C can contribute to Q_P since the enclosed charge due to A is zero whatever its state of polarization. The charge crossing S from inside to outside will, by (11.5), be the surface integral of **P** over S_B and S_C. However, since **P** is zero over the rest of S, the total polarization charge leaving the enclosed volume can be written as $\oint \mathbf{P}\cdot d\mathbf{S}$ (remember that outward directions are positive for closed surfaces). It follows that the ΣQ_P left inside S when the polarization takes place will be of opposite sign, so that it can be replaced in (11.12) by $-\oint \mathbf{P}\cdot d\mathbf{S}$. A slight rearrangement then gives

$$\oint_S (\varepsilon_0 \mathbf{E} + \mathbf{P}) \cdot d\mathbf{S} = \Sigma Q_c \tag{11.13}$$

and this is the new form taken by Gauss's law when dielectrics are present.

The electric displacement or D-field The quantity $\varepsilon_0\mathbf{E}+\mathbf{P}$ occurs frequently in dielectric problems once the **E**-field is defined as in (11.9) and it is given a special name, the electric displacement, and a symbol **D** so that

$$\boxed{\mathbf{D} = \varepsilon_0 \mathbf{E} + \mathbf{P} \qquad \text{(definition of } \mathbf{D}\text{)}} \tag{11.14}$$

Its **SI unit** is the $\mathrm{C\,m^{-2}}$.

Gauss's law (11.13) then becomes

Gauss's law for **D**:

$$\boxed{\oint_S \mathbf{D} \cdot d\mathbf{S} = \Sigma Q_c} \tag{11.15}$$

or, in words, *the outward flux of* **D** *over any closed surface S equals the algebraic sum of the conduction charges enclosed by S* (polarization charges do not contribute to the right-hand side). This shows that, while both conduction and polarization charges are sources of **E**-fields, the only sources of **D**-fields are conduction charges.

The differential form of (11.15), obtained by the same method as in Sec. 4.5, illustrates the same point:

$$\boxed{\operatorname{div} \mathbf{D} \equiv \nabla \cdot \mathbf{D} = \rho_c} \tag{11.16}$$

where ρ_c is the volume density of conduction charges. The form of (11.15) and (11.16) shows that the **D**-field due to specified distribution of conduction charge can be obtained by the same methods as those for the **E**-fields in Chapters 3 and 4. The expressions for **D** can in all cases be obtained by multiplying the **E**-field by ε_0 since no dielectrics were present, $\mathbf{P}=0$ and $\mathbf{D}=\varepsilon_0\mathbf{E}$. For instance, the **D**-field at a distance r from a point charge Q is $Q/(4\pi r^2)$.

'Displacement' is not a happy choice of name. It originated historically from the idea that it was the result of a movement of charge in the dielectric (giving **P**) added to a similar state of strain in space (giving $\varepsilon_0\mathbf{E}$). We can give another physical interpretation later in Sec. 11.8.

Relation between susceptibility and relative permittivity By substituting (11.10)—$\mathbf{P}=\varepsilon_0\chi_e\mathbf{E}$—into (11.14) we obtain:

$$\mathbf{D} = \varepsilon_0(1+\chi_e)\mathbf{E} \tag{11.17}$$

Consider now any system of conduction charges. Whether these are *in vacuo* or in an infinite LIH medium, **D** at any point is the same since it depends only on conduction charges. In that case, (11.17) tells us that **E** at any point will fall to $1/(1+\chi_e)$ of its vacuum value when the dielectric is introduced. In Sec. 11.1, however, we saw that **E** drops to $1/\varepsilon_r$ of its vacuum value, so we are led to a general definition

$$\varepsilon_r = 1 + \chi_e \qquad \text{(definition of } \varepsilon_r\text{)} \tag{11.18}$$

which gives the relative permittivity at any point and not just its mean value over a volume. We can now write the relation between **D** and **E** in (11.17) as

$$\mathbf{D} = \varepsilon_r \varepsilon_0 \mathbf{E} \tag{11.19}$$

and that between **P** and **E**, (11.10), as

$$\mathbf{P} = \varepsilon_0(\varepsilon_r - 1)\mathbf{E} \tag{11.20}$$

If we agree to adopt (11.18) as a general definition not restricted to LIH media, then (11.19) and (11.20) become general as well, although they may not be as simple as they appear (see next section).

Circuital law and Faraday's law for E We saw in Chapter 4 that charges make no contribution to the line integral of **E** round a closed path, or to the curl of **E**. Because the **E**-field in dielectrics is defined by (11.9), and thus still arises from charges, it will still obey the circuital law $\oint \mathbf{E}\cdot d\mathbf{L} = 0$ in the absence of time variations and $\oint \mathbf{E}\cdot d\mathbf{L} = -\partial\Phi/\partial t$ more generally, so that no modification is necessary due to the introduction of dielectric materials.

11.4 LIH and non-LIH dielectrics

Non-LIH dielectrics None of the relations and definitions in Secs 11.2 and 11.3 are limited to LIH dielectrics, but it must be emphasized that χ_e and hence ε_r are not necessarily constants but may vary from one point to another either because of non-homogeneity or because non-linearity occurs and **E** is not uniform: because of this and because ε_r also varies with physical conditions, the term 'dielectric constant' is not favoured. Moreover, if the dielectric is anisotropic, (11.10) is an abbreviation for

$$P_x = \varepsilon_0 \chi_{xx} E_x + \varepsilon_0 \chi_{xy} E_y + \varepsilon_0 \chi_{xz} E_z$$

with two similar expressions for P_y and P_z. Equation (11.19) is similar and both χ_e and ε_r are *tensors* rather than scalars: in this case **D**, **E** and **P** are not necessarily in the same direction. Thus, while (11.19) is sometimes used as a definition of **D** in preference to (11.14), it is deceptively simple: only in LIH dielectrics are χ_e and ε_r scalar constants under constant physical conditions.

Anisotropy is confined to single crystals whose symmetry is tetragonal or lower and to substances which are strained: we shall generally assume isotropy. **Non-linearity** would occur at very high electric fields when saturation sets in or at low fields in substances known as *ferroelectric* (only by analogy with *ferromagnetic*—ferroelectricity has nothing to do with iron). **Non-homogeneity** is caused by variation in composition, density and structure of the dielectric.

Polarization charge in LIH dielectrics In Sec. 11.2, a polarized dielectric was seen to be equivalent to a set of surface and volume charges denoted by σ_P and ρ_P. We

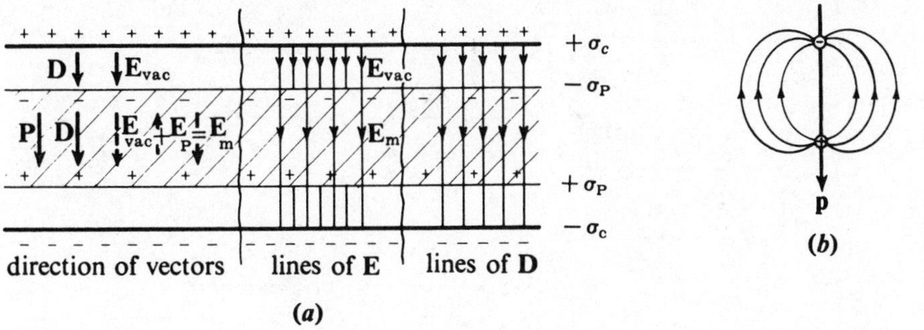

Figure 11.3 (a) Vectors in an LIH dielectric between the plates of a parallel plate capacitor; (b) an elementary dipole showing flux and field in a direction opposite to that of the moment **p**.

now show that *in LIH dielectrics there is no volume distribution of charge*, and we do this by consideration of Eq. (11.7): $Q_P = \oint \mathbf{P} \cdot d\mathbf{S}$, giving the resultant charge inside a surface S wholly within a polarized dielectric. Now **P** can be written as $\chi_e \mathbf{D}/\varepsilon_r$ using (11.10) and (11.19), and in this expression χ_e/ε_r is a scalar constant independent of position because the dielectric is LIH. Q_P is thus given by $-\chi_e/\varepsilon_r$ times $\oint \mathbf{D} \cdot d\mathbf{S}$. However, this surface integral is zero by (11.15) because there is no conduction charge within S and hence Q_P is zero. It follows that the equivalent distribution to a polarized LIH dielectric consists solely of surface charges whose density σ_P is given by the normal component of **P**. Because $\mathbf{P} = \varepsilon_0(\varepsilon_r - 1)\mathbf{E}$ according to (11.20), we have

$$\sigma_P = P_n = \varepsilon_0(\varepsilon_r - 1)E_m \tag{11.21}$$

in which E_m is the normal component of the **E**-field *in the medium* at the surface.

We can now see why the presence of a dielectric always reduces the **E**-fields and potentials due to a fixed system of conduction charges. Figure 11.3a represents a slab of dielectric placed between the plates of a parallel plate capacitor leaving a gap in which the fields have their vacuum values, $D = \sigma_c$, $E_{vac} = \sigma_c/\varepsilon_0$. Inside the dielectric, D is unchanged because it is unaffected by polarization charges. **E**, on the other hand, is reduced because it is the sum of $\mathbf{E}_{vac}$ and $\mathbf{E}_P$ according to the definition (11.9), and $\mathbf{E}_P$ is in the opposite direction to $\mathbf{E}_{vac}$: although the elementary dipoles in the dielectric point in the direction of **P**, Fig. 11.3b shows that their **E**-flux will be in the opposite direction. Now $\mathbf{E}_P$ will be given by σ_P/ε_0 so that the field in the medium, E_m, is

$$E_m = E_{vac} + E_P = (\sigma_c - \sigma_P)/\varepsilon_0 = \sigma_c/\varepsilon_0 - (\varepsilon_r - 1)E_m$$

using (11.21). This gives $E_m = \sigma_c/\varepsilon_r\varepsilon_0$ and justifies the expression in (11.3) and similar expressions.

Introduction of a dielectric into a capacitor Finally, we consider the process of *introducing* a dielectric into the space between the plates of a capacitor as in Fig. 11.4a and b. Here, we could isolate the capacitor as in Fig. 11.4a, thus keeping the charge Q constant. In this case we have the same arrangement as in Fig. 11.3a without the gap, and we have seen that the values of V and E fall by a factor ε_r. Alternatively, as in Fig. 11.4b, the dielectric could be introduced while keeping the plates at a constant potential difference by batteries. Since the definition (11.1) does not depend on the way the dielectric is introduced, the capacitance still increases by ε_r. If V is constant, this must mean that Q increases to $\varepsilon_r\sigma_c$, the extra

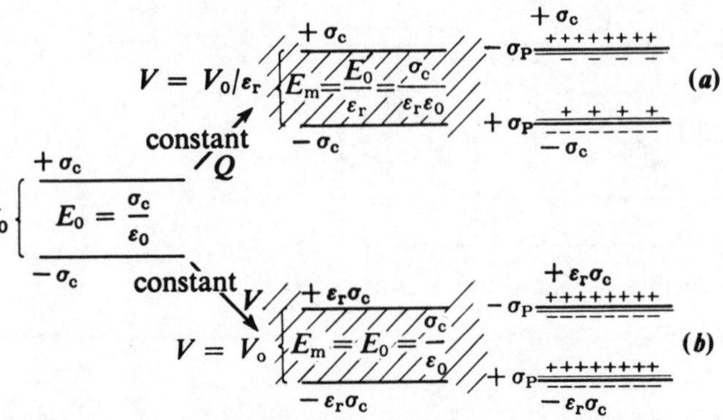

Figure 11.4 The effect of introducing a dielectric into a capacitor (a) at constant charge, (b) at constant potential.

charge density $(\varepsilon_r - 1)\sigma_c$ being supplied by the batteries. The reader should be able to show that the value of σ_P on the dielectric is then exactly what is needed to keep the E-field constant at σ_c/ε_0.

11.5 Boundary conditions

Values of D and E across a boundary The solutions to many problems involving conducting and insulating media are determined by what happens at the various boundaries. Here, we shall use the general laws derived so far to find what conditions must be obeyed by **D**- and **E**-fields on the two sides of any boundary.

Figure 11.5 shows the surface between two media labelled 1 and 2, either of which may be conductors or insulators. Along this surface there resides in general a conduction charge of surface density σ_c and a polarization charge. The generalized Gauss's law (11.15) applied to the volume in Fig. 11.5a generated by the lines of **D** forming the curved sides and having elementary surfaces $d\mathbf{S}_1$ and $d\mathbf{S}_2$ parallel to the surface S gives

$$\mathbf{D}_1 \cdot d\mathbf{S}_1 - \mathbf{D}_2 \cdot d\mathbf{S}_2 = \sigma_c \, dS$$

As the heights of the cylinders shrink, both $d\mathbf{S}_1$ and $d\mathbf{S}_2$ tend to $d\mathbf{S}$ and in the limit

$$\boxed{D_{1n} - D_{2n} = \Delta D_n = \sigma_c} \tag{11.22}$$

where D_n is the component of **D** normal to S. Thus we can say that in general *the normal component of* **D** *is discontinuous by* σ_c *across any surface.*

In Fig. 11.5b, the circuital law for **E** in (11.11) is applied to a closed path consisting of two elements $d\mathbf{L}_1$ and $d\mathbf{L}_2$ as shown and completed by two small portions at the ends. The contributions of the ends to $\oint \mathbf{E} \cdot d\mathbf{L}$ vanish in the limit as long as **E** is finite, so that

$$E_{1t} \, dL_1 - E_{2t} \, dL_2 = 0$$

and hence, because $dL_1 = dL_2$,

$$\boxed{E_{1t} = E_{2t} \quad \text{or} \quad \Delta E_t = 0} \tag{11.23}$$

where E_t is the component of **E** tangential to the surface. Thus *the tangential component of* **E** *is continuous across any surface.* This conclusion is unaffected if

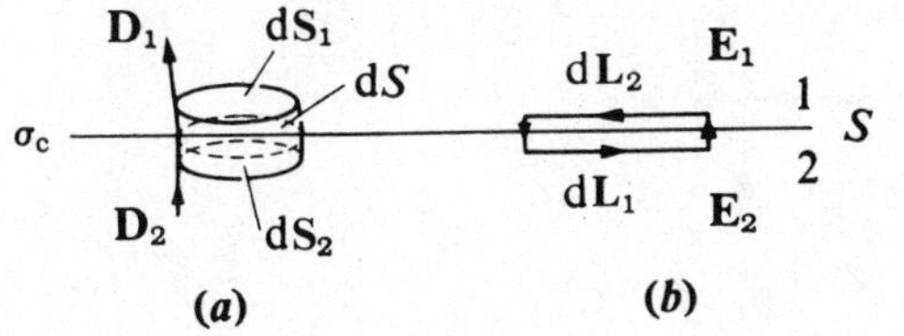

Figure 11.5 Derivation of boundary conditions for (a) the normal component of **D**, (b) the tangential component of **E**.

time variations occur, even though the circuital law becomes $\oint \mathbf{E}\cdot d\mathbf{L} = -\partial\Phi/\partial t$. The magnetic flux Φ is that across the small rectangle in Fig. 11.5b and in the limit, as long as any **B**-field remains finite, Φ will tend to zero with the area.

Worked example *capacitors with more than one dielectric* An application of (11.22) and (11.23) occurs in capacitors where the space between the plates contains more than one dielectric, with the bounding surfaces being either perpendicular or parallel to the lines of the fields.

When the surfaces are perpendicular to the lines of **D** and **E**, (11.22) shows that **D** will be the same on both sides of any one surface while **E** must therefore change inversely as the ε_r's. In fact the values of **E** and therefore of the potential differences between the surfaces of the dielectrics must each be $1/\varepsilon_r$ of the values they would have *in vacuo*. If the potential differences are V_1, V_2, etc., the total capacitance C will be $Q/(V_1+V_2+\cdots)$ and hence $1/C = 1/C_1 + 1/C_2 + \cdots$ where $C_1 = Q/V_1$, etc., showing that C may be obtained by treating the single capacitor as a set in series.

Figure 11.6 illustrates the case of a parallel plate capacitor with a dielectric slab of thickness t. For this the capacitance is given by

$$\frac{1}{C} = \frac{x-t}{\varepsilon_0 A} + \frac{t}{\varepsilon_r \varepsilon_0 A}$$

or

capacitor of Fig. 11.6:
$$C = \frac{\varepsilon_0 A}{x - t(1 - 1/\varepsilon_r)}$$

showing that the effect of the slab is the same as that of reducing the distance apart of the plates *in vacuo* by $t(1-1/\varepsilon_r)$.

A similar argument applies when boundaries are parallel to the field lines. The **E**-field is then continuous and the total capacitance is obtained by treating the system as a set of capacitors in parallel. Figure 11.7 shows this for a parallel plate capacitor. Note that in both Figs 11.6 and 11.7, the conduction charge density σ_c is not necessarily that which the plates would possess if the dielectric were withdrawn, since what then happens depends on whether or not the plates are isolated: the figures show merely a static situation.

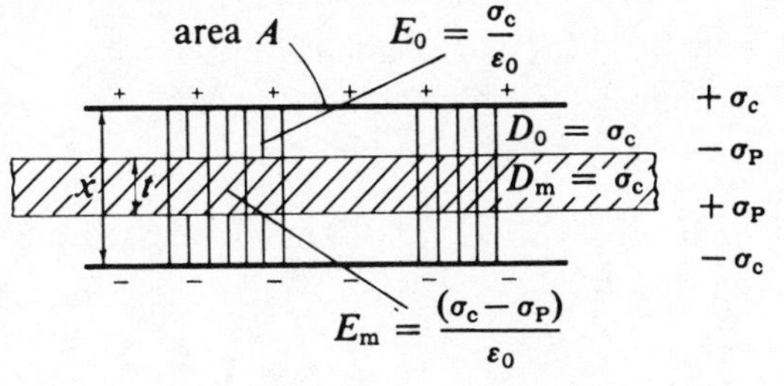

Figure 11.6 Capacitor with dielectric boundaries perpendicular to directions of **D** and **E**.

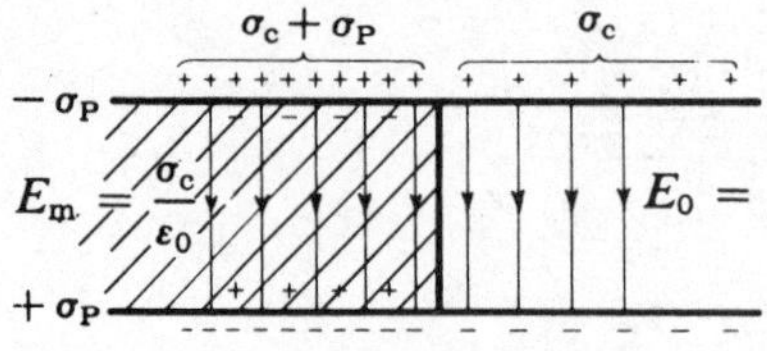

Figure 11.7 Capacitor with dielectric boundaries parallel to directions of **D** and **E**.

11.6 Electric energy with dielectrics

In Secs 5.5 and 5.6 it was shown that the work done in assembling a set of charges could be expressed as $\Sigma\frac{1}{2}QV$ or $\int\frac{1}{2}\varepsilon_0 E^2\,d\tau$ in the absence of dielectrics. Similar arguments can be used to obtain more general expressions, but some simplifying assumptions will be made.

Firstly, while we may not always wish to insist on linearity, ε_r must be single valued so that any integrals have a unique magnitude: our treatment will not necessarily apply when hysteresis occurs in ferroelectrics.

Secondly, we must treat systems including dielectrics as thermodynamic rather than purely mechanical so that instead of equating work done on a system W to the increase in mechanical energy U, we should specify the conditions more precisely. For instance, for an adiabatic change, $dW = dU$, the increase in internal energy, while for a reversible isothermal change, $dW = dF$, the increase in Helmholtz free energy. Because ε_r varies with temperature and we wish to avoid this complication, we assume that all processes are carried out reversibly and isothermally so that strictly $dW = dF$: we shall retain the symbol U_E instead of F since it is still electric in origin and plays the same part as the corresponding quantity in Chapter 5.

Finally, stresses in a dielectric caused by polarization will normally strain the material (electrostriction) and cause changes in ε_r: again we wish to avoid this and shall assume that all fluids are incompressible.

Work done in charging a capacitor and a conductor As in Sec. 5.5, the work done by an external agency in transferring a charge dq from one plate of a capacitor to the other is, in the limit,

$$dW = v\,dq \tag{11.24}$$

where v is the potential difference. The total work in charging to Q is

$$W = \int_0^Q v\,dq$$

and if all dielectrics are linear, $q = vC$ at all stages and

$$\text{capacitor or conductor:}\quad \boxed{U_E = W = \tfrac{1}{2}QV = \tfrac{1}{2}CV^2 = \tfrac{1}{2}Q^2/C} \tag{11.25}$$

as before, Q being the conduction charge. The same applies to an isolated conductor.

Energy of charged conductors in terms of Q and V (linear dielectrics) For a general collection of charges we cannot use quite the same argument as in Sec. 5.5 since Coulomb's law does not give the complete force on a conduction charge when dielectrics are present. Consider, however, three conduction charges Q_1, Q_2 and Q_3 and let the potential at the point Q_1 due to Q_2 alone be V_{12} with corresponding potentials for all the others. Any dielectrics are assumed to remain permanently in position. The work done in assembling Q_1 and Q_2 only is given either by Q_1V_{12} or by Q_2V_{21}, which must therefore be equal to each other and to $\frac{1}{2}Q_1V_{12}+\frac{1}{2}Q_2V_{21}$. In a similar way the addition of Q_3 will entail work $Q_3V_{31}+Q_3V_{32}$ which can also be written as four terms. The total work is thus

$$\begin{aligned} W &= \tfrac{1}{2}Q_1(V_{12}+V_{13})+\tfrac{1}{2}Q_2(V_{21}+V_{23})+\tfrac{1}{2}Q_3(V_{31}+V_{32}) \\ &= \tfrac{1}{2}Q_1V_1+\tfrac{1}{2}Q_2V_2+\tfrac{1}{2}Q_3V_3 \end{aligned}$$

where V_1 is the potential at Q_1 due to all the other charges, etc. This can be generalized for any number of charges to

set of charges: $$\boxed{U_E = W = \sum_i \tfrac{1}{2}Q_iV_i} \tag{11.26}$$

and, as in Sec. 5.5, to a collection of conductors:

set of conductors: $$U_E = W = \sum_A \tfrac{1}{2}Q_AV_A \tag{11.27}$$

Energy in terms of field quantities The argument of Sec. 5.6 can be taken over in the presence of LIH dielectrics in which the directions of **D** and **E** are the same at any point. Referring back to Fig. 5.15 we now apply the generalized Gauss's law to the volume bounded by dQ, dS and the lines of force and obtain $D\,dS=dQ$ so that

$$dU_E = \tfrac{1}{2}D\,dS\,V$$

The relation $dQ=D\,dS$ applies to all the equipotentials cut by the tube and thus

$$dU_E = \tfrac{1}{2}D\,dS\int_{S'}^{\infty} E\,dr = \int_{S'}^{\infty} \tfrac{1}{2}DE\,dS\,dr$$

Hence the work or free energy is

electric energy: $$\boxed{U_E = \int_\tau \tfrac{1}{2}\mathbf{D}\cdot\mathbf{E}\,d\tau = \int_\tau \tfrac{1}{2}\varepsilon_r\varepsilon_0E^2\,d\tau} \tag{11.28}$$

Further, Eq. (11.24) gave us the work done in increasing the amount of charge by dq at a potential v as $v\,dq$. This can be transformed in exactly the way U_E was, in obtaining (11.28), so that (11.24) can also be expressed as

$$\boxed{dW = \int_\tau \mathbf{E}\cdot d\mathbf{D}\,d\tau} \tag{11.29}$$

11.7 Forces and changes in energy

As long as there is no contact between conductors and dielectrics the expressions for forces on *conductors* obtained in Sec. 5.7 are still valid since the same arguments can be taken over. The forces on any *dielectrics* can also be obtained from the energy. For instance, at constant Q, if a dielectric has a force $\mathbf{F}$ acting on it internally and is kept in equilibrium by an external force $\mathbf{G}=-\mathbf{F}$, the same arguments can be used to obtain (5.22) and thus (5.23) for the dielectric in a capacitor:

$$\text{force on dielectric:} \qquad F_x=-\left(\frac{\partial U_\mathrm{E}}{\partial x}\right)_Q=\tfrac{1}{2}V^2\frac{\partial C}{\partial x} \tag{11.30}$$

and any external work done on the system results in an equal increment of the free energy.

At constant V, as before,

$$\text{force on dielectric:} \qquad F_x=+\left(\frac{\partial U_\mathrm{E}}{\partial x}\right)_V=\tfrac{1}{2}V^2\frac{\partial C}{\partial x} \tag{11.31}$$

and of the energy supplied by the battery, half goes in doing external work and half in the increase of free energy (see Problem 11.13).

Equation (11.30) shows that the force on a dielectric is always such as to try to decrease U_E when the conductors are isolated so that the conduction charges Q are constant. If, *in addition*, no redistribution of Q on the conductors occurs, then $\mathbf{D}$ will also be unchanged when the dielectric moves. In that case, we write U_E as $\int \frac{1}{2}D^2\,\mathrm{d}\tau/\varepsilon_\mathrm{r}\varepsilon_0$, from which the reader should be able to argue that U_E will be minimized by the motion of the dielectric into regions of greater $\mathbf{D}$ (since ε_r is always greater than 1). A particular case of this is the attraction between a point charge and a dielectric body first encountered in Sec. 1.1.

Force on the surface of a charged conductor If there is no contact between dielectric and conductor the analysis is the same as in Sec. 5.7 and the outward force is $\sigma_\mathrm{c}^2/2\varepsilon_0$ per unit area. When a solid dielectric touches a charged conducting surface the force is indeterminate because the form of contact is indefinite. Only for a fluid can a calculation be performed, and this case we now discuss.

If we use the energy method of Sec. 5.7, we again find that the force per unit area is equal to U_E outside the conducting surface, i.e.

$$\text{Outward pressure on the surface of a charged conductor in contact with a fluid dielectric}=\tfrac{1}{2}DE=\sigma_\mathrm{c}^2/2\varepsilon_\mathrm{r}\varepsilon_0 \tag{11.32}$$

since $E=\sigma_\mathrm{c}/\varepsilon_\mathrm{r}\varepsilon_0$ and $D=\sigma_\mathrm{c}$.

This result can be confirmed for a parallel plate capacitor by using (11.30). There is, on the other hand, the first method of Sec. 5.7: if there were a small gap left between the fluid dielectric and the conducting surface, that method would still yield an outward pressure of $\sigma_\mathrm{c}^2/2\varepsilon_0$. How is it that the mere contact of the fluid changes this? An analysis of the forces acting on a polarized dielectric reveals that there are two sources of hydrostatic pressure acting outwards from the dielectric surface which reduce the pressure on the conductor to (11.32). The first is due to a force on every element of volume $\mathrm{d}\tau$ given by $P\,\mathrm{d}\tau\,\mathrm{d}E/\mathrm{d}x$, using (3.41), where x is the direction of both $\mathbf{P}$ and $\mathbf{E}$: it can be written as $\frac{1}{2}\varepsilon_0\chi_e(\partial E^2/\partial x)$ per unit volume and it produces a hydrostatic pressure gradient of the same value (Fig. 11.8a). There is thus an excess pressure inside a polarized dielectric of $\frac{1}{2}\varepsilon_0\chi_e E^2$ compared with a region where $E=0$ (Fig. 11.8b). The second source of pressure is the unbalanced surface polarization charge of density σ_P and the force on this is obtained by a method similar to that used in Sec. 5.7 for the force on a charged conducting surface. Figure 5.17 is applicable except that the inside field is E_m and the outside field E_0. The integration of $\varepsilon_0 E'\,dE'$ is, however, between the limits 0 and $E_0 - E_m$ because a force due to E_m acts on the negative ends of the dipoles whose positive ends are in the surface layer. The outward pressure is thus $\frac{1}{2}\varepsilon_0(E_0-E_m)^2$ and, because $E_0=\varepsilon_r E_m$ and $\sigma_P = P = \varepsilon_0(\varepsilon_r - 1)E_m$, this becomes $P^2/2\varepsilon_0$. Since both forces oppose the original $\sigma^2/2\varepsilon_0$, the resultant pressure is

$$\frac{D^2}{2\varepsilon_0} - \frac{\varepsilon_0\chi_e E^2}{2} - \frac{P^2}{2\varepsilon_0}$$

using $D=\sigma_c$. This reduces to (11.32), surprisingly.

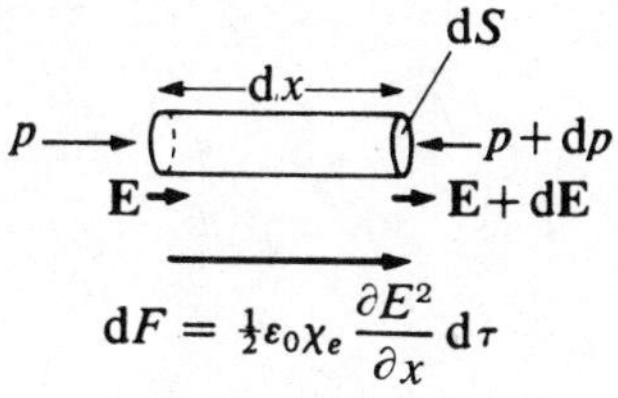

(a)

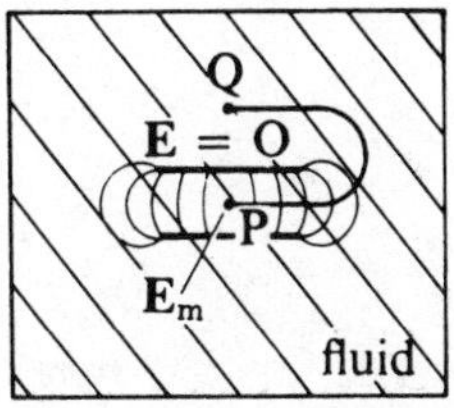

(b)

Figure 11.8 (a) Equilibrium of an element of material in a fluid dielectric: the pressure difference dp is dF/dS and the expression for dF with $d\tau = dx\,dS$ means that $dp/dx = \frac{1}{2}\varepsilon_0\chi_e(\partial E^2/\partial x)$; (b) integrating from P to Q gives an excess pressure $\frac{1}{2}\varepsilon_0\chi_e E^2$ between the plates.

11.8 Electric fields within charge distributions

In Chapter 4 we shelved the question of what was meant by the electric field strength $\mathbf{E}$ within a distribution of charge. The problem has arisen once more in dielectric theory where the definitions of $\mathbf{E}$ and $\mathbf{D}$ in (11.9) and (11.14), respectively, do not answer the question. We wish to know how, if at all, these fields are related to the force per unit charge inside a material.

It is important to realize that calculations of fields due to, and forces on, charge distributions are performed on a *model* and the results compared with experiment. Good agreement means that the model is a satisfactory one as far as those experiments go, while complete disagreement means that the model must be discarded (cf. the Thomson and Rutherford atoms in Sec. 4.2). It more often happens, however, that different models are required to account for different sets of experimental results or that a crude model is sufficient for one set but that a more detailed model must be used for another. Thus we found in Chapter 5 that our model of a charged conductor as a body carrying a continuous distribution of charge on its surface was quite adequate to account for all the results connected with capacitance, although we realized that other experiments require us to assume the existence of discrete charges.

When we are only interested in macroscopic or large-scale phenomena it is often possible to adopt a perfectly adequate model in which only continuous distributions of charge exist: the reason for this was discussed in Sec. 1.5. Only one thing about this need worry us: even if we have only continuous distributions and no point charges, is it not possible that $\mathbf{E}$ and V become infinite within such distributions? However, we have discussed this point to our satisfaction in Sec. 4.2.

The next question concerns the meaning of $\mathbf{E}$. We know that *outside* conductors and dielectrics, the models we have adopted of smoothed-out distributions are adequate because we can measure $\mathbf{E}$ and confirm our results. *Inside* conductors and insulators, however, we have an apparent contradiction: on the one hand we have that $\mathbf{E}$ in conductors carrying no current is zero and in

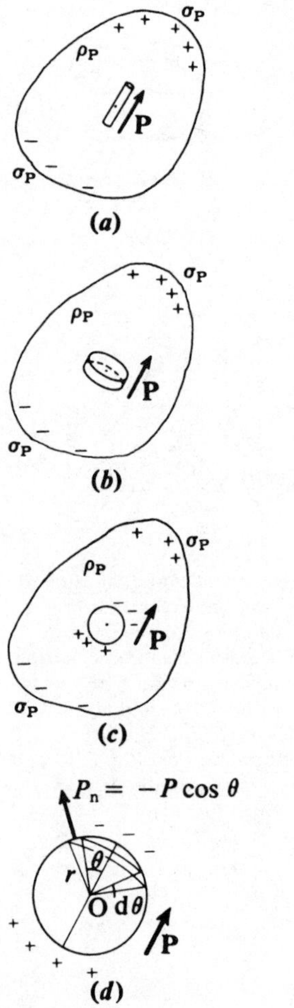

Figure 11.9 Cavities in dielectrics. The force per unit charge is given by (a) **E** in a needle-like cavity parallel to **P**, (b) $\mathbf{D}/\varepsilon_0$ in a disc-like cavity perpendicular to **P**, (c) $\mathbf{E}+\mathbf{P}/3\varepsilon_0$ in a spherical cavity. (d) The polarization charges on the inner surface of a spherical cavity.

dielectrics is the field due to external sources plus that due to the equivalent σ_P and ρ_P: these fields we shall call the **macroscopic** electric field. On the other hand, we know that matter is discrete and that **E** *defined as force per unit charge* within any material will fluctuate violently from point to point and from time to time: this we shall call the **microscopic** field, denoted by **f**.

We should expect the macroscopic field to be the space and time average of **f**, such as might be experienced by a charge moving rapidly through the material. Another way of imparting physical significance to the macroscopic field in a dielectric is to realize that, if we scoop out a cavity within the material which is macroscopically small but large enough for a point within it to be counted as *outside* the dielectric, we shall obtain a force per unit charge which does not fluctuate like **f**. However, the force now depends on the shape of the cavity since extra polarization charges will appear on its inner surface.

If such a cavity is needle-like with its axis in the direction of **P** then the polarization charges will appear at the ends and can be made negligibly small (Fig. 11.9a). The force per unit charge is then due only to σ_P, ρ_P and any external field, and is thus just **E**. A disc-like cavity with its plane perpendicular to **P** as in Fig. 11.9b carries a surface density of polarization charge equal to $\pm\mathbf{P}$ on its flat faces which will produce an extra force $\mathbf{P}/\varepsilon_0$ per unit charge by (4.12): the total force per unit charge in such a cavity is thus $\mathbf{E}+\mathbf{P}/\varepsilon_0$ or $\mathbf{D}/\varepsilon_0$.

In a spherical cavity of radius r, the polarization charges appearing on the inner surface are of surface density $-P_n$ or $-P\cos\theta$ (Fig. 11.9c and d). According to Problem 3.7, a ring of charge like the elementary strip in the figure produces an electric field at O of $-(Q\cos\theta)/(4\pi\varepsilon_0 r^2)$ where Q is the total charge—in this case $(-P\cos\theta\times 2\pi r\sin\theta)r\,d\theta$. Thus the field due to the elementary strip is $+P\cos^2\theta\sin\theta\,d\theta/2\varepsilon_0$ and due to the whole surface is

$$\int_{\theta=0}^{\theta=\pi}\frac{-P\cos^2\theta}{2\varepsilon_0}\,d(\cos\theta)=\frac{P}{3\varepsilon_0}$$

and the total force in the cavity is

$$\mathbf{E}+\frac{\mathbf{P}}{3\varepsilon_0}\text{ per unit charge}$$

We therefore have the following results:

Force per unit charge in needle-like cavity along **P**	$=\mathbf{E}$	
Force per unit charge in disc-like cavity perpendicular to **P**	$=\mathbf{D}/\varepsilon_0$	(11.33)
Force per unit charge in spherical cavity	$=\mathbf{E}+\mathbf{P}/3\varepsilon_0$	

which enable us to give some physical meaning to **E**- and **D**-fields within dielectrics.

If we enquire further as to the field experienced by an actual molecule within the dielectric we are asking for a microscopic theory, and this we consider below in Sec. 11.9.

The status of Coulomb's law in a dielectric Mention must be made of the 'law' $F=Q_1Q_2/(4\pi\varepsilon_r\varepsilon_0 r^2)$ purporting to give the force between point charges in a medium and sometimes used to define ε_r. This is a complex matter which can only be properly discussed from a more advanced viewpoint, but if we split it into its two components $E=Q/(4\pi\varepsilon_r\varepsilon_0 r^2)$ and $\mathbf{F}=Q\mathbf{E}$ we can see that it will be of restricted validity. The first component only applies in an infinite LIH medium (Secs 11.1 and 11.4), while the discussions in this section above and in the following section are relevant to the validity of $\mathbf{F}=Q\mathbf{E}$, for it depends on what we mean by Q. If we mean it to be the limiting case of a small charged conductor immersed in a fluid, we must take account of the extra forces giving rise to hydrostatic pressure, while if we mean it to be an elementary charge or atom, we must find **F** by methods to be discussed in the next section. It will be clear that $F=Q_1Q_2/(4\pi\varepsilon_r\varepsilon_0 r^2)$ can give little more than an order of magnitude in most cases and should never be used as a basic law, while the view that the force between charges is in some way modified by the medium can be misleading.

11.9 The approach to microscopic theory

We know from evidence outside the field of electricity that, on an atomic scale, insulators are built up from groups of atoms or ions forming what we shall call a 'molecule', although strictly this term should be kept for a more restricted group of materials. Thus, in an inert gas our molecule is a single atom, in a gas such as carbon dioxide it is a single CO_2 group, while in a solid such as potassium chloride we shall take it as a single KCl group even though the structure is an ionic framework in which molecules proper do not exist.

Two specifically microscopic quantities are introduced: firstly, the **local field** $\mathbf{E}_L$ experienced by a single molecule and, secondly, the **polarizability** α defined as the *mean electric dipole moment per molecule per unit field.* The electric dipole moment of a molecule in a polarized dielectric is thus $\alpha\mathbf{E}_L$ and because there are $N'_A\rho/M$ molecules per unit volume (Problem 1.7), it follows from (11.4) that

$$\mathbf{P}=\frac{N'_A\rho\alpha}{M}\mathbf{E}_L \tag{11.34}$$

which relates the microscopic quantities to a macroscopic **P**. Because measurements of ε_r are usually made, it is better to use (11.20) and obtain

$$\boxed{\varepsilon_r=1+\frac{N'_A\rho\alpha}{\varepsilon_0 M}\frac{\mathbf{E}_L}{\mathbf{E}}} \tag{11.35}$$

Dielectric theory is concerned with the value of $\mathbf{E}_L/\mathbf{E}$, because only then can α be calculated and information about the molecules be obtained.

The local field Only for very dilute gases should we expect $\mathbf{E}_L$ to be equal to **E**. One method of taking into account the molecules of the material is due to Lorentz: imagine a macroscopically small sphere described about the molecule in question so that all the material outside it gives the field $\mathbf{E}+\mathbf{P}/3\varepsilon_0$ and all the molecules inside it must be treated individually and their effect added to that field. Lorentz showed that for material with a cubic lattice, the contribution of the molecules inside the sphere would be zero and that it would be very small for fluids in which the random motion produced isotropy. For these materials (11.35) becomes

$$\varepsilon_r = 1 + \frac{N'_A \rho \alpha}{\varepsilon_0 M}\left(1 + \frac{\varepsilon_r - 1}{3}\right)$$

using $\mathbf{E}_L = \mathbf{E} + \mathbf{P}/3\varepsilon_0$ and $\mathbf{P} = \varepsilon_0(\varepsilon_r - 1)\mathbf{E}$. This gives

$$\alpha = \frac{3\varepsilon_0 M}{N'_A \rho} \frac{\varepsilon_r - 1}{\varepsilon_r + 2} \tag{11.36}$$

known as the **Clausius–Mossotti formula**. This approximation is found to be accurate enough in many cases for α to be calculated from ε_r.

The origin of the polarizability α There are three ways in which α may arise, two involving distortion of the molecules and one involving orientation. If the molecule has no permanent electric dipole moment of its own, it is said to be *non-polar*. However, the application of an electric field can still *induce* a dipole moment because the positive and negative charges move in opposite directions. Such a distortion may be due either to the relative motion of nuclei and extra-nuclear electrons (this is called **electronic polarization**) or to the relative motion of positive and negative ions as a whole in ionic solids (**ionic polarization**). These are illustrated in Fig. 11.10a and b, respectively.

Both of the above effects will also occur in *polar* substances, i.e. those in which the molecules have a permanent dipole moment. However, a further effect can now occur when an electric field is applied: the dipole moments will tend to turn into the direction of **E** and produce an overall polarization (**orientational polarization**). Unlike the other two types, this one is temperature-dependent since the randomizing effect of thermal agitation on the dipole directions will be greater at higher temperatures. Thus, if α varies with temperature, the structure is known to be polar (Fig. 11.10d) and although this information is qualitative it is nonetheless valuable.

Problems 11.14 and 11.15 show that the electronic polarizability of a single atom should be of the order of $3\varepsilon_0$ multiplied by the volume of the atom, and experimental measurements confirm this.

11.10 Dielectric properties

Measurement of ε_r and typical results Measurements of relative permittivity at lower frequencies use the a.c. bridge and resonance methods described in Sec.

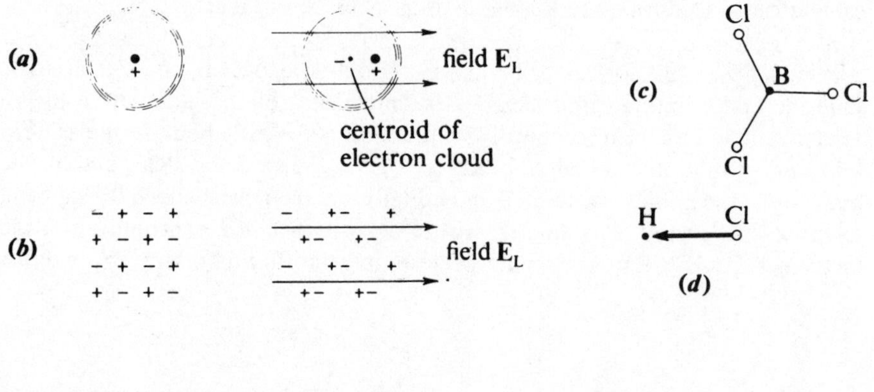

Figure 11.10 (a) Electronic polarization; (b) ionic polarization; (c) BCl_3, a non-polar molecule; (d) HCl, a polar molecule.

Table 11.1 Dielectric properties of some representative materials

Gases	Relative permittivity at STP			Dielectric strength
Hydrogen	1.00027			$2.0\,\mathrm{kV\,mm^{-1}}$
Dry air	1.00058			$3.0\,\mathrm{kV\,mm^{-1}}$
Carbon dioxide	1.00099			$2.9\,\mathrm{kV\,mm^{-1}}$
Liquids and solids† (20°C)	**Relative permittivity at low frequencies**	**Power factor**	**Conductivity ($\mathrm{S\,m^{-1}}$)**	**Dielectric strength ($\mathrm{kV\,mm^{-1}}$)**
Polythene	2.3	0.0002	10^{-11}	20
Paper	3.7	0.009	10^{-10}	16
Mica	6.0	0.0002	10^{-14}	100
Ethanol	26	Large	3×10^{-4}	—
Water	81	Large	2×10^{-4}	—
Barium titanate‡	4000	0.02	—	—

† Approximate values only—materials vary in properties from specimen to specimen in many cases.
‡ Anisotropic. The quoted ε_r is that for a single crystal along the *a* axis.

10.11 to obtain C_m/C_0 [Eq. (11.1)]. At microwave frequencies, waveguides and cavities can be used. At higher frequencies still, in the infrared and visible regions of the electromagnetic spectrum, the relation between refractive index and ε_r (Sec. 14.1) can be used.

Table 11.1 gives some values of ε_r for common materials at low frequencies together with some other relevant properties. Figure 11.11 illustrates the type of variation of ε_r with frequency encountered in various types of dielectric. The high value at low frequencies is typical of a polar material and is the result of orientational polarization. The rotation of the polar molecules into the direction of a static field is such that the polarization increases exponentially with time. This means that a relaxation time or time constant can be defined for the process just as in a *CR* circuit. In an oscillating field the molecules will thus never be quite in phase with the reversals in direction of the applied field and at high frequencies will not follow them at all. The orientation contribution thus disappears. The variations in the infrared and visible regions are due to resonant response of the vibrating ionic or atomic dipoles to the incoming radiation.

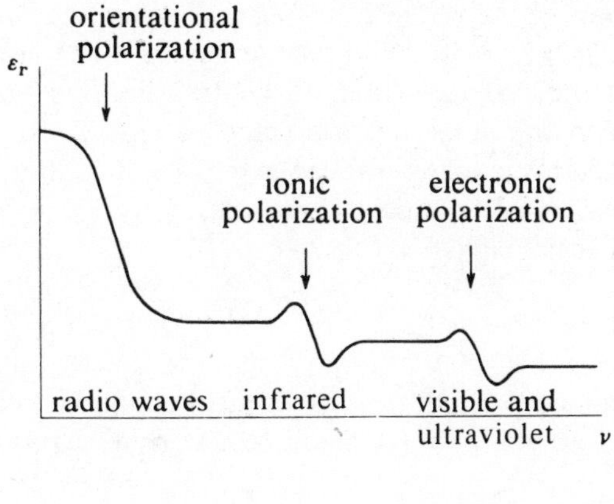

Figure 11.11 The variation of relative permittivity with frequency of the applied electric field.

Complex permittivity In a capacitor with an ideal dielectric, we have seen that **I** and **V** are in quadrature and no loss of energy occurs. From (10.30):

$$\mathbf{I} = j\omega\varepsilon_r C_0 \mathbf{V} \tag{11.37}$$

where C_0 is the vacuum capacitance. For a variety of reasons, a real dielectric passes a component of current in phase with **V** (e.g. due to conduction current or the dielectric relaxation time just mentioned), and a real term must be added to (11.37). This can be allowed for, whatever the cause, by using a complex relative permittivity ε_r equal to $\varepsilon_r' - j\varepsilon_r''$. Thus (11.37) becomes

$$\mathbf{I} = j\omega\varepsilon_r C_0 \mathbf{V} = j\omega\varepsilon_r' C_0 \mathbf{V} + \omega\varepsilon_r'' C\mathbf{V} \tag{11.38}$$

The ε_r'' component is responsible for the total loss of energy in the dielectric and $\tan\delta = \varepsilon_r''/\varepsilon_r'$ is a measure of this [see Eq. (10.90)].

Other dielectric properties The arrangements of atoms in the crystal structures of some materials give them dielectric properties which are unusual but in many cases have important applications. We can do no more here than give a brief description of these, and fuller accounts may be found in many solid-state texts (e.g. Kittel, 1986) or in books dealing with electrostatic applications (e.g. Cross, 1987).

1. *Ferroelectrics.* A small class of materials exhibit ferroelectricity, in which a graph of **D** against **E** is not a straight line but has the same general shape as the ferromagnetic hysteresis loops in the next chapter. Moreover, their relative permittivities are very high (see barium titanate in Table 11.1). If their hysteresis loop is very square they can be used as binary elements in memory devices.
2. *Electrets.* Ferroelectric materials and certain others are capable of remaining permanently polarized even when an applied electric field is removed to form electrets (the word is coined by analogy with 'magnet'). Electrets are used in electrostatic microphones.
3. *Piezoelectrics.* Some anisotropic materials polarize on the application of a mechanical stress, an effect known as piezoelectricity. The inverse effect also occurs: they are strained mechanically when polarized by an electric field (this is quite distinct from electrostriction—Sec. 11.6). Quartz is the most important example of a piezoelectric material and a suitably cut crystal will polarize along one direction if compressed along another at right angles. Mechanical vibrations can be converted into electric ones in this way by incorporating the quartz in a capacitor, and the mechanical resonance that occurs is so sharp that the system is used to control the frequency of electronic oscillators. There are numerous other applications.
4. *Pyroelectrics.* Some piezoelectric materials also exhibit polarization when their temperature changes, an effect known as pyroelectricity: the property has an obvious use in thermometry.

11.11 Conclusion

The presence of dielectric materials alters the general field equations through their effect on the **E**-field. However, by introducing a new field quantity, the electric displacement or **D**-field, the simple form of Gauss's law and the circuital law have been retained:

$$\oint_S \mathbf{D}\cdot d\mathbf{S} = \Sigma Q_c; \qquad \oint_L \mathbf{E}\cdot d\mathbf{L} = -\partial\Phi/\partial t$$

together with their differential forms:

$$\nabla\cdot\mathbf{D} = \rho_c; \qquad \nabla\times\mathbf{E} = -\partial\mathbf{B}/\partial t$$

The **B**-field equations are unchanged, but we now go on to see how they might be affected by the presence of magnetic materials.

11.12 Revision summary

General field equations By defining the polarization **P** of a dielectric as the electric moment per unit volume (**SI unit,** $\mathrm{C\,m^{-2}}$), we find that any piece of the material can be represented by Poisson's equivalent distribution. This is a set of surface and volume charges of densities σ_P and ρ_P given by

$$\sigma_P = P_n; \qquad \rho_P = -\nabla\cdot\mathbf{P} \qquad (11.6)\ (11.8)$$

which can replace the actual polarized dielectric. The **E**-field at any point is defined as the sum of any applied field and that due to σ_P and ρ_P. If we also use a **D**-field, with an **SI unit** also $\mathrm{C\,m^{-2}}$, defined by

$$\mathbf{D} = \varepsilon_0\mathbf{E} + \mathbf{P} \qquad (11.14)$$

then, in the absence of time variations, the circuital law for **E** and Gauss's law take the forms

$$\oint_L \mathbf{E}\cdot d\mathbf{L} = 0; \qquad \oint_S \mathbf{D}\cdot d\mathbf{S} = \Sigma Q_c \qquad (11.15)$$

where Q_c includes only conduction and convection charges. The differential form of these is

$$\nabla\times\mathbf{E} = 0; \qquad \nabla\cdot\mathbf{D} = \rho_c \qquad (11.16)$$

When time-varying **B**-fields exist, the circuital law becomes $\oint \mathbf{E}\cdot d\mathbf{L} = -\partial\Phi/\partial t$ with $\nabla\times\mathbf{E} = -\partial\mathbf{B}/\partial t$ as in Chapter 9.

From these laws, it follows that across any boundary between two media:

$$D_{\text{normal}} \text{ is discontinuous by } \sigma_c; \qquad E_{\text{tangential}} \text{ is continuous} \qquad (11.22)\ (11.23)$$

where σ_c is the surface density of any conduction charges along the boundary.

Properties of dielectrics The relationship between **P** and **E** at any point is expressed in terms of the electric susceptibility χ_e:

$$\mathbf{P} = \varepsilon_0\chi_e\mathbf{E} \qquad (11.10)$$

A relative permittivity is defined by

$$\varepsilon_r = 1 + \chi_e \qquad (11.18)$$

and this quantity then has all the properties of the ε_r defined by C_m/C_0 as in Eq. (11.1). Moreover there is then the relation

$$\mathbf{D}=\varepsilon_r\varepsilon_0\mathbf{E} \tag{11.19}$$

In LIH dielectrics only, χ_e and ε_r are scalar constants and no volume polarization charges ρ_P exist. It follows that an infinite LIH medium reduces the **E**-field and potential of specified charges to $1/\varepsilon_r$ of their vacuum value.

Many of the expressions giving forces and energy in Chapter 5 are unaffected, but a system of charged conductors in the presence of dielectrics has an energy that can be expressed in terms of

$$u_E=\tfrac{1}{2}\mathbf{D}\cdot\mathbf{E} \qquad \text{from (11.28)}$$

giving the energy density or energy per unit volume. It follows that the increment of energy when charges are changed by small amounts in a system can be expressed as

$$\mathrm{d}W=\mathrm{d}U_E=\int_\tau \mathbf{E}\cdot\mathrm{d}\mathbf{D}\,\mathrm{d}\tau \tag{11.29}$$

Commentary

C11.1 On Definitions of D and $\boldsymbol{\chi_e}$ The substitution $\varepsilon_0\rightarrow 1/4\pi$ used in Chapters 2–5 to convert formulae to the CGS e.s.u. form does not work for relations involving **D** and χ_e. The reason is that the definitions differ by a factor of 4π. Compare

SI definitions:	$\mathbf{D}=\varepsilon_0\mathbf{E}+\mathbf{P}$;	$\chi_e=\mathbf{P}/\varepsilon_0\mathbf{E}$
CGS definitions:	$\mathbf{D}=4\pi(\varepsilon_0\mathbf{E}+\mathbf{P})$;	$\chi_e=\mathbf{P}/4\pi\varepsilon_0\mathbf{E}$

It follows that the substitutions needed to convert all formulae in this chapter to CGS e.s.u. are

$$\varepsilon_0\rightarrow 1/4\pi; \qquad \mathbf{D}\rightarrow\mathbf{D}/4\pi; \qquad \chi_e\rightarrow 4\pi\chi_e \tag{11.39}$$

so that, for instance, $\mathbf{D}=\mathbf{E}+4\pi\mathbf{P}$, $\varepsilon_r=1+4\pi\chi_e$, $\mathbf{P}=\chi_e\mathbf{E}$.

Problems

Section 11.1

11.1 Show that a surface density of charge given by $\varepsilon_0(\varepsilon_r-1)E_m$ is alone sufficient to account for the appearance of ε_r in the denominators of Eqs (11.3). E_m is the E-field in the medium.

Section 11.2

11.2 An LIH dielectric sphere has a uniform polarization P. Find an expression for the surface density of polarization charge σ_P at any point on the surface. Show that the 'depolarizing' field produced at the centre of the sphere by σ_P is $P/3\varepsilon_0$.

11.3* If you are familiar with vector field theory, derive the Poisson distribution as follows. Write the potential V at a point due to a polarized dielectric as the volume integral over the dielectric of elementary dipole potentials due to each $\mathbf{P}\,\mathrm{d}\tau$. Express the potential due to a dipole (3.38) as $-(1/4\pi\varepsilon_0)\mathbf{p}\cdot\nabla(1/r)$ and transform it into two terms using general vector relation (B.20) of Appendix B.11. Transform one term, after integration, from a volume to a surface integral using Gauss's divergence theorem. Interpret the results using Eqs (3.17).

Section 11.4

11.4 A 100 pF parallel plate capacitor is charged to a potential difference of

50 V. The space between the plates is then completely filled with an insulating liquid of $\varepsilon_r = 3$, a 50 V battery being permanently connected across the plates. Find the new charge. If the filling had been carried out with the plates isolated from the battery, what would the new charge and potential difference be?

11.5 The capacitor of Problem 11.4 is connected in parallel with one of 200 pF, the combination is charged to a potential difference of 50 V and the battery is removed. If the first capacitor is now filled with the insulating liquid, what are the final charges and potential differences?

Section 11.5

11.6 A rectangular box has two opposite faces of conducting material so that it forms a parallel plate capacitor. If it is one-third filled with an insulating liquid of $\varepsilon_r = 4$, find the ratio of the capacitance when the conducting faces are horizontal to that when they are vertical. (Neglect edge effects.)

11.7 A long cylindrical capacitor with radii a and $4a$ has its inner conductor covered with a cylindrical sleeve of dielectric with $\varepsilon_r = 2$ and whose outer radius is $2a$. Find the capacitance per unit length.

11.8 If the total plate area of the capacitor in Fig. 11.7 is A, show that the effect of inserting the dielectric to cover an area A_1 has the same effect on the capacitance as extending the area *in vacuo* by $(\varepsilon_r - 1)A_1$.

11.9 Show that the lines of **E** and of **D** in crossing a plane boundary from one LIH dielectric of relative permittivity ε_1 to a second with ε_2 are refracted; and that the law governing the refraction is $\varepsilon_1/\varepsilon_2 = (\tan\theta_1)/\tan\theta_2$, where θ_1 and θ_2 are the angles between the lines and the normal to the boundary.

11.10 Assuming that the condition (11.23) applies in current-carrying conductors, find how the lines of flow of a current bend in crossing the boundary between conductors of conductivity σ_1 and σ_2. Show that for a metal in contact with an electrolyte, the **E**-field in the electrolyte is likely to be normal to the boundary.

Section 11.6

11.11 The plates of a parallel plate capacitor are of area A, distance apart x and are at a potential difference V. A slab of dielectric of uniform thickness t and relative permittivity ε_r is inserted between the plates. Find the change in electric energy of the capacitor if the plates are (a) isolated and (b) maintained at their initial potential difference by a battery.

11.12 A long cylindrical capacitor has radii a and b ($b > a$) and the space between the cylinders is filled with a dielectric of relative permittivity ε_r and dielectric strength K. What is the maximum energy which can be stored per unit length of the capacitor?

Section 11.7

11.13 A slab of dielectric is of such shape and size as to fill the space between the rectangular plates of a parallel plate capacitor except for negligible air gaps. The slab is withdrawn in a direction parallel to one side of the plates whose length is a. Find the force on the dielectric when only part of it is still between the plates, if the potential difference is maintained at V throughout and if the vacuum capacitance is C.

Section 11.9

11.14 Orders of magnitude of atomic polarizability can be obtained from simple models as in this problem and the next. Assume that an atom consists of a nucleus with charge $+Q$ and an electron with charge $-Q$ orbiting in a circle of radius a centre at the nucleus. Show that if an electric field **E** is applied in a direction perpendicular to the plane of the orbit and displaces the nucleus by an amount small compared with a, the induced dipole moment is $4\pi\varepsilon_0 a^3 E$ and the polarizability $4\pi\varepsilon_0 a^3$.

11.15 If the atomic model is taken instead as a nucleus of charge $+Q$ at the centre of a spherical volume of radius a carrying the total electronic charge $-Q$, show that the atomic polarizability is again $4\pi\varepsilon_0 a^3$.

Section 11.10

11.16 A parallel plate capacitor is filled with an imperfect dielectric of relative permittivity ε_r and conductivity σ. What is the time constant of any discharge which takes place? Estimate its value for a poor conductor from the data in Table 11.1.

11.17 An electret is in the form of a thin sheet with **P** perpendicular to the faces. Show that if edge effects are neglected, the **D**-field is everywhere zero.

12

Magnetic materials

The effects of material media on the magnetic laws of Chapters 7, 8 and 9 bear some resemblance to the effects of dielectrics described in the last chapter and many similarities of treatment will occur. Thus we begin by seeing in Sec. 12.1 how far a description in terms of a relative permeability only will take us. We find that magnetization, like polarization, is a more fruitful concept and leads to the field theory of Secs 12.2–12.5. There are, however, two complicating factors that make the treatment differ from that of dielectrics. One is the existence of two possible models for the elementary dipoles constituting a magnetic material, one based on the current loop and the other on the monopole. We prefer the former, but discuss the use of the monopole in Sec. 12.8. The other complicating factor is the existence of the small but technically important class of ferromagnetic materials with their permanent magnetism and their hysteresis. These require separate treatment in Secs 12.6 and 12.7. We end with three sections on models of magnetic materials and on measured properties, relating the continuum approach of field theory to atomic aspects.

We shall find it useful in this chapter to make greater use of a terminology previously introduced only in passing (in the revision summaries of Secs 4.10

and 8.9). The two important properties of vector fields are the flux over a closed surface and circulation round a closed path. The enclosed volumes and surfaces can be shrunk so that the flux per unit volume and circulation per unit area reduce respectively to the divergence and curl at a point. Anything that gives rise to divergence at a point is said to be a **source** of the field, e.g. charges are sources of **E** (negative charges are often called **sinks**), while the **B**-field has no sources. Anything that gives rise to curl at a point is said to be a **vortex** of the field. Thus, electric currents are vortices of **B** while electrostatic **E**-fields have no vortices since **curl** $\mathbf{E}_Q$ is always zero.

12.1 Relative permeability of isotropic, homogeneous media

The self-inductance L of a circuit was shown in Chapter 9 to depend only on geometry provided the circuit was *in vacuo*. As soon as any material media are present, however, we find experimentally that L changes even if the geometry stays the same. The *relative permeability* μ_r of a medium is defined by

$$L_m/L_0 = \mu_r \qquad \text{(definition of mean } \mu_r\text{)} \tag{12.1}$$

where L_0 is the self-inductance of a circuit *in vacuo* and L_m is that of the same circuit with the medium filling the whole of space in which a magnetic field exists: effectively an infinite medium. The μ_r defined like this is clearly an average in some sense over the whole volume of material and we shall eventually need a definition giving μ_r at any point.

Because the magnetic flux $\Phi = LI$, it follows that if the same current I in a circuit gives a self-flux Φ_0 *in vacuo* and Φ_m in an infinite medium,

$$\Phi_m/\Phi_0 = \mu_r \tag{12.2}$$

as well. The same relation clearly applies to the mean **B**-field $\langle B \rangle$ across any circuit:

$$\langle B \rangle_m / \langle B \rangle_0 = \mu_r \tag{12.3}$$

and both (12.2) and (12.3) are alternative definitions of the mean μ_r.

Measurements of μ_r by methods to be described later show that all substances are magnetic and that the following classes of material exist:

1. *Diamagnetic*, with $(\mu_r - 1)$ small, negative, independent of **B** (linear) and independent of temperature.
2. *Paramagnetic*, with $(\mu_r - 1)$ small, positive, linear and decreasing with increasing temperature.
3. *Ferromagnetic*, with μ_r large and positive, non-linear and dependent on previous history (showing hysteresis, Sec. 12.6). These materials are metallic and become paramagnetic above a temperature known as the **Curie point**. We distinguish between ideally *soft* ferromagnetics, which retain no magnetism when the external field is removed, and *hard* ferromagnetics, which may form permanent magnets.
4. *Antiferromagnetic*, with $(\mu_r - 1)$ small and positive, but showing non-linearity and hysteresis effects, and having a temperature similar to the Curie point known as the **Néel point**.

5. *Ferrimagnetic*, with μ_r as for a ferromagnetic but non-metallic. A class consisting of mixed oxides of iron known as *ferrites*.

Single crystals may be anisotropic, with different μ_r's in different directions, and non-homogeneity may also occur, but *we shall restrict the treatment entirely to isotropic homogeneous (IH) materials.*

Effect of μ_r on magnetic B-field The definitions of μ_r tell us that when the vacuum round a circuit in which a current I is maintained constant is completely filled with a medium, the inductance, flux and mean **B**-field are all increased by μ_r. It does not follow, however, that the **B**-field *at any point* due to these currents will also change by the same factor. Nevertheless we make an assumption, justified in Sec. 12.3, that this is indeed the case for IH media. This would then mean that all the expressions for **B** in Chapters 7 and 8 due to specified currents I must be modified by the addition of a μ_r in the numerator:

$$\left.\begin{aligned} &\text{magnetic dipole:} && B_r = \frac{\mu_r \mu_0 2IA\cos\theta}{4\pi r^3} \\ &\text{infinite solenoid:} && B = \mu_r \mu_0 n I \\ &\text{current element:} && \mathrm{d}\mathbf{B} = \frac{\mu_r \mu_0 I\,\mathrm{d}\mathbf{l}\times\mathbf{r}}{4\pi r^3} \end{aligned}\right\} \qquad (12.4)$$

etc. The effect on the monopole formulae is discussed in Sec. 12.8. While Eqs (12.4) give **B** measured as a flux density, they do not necessarily give the force per unit current element or the couple per unit dipole within the medium. This will be discussed in Sec. 12.9.

Some explanation of the increase in the **B**-field is given by the magnetization of the material. It is as if the effective currents were increased from I to $\mu_r I$ when an infinite medium is introduced. If we use the parallel case of the polarized dielectric to guide us, we see that currents of $(\mu_r - 1)I$ flowing in the surface of the media adjacent to the conduction currents I would have the desired effect. These current, known as **Amperian currents**, play a similar role to the polarization charges in a dielectric. We distinguish them from conduction currents I_c by using the subscript M: I_M.

We shall justify this model in later sections, but Fig. 12.1 illustrates the idea for an infinite solenoid. Here there is a current per unit length of solenoid, J_{sc} (see Sec. 1.7 for surface current density J_s in general) and the **B**-field *in vacuo* is thus $\mu_0 J_{sc}$, as in Fig. 12.1a, since $J_{sc} = nI_c$. With the medium filling the solenoid, the **B**-field becomes $\mu_r \mu_0 J_{sc}$ as in Fig. 12.1b. In Fig. 12.1c, we show that Amperian currents flowing in the surface of the medium would have the same effect if their surface density were J_{sM} given by

$$J_{sM} = (\mu_r - 1) J_{sc} = \frac{(\mu_r - 1) B_m}{\mu_r \mu_0} \qquad (12.5)$$

where B_m is the **B**-field in the medium.

Practical application *the displacement transducer* Figure 12.1d represents a transformer with a single primary winding and two secondary windings, the latter so connected that the induced electromotances oppose each other. If they are

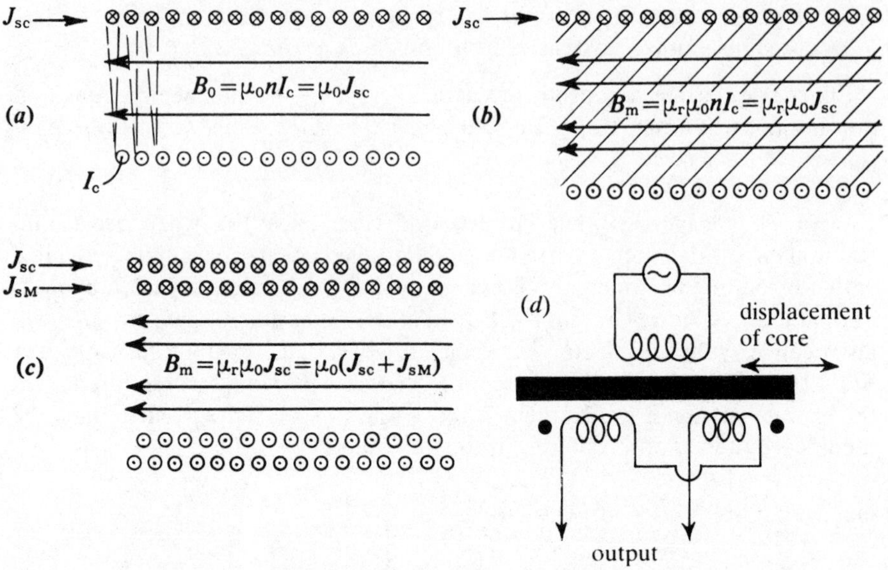

Figure 12.1 (a) An ideal current-carrying solenoid *in vacuo*; (b) an ideal solenoid with magnetic core; (c) surface currents on the core of density J_{sM} producing the same **B**-field as in (b); (d) principle of a displacement transducer.

completely symmetrical, the output in the absence of a core will be zero. It will still be zero if an iron core in the shape of a rod is situated symmetrically with respect to the windings as shown. If, however, the rod is displaced parallel to its axis in either direction, the two mutual inductances are no longer equal and an output is produced whose magnitude is related to the distance moved by the rod. This forms a **transducer**, the name given to a device that converts one physical magnitude into another, more easily measurable, magnitude, usually electrical. The phase of the output changes as the rod moves through the central position, so that a detector sensitive to phase will also yield the *direction* of any displacement. The main advantage of such a device is the absence of any electrical contacts on the movable element.

12.2 Magnetization and B-field in a medium

Magnetization: concept and definition We now look for a model of a magnetic material which will enable us to explain its macroscopic behaviour. Here we are guided by elementary experiments with pieces of soft iron which behave as magnets when placed in a magnetic field and thus can be considered (like a polarized dielectric) as a collection of elementary dipoles. Each small volume $d\tau$ of a magnetized material will possess a magnetic dipole moment $\mathbf{dm}$. If the *magnetization* $\mathbf{M}$ is defined as the *magnetic dipole moment per unit volume*, then

$$\text{Magnetic moment of volume } d\tau,\ \mathbf{dm}=\mathbf{M}\,d\tau \quad \text{(definition of } \mathbf{M}) \tag{12.6}$$

The **SI unit** of **M** is the $\mathrm{A\,m^{-1}}$.

The polarization **P** was interpreted as arising from small displacements of the bound charges whose existence we inferred in dielectrics. We have two alternatives for the origin of **M**: one that it is due to small displacements of monopoles (cf. **P**), the other that it arises from the alignment of current loops. No distinction between the two arises if points outside the medium are considered since we have seen in Chapter 7 that the **B**-field at some distance from them is identical. Figure 12.2 shows, however, that the flux of the permanent dipole is in the opposite direction to its moment while that of the current loop is in the same direction, so that the explanation of the observed flux changes on magnetization offered by the two models is bound to differ. We prefer the current-loop model because there is evidence that magnetism is associated with angular momentum which must arise with any current and because it accounts naturally for the observed changes in **B**. We shall see that the monopole model is the more natural when changes in the magnetic field strength **H** are to be explained.

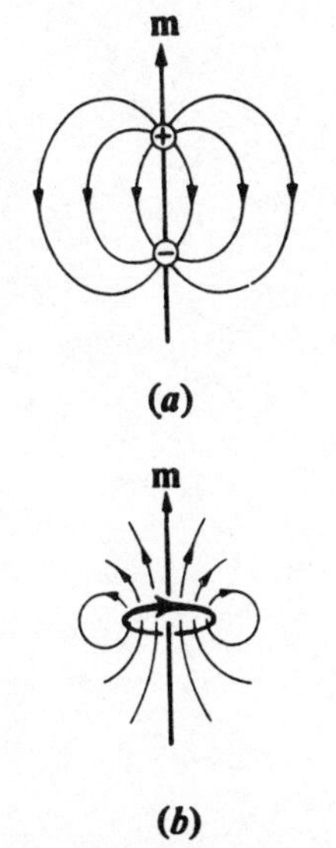

Figure 12.2 Contrast between the internal magnetic fields of (a) a two-pole dipole and (b) a current-loop dipole.

Surface and volume Amperian currents Consider an element of material in isolation with its length $\mathbf{d}l$ in the same direction as **M** and a cross-section $\mathrm{d}S$ as in Fig. 12.3a. From the definition of **M**, the magnetic dipole moment of this element is $\mathbf{M}\,\mathrm{d}l\,\mathrm{d}S$. Now suppose that the equivalent set of Amperian currents has a surface current density J_s and thus a total current $J_s\,\mathrm{d}l$. These would form a current-loop dipole of moment $J_s\,\mathbf{d}l\,\mathrm{d}S$. These two expressions for the moment must be equal so, *for an element in the direction of* **M**,

$$\text{Surface density of Amperian currents} = M \qquad (12.7)$$

Should **M** be uniform throughout a large block of material, then the J_s on the surface of one element will be equal and opposite to those on the adjacent elements and thus will produce no magnetic effect. Only the J_s on the outer surface of the block will remain unbalanced and these will have a value given by (12.7) provided **M** is parallel to that surface or, put another way, provided **M** is perpendicular to the normal to the surface denoted by the unit vector $\hat{\mathbf{n}}$. However, for outer surfaces, **M** and $\hat{\mathbf{n}}$ will not in general be perpendicular but will be at some angle θ to each other as in Fig. 12.3b. The figure shows that in that case the surface current density decreases to $J_s \sin\theta$ or, since $J_s = M$, to $M\sin\theta$. We denote the equivalent Amperian surface current density on external surfaces by J_{sM}, so that we have

$$\boxed{\text{Surface current density of Amperian currents, } J_{sM} = M\sin\theta = |\mathbf{M}\times\hat{\mathbf{n}}|} \qquad (12.8)$$

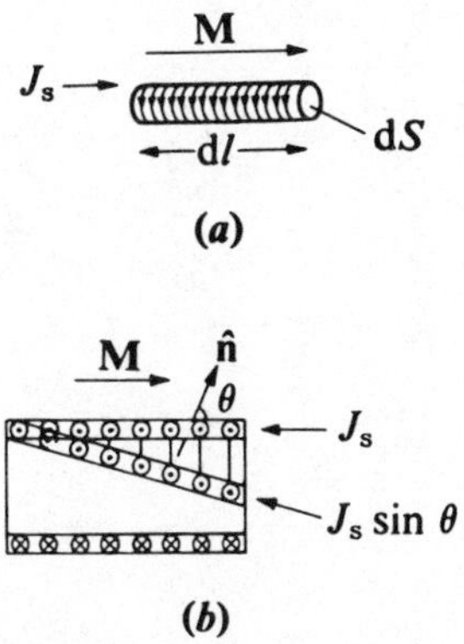

Figure 12.3 (a) An element of magnetized material and its equivalent Amperian surface current density. (b) If the outer surface of an element is not parallel to **M**, the total current remains the same but is spread over a longer distance: the surface current density therefore decreases to $J_s \sin\theta$.

If **M** is not uniform, the J_s from one element will not balance that from an adjacent one and a volume density of Amperian currents $\mathbf{J}_M$ may also occur. The form of $\mathbf{J}_M$ is not necessary for our treatments since we shall show that it is zero within LIH media carrying no conduction currents. (Its value could be obtained by taking a closed path within a magnetic material. This path will link a total Amperian current $I_M = \oint \mathbf{M}\cdot\mathrm{d}\mathbf{L}$ by a similar argument to that in the next section. Shrinking this path to a point leads to the equivalent differential form $\mathbf{J}_M = \nabla\times\mathbf{M}$.) The J_{sM} and $\mathbf{J}_M$ play the same part as Poisson's equivalent distribution for dielectrics.

B-field in the magnetic material Suppose now that a field $\mathbf{B}_0$ is maintained over a region of free space by steady external conduction currents. Let a piece of magnetic material be introduced into this space so that it becomes magnetized. The Amperian currents produce an additional field $\mathbf{B}_M$ outside the material as if it arises from the J_{sM} and $\mathbf{J}_M$ above. We now **define** the total **B**-field everywhere, including the interior of the medium, as

$$\boxed{\mathbf{B} = \mathbf{B}_0 + \mathbf{B}_M} \tag{12.9}$$

so that it has the same properties as the **B**-fields of conduction currents alone.

12.3 Ampère's circuital law, H-field, and susceptibility

Ampère's circuital law with magnetic materials In Chapter 8 we saw that the properties of **B**-fields due to currents could be summarized by the circuital law

$$\oint_L \mathbf{B} \cdot d\mathbf{L} = \mu_0 I \tag{12.10}$$

and Gauss's law

$$\oint_S \mathbf{B} \cdot d\mathbf{S} = 0 \tag{12.11}$$

These will be unaffected by the presence of magnetic materials if the definition (12.9) is used, except that the I in (12.10) must now include both conduction currents I_c and equivalent Amperian currents I_M:

$$\oint_L \mathbf{B} \cdot d\mathbf{L} = \mu_0 I_c + \mu_0 I_M \tag{12.12}$$

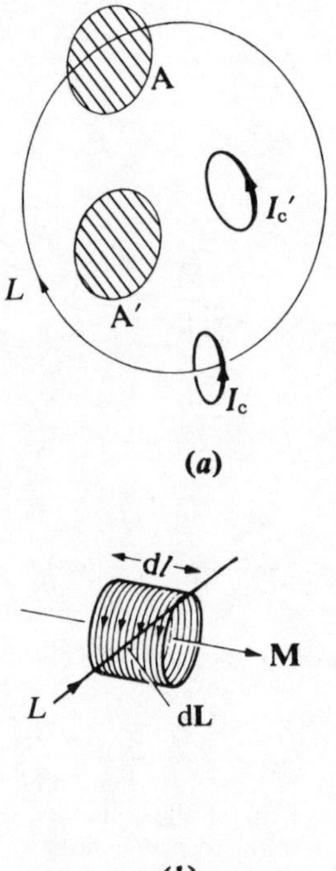

Figure 12.4 (a) Derivation of the generalized circuital law for **B**; (b) an element of path **dL** intersecting Amperian currents of amount $\mathbf{M} \cdot d\mathbf{L}$.

Apply this to the typical path L shown in Fig. 12.4a which links conduction currents and passes through pieces of magnetized material like A. Contributions to the right-hand side of (12.12) are only made by currents linked by L so that I_c' and A' will not contribute. The contribution of A to I_M can be obtained by considering Fig. 12.4b in which **dL** is an element of the path L at a region of A where the magnetization **M** makes an angle α with **dL**. The Amperian currents linked by this part of L are those in the element of material shown, amounting to $J_s\, dl$ or $J_s\, dL \cos\alpha$. Since $J_s = M$, by (12.7)

$$dI_M = J_s\, dl = M\, dL \cos\alpha = \mathbf{M} \cdot d\mathbf{L}$$

and I_M of (12.12) is therefore the line integral of **M** round C. Hence, with a slight rearrangement,

$$\oint_L \left(\frac{\mathbf{B}}{\mu_0} - \mathbf{M} \right) \cdot d\mathbf{L} = I_c \tag{12.13}$$

and this is the new form taken by Ampère's circuital law when magnetic materials are present.

The magnetic field strength or H-field The quantity $\mathbf{B}/\mu_0 - \mathbf{M}$ occurs frequently in problems connected with magnetic materials. It is often called the magnetic field strength and given a special symbol **H**, although we shall refer to it simply as the **H**-field:

$$\mathbf{H} = \frac{\mathbf{B}}{\mu_0} - \mathbf{M} \qquad \text{(definition of } \mathbf{H}\text{)} \tag{12.14}$$

Its **SI unit** is the A m^{-1}. Equation (12.13) now becomes

circuital law for **H**:
$$\oint_L \mathbf{H} \cdot d\mathbf{L} = I_c \tag{12.15}$$

In Eq. (8.1) the magnetomotance was defined *in vacuo* as the line integral round a closed path of $\mathbf{B}/\mu_0$. We now generalize this to

$$\text{Magnetomotance } \mathscr{H} = \oint_L \mathbf{H} \cdot d\mathbf{L} \qquad \text{(definition of } \mathscr{H}\text{)} \tag{12.16}$$

which includes the previous definition as a special case and allows us to quote the general circuital law (12.15) as 'the magnetomotance round a closed path is equal to the total conduction current linking the path'.

The general laws are now (12.11) and (12.15). The former tells us that there are still no *sources* of **B**. Equations (12.12) and (12.15) show that *all* currents are *vortices* of **B** but only conduction currents are vortices of **H**. This is brought out by the differential forms of the general laws, which take the form

$$\text{div}\,\mathbf{B} \equiv \nabla \cdot \mathbf{B} = 0; \qquad \textbf{curl}\,\mathbf{H} \equiv \nabla \times \mathbf{H} = \mathbf{J}_c \tag{12.17}$$

where $\mathbf{J}_c$ is the volume current density of conduction or convection charges.

We ask finally, what of the *sources* of **H**? If **B** in Gauss's law (12.11) is replaced by $\mu_0(\mathbf{H}+\mathbf{M})$, we find that

$$\oint_S \mathbf{H} \cdot d\mathbf{S} = -\oint_S \mathbf{M} \cdot d\mathbf{S} \tag{12.18}$$

or, in differential form, $\nabla \cdot \mathbf{H} = -\nabla \cdot \mathbf{M}$. These mean that the sources of **H** are the sinks of **M**, i.e. the disappearance of **M** gives rise to a flux of **H**. We shall see in Secs 12.4 and 12.8 that these sources can be identified with poles.

Magnetic susceptibility and permeability The magnetic susceptibility χ_m at a point is defined by

$$\mathbf{M} = \chi_m \mathbf{H} \qquad \text{(definition of } \chi_m\text{)} \tag{12.19}$$

and is thus dimensionless (but see Comment C12.3). For an IH medium, χ_m is independent of position and of the direction of **H** and is thus a scalar. Only for linear media is it independent of the magnitude of **H**.

If we consider infinite LIH media for the moment and make the substitution of $\mathbf{M} = \chi_m \mathbf{H}$ into (12.18), we see that there are no sources of **H** or **M** anywhere since χ_m can be taken outside the integral on the right and hence $\oint \mathbf{H} \cdot d\mathbf{S} = 0$. It then follows that **H** can only arise from its vortices, i.e. the conduction currents according to (12.15). It is clear from this that the **H**-field due to distributions of conduction current considered in Chapters 7 and 8 will be given by the same expressions as for **B** but divided by μ_0. Thus

$$\left.\begin{array}{ll}\text{magnetic dipole:} & H_r=\dfrac{2IA\cos\theta}{4\pi r^2}\\[2ex] \text{toroid and infinite solenoid:} & H=nI\\[2ex] \text{current element:} & \mathrm{d}\mathbf{H}=\dfrac{I\,\mathrm{d}\mathbf{l}\times\mathbf{r}}{4\pi r^3}\end{array}\right\}\qquad(12.20)$$

and *these will be unaffected by the presence of an infinite LIH medium.* On the other hand, we saw in Sec. 12.1 that the corresponding expressions for **B** increased by μ_r times. We use this to seek a relation between χ_m and μ_r.

If we put $\mathbf{M}=\chi_m\mathbf{H}$ into (12.14) written as $\mathbf{B}=\mu_0(\mathbf{H}+\mathbf{M})$, we obtain

$$\mathbf{B}=\mu_0(1+\chi_m)\mathbf{H}$$

From the experience with infinite LIH media, this suggests that we take as a general definition of μ_r at a point:

$$\boxed{\mu_r=1+\chi_m \qquad \text{(definition of } \mu_r\text{)}} \qquad (12.21)$$

which includes the definition of Sec. 12.1 as a special case. We see that we expect diamagnetic materials to have small negative susceptibilities and paramagnetic materials small positive ones (see Table 12.2, page 326). It also follows from (12.21) that

$$\boxed{\mathbf{B}=\mu_r\mu_0\mathbf{H}} \qquad (12.22)$$

Although these are general relations, χ_m and μ_r are only simple scalar constants in LIH media. Moreover, it cannot be too strongly emphasized that (12.22) does not imply that **H** is completely independent of the presence of media and that **B** increases μ_r times: that is only true for infinite LIH media, as we have shown. Finite media are considered in the next section.

Finally, we should look again at (12.5) in Sec. 12.1 and justify it in the new context. In an LIH medium carrying no conduction current, $\oint \mathbf{H}\cdot\mathrm{d}\mathbf{L}=0$, and because $\mathbf{M}=\chi_m\mathbf{H}$, $\oint \mathbf{M}\cdot\mathrm{d}\mathbf{L}=0$ as well. By our argument in Sec. 12.2, this means that there are no volume Amperian currents but only surface ones. In the special case of (12.5), the J_{sM} ought to be simply given by M as in (12.7). Now $M=B/\mu_0-H=B(1-1/\mu_r)/\mu_0$ using (12.22). This is the same as (12.5).

12.4 Boundaries and finite media

Values of B and H across a boundary At any boundary we shall have in general both a conduction surface current J_{sc} and an Amperian surface current J_{sM}. Applying Gauss's law $\oint \mathbf{B}\cdot\mathrm{d}\mathbf{S}=0$ to the cylinder of Fig. 12.5a as we did for Eq. (11.22), we obtain the rule that

$$\boxed{B_{1n}-B_{2n}=\Delta B_n=0} \qquad (12.23)$$

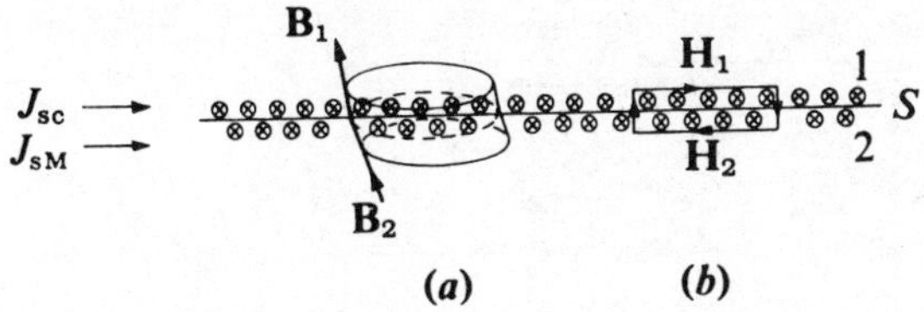

Figure 12.5 Derivation of boundary conditions for (a) the normal component of **B**, (b) the tangential component of **H**.

where B_n is the component of **B** normal to S. Thus *the normal component of* **B** *is continuous across any surface.*

If the generalized circuital law $\oint \mathbf{H} \cdot d\mathbf{L} = I_c$ is applied to the path in Fig. 12.5b as we did for Eq. (11.23), we obtain

$$\boxed{H_{1t} - H_{2t} = \Delta H_t = J_{sc}} \tag{12.24}$$

where H_t is the component of **H** tangential to S. Thus *the tangential component of* **H** *is discontinuous by* J_{sc} *across any surface.*

The effect of finite magnetic media When a finite piece of magnetic material is placed in the field of conduction currents, it may not fill the whole of the region in which there is a **B**-field. The first task in that case is to find whether there are any sources of **H**, and this depends on the relative directions of the **H**-field and the boundaries of the medium.

When the boundaries are **parallel** to the direction of **H**, the fact that there are no surface conduction currents means that **H** is continuous across the boundary, from (12.24) with $J_{sc} = 0$. The absence of a discontinuity of **H** means that there are no sources of **H** along the boundary. The **H**-field is therefore determined entirely by its vortices according to the circuital theorem and **B** is given by $\mu_r \mu_0 \mathbf{H}$. Example 1 below illustrates such a case.

When the boundaries are **perpendicular** to **B**, (12.23) shows that **B** is continuous across them, so that **H** must change abruptly by a factor $1/\mu_r$. There are therefore sources of **H** at such boundaries, as is otherwise clear because lines of **M** must begin and end at them to form sources and sinks (see Example 2 below). We shall see that for permanent magnets it is sometimes useful to interpret these sources as poles, but for para-, dia- and soft ferromagnetic materials even apparently simple examples are difficult to solve.

Worked example 1 *boundaries parallel to lines of* **H** Consider the ideal solenoid in Fig. 12.6 through the centre of which a rod of cross-section a ($< A$) and infinite length is inserted. The **H**-field is nI everywhere inside the solenoid so that in the material $B = \mu_r \mu_0 nI$ while in the space between the material and the solenoid $B = \mu_0 nI$. The flux linkage per turn is thus $\mu_r \mu_0 nIa + \mu_0 nI(A - a)$. The self-inductance is then easily seen to be $\mu_0 n^2[A + a(\mu_r - 1)]$ per unit length.

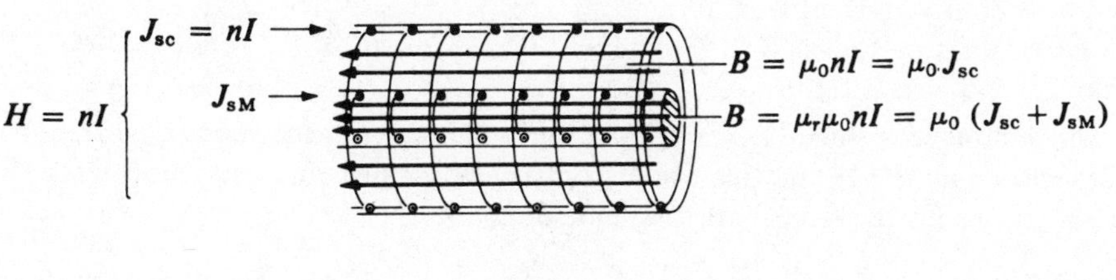

Figure 12.6 Example of a finite medium with boundaries parallel to **H**: an ideal current-carrying solenoid with a magnetic core.

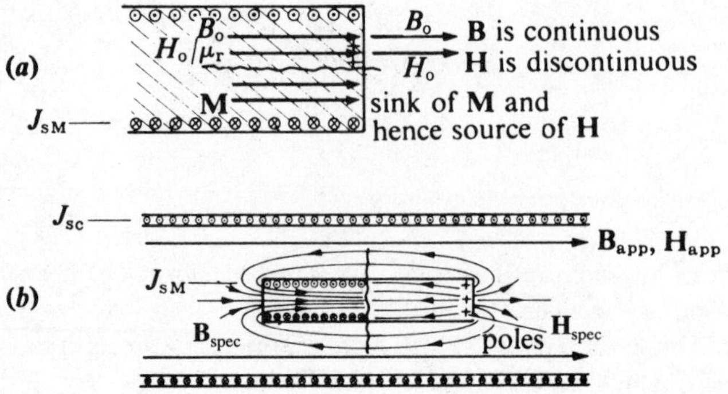

Figure 12.7 (a) A boundary perpendicular to **B** and **H**; (b) sources and vortices of **B** and **H** in a magnetized rod within a current-carrying solenoid. For clarity, the fields of the solenoid ($\mathbf{B}_{app}$, $\mathbf{H}_{app}$) and of the specimen ($\mathbf{B}_{spec}$, $\mathbf{H}_{spec}$) are drawn separately; the resultant fields are obtained by superposition. The left-hand end of the rod shows the **B**-field and its vortices; the right-hand end shows the **H**-field and its sources. The two versions are equivalent.

Worked example 2 *boundaries perpendicular to lines of* **B** The situation at such a boundary is illustrated in Fig. 12.7a, which shows the presence of sources of **H** and sinks of **M** as discussed above. While completely uniform magnetization is not strictly possible, consider a long thin cylinder inside an ideal current-carrying solenoid as in Fig. 12.7b. It will be approximately true that **M**, and hence **H** and **B**, will be approximately uniform within such a rod. The total **B**-field is the sum of that due to the solenoid, $\mathbf{B}_{app}$, and that due to the Amperian currents in the surface of the rod, $\mathbf{B}_{spec}$. The **H**-field will be the sum of $\mathbf{H}_{app}$ and **H** due to sources located at the ends of the specimen, $\mathbf{H}_{spec}$. It does not matter, for points outside the specimen, whether the **B**-field or **H**-field is calculated, but inside the material we see that if χ_m is positive $\mathbf{H}_{spec}$ opposes $\mathbf{H}_{app}$. For para- and diamagnetics, $\mathbf{H}_{spec}$ is so small that the resultant **H** is in the same direction as the resultant **B**: this is not so in ferromagnetics, as we see later.

12.5 Magnetic energy and forces

The same conditions must apply to magnetic materials as were specified for dielectrics at the beginning of Sec. 11.6, with the sole exception that hysteresis effects must now be considered at some time. In this section we ignore them.

Magnetic energy excluding non-linear materials If all the materials present are para- or diamagnetic, the arguments of Sec. 9.9 are unaffected since the values of L and M include their effects. Thus

$$U_M = \sum_i \tfrac{1}{2} L_i I_i^2 + \sum_i \sum_j M_{ij} I_i I_j = \sum_i \tfrac{1}{2} I_i \Phi_i \tag{12.25}$$

still. To express U_M in terms of field quantities the argument of Sec. 9.10 can be used except that we must now use the more general expressions:

$$U_M = \int_\tau \tfrac{1}{2} BH \, d\tau = \int_\tau B^2/(2\mu_r\mu_0) \, d\tau = \int_\tau \tfrac{1}{2}\mu_r\mu_0 H^2 \, d\tau \tag{12.26}$$

which replaces (9.44). The energy per unit volume, or energy density, u_M can be expressed as $\tfrac{1}{2}\mathbf{B}\cdot\mathbf{H}$ and this would apply even in anisotropic materials where **B** and **H** are not necessarily in the same direction.

Magnetic energy including non-linear materials The argument of Sec. 9.9 must now take account of the fact that the L's and M's are not constant and that $M_{12} \neq M_{21}$ in general. This last fact is exemplified in an ideal transformer consisting of two solenoids of the same length l and cross-section A wound on the same infinite soft ferromagnetic core so that the primary and secondary differ only in the number of turns per unit length (n_p and n_s). A current I in the primary gives $M_{12} = \mu_{rp}\mu_0 n_p n_s l A$, where μ_{rp} is the value of μ_r for $H = n_p I$, while a current I in the secondary gives $M_{21} = \mu_{rs}\mu_0 n_p n_s l A$, where μ_{rs} is for a field $n_s I$. If the core is non-linear, $\mu_{rp} \neq \mu_{rs}$ in general, and $M_{12} \neq M_{21}$.

Taking non-linearity into account, we now find the increment in U_M for small changes of current. For instance, with two circuits carrying currents I_1 and I_2 and with self- and mutual inductances L_1, L_2, M_{12} and M_{21}, the increment in U_M when I_1 increases by dI_1 with I_2 constant is $L_1 I_1\, dI_1 + M_{12} I_2\, dI_1$ by the same method as in Sec. 9.9. If now I_2 is increased by dI_2 keeping I_1 constant, we find in the limit that

$$dU_M = L_1 I_1\, dI_1 + L_2 I_2\, dI_2 + M_{12} I_2\, dI_1 + M_{21} I_1\, dI_2$$

and because $\Phi_1 = L_1 I_1 + M_{21} I_2$, etc.,

$$d\,U_M = I_1\, d\Phi_1 + I_2\, d\Phi_2$$

and *not* $\Phi_1\, dI_1 + \Phi_2\, dI_2$ unless $M_{12} = M_{21}$. Thus for any number of circuits,

$$d\,U_M = \Sigma I_i\, d\Phi_i \tag{12.27}$$

which is not equal to $\Sigma \Phi_i\, dI_i$ unless all materials are linear. Equation (12.27) can be integrated to give (12.25) only for linear materials.

The argument of Sec. 9.10, applied to obtain U_M in terms of fields, now gives

$$\boxed{dU_M = \int_\tau \mathbf{H} \cdot d\mathbf{B}\, d\tau} \tag{12.28}$$

so that the increment in energy density is $\mathbf{H} \cdot d\mathbf{B}$ (and not $\mathbf{B} \cdot d\mathbf{H}$ unless linear). By writing $\mathbf{H} \cdot d\mathbf{B} = \mu_0 \mathbf{H} \cdot d\mathbf{H} + \mu_0 \mathbf{H} \cdot d\mathbf{M}$, it can be seen that if the materials are so weakly magnetic that $\mathbf{H}_{spec}$ is negligible or if the specimen is of such a shape that no poles exist, the first term represents work which would be done in establishing the fields with no material present and the second can therefore properly be associated with the work done in magnetizing the specimen.

Forces between currents in the presence of magnetic materials Provided there is no contact between the materials and the current-carrying conductors the forces on the latter will still be given by

$$F_x = (\partial\, U_M/\partial x)_I = I(\partial \Phi/\partial x)_I \tag{12.29}$$

as in Sec. 9.11. The forces between currents are thus only likely to be ($\mu_r \times$ the vacuum value) in an infinite LIH medium.

Forces on linear magnetic materials Like the forces on dielectrics, these calculations are very complex and often of questionable validity unless simplifying assumptions are made which can be reflected in practical situations. Fortunately, the problems of interest refer to weakly magnetic materials in which the fields due to the specimen are so feeble that they may be neglected in comparison with the applied fields. It must be emphasized that many of the formulae to be developed do not apply unless this is so.

We can use either energy methods or force methods as we did with dielectrics. Using the force method first, consider a small volume $\mathrm{d}\tau$ of material which has a magnetic moment $\mathbf{M}\,\mathrm{d}\tau$ or $\chi_\mathrm{m}\mathbf{H}\,\mathrm{d}\tau$. In a non-uniform field it will experience a force given by (8.27) whose x component is

$$\begin{aligned} \mathrm{d}F_x &= \chi_\mathrm{m}(H_x\,\partial B_x/\partial x + H_y\,\partial B_x/\partial y + H_z\,\partial B_x/\partial z)\,\mathrm{d}\tau \\ &= \mu_\mathrm{r}\mu_0\chi_\mathrm{m}(H_x\,\partial H_x/\partial x + H_y\,\partial H_x/\partial y + H_z\,\partial H_x/\partial z)\,\mathrm{d}\tau \qquad (12.30) \end{aligned}$$

If no conduction currents flow in the material then $\oint \mathbf{H}\cdot\mathrm{d}\mathbf{L}=0$ so that $\nabla\times\mathbf{H}=0$ and $\partial H_x/\partial y = \partial H_y/\partial x$, etc. Hence:

$$\begin{aligned} \mathrm{d}F_x &= \mu_\mathrm{r}\mu_0\mu_\mathrm{m}(H_x\,\partial H_x/\partial x + H_y\,\partial H_y/\partial x + H_z\,\partial H_z/\partial x)\,\mathrm{d}\tau \qquad (12.31) \\ &= \tfrac{1}{2}\mu_\mathrm{r}\mu_0\chi_\mathrm{m}(\partial H^2/\partial x)\,\mathrm{d}\tau = \tfrac{1}{2}\chi_\mathrm{m}(\partial B^2/\partial x)\,\mathrm{d}\tau/\mu_\mathrm{r}\mu_0 \qquad (12.32) \end{aligned}$$

If immersed in a fluid medium with a comparable susceptibility the hydrostatic pressures make the calculation very complex. If, however, we can assume that the weakly magnetic material and fluid do not alter **H** from its vacuum value by more than a negligible amount, we can assert that were the material to be removed and replaced by the fluid, the force on the same volume of *fluid* would be given by (12.32) with χ_m replaced by χ'_m, this being the susceptibility of the fluid. Since we know that the hydrostatic pressures keep the fluid in equilibrium, we conclude that the forces due to these pressures are opposite to those on the volume $\mathrm{d}\tau$ but equal in magnitude. Since these pressures are determined by the fields, they are the same when the material fills the volume and so we finally have

$$\mathrm{d}F_x = \tfrac{1}{2}\mu_0(\chi_\mathrm{m} - \chi'_\mathrm{m})(\partial H^2/\partial x)\,\mathrm{d}\tau \qquad (12.33)$$

taking $\mu_\mathrm{r} \approx \mu'_\mathrm{r} \approx 1$. For a finite volume this must be integrated.

To use the energy method, let U'_M be the energy of the fluid before the magnetic material is inserted. Then with the volume $\mathrm{d}\tau$ of material

$$U_\mathrm{M} = U'_\mathrm{M} + \int_\tau \tfrac{1}{2}\mu_\mathrm{r}\mu_0 H^2\,\mathrm{d}\tau - \int_\tau \tfrac{1}{2}\mu'_\mathrm{r}\mu_0 H'^2\,\mathrm{d}\tau$$

and if we again assume that $\mathbf{H}\approx\mathbf{H}'$, this becomes

$$U_\mathrm{M} = U'_\mathrm{M} + \int_\tau \tfrac{1}{2}(\mu_\mathrm{r} - \mu'_\mathrm{r})\mu_0 H^2\,\mathrm{d}\tau$$

and hence

$$F_x = \int_\tau \tfrac{1}{2}\mu_0(\chi_\mathrm{m} - \chi'_\mathrm{m})(\partial H^2/\partial x)\,\mathrm{d}\tau \qquad (12.34)$$

since $\mu_\mathrm{r} = 1 + \chi_\mathrm{m}$. This is the same as (12.33) integrated over τ.

These forces are always such as to cause paramagnetic bodies to move into regions of stronger field irrespective of sign and diamagnetic bodies to move into weaker ones, although a paramagnetic immersed in a fluid of greater positive susceptibility will behave as a diamagnetic. Thus in Fig. 12.8a the diamagnetic will set across the field (this is the usual case considered because the non-uniformity is of the type commonly found between the poles of magnets, but note

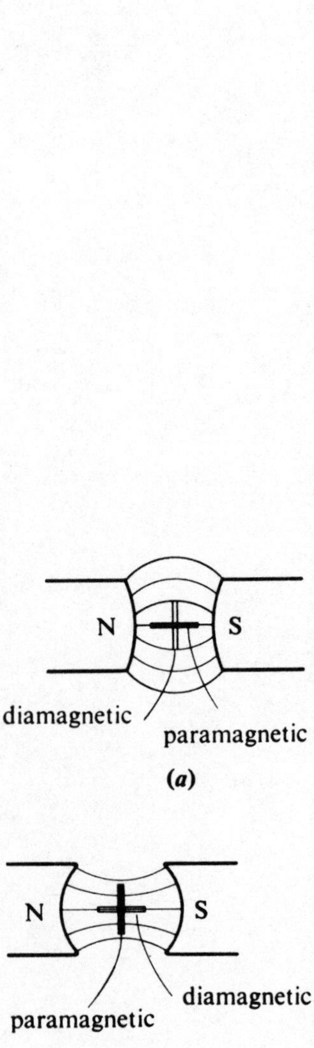

Figure 12.8 Equilibrium positions of paramagnetic and diamagnetic specimens in two different types of non-uniform magnetic field.

what happens in Fig. 12.8b). It can be shown that elongated para- or diamagnetic specimens in a *uniform* field experience feeble couples turning them *along* lines of force.

12.6 Ferromagnetic materials: hysteresis

Hysteresis loops The properties of ferromagnetic materials are commonly displayed by curves of B against H which may be obtained experimentally as explained in Sec. 12.10. In our discussion of the results we shall assume that the specimen is in the form of an anchor ring magnetized by a current in a toroidal coil wound on the ring: since there are no poles, H is simply nI and B is measured by the flux change through a second winding.

If initially unmagnetized, an increase of H gives a B–H curve following OXYZ, the virgin curve (Fig. 12.9a). Up to a point X low on the curve a reduction of H causes the curve to be retraced: OX is reversible. Beyond X, a reduction of H causes a path such as that from Y to be followed and, if H is continuously run through a cycle from Y to Y′ and back, the small loop YY′, known as a **hysteresis loop**, is followed. The largest of these loops, ZZ′, is reproducible for a given specimen and is known as **the** hysteresis loop. If at any point such as W′ small reversals of H are made, subsidiary loops like WW′ are traced.

The intercepts on the B-axis denote flux when $H=0$ and thus indicate the possibility of permanent magnets: OR is called the **remanence**, B_r. The intercepts on the negative H axis give the reverse fields needed to demagnetize the material, OC being known as the **coercivity**, H_c. We shall see that the so-called demagnetization curve RC contains the useful information about a material intended for use as a permanent magnet.

The value of $\mu_r = B/\mu_0 H$ will vary considerably over the virgin curve, increasing to a maximum and then falling almost to 1. What is often more important is the *differential* or *incremental* μ_r defined for a small range as $\Delta B/\mu_0 \Delta H$. Clearly μ_r and μ_{inc} will only be equal for the initial part of the virgin curve, but when a

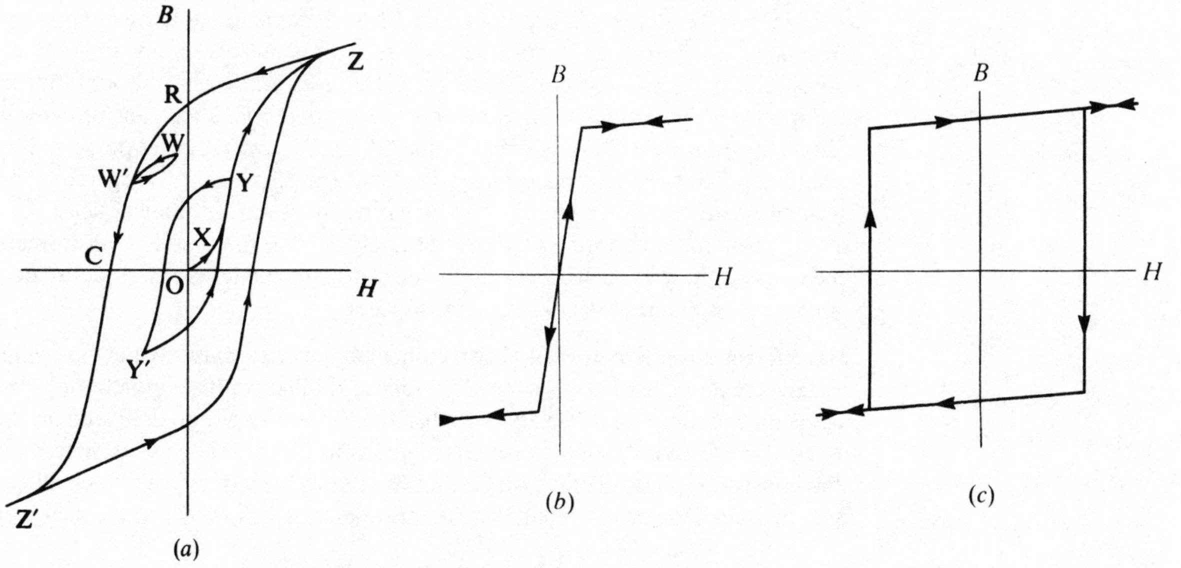

Figure 12.9 (a) General hysteresis loops; (b) hysteresis 'loop' of an ideally soft ferromagnetic; (c) hysteresis loop of an ideally hard ferromagnetic.

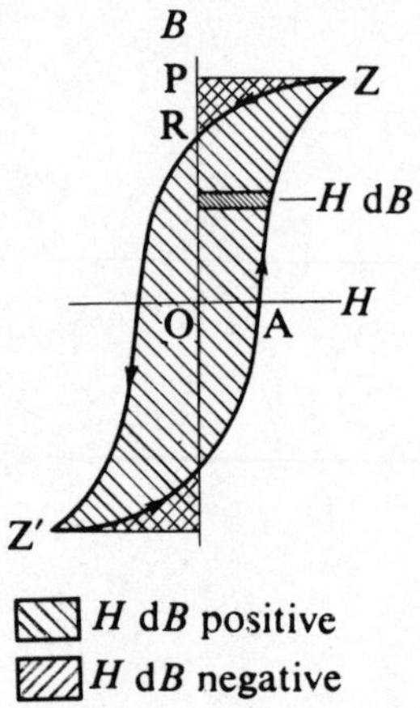

Figure 12.10 Energy loss in the hysteresis loop.

ferromagnetic forms the core of two coils, one carrying d.c. and the other a.c., the loop WW′ of Fig. 12.9a is typical of that which is followed and μ_{inc} is the slope of WW′ divided by μ_0. The value of μ_{inc} determines the effective inductance and reactance and, since it varies with the operating point on the hysteresis loop and thus with d.c., the latter can be used to control the impedance of a circuit. Such a device is known as a *saturable reactor.*

Energy loss in the hysteresis loop We have seen that the energy input needed to increase currents by small amounts can be expressed as $H\,dB$ per unit volume, even when non-linear materials are present. If the currents decrease, this amount of energy is recovered so that, without hysteresis, there is no net gain or loss of energy in a complete cycle. Figure 12.10 shows, however, that because the shaded element represents the energy input for a small portion of the loop, the area OAZP gives the energy input for the portion AZ of the loop, while RPZ gives the recovered energy for the portion ZR. Thus OAZR is lost, and when continued round the loop the total energy loss per unit volume per cycle is found to be just the area of the hysteresis loop, which may be written $\oint H\,dB$.

Magnetization loop A graph of M against H shows similar features but becomes horizontal at the extremities due to saturation. (Because $B=\mu_0 H+M$, the value of B is still increasing at saturation, unlike M.) We can also write

$$\oint H\,dB=\oint \mu_0 H\,dH+\oint \mu_0 H\,dM$$

where the left-hand side is the area of the B–H loop, and the right-hand side is the sum of $\mu_0 \times$ the area of the H–H loop (zero) and of $\mu_0 \times$ the area of the M–H loop. Hence the hysteresis loss per unit volume per cycle is also $\mu_0 \times$ the area of the M–H loop.

Soft magnetic materials In Sec. 12.1 an ideally soft ferromagnetic was defined as one having zero remanence. In addition, its magnetization curve should be linear up to saturation, so that the hysteresis 'loop' of an ideally soft ferromagnetic would have the shape illustrated in Fig. 12.9b. In practice, soft ferromagnetics have a small coercivity, a low loss because of a narrow loop and a high permeability. Apart from iron itself, used in electromagnets, the material in common use, particularly for large motors, generators and transformers, is silicon–iron (up to 4% Si). Higher permeability and lower loss are possessed by nickel–iron alloys containing between about 40 and 80 per cent Ni (e.g. mumetal, permalloy, supermalloy), but because they generally saturate at lower fields than silicon–iron and are more costly, they are only used in the smaller transformers and chokes: they find extensive other uses, e.g. as magnetic screens and, because of their large magnetostriction, in transducers.

Hard ferromagnetic materials Permanent magnet materials should have high remanence and coercivity (see next section) and ideally a hysteresis loop of the shape indicated in Fig. 12.9c. Beyond this, the properties required depend on the particular use and size of the magnet in question. The largest class of material is that consisting of alloys of Fe, Al, Co and Cu, known generically as Alnico alloys, but the development of cobalt–rare earth materials has shown that markedly

Table 12.1. Properties of some representative ferromagnetics

	Initial	Maximum		
Soft	μ_r	μ_r	$H_c\,\mathrm{A\,m^{-1}}$	B_{sat}(T)
3% Si–Fe	1.5×10^3	4.0×10^4	8.0	2.0
Mn–Zn ferrite	1.0×10^3	1.5×10^3	0.8	0.2
Mumetal	2.0×10^4	1.0×10^5	4.0	0.6
Supermalloy	1.0×10^5	1.0×10^6	0.2	0.8
Hard		$(BH)_{max}\,\mathrm{J\,m^{-3}}$	$H_c\,\mathrm{A\,m^{-1}}$	B_r(T)
5% Chromium steel		2.4×10^3	5×10^3	0.94
Alnico (high remanence)		1.8×10^4	8×10^4	0.62
Cobalt–samarium, Co_5Sm		1.5×10^5	1×10^6	1.50
Fe–Nd–B alloy		2.7×10^5	1×10^6	1.30

better characteristics are possible, while iron–neodymium–boron alloys discovered in the mid-1980s are even more promising.

Ferrites These are a widely used and technically important class of material whose magnetic properties arise from ferrimagnetism. They have a very high resistivity so that the Joule heating losses are small, they are cheap to produce and they can be fabricated in a wide variety of shapes. Moreover, ferrites can be produced with magnetic properties ranging from very soft to very hard depending on their chemical composition. Soft ferrites, with a general formula $Fe_2O_3.MO$ where M is a divalent metal, are of great use in high-frequency transformers, filters and choke coils. A squarer hysteresis loop can be obtained with Mg–Mn ferrites and these are used as computer memory stores. Hard ferrites, with a general formula $6Fe_2O_3.MO$, where M is barium or strontium, are used for permanent magnets which can be much smaller than the older bulky metallic magnets for a given field. Finally, there are what might be called semi-hard materials, such as γ-Fe_2O_3 and CrO_2, which are used in particulate form on an oxide-coated plastic tape for magnetic recording.

Values of the important constants for a few representative ferromagnetics are given in Table 12.1, but they should be taken as typical rather than definitive since the materials themselves vary in composition and physical constitution.

12.7 The magnetic circuit and the production of magnetic fields

The magnetic circuit Suppose we have a toroidal coil wound closely on an anchor ring formed of sections of various lengths as in Fig. 12.11. If the ring is ferromagnetic there will be little flux leakage, and if it is also thin compared with its diameter both B and H will be approximately constant across any cross-section. The magnetomotance round the ring is

$$\mathscr{H} = \Sigma H_i l_i$$

the summation occurring over the various sections each with their own μ_r, H, and in general with differing cross-sections A and flux densities B (the figure does not

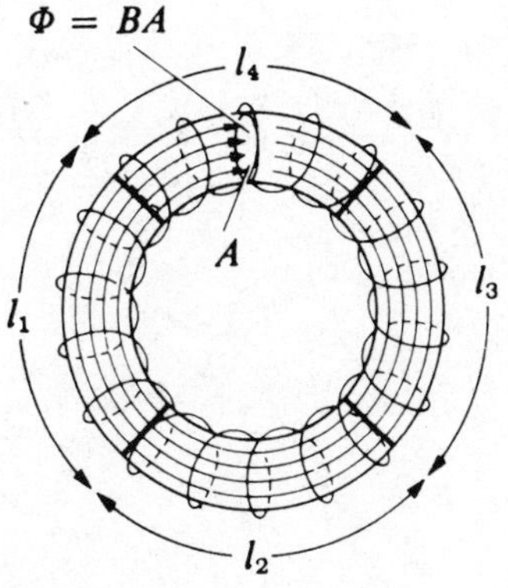

Figure 12.11 A simple magnetic circuit.

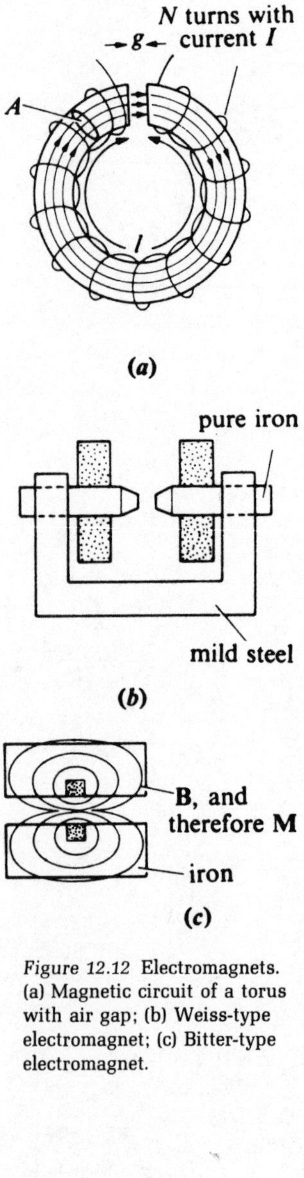

Figure 12.12 Electromagnets. (a) Magnetic circuit of a torus with air gap; (b) Weiss-type electromagnet; (c) Bitter-type electromagnet.

show this). For each length, $Hl = Bl/\mu_r\mu_0 = \Phi \times l/(\mu_r\mu_0 A)$, where Φ is the constant flux in the tube defined by the material. Thus

magnetomotance: $$\mathscr{H} = \Phi \sum_i \frac{l}{\mu_r\mu_0 A} \tag{12.35}$$

The quantity $l/(\mu_r\mu_0 A)$, known as the **reluctance** of the length l and denoted by $\mathscr{R}$, is analogous to the resistance of a simple series electric circuit for which

electromotance: $$\mathscr{E} = I\Sigma \frac{l}{\sigma A}$$

σ being the conductivity [cf. Eq. (6.32)]. Because of the analogy between this equation and (12.35), it follows that reluctances in series and parallel will combine according to the same laws as resistances.

The equation

magnetic circuit: $$\boxed{\mathscr{H} = \mathscr{R}\Phi} \tag{12.36}$$

is not so exact in practice as $\mathscr{E} = RI$ because the assumptions made above do not hold exactly and because μ_r is not the constant we have assumed. The concept of a magnetic circuit does, however, yield useful results provided any air gaps are small.

Production of magnetic fields The variety in the type of magnetic field required for physical investigation is such that their production is a highly technical matter which can only be touched on here. **Permanent magnets** have the advantage of constancy without power input but cannot produce flux densities even with the best modern materials greater than about 1 T (10 kG—see Table 12.1). **Electromagnets** with iron cores are useful up to about 3 T and have been constructed to produce as much as 7 T with the help of over 100 kW input power and a weight of over 35 tonne. Beyond 2–3 T, an iron core saturates and any additional flux density is produced by the current while the contribution from the core diminishes: the cost of the core is often not justified by its contribution and for the highest fields *air-cored coils* are used. The great problem here, the dissipation of heat produced by the large currents, is solved by using water-cooling (up to 22 T but more generally around 10 T), superconducting alloys (up to about 10 T) or pulses of current giving transient fields where they can be used (up to 100 T for tens of μs or to 1000 T for one μs).

Comment C12.1

We now look in more detail at electromagnets and permanent magnets using the concept of the magnetic circuit developed above.

Electromagnets The torus with air gap shown in Fig. 12.12a is a typical electromagnet. If the gap is of length g and the rest of the torus of length l and relative permeability μ_r, the reluctance of the circuit is $l/\mu_r\mu_0 A + g/\mu_0 A$ while the magnetomotance is NI. The flux is therefore

$$\Phi = \frac{\text{magnetomotance}}{\text{reluctance}} = \frac{\mu_0 NIA}{l/\mu_r + g}$$

and if the subscript g refers to the gap and m to the material

$$B_m = B_g = \frac{\mu_0 NI}{l/\mu_r + g}; \qquad H_g = \frac{NI}{l/\mu_r + g}; \qquad H_m = \frac{NI}{l + \mu_r g} \tag{12.37}$$

compared with the following values for a complete torus of length $l+g$:

$$B = \mu_r \mu_0 NI/(l+g); \qquad H = NI/(l+g) \tag{12.38}$$

These relations show that for a gap g small compared with l, the opening of the gap in the torus reduces B to $1/(1+\mu_r g/l)$ of its original value, e.g. to 1% if $g/l = \frac{1}{50}$ and $\mu_r = 5000$. It is also clear that the large reluctance of the gap causes most of the magnetomotance to be concentrated there, giving an H_g larger than H_m.

The Weiss type of electromagnet shown in Fig. 12.12b is very common, and the use of conical pole pieces concentrates the flux over a smaller area, thus increasing B_g and the reluctance of the gap. The Bitter type of magnet (Fig. 12.12c) uses magnetic material more economically than any other type: in it, a single coil, producing a nearly dipolar field, is embedded in a shell of soft ferromagnetic material which is magnetized along the lines of force shown. Each magnetized element then produces its maximum field at O (see Problem 12.11).

Permanent magnets If we could freeze in the magnetization of the gapped torus of Fig. 12.12a while removing the magnetizing coil, we should have the situation of Fig. 12.13 in which the sources and sinks of H (like those of M) occur at the sides of the gap: this is the situation in a permanent magnet. Because the magnetomotance is now zero,

$$H_g g + H_m l = 0 \tag{12.39}$$

and H_m is in the opposite sense to H_g as we should expect: H_m is known as the demagnetizing field. The flux density is continuous, however, and is $\mu_0 H_g$ all round the circuit.

Consider the common form of magnet shown in Fig. 12.13b in which part of the circuit is of soft ferromagnetic material of negligible reluctance so that (12.39) still applies but in which the pole pieces define a gap of cross-section a. Then

$$\Phi = B_g a = B_m A$$

and since $H_g g = -H_m l$, the product of the two equations gives

$$B_g H_g \tau_g = -B_m H_m \tau_m$$

where τ_g and τ_m are the volumes of gap and permanent magnet. This equation shows us that to produce a given flux density $B_g (= \mu_0 H_g)$ in a given volume τ_g, the smallest volume of material is required when the product $B_m H_m$ is a maximum. Now we have already seen that (12.39) means that H_m is negative if the direction of B is positive so that the state of the ferromagnetic material is given by some point on the demagnetization curve. Figure 12.13c shows this curve together with the product BH plotted against B and we now see that the point P corresponding to $(BH)_{max}$ is the most economical to work at. We also see that for large B_g we require large $(BH)_{max}$ which entails both a high B_r and a high H_c.

A 'keeper' of soft iron placed in the gap has a negligible reluctance and reduces $H_g g$ almost to zero, so that by (12.39), H_m is also reduced almost to zero. This

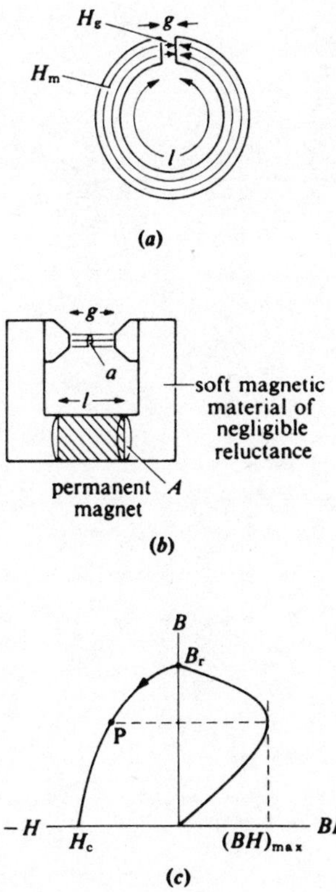

Figure 12.13 Permanent magnets. (a) Magnetic circuit; (b) use of pole pieces; (c) demagnetization curves.

takes the permanent magnet from the working point P on the hysteresis loop back to a point just below B_r, the demagnetizing field being zero (see the loop WW′ in Fig. 12.9).

For an up-to-date treatment of permanent magnets, see McCaig and Clegg (1987).

The ideal permanent bar magnet A true or ideal permanent magnet has a magnetization **M** independent of **H** so that the relations (12.7)–(12.18) apply, but not (12.19)–(12.22). Consider such a magnet as in Fig. 12.14 in the form of a uniformly magnetized bar. The **B** field is due simply to the Amperian currents in the surface and is similar to that of a finite solenoid. We have seen from (12.18) that sources of **H** are sinks of **M** and vice versa so that the **H** field is as shown, the relation $\mathbf{B}=\mu_0(\mathbf{H}+\mathbf{M})$ applying everywhere. Notice that inside the magnet **H** is in opposition to **B** along the axis and that at a general point **B**, **H** and **M** are all in different directions.

In practice the hysteresis curve shows us that **M** cannot be completely independent of **H** and so the magnetization cannot remain uniform. This is obvious when it is realized that the direction of **H** near the ends would not be parallel to **M** because the sheets of poles have a finite area (cf. the electric field in Fig. 5.4b). Because of the non-uniformity of **M** thus induced, the **H** field is in turn modified so that a complete solution is very difficult.

12.8 Magnetic poles

We have already discussed the monopole concept and its limitations in Chapter 7. For ideal permanent magnets in the absence of conduction currents it is a useful concept because of its similarity to electric charge, and it is possible to develop the theory of magnetization in terms of it. For instance, the dipole moment per unit

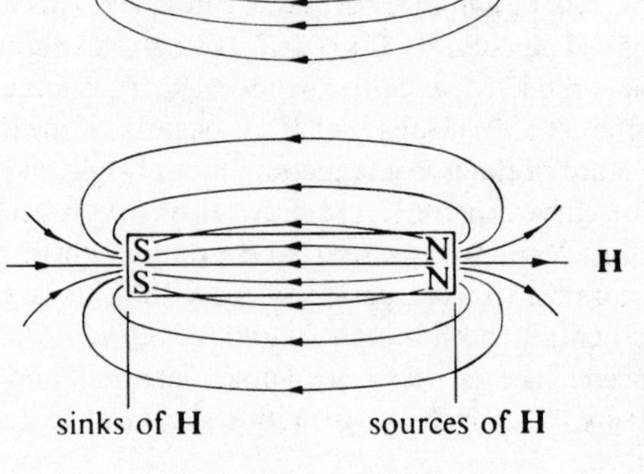

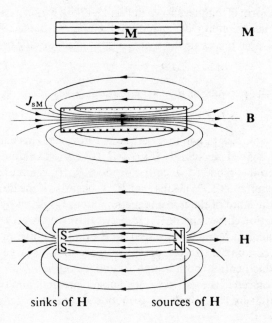

Figure 12.14 **M**, **B** and **H** in an ideal bar magnet.

volume **M** would be equivalent to a surface and volume distribution of poles, the volume distribution being absent if **M** is uniform or in LIH materials. At a surface, the pole strength per unit area would be equal to the normal component of **M** and thus gives rise to sources of **H** which we should have defined *in vacuo* in Chapter 7 by $\mathbf{H}=\mathbf{B}/\mu_0$ so that Gauss's law would become $\oint_S \mathbf{H}\cdot d\mathbf{S}=\Sigma P$. Since *all* poles P are due to magnetization we should have, over the surface S, $\Sigma P=-\oint \mathbf{M}\cdot d\mathbf{S}$ and hence $\oint (\mathbf{H}+\mathbf{M})\cdot d\mathbf{S}=0$. We should define **B** in a medium by $\mathbf{B}=\mu_0(\mathbf{H}+\mathbf{M})$ so that $\oint \mathbf{B}\cdot d\mathbf{S}=0$, which is (12.11), while (12.10) would be unaffected since poles are not vortices.

In a bar magnet the essential non-uniformity of **M** in practice, discussed in the previous section, means that there must be a volume distribution of poles near the ends so that only in the limit of a long thin magnet would the resultant pole strength appear to be situated *at* the ends.

We see that on the pole picture the obvious vector to use is **H**, which has poles as sources. In our methods we can use whichever concept is most useful: the **B**-field is given by conduction and Amperian currents as vortices and by no sources, while the **H**-field has conduction current as vortices and poles as sources.

The inverse square law between poles in a medium Insofar as single poles cannot be isolated or located with any accuracy it seems fruitless to discuss further a law of force between them. Nevertheless the question is asked as to whether the μ_r occurs in the numerator or the denominator when two point poles are immersed in a medium. All the difficulties arising in the case of Coulomb's law in a medium apply here and in addition there is no way of distinguishing the poles of the medium from those of a magnet since only the discontinuity in **M** gives pole strength. Experimental tests are also impossible because of the absence of a conservation law applying to poles of a single sign and, because no magnet is exactly permanent, the pole strength would always be uncertain.

12.9 Magnetic fields within materials

The discussion here follows the same lines as in Sec. 11.8 and need not therefore be so extensive. Our model of a magnetic material has been that of a smoothed-out distribution of dipoles which can be shown to be equivalent to surface and volume Amperian currents (**B**-field) or surface and volume poles (**H**-field) as far as points outside the material and far enough away from it are concerned. Inside materials we have defined **B** and **H** in such a way that they are still the flux density and field strength due to the surface and volume distributions, but we should not expect the *actual* couple per unit dipole or force per unit charge per unit velocity to be equal to **B** since we know that on a microscopic scale the smoothed-out model is not good enough.

As with **E**, there are two cases in which the couple or force would be expected to be **B**. One would think, first, that the space and time average of the microscope **B** would be equal to the macroscopic **B** and that therefore a charge or dipole moving with a high velocity would experience forces and couples appropriately. This has been shown to be so within a few per cent for mesons and neutrons in iron. The second case is one in which a cavity is scooped out of the material so that the **B**-detecting entity can be placed in it and be 'outside' the material once more. As in the dielectric, there are now surface distributions of Amperian currents or poles produced on the interior surface of the cavity: the reader should have no difficulty in showing that the couple on a dipole in a disc-like cavity perpendicular to **M** is $\mathbf{m}\times\mathbf{B}$, in a needle-like cavity along **M** is $\mu_0\mathbf{m}\times\mathbf{H}$, and in a spherical cavity is $\mu_0\mathbf{m}\times(\mathbf{H}+\frac{1}{3}\mathbf{M})$. (Note particularly that the definition of **B** by $\mathbf{T}=\mathbf{m}\times\mathbf{B}$ in Sec. 7.8 applied to *a vacuum*.)

12.10 Measurement of magnetic permeability and susceptibility

Methods of measuring μ_r and χ_m fall into two broad classes, one depending on *forces* on magnetic materials, the other on *induction* of electromotances in nearby circuits.

Force methods These are of most use for *paramagnetic* and *diamagnetic* substances, where the methods are based on the formulae developed in Sec. 12.5 giving the susceptibility χ_m directly, in terms of the forces in non-uniform magnetic fields. In *Faraday*'s method (also known as *Curie*'s) only a small sample is used and the force given by (12.33), in the form

$$dF_x=\mu_0(\chi_m-\chi_m')\left(H_x\frac{\partial H_x}{\partial x}+H_y\frac{\partial H_y}{\partial x}+H_z\frac{\partial H_z}{\partial x}\right)d\tau$$

is measured. The non-uniformity of **H** is generally such that one term of the three predominates and only dF_x is measured, not dF_y or dF_z. The forces are very small and have led to the development of several delicate balances for their measurement.

In *Gouy*'s method the specimen is in the form of a long rod of uniform cross-section A perpendicular to the direction x in which the force is to be measured and, by (12.34), since $d\tau=A\,dx$,

$$F_x=\frac{1}{2}\int_0^l \mu_0(\chi_m-\chi_m')A\frac{\partial H^2}{\partial x}dx$$

If the rod has one end in a field H_0 and the other in the zero field as in Fig. 12.15,

$$F_x=-\tfrac{1}{2}\mu_0(\chi_m-\chi_m')H_0^2A \qquad \text{downwards if } \chi_m>\chi_m'$$

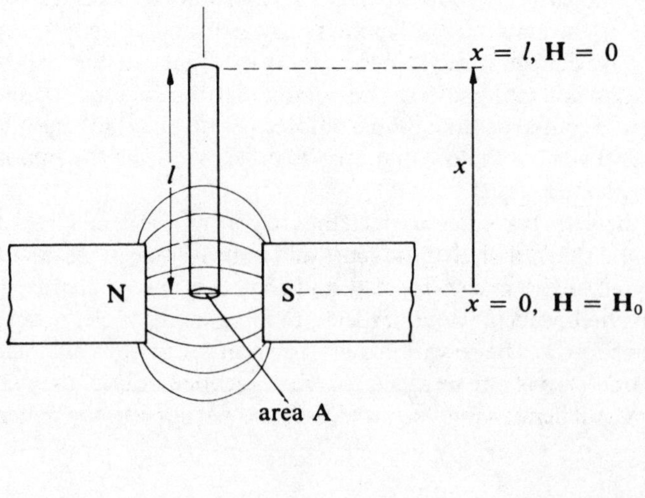

Figure 12.15 Gouy's method for the measurement of magnetic susceptibility.

and this is generally rather greater than the forces in the Faraday method. For liquids, the specimen is placed in a U-tube and the force on the liquid in one arm balanced by that due to hydrostatic pressure from the resultant difference in levels (*Quincke*'s method). For gases, χ_m' is determined by using a liquid of known χ_m.

Induction methods These are of most use for *ferromagnetic* substances where a complete B–H or M–H hysteresis loop is required, from which the μ_r can be obtained if required. We deal first with methods yielding B–H loops. If the specimen can be so formed that it has no boundaries crossing field lines, there will be no sources of H ('poles') and therefore no demagnetizing field. This is the situation with a toroidal ring specimen (Fig. 12.16a) and the H-field arises only from the magnetizing windings and is given by nI. In the case of a bar specimen (Fig. 12.16b), the magnetic circuit is completed by using a yoke as shown. The H-field is measured this time by the induced electromotances in small coils placed near and parallel to the surface but linking no material (the magnetizing coil here is independent of the H-coil). The B-field is measured in each case by a sensing or search coil wound so that the flux in the specimen links it. If the magnetizing field is altered, the flux linkage NAB changes and induces an electromotance $\mathscr{E}$ given by $NA\,\mathrm{d}B/\mathrm{d}t$. The change in B is thus given by $\int \mathscr{E}\,\mathrm{d}t/NA$. Older methods performed the integration by allowing $\mathscr{E}$ to produce a current I and thus a total charge $\int I\,\mathrm{d}t$ measured by a ballistic galvanometer. Present-day methods use electronic integrating circuits: e.g. the voltage $\mathscr{E}$ can be used to generate a set of pulses whose frequency is proportional to $\mathscr{E}$. The integral $\int \mathscr{E}\,\mathrm{d}t$ over a certain time interval is then simply the number of counted pulses in that time. Such a method can be made automatic, or the loop can be plotted step by step. In the latter case it is important to keep on the hysteresis loop by a proper sequence of operations.

M–H loops (as well as susceptibilities of paramagnetics and diamagnetics) can

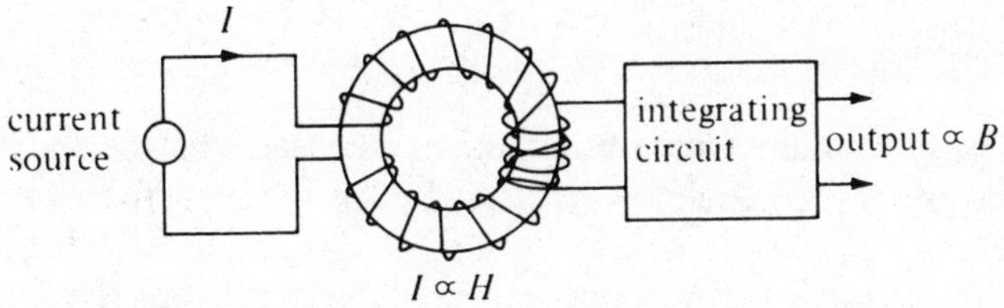

(a)

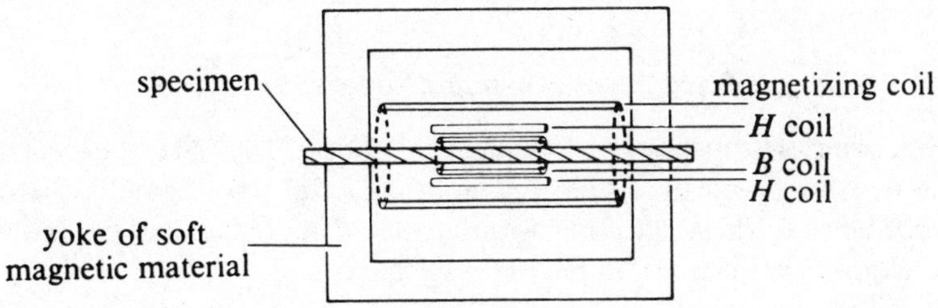

(b)

Figure 12.16 Permeameters with specimens in the form of (a) a torus; (b) a rod.

Table 12.2. Susceptibilities of some diamagnetic and paramagnetic materials at 20°C

	Susceptibility†, χ_m	Mass susceptibility†, χ_ρ ($m^3 kg^{-1}$)
Diamagnetic		
Bismuth	-16.7×10^{-5}	-1.69×10^{-8}
Copper	-0.92×10^{-5}	-0.10×10^{-8}
Water	-0.91×10^{-5}	-0.91×10^{-8}
Hydrogen	-0.22×10^{-8}	-2.5×10^{-8}
Paramagnetic		
Aluminium	2.2×10^{-5}	0.82×10^{-8}
Platinum	2.6×10^{-5}	1.22×10^{-8}
Nitric oxide	0.08×10^{-5}	59.3×10^{-8}
Oxygen	0.19×10^{-5}	133×10^{-8}

† To convert χ_m to unrationalized values (usually called e.m.u.) divide by 4π: to convert χ_ρ to CGS unrationalized values, multiply by $1000/4\pi$.

be measured by making a small ellipsoidal or spherical specimen for which a demagnetizing factor can be calculated. It is then suspended in a magnetizing field and oscillated rapidly at right angles to the field. Electromotances are induced in a coil round the specimen which are proportional to its magnetic moment and thus to M.

Results Some typical susceptibilities are shown in Table 12.2. Diamagnetic susceptibilities are independent of temperature and their ratio to the density (mass susceptibility χ_ρ—see next section) are all of the same order of magnitude. Paramagnetic materials have χ_ρ's differing more widely in magnitude and, except for paramagnetic metals, depending on temperature according to the Curie–Weiss law:

paramagnetics:
$$\chi_m = \frac{C}{T-\theta_p} \tag{12.40}$$

where T is the absolute temperature and θ_p a constant for the material known as its paramagnetic Curie point, which may be positive, negative or zero (in the last case the law is known as Curie's law). Ferromagnetic substances become paramagnetic above a critical temperature θ_f known as the ferromagnetic Curie point and then obey the Curie–Weiss law. θ_f and θ_p are not equal but differ by several degrees for most substances.

12.11 The approach to microscopic theory

We introduce two microscopic quantities as in Sec. 11.9: firstly the local magnetic field $\mathbf{H}_L$ experienced by a single atom or molecule and secondly the magnetic polarizability α_m defined as the mean magnetic dipole moment per molecule per unit magnetic **H**-field. As in Sec. 11.9 we have

$$\mathbf{M} = \frac{N'_A\rho\alpha_m}{M}\mathbf{H}_L; \qquad \chi_m = \frac{N'_A\rho\alpha_m}{M}\frac{\mathbf{H}_L}{\mathbf{H}} \tag{12.41}$$

The mass susceptibility, $\chi_\rho = \chi_m/\rho$, and the molar susceptibility, $\chi_M = \chi_m M/\rho$, are also used so that, for instance, $\chi_M = N'_A \alpha_m \mathbf{H}_L/\mathbf{H}$.

Local field As in a dielectric, the couple exerted on a molecular dipole will not be that due only to the macroscopic **B**. However, the susceptibilities of paramagnetic and diamagnetic materials are so small that we might hope to find the assumptions that $\mathbf{H}_L = \mathbf{H}$ and $\mathbf{B}_L = \mu_0 \mathbf{H}$ are sufficient. When they prove inadequate, we shall assume that $\mathbf{H}_L$ is given by a **molecular field** proportional to **M** in addition to **H**, i.e. that $\mathbf{H}_L = \mathbf{H} + \lambda \mathbf{M}$ where λ is a constant for the material. The Lorentz method as used in the dielectric case would yield $\lambda = \frac{1}{3}$.

Diamagnetism We start with a material in which the atoms or ions have no resultant moment, usually because the electronic structure consists only of completed shells. In that case the only effect of applying a magnetic **B**-field is to cause a precession of the orbital electrons as discussed in Sec. 8.7. The Larmor frequency of precession is given by $\omega_L = -eB/2m$, since $\gamma = e/2m$ for orbital electrons, and is in such a direction as to provide moments and fields opposing the applied field. The precession is equivalent to a rotation of each electron with an additional angular velocity ω_L at its distance r_{xy} from the nucleus (taking the xy plane as normal to the direction of **B**). The additional angular momentum produced is thus $mr_{xy}^2\omega_L$. From (8.29), the additional magnetic moment per electron is then $e\langle r_{xy}^2\rangle\omega_L/2$ or $-e^2\langle r_{xy}^2\rangle B/4m$. Since the electron in fact moves in three dimensions we prefer to express this in terms of $r^2 = x^2 + y^2 + z^2$ whose mean is $\frac{3}{2}$ times that of $r_{xy}^2 = x^2 + y^2$ for one of a number of randomly oriented orbits. Summing over all electrons in the atom or ion, we have $\alpha_m = -(\mu_0 e^2 6m)\Sigma\langle r^2\rangle$. Hence

diamagnetics: $$\chi_m = -\frac{N'_A \rho \mu_0 e^2}{6Mm}\Sigma\langle r^2\rangle \qquad (12.42)$$

This gives good agreement with measured values using as r the radius of atoms as given by independent experiments. The effect is internal to each atom and is thus independent of temperature.

All the above would clearly also apply to atoms or ions which had a resultant magnetic moment. Diamagnetism is thus a universal effect but, like the distortion term in a dielectric, is usually swamped by the effect of a permanent moment when present.

Paramagnetism An intrinsic moment is possessed by an atom or ion when one or more of its electron shells are incomplete (apart from the outer shells providing the conduction electrons if any). The value of the intrinsic moment and its orientation with respect to **B** are given by quantum-mechanical methods and result in the expressions of Sec. 8.7: the contribution of the total moment to the magnetization is $gm_J\mu_B$, where μ_B is the Bohr magneton and m_J may take values 0, ± 1, etc., or $\pm\frac{1}{2}$, $\pm\frac{3}{2}$, etc., according to the direction in space. In the absence of an external field, the distribution amongst the various possible alignments is such that there is no resultant moment, but the application of **B** causes a partial alignment with a distribution governed by the Maxwell–Boltzmann law. Changes of alignment may take place through interactions between one atom and another so that paramagnetism will depend on temperature. A complete calculation shows that for moderate fields, the value of α_m is $\mu_0 \mathbf{m}^2/(3kT)$, where $\mathbf{m}$ is the total atomic magnetic moment and k is Boltzmann's constant. Substitution in (12.41) with $\mathbf{H}_L = \mathbf{H}$ yields Curie's law.

To account for the Curie–Weiss law, we assume a molecular field $\lambda\mathbf{M}$ is contributed by interaction with other atoms. If we write C for the expression $\mu_0 N'_A \rho \mathbf{m}^2/(3kM)$, then (12.41) yields $\chi_m = C(1+\lambda_m)/T$ or

paramagnetics: $$\chi_m = \frac{C}{T - \lambda C} \tag{12.43}$$

This is the Curie–Weiss law with $\theta_p = \lambda C$.

Ferromagnetism There are two separate aspects of ferromagnetism: the large magnetization, which can occur in the absence of any external field, and the non-linearity and hysteresis when the external field varies.

The first question to be settled in connection with the high value of magnetization is the nature of the atomic contribution: is it, as in paramagnetism, a combination of orbital and spin moments of the electrons? A clue is provided by **gyromagnetic experiments**, which not only confirm that magnetization is related to angular momentum but allow the value of $\gamma(=\mathbf{m}/\mathbf{L}$, Sec. 8.7) to be determined for the magnetic carriers. Two types of experiment have been performed. In one, the rotation of a cylindrical rod about its axis aligns the angular momenta and thus magnetizes it—this is the **Barnett effect**. The **Einstein–de Haas effect** is the reverse: if a rod is magnetized there will be an accompanying change $\Delta\mathbf{L}$ in the total internal angular momentum of the carriers due to their alignment. Because there is no external torque on the system, angular momentum is conserved and the rod as a whole must therefore rotate with angular momentum $-\Delta\mathbf{L}$. Both effects show that γ for ferromagnetics has a value almost equal to e/m and not to $e/2m$ as it would if only the orbital motion of the electrons were involved. The concept of electron spin with a g-factor of 2 has to be invoked to explain optical spectra, so that gyromagnetic experiments show that such spins are almost entirely responsible for ferromagnetism.

Now it is clear from the derivation of the Curie–Weiss law that the assumption of a molecular field leads to the prediction that the susceptibility may become infinite as T is lowered to θ_p, i.e. there may be a finite $\mathbf{M}$ even when the applied $\mathbf{H}$ is zero. This is **spontaneous magnetization**. Measurements of the Curie point in ferromagnetic metals show that λ would be of the order of several thousands, so that there must be an enormous interaction between spins causing their alignment over large regions. Such interactions, due to *exchange forces*, are accounted for by a quantum-mechanical calculation and the reader should consult solid-state texts for further details.

The spontaneous magnetization does not show itself on a large scale, however, until a field is applied and even then shows hysteresis when the field is varied. This is now known to be due to the existence of small **domains** of the material each spontaneously magnetized to saturation in a definite crystalline direction, adjacent domains having different directions. These domains (Fig. 12.17), of size about 20 μm, are separated by walls about 5 nm thick in which the direction of the spins is changing. Domains are formed in the first place even in a single crystal

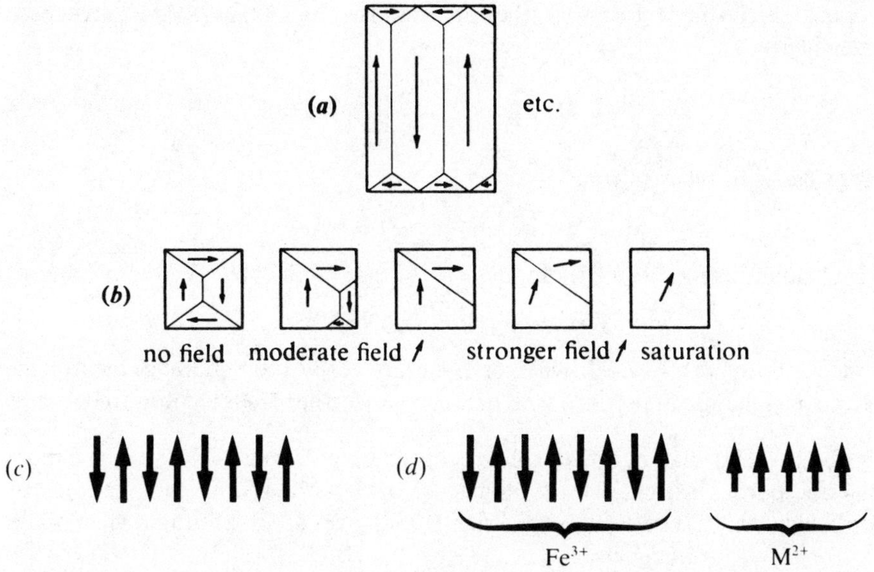

Figure 12.17 (a) Ferromagnetic domains in a typical closure pattern; (b) ferromagnetic domains and the effect of applying an external magnetic field; (c) schematic diagram of spin arrangement in an antiferromagnetic material; (d) spins in a ferrimagnetic material.

because the energy is lowered thereby, but although smaller domains mean less energy the wall area increases and the process ceases when the decrease in energy due to a splitting into smaller domains equals the increase due to the resultant increase in wall area.

The application of an increasing magnetic field causes (a) initially, reversible wall movements allowing domains more favourably magnetized to grow at the expense of others, (b) for moderate fields, irreversible wall movements of the same type but hindered by impurities or defects in the crystal structure, and (c) for high fields, rotation of magnetization from easy directions towards that of the field. Each of these stages accounts for part of the hysteresis loop and the irreversible step (b) can be detected audibly by electromotances induced in loudspeaker coils wound on a specimen as a wall jumps irreversibly across a defect (the **Barkhausen effect**). The patterns of domains on the surfaces of materials can be made visible under a microscope by fine ferromagnetic powders and the wall movements can be watched as a magnetic field is applied.

Antiferromagnetism and ferrimagnetism Antiferromagnetics are materials with highly aligned spins but with an interaction causing adjacent atoms to be antiparallel. There are thus two sets with equal spins in opposite directions giving a very small resultant moment even in an external field. Ferrimagnetics are similar, but the two sets of antiparallel spin moments are not equal in magnitude and there is a much larger magnetization more like that of ferromagnetics. The arrangement of spins in antiferromagnetic and ferrimagnetic materials is illustrated schematically in Fig. 12.17c and d.

12.12 Conclusion

The presence of magnetic materials alters the general field equations through their effect on the **B**-field. However, by introducing a new field quantity, the

H-field, the simple forms of the circuital law and Gauss's law have been retained:

$$\oint_L \mathbf{H}\cdot d\mathbf{L} = I_c; \qquad \oint_S \mathbf{B}\cdot d\mathbf{S} = 0$$

together with their differential forms:

$$\nabla\times\mathbf{H} = \mathbf{J}_c; \qquad \nabla\cdot\mathbf{B} = 0.$$

The equations for **D** and **E** obtained at the end of Chapter 11 are unchanged:

$$\nabla\times\mathbf{E} = -\partial\mathbf{B}/\partial t; \qquad \nabla\cdot\mathbf{D} = \rho_c,$$

but we now go on to ask whether these are really the definitive form of the electromagnetic equations or whether there are further modifications to be made.

References For theories of magnetic properties, see Crangle (1977) or, at a more advanced level, Kittel (1986). Crangle also deals with experimental methods, as do Kalvius and Tebble (1979). Carden (1976) gives an interesting review of the history of electromagnets.

12.13 Revision summary

General field equations By defining the magnetization of a material, **M**, as the magnetic dipole moment per unit volume, with **SI unit** the A m^{-1}, we find that any piece of the material can be represented by Amperian currents. These consist of surface currents of density $J_{sM} = M \sin\theta$, where θ is the angle between **M** and the normal to the surface, together with volume currents of density $\mathbf{J}_M = \nabla\times\mathbf{M}$. In LIH materials, $\mathbf{J}_M = 0$.

The **B**-field at any point is defined as the sum of any applied field and that due to J_{sM} and $\mathbf{J}_M$. If we also define an **H**-field (**SI unit** also the A m^{-1}) by

$$\mathbf{H} = \mathbf{B}/\mu_0 - \mathbf{M} \qquad (12.14)$$

then Ampère's circuital law and Gauss's law in the presence of magnetic materials take the forms

$$\oint_L \mathbf{H}\cdot d\mathbf{L} = I_c; \qquad \oint_S \mathbf{B}\cdot d\mathbf{S} = 0 \qquad (12.15)$$

where I_c includes only conduction currents. The differential forms are

$$\nabla\times\mathbf{H} = \mathbf{J}_c; \qquad \nabla\cdot\mathbf{B} = 0 \qquad (12.17)$$

Thus we find that **B** has no sources, but has as its vortices both conduction and Amperian currents. On the other hand, **H** has only conduction currents as vortices but does have sources. The latter are the same as the sinks of **M** and can be thought of as magnetic poles.

From these laws it follows that across any boundary between two media:

B_{normal} is continuous; $H_{\text{tangential}}$ is discontinuous by J_{sc}(12.23) (12.24)

Properties of magnetic materials The relationship between **M** and **H** at any point is expressed in terms of magnetic susceptibility χ_m, where

$$\mathbf{M} = \chi_m \mathbf{H} \qquad (12.19)$$

A relative permeability μ_r is defined by

$$\mu_r = 1 + \chi_m \tag{12.21}$$

and this then has all the properties of the μ_r introduced as L_m/L_0 in (12.1). Moreover, there is then the relation

$$\mathbf{B} = \mu_r \mu_0 \mathbf{H} \tag{12.22}$$

In LIH media only, μ_r and χ_m are scalar constants. If such a medium fills the space completely round a system of conduction currents, it increases **B** everywhere by μ_r and leaves **H** unaffected. No such simple results occur when media are finite.

In practice, LIH materials are either paramagnetic, with χ_m small and positive, or diamagnetic, with χ_m small and negative. Ferromagnetic materials have large positive μ_r and χ_m, are non-linear and show hysteresis.

Expressions for forces and energy from Chapter 9 are largely unaffected, but a system of currents has an energy that can be generalized to

$$u_M = \tfrac{1}{2}\mathbf{B} \cdot \mathbf{H} \qquad \text{from (12.26)}$$

which gives the energy density per unit volume. It follows that the increment of energy when currents are increased by small amounts can be expressed as

$$dU_M = \int_\tau \mathbf{H} \cdot d\mathbf{B}\, d\tau \tag{12.28}$$

of which $\mu_0 \mathbf{H} \cdot d\mathbf{M}$ per unit volume is associated with the magnetization of materials.

Commentary

C12.1 On Intense Magnetic Fields B-fields of 5–10 T and greater are needed in such areas as the production of low temperatures by nuclear demagnetization; the containment of plasmas; the investigation of properties of solids, in which the atomic interaction fields are of the order of 10 T so that external fields of this magnitude will produce first-order effects. The methods used to produce such fields use principles of classical physics and the same applies to the solution of the problems arising, such as the dissipation of heat, the reduction of cost and above all the enormous Lorentz forces. These forces, between current-carrying conductors, are comparable with interatomic forces and can cause plastic flow in the materials normally used. A general survey is given in Parkinson and Mulhall (1967).

The most intense fields of 100 T and over fall into a class by themselves since they are only likely ever to be produced for very short times. The methods used involve the discharge of capacitor banks or implosion. In the latter, a magnetic flux established within a cylindrical tube of a good conductor is 'compressed' by detonating explosives placed uniformly round the outside. The currents induced in the imploding cylinder are sufficient to increase the field to very high values, although the specimen being investigated, and the apparatus, are invarably destroyed in the process. For further details see Herlach (1968, 1985).

C12.2 On Modern Magnetometry The magnitudes of **B**-fields which need to be measured cover an enormous range as indicated by the following set of typical values:

Interplanetary field	5×10^{-9} T (50 μG or 5γ)
Terrestrial field	4×10^{-5} T
Aircored solenoid (1 turn/mm, 1 A)	1×10^{-3} T (10 G)
Good permanent magnet	1 T
Largest steady laboratory fields	25 T (water-cooled coil)
Pulsed fields	50 T (10 ms)
Pulsed fields	150 T (3 μs)
Largest flux-compression fields	10^3 T (<1 μs)
White dwarf star	10^3–10^4 T (measured by Zeeman effect)
Neutron star	10^8 T (estimated)

The following methods are in use for the measurement of **B**-fields:

1. Induction techniques using sensing coils or probes as in Sec. 9.4. For low fields, ferrite cores are added to increase sensitivity.
2. Proton precession using the resonance or induction techniques of Sec. 8.7 (i.e. methods for γ_p in reverse). This is used over a wide range of **B**-fields and is very precise because of the sharp resonance and the precision with which frequency can be measured.
3. Zeeman effect in the rubidium magnetometer. The Zeeman splitting varies with **B** and thus changes the resonance absorption of light in rubidium vapour.
4. Fluxgate method. When an auxiliary field of frequency ω is applied to a *non-linear* magnetic material, the variations of **B** have symmetrical but opposing half-cycles and contain frequencies ω, 3ω, 5ω, but no even harmonics. If the unknown field is added to the original variations the symmetry is upset and even harmonics appear in the flux variations. The second harmonic is usually picked out and its amplitude is related to that of the unknown **B**.
5. Hall probes. These use the Hall effect as explained in Sec. 7.6.
6. Magnetoresistive probes. Using the variation of resistivity with applied magnetic field (Sec. 6.10).
7. SQUIDs (superconducting quantum interference devices) using the Josephson effect in superconducting loops. These can detect and measure magnetic flux in units of $h/2e$, the flux quantum, i.e. 2×10^{-15} Wb. Modern SQUIDs are capable of detecting fields as low as 10^{-10} T (from the heart) and even 10^{-13} T (from the brain).

For further details of the use of methods 1 and 2 in palaeomagnetism see Aitken (1962) and of methods 2 and 3 in space probes see Hall (1967). A survey of the many applications of SQUIDs can be found in Dickens (1987).

C12.3 On the Earth's Magnetic Field and Palaeomagnetism The origin of the earth's magnetic field, and indeed of other planetary and stellar fields, is still the subject of theoretical investigation. There is general agreement that the source must be convective motion of conducting fluids in the earth's outer core, and that the field is produced by a dynamo effect (motional electromotance). Faraday's disc of Fig. 9.3a can be used to illustrate the basic principle proposed: if the current produced in the circuit were used to generate the magnetic field across the disc, the system would become a self-excited generator. However, adapting this idea to the motion of a spherical volume of fluid is no easy matter and no agreed solution has yet been found. Gubbins (1984) surveys the current position as regards the earth's field, while Parker (1983) and Stevenson (1983) deal with planetary magnetic fields in general, and Hones (1986) describes the interaction of the solar wind with the earth's field.

The earth's field magnetizes various forms of rock in the direction of the field prevailing at the time they were laid down. Measurements of the resultant magnetization, coupled with geological evidence and radioactive dating, have yielded a great deal of information about past variations in the terrestrial field (palaeomagnetism) and about the sea-floor spreading and continental drift that are involved in plate tectonics. There is abundant evidence for frequent reversals in the earth's polarity (Hoffman, 1988), so that any model for the source of the field must allow for this possibility. A good general description of these matters is to be found in Gass, Smith and Wilson (1971).

C12.4 On definitions of H and χ_m The substitution $\mu_0 \to 4\pi$ used in Chapters 7–9 to convert formulae to CGS e.m.u. does not work for relations involving **H** and χ_m. The reason is that the definition of **H** differs by a factor of 4π (compare Comment C11.1):

SI definition: $\mathbf{H} = \mathbf{B}/\mu_0 - \mathbf{M}$

CGS definition: $\mathbf{H} = 4\pi(\mathbf{B}/\mu_0 - \mathbf{M})$

It follows that the substitutions needed to convert all the formulae in this chapter to CGS e.m.u. are (since $\chi_m = \mathbf{M}/\mathbf{H}$)

$$\mu_0 \to 4\pi; \qquad \mathbf{H} \to \mathbf{H}/4\pi; \qquad \chi_m \to 4\pi\chi_m \tag{12.44}$$

so that, for instance,

$$\mathbf{B} = \mathbf{H} + 4\pi\mathbf{M}; \qquad \mathbf{M} = \chi_m\mathbf{H}; \qquad \mu_r = 1 + 4\pi\chi_m \text{ (CGS e.m.u)} \tag{12.45}$$

There have been suggestions that the **H**-field should be defined by $\mathbf{H}' = \mathbf{B} - \mu_0\mathbf{M}$, giving **B** and **H**′ the same units and making the substitutions for **H** and χ_m of (12.44) unnecessary.

C12.5 On Sommerfeld and Kennelly magnetization Because **m** is defined differently in the S and K versions of SI units (by a factor μ_0—see Comment C7.3), the magnetization is also different. We have used the S version throughout the chapter, but we summarize here the alternative for those who may wish to consult texts using it.

The K version defines the magnetic moment per unit volume, **I**, in T so that in the derivation of the surface density of Amperian current, the moment of the element in Fig. 12.3 is $\mu_0 J_s \, dS \, d\mathbf{l}$. This is therefore $\mu_0\mathbf{I}$ so that $\mathbf{I}/\mu_0$ must replace **M** throughout. Thus in the derivations of Sec. 12.3, the circuital law (12.10) becomes $\oint \mathbf{B} \cdot d\mathbf{L} = \mu_0 I_c + \mu_0 I_M$ in which $I_M = \mathbf{I} \cdot d\mathbf{L}/\mu_0$ and hence (12.12) becomes $\oint (\mathbf{B}/\mu_0 - \mathbf{I}/\mu_0) \cdot d\mathbf{L} = I_c$.

H is now defined as

$$\mathbf{H} = \mathbf{B}/\mu_0 - \mathbf{I}/\mu_0 \qquad \text{or} \qquad \mathbf{B} = \mu_0\mathbf{H} + \mathbf{I}$$

so that the generalized circuital law (12.15) follows.

The magnetic susceptibility χ_m is defined by

$$\mathbf{I} = \mu_0\chi_m\mathbf{H}$$

The K version follows more closely the dielectric one and is thus more appropriate when the pole model is used. To convert Kennelly formulae to CGS e.m.u., the substitution $\mathbf{I} \to 4\pi\mathbf{I}$ must be made in addition to (12.44) so that, for instance,

$$\mathbf{B} = \mathbf{H} + 4\pi\mathbf{I}; \qquad \mathbf{I} = \chi_m\mathbf{H}; \qquad \mu_r = 1 + 4\pi\chi_m \text{ (CGS e.m.u.)}$$

The conversion from the S version in this chapter to the K version is achieved by the substitution $\mathbf{M} \to \mathbf{I}/\mu_0$.

I is sometimes called the *magnetic polarization* to distinguish it from **M**.

Problems

Section 12.1

12.1 A ferromagnetic rod of relative permeability 10^4 is pushed into a long solenoid. If the rod fills the whole of space in which any appreciable magnetic field occurs, find its effect on (a) the time constant of the solenoid, (b) the resonant frequency when connected to a capacitor and (c) the $\hat{Q}$-factor of the circuit in (b) assuming no losses in the rod.

12.2 A long solenoid with 15 turns per cm each carrying 0.1 A is wound on an even longer iron core ($\mu_r = 1000$). Compare the magnitudes of conduction and Amperian surface current densities. What does the Amperian surface current density become if the rod is of copper ($\mu_r - 1 = -10^{-5}$)?

Section 12.2

12.3 An anchor ring of cross-section 2 cm^2, mean radius 20 cm and $\mu_r = 1500$ is closely and uniformly wound with 2000 turns of wire. Calculate the self-inductance of the toroidal coil. If a current of 0.1 A is passed through the wire, find the mean **B**-field and the magnetization in the anchor ring.

12.4* Show that, provided $\mathbf{M} = \sigma\omega a$, identical magnetic **B**-fields are produced by (a) a ferromagnetic sphere with uniform magnetization **M** and (b) a non-ferromagnetic conducting sphere carrying a uniform surface density of charge σ and rotating about a diameter with angular velocity ω.. The radius of both spheres is a.

Section 12.3

12.5 Find the mean **H**-field in the anchor ring of Problem 12.3. What happens if 500 of the 2000 turns are reverse wound?

Section 12.4

12.6 The inner conductor of a coaxial cable has a radius a and is coated with a sleeve of non-conducting ferromagnetic material whose external radius is $2a$. If the inner radius of the outer conductor is $4a$ and the relative permeability of the ferromagnetic is μ_r, find the self-inductance per unit length of the cable, neglecting the magnetic flux in the material of the conductors.

12.7 Derive the condition governing the refraction of lines of **B** and **H** across a plane boundary between two LIH media. Explain why such refraction is in practice extremely small.

Section 12.7

12.8 Show that the increment in magnetic energy per unit volume in the magnetic circuit of Fig. 12.11 for an increment in current may be written as $\mathscr{H}\,\mathrm{d}\Phi$.

12.9 What is the **B**-field in a gap of width 5 mm opened in the anchor ring of Problem 12.3?

12.10 Find the magnetic flux through the centre arm of the magnetic circuit shown in Fig. 12.18 if the cross-section of all arms is 20 cm^2, if the relative permeability is 1000 and the magnetizing coil is of 1000 turns carrying 0.2 A.

12.11* Show that the **B**-field at the origin produced by a small magnetic dipole situated at the point (r, θ) and making an angle α with r is a maximum when $\tan\alpha = \frac{1}{2}\tan\theta$. Hence explain the principle of the Bitter magnet of Fig. 12.12c.

Section 12.8

12.12 Show that, inside a uniformly magnetized permanent magnet, **B** and **H** are always in opposite directions.

Section 12.11

12.13* Show that, when a uniform **B**-field is established normal to the plane of a circular electronic orbit, the change in angular velocity induced is $\omega_L = eB/2m$. Show also that the additional force produced by the **B**-field is just enough to

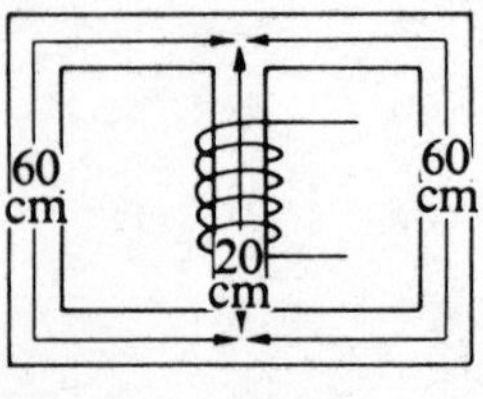

Figure 12.18 Magnetic circuit for Problem 12.10.

provide the additional acceleration so that the orbit stays the same size. (Assume that ω_L is much less than the angular velocity of the electron before the application of **B**.)

12.14 A ferromagnetic rod is suspended, already magnetized to saturation along its axis. Estimate an order of magnitude for the change in angular momentum per cm^3 of the rod when its magnetization is completely reversed.

13

Maxwell's equations and their solutions

Our study of electromagnetism has now reached the point at which we take stock of the laws established in previous chapters and see whether they need any further generalization. One of the merits of using field quantities (**E** and **B**, **H** and **D**) to express the laws is that an inconsistency is detected which can only be resolved by the addition of the so-called displacement current. This addition, first suggested by Maxwell, completes the set of equations known by his name and has far-reaching consequences.

The chapter falls broadly into three parts. In the first two sections we discuss the introduction of displacement current and the forms taken by Maxwell's equations. Then in Secs 13.3–13.7, we look at solutions to the equations: first relating them to previous work in this book and then seeing the effect of the new displacement current term in predicting the way field changes are propagated (Secs 13.6 and 13.7). Finally, the last three sections deal with some fundamental aspects: those of energy, momentum and the relationship with special relativity.

13.1 Completion of Maxwell's equations: displacement current

Summary of electromagnetic laws so far The most general forms taken by the laws of electromagnetism are those expressing the relations of the fields to each other and to the charges and currents producing them. We began by defining two basic fields *in vacuo*, the **E**-field as the force per unit stationary charge and the **B**-field as the force per unit current element, which together lead to the Lorentz force law:

$$\mathrm{d}\mathbf{F} = (\rho\mathbf{E} + \mathbf{J} \times \mathbf{B})\,\mathrm{d}\tau \qquad \text{or} \qquad \mathbf{F} = Q(\mathbf{E} + \mathbf{v} \times \mathbf{B}) \tag{13.1}$$

Inside materials, **E** and **B** are defined by processes described in Chapters 11 and 12, and two further fields introduced by $\mathbf{D} = \varepsilon_0\mathbf{E} + \mathbf{P}$ and $\mathbf{H} = \mathbf{B}/\mu_0 - \mathbf{M}$.

The laws governing the two electric field quantities, **E** and **D**, and the two magnetic field quantities, **B** and **H**, can be expressed in either integral or differential form:

Integral forms **Differential forms**

$$\oint_S \mathbf{D} \cdot \mathrm{d}\mathbf{S} = \sum_\tau Q_c \text{ or } \int_\tau \rho_c\,\mathrm{d}\tau \qquad \nabla \cdot \mathbf{D} = \rho_c \tag{13.2}$$

$$\oint_L \mathbf{E} \cdot \mathrm{d}\mathbf{L} = -\partial\Phi_S/\partial t \text{ or } \int_S (-\partial\mathbf{B}/\partial t) \cdot \mathrm{d}\mathbf{S} \qquad \nabla \times \mathbf{E} = -\partial\mathbf{B}/\partial t \tag{13.3}$$

$$\oint_S \mathbf{B} \cdot \mathrm{d}\mathbf{S} = 0 \qquad \nabla \cdot \mathbf{B} = 0 \tag{13.4}$$

$$\oint_L \mathbf{H} \cdot \mathrm{d}\mathbf{L} = \sum_S I_c \text{ or } \int_S \mathbf{J}_c \cdot \mathrm{d}\mathbf{S} \qquad \nabla \times \mathbf{H} = \mathbf{J}_c \tag{13.5}$$

In any integral equation, a volume τ is always bounded by a closed surface S, and an open surface S by a closed path L. The subscript c means that only conduction and convection charges and currents are included.

Finally, we have the relation between charge and current expressing the conservation of charge in an equation of continuity:

$$\oint_S \mathbf{J} \cdot \mathrm{d}\mathbf{S} = -\partial Q/\partial t \qquad \text{or} \qquad \nabla \cdot \mathbf{J} = -\partial\rho/\partial t \tag{13.6}$$

We have developed all these equations through a process of greater and greater generalization, starting with purely static fields *in vacuo* and working through to time-varying **B**-fields and fields within media. The question we now ask is whether any further terms need to be added if, for instance, the **E**- or **D**-field varies in time. We resolve this by examining, as Maxwell did, a discrepancy in the laws stated above.

Displacement current: argument from integral form of circuital law for H Consider a parallel plate capacitor as in Fig. 13.1† being charged by a conduction current I_c. The **H**-field round a path L satisfies Eq. (13.5), i.e. $\oint \mathbf{H} \cdot \mathrm{d}\mathbf{L} = I_c$, where I_c is the current encircled by the path L. If we ask the question

† Objections to this arrangement are sometimes made that the current takes some finite time to spread over the capacitor plates and that therefore the **D**-field between them is not uniform (e.g. Catt, Davidson and Walton, 1978). To counter this, one may take the 'plates' as no larger than the cross-sections of the wire carrying I_c and make their separation small enough for leakage of the **D**-field to be negligible. However, in any case the argument is only used to suggest the generalization which follows.

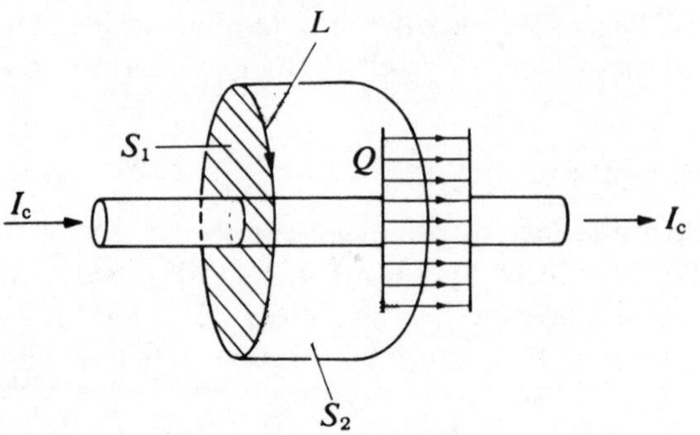

Figure 13.1 Application of the circuital law for **H** to a conduction current not flowing in a complete circuit.

'how do we decide what *encirclement* means?', we should reply in the case of a steady current that *any* surface S bounded by L should be taken and the current across it found. In Fig. 13.1, however, while S_1 is certainly crossed by I_c, S_2 has no conduction current across it and so we have both

$$A: \oint_L \mathbf{H}\cdot d\mathbf{L} = I_c \quad \text{if } S_1$$
$$B: \oint_L \mathbf{H}\cdot d\mathbf{L} = 0 \quad \text{if } S_2 \qquad (13.7)$$

We are in no doubt that at some distance from the capacitor the correct expression is very close to A because we can measure **H** and confirm it and also because the highly respected current-element formula of Sec. 7.7 would give us the same result. We are equally sure therefore that **H** does not satisfy B. There is an inconsistency here which is due to the time variations: the presence of the capacitor means that we have a non-steady situation with charges building up on the plates. We seek a way of generalizing the circuital law so that the right-hand side is the same no matter how the surface S is chosen, always assuming that it is bounded by L.

Some indication of the solution is provided by the special case of Fig. 13.1. While S_2 does not have I_c flowing across it, it *does* have a flux of **D** *which changes in time as long as I_c flows*. Moreover, because $D = \sigma = Q/A$, where A is the area of the plates, $I_c = dQ/dt = d(DA)/dt$. So I_c is equal to the rate of change of **D**-flux over S_2. If now we define a quantity we call the *displacement current* I_d by

$$I_d = d(DA)/dt \qquad (13.8)$$

and amend the right-hand side of (13.7) to $I_c + I_d$, the discrepancy is resolved: over S_2, $I_c = 0$, while over S_1, $I_d = 0$, so that $\oint \mathbf{H}\cdot d\mathbf{L}$ is the same for both.

Generalization Guided by the special case, we define displacement current generally by

$$I_d = \frac{\partial}{\partial t}\int_S \mathbf{D}\cdot d\mathbf{S} = \int_S \frac{\partial \mathbf{D}}{\partial t}\cdot d\mathbf{S} \qquad \text{(definition of } I_d \text{ over } S) \qquad (13.9)$$

while $\partial\mathbf{D}/\partial t$ is the *displacement current density* $\mathbf{J}_d$. We also generalize the situation of Fig. 13.1 into that of Fig. 13.2, where both S_1 and S_2 have conduction and

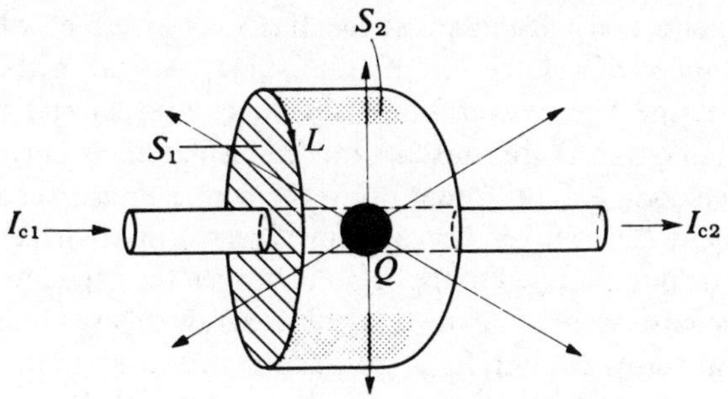

Figure 13.2 Generalization of displacement current. The circuital law for **H** applied round L will give different results if I_{c_1} crossing S_1 is different from I_{c_2} crossing S_2, unless the displacement current arising from the time variation of **D** across the surfaces is included as well. **D** varies because the unequal currents cause Q to vary.

displacement currents across them. The current entering the volume bounded by S_1 and S_2 is different from that leaving it, so that the charge Q within the volume changes according to

$$I_{c1} - I_{c2} = \frac{dQ}{dt} \tag{13.10}$$

If Gauss's law for **D** is applied to the same volume, we have

$$\int_{S_1} \mathbf{D} \cdot d\mathbf{S} + \int_{S_2} \mathbf{D} \cdot d\mathbf{S} = Q$$

Taking the time rate of change of both sides of this equation gives

$$-I_{d1} + I_{d2} = dQ/dt \tag{13.11}$$

the negative sign occurring because I_d must be related to the sense of L by the right-hand screw rule. Equations (13.10) and (13.11) give

$$I_{c1} + I_{d1} = I_{c2} + I_{d2}$$

and the sum of conduction and displacement currents is independent of S. It is therefore postulated that the required generalization of (13.5) is

$$\oint_C \mathbf{H} \cdot d\mathbf{L} = I_c + I_d \quad \text{or} \quad \nabla \times \mathbf{H} = \mathbf{J}_c + \mathbf{J}_d \tag{13.12}$$

Displacement current: argument from equation of continuity It is easily shown (using, for example, Cartesian coordinates) that $\nabla \cdot \nabla \times \mathbf{A}$ or div **curl A** is zero for *any* vector field **A**. If we take the divergence of both sides of (13.5b), we deduce that $\nabla \cdot \mathbf{J}_c = 0$. Now this is certainly valid for steady currents, but it violates the conservation of charge as expressed in the equation of continuity (13.6). This means that when time variations occur in currents and **H**-fields, (13.5) and (13.6) are inconsistent.

Since we would wish to preserve the conservation of charge, we keep (13.6) but write it in a different way using $\nabla \cdot \mathbf{D} = \rho$. Replacing ρ by $\nabla \cdot \mathbf{D}$ in (13.6) enables it to be written as

$$\nabla \cdot (\mathbf{J}_c + \partial \mathbf{D}/\partial t) = 0 \tag{13.13}$$

To make (13.5) consistent with this, we must clearly replace $\mathbf{J}_c$ by $\mathbf{J}_c + \partial \mathbf{D}/\partial t$, which is the same as (13.12) with $\mathbf{J}_d = \partial \mathbf{D}/\partial t$.

Displacement current: effect on calculations so far If (13.12) is correct, why have we been able to obtain verifiable results in this subject so far without the displacement current term? The answer for steady currents is that, referring to Fig. 13.2, $I_{c1}=I_{c2}$ so that Q and $\mathbf{D}$ are constant and the displacement currents are zero. In a.c. circuits we confined our attention to the connecting wires so that, referring to Fig. 13.1, only I_c was taken into account: this will give correct results as long as the electric field is confined to the region between the capacitor plates (for then $I_c=I_d$ and we have seen that *either* may be used) and also as long as the variations in current are slow enough for I_c to be the same at all points in the connecting wires (the assumption of quasi-steady conditions). We therefore expect any effects due to displacement current to become noticeable when high-frequency electric fields occur.

Problem 13.2 is worth tackling at this stage because it shows clearly the point that the current element formula using conduction currents *alone* will give the correct **H**- and **B**-fields due to any system of varying currents. However, when using the circuital law, both conduction and displacement currents over the chosen surface S must be included. A mistake which must be avoided can be illustrated using Fig. 13.1. In applying the circuital law round L do not include both the I_c across S_1 and the I_d across S_2, but only one or the other. In practice, of course, the **H**-field given by A of (13.7) will not be quite correct but will give a value slightly higher than it really is. Three different approaches show that the **H**-field in Fig. 13.1 is slightly smaller than the value $I_c/2\pi r$ for an infinite current. Firstly, the circuital law using S_1: there is a small amount of displacement current crossing S_1 in opposition to I_c because of leakage fields from the capacitor. Secondly, the circuital law using S_2: the same leakage fields mean that the displacement current across S_2 is not quite equal to I_c. Thirdly, the current element formula using only I_c must take into account the missing element between the capacitor plates.

13.2 Forms of Maxwell's equations

The set of equations now developed is known as Maxwell's equations, and we adopt the hypothesis that they need no further generalization. We set them out in their integral and differential forms.

Integral forms in which surfaces S are bounded by closed paths L, volumes τ by closed surfaces S, while Q, I or ρ, $\mathbf{J}$ include conduction and convection charges only:

---

a. Coulomb's law $\rightarrow \oint_S \mathbf{D}\cdot d\mathbf{S}=\sum_\tau Q$ or $\int_\tau \rho\, d\tau$

b. Coulomb's and Faraday's laws

$$\rightarrow \oint_L \mathbf{E}\cdot d\mathbf{L}=-\partial\Phi_S/\partial t \quad \text{or} \quad \int_S -\frac{\partial \mathbf{B}}{\partial t}\cdot d\mathbf{S}$$

c. Ampère's law and absence of poles

$$\rightarrow \oint_S \mathbf{B}\cdot d\mathbf{S}=0$$

d. Ampère's law and displacement current

$$\rightarrow \oint_L \mathbf{H}\cdot d\mathbf{L}=I_c+I_d \quad \text{or} \quad \int_S (\mathbf{J}+\partial\mathbf{D}/\partial t)\cdot d\mathbf{S}.$$

(13.14)

Differential forms These are given in terms of the vector operators $\nabla\cdot$ (divergence) and $\nabla\times$ (curl) but the Cartesian forms should be well known to the reader (see Appendix B.9):

Scalar	Vector	
$\nabla\cdot\mathbf{D}=\rho$	$\nabla\times\mathbf{E}+\dot{\mathbf{B}}=0$	a b
		(13.15)
$\nabla\cdot\mathbf{B}=0$	$\nabla\times\mathbf{H}-\dot{\mathbf{D}}=\mathbf{J}$	c d

Equations (b) and (c) are homogeneous differential equations, while (a) and (d) are inhomogeneous or source equations, containing ρ and J. The conservation of charge in the form of the equation of continuity now *follows from* the inhomogeneous equations and does not need to be independently postulated.

Supplementary equations Maxwell's equations need to be supplemented by three equations embodying definitions of the fields. These are the Lorentz force law (13.1):

$$\mathbf{F}=Q(\mathbf{E}+\mathbf{v}\times\mathbf{B}) \quad \text{or} \quad d\mathbf{F}=(\rho\mathbf{E}+\mathbf{J}\times\mathbf{B})\,d\tau$$

and the definitions of **D** and **H**:

$$\mathbf{D}=\varepsilon_0\mathbf{E}+\mathbf{P} \quad \text{and} \quad \mathbf{H}=\mathbf{B}/\mu_0-\mathbf{M} \tag{13.16}$$

When particular media are specified, they will obey constitutive equations such as those for LIH dielectrics, magnetic media, and conductors, namely:

$$\mathbf{D}=\varepsilon_r\varepsilon_0\mathbf{E}; \quad \mathbf{H}=\mathbf{B}/(\mu_r\mu_0); \quad \mathbf{J}=\sigma(\mathbf{E}+\mathbf{e}) \tag{13.17}$$

where **e** represents the contributions of electromotances not included in the Maxwell **E**-field.

Boundary conditions These were derived in Chapters 11 and 12 using the integral forms of Maxwell's equations but without the displacement current. However, we saw in Sec. 11.5 that the $\partial\mathbf{B}/\partial t$ term made no difference to the static boundary condition on the tangential component of **E**, denoted by E_t. For exactly the same reason, the $\partial\mathbf{D}/\partial t$ term makes no difference to the boundary condition (12.24) on H_t. Hence, we have quite generally:

	Normal	Tangential	
Electric	D_n discontinuous by σ_c	E_t continuous	a b
			(13.18)
Magnetic	B_n continuous	H_t discontinuous by J_{sc}	c d

where σ_c is the surface density of conduction charge and J_{sc} is the surface density of conduction current.

13.3 Solutions of Maxwell: specifying the conditions

All the problems that have been solved in previous chapters provide solutions to Maxwell's equations in various restricted forms. The restrictions arise from specifications made in the problems (e.g. that no time variations occur) and only when these are clear can a decision be made about the method to be adopted for solution. In this section we survey the types of specification that can be made, the most important of which is the type of time variation occurring since this has most influence on the method available.

Time variations *Steady conditions*, i.e. no time variations. See Chapters 1–8 and the further discussion in Sec. 13.4.
Quasi-steady conditions, i.e. $\partial\mathbf{D}/\partial t$ negligible. See Chapters 9 and 10 and the discussion in Sec. 13.5.
Otherwise the complete equations are needed. See Sec. 13.6 onwards.

Sources *No sources.* $\rho=0$, $\mathbf{J}=0$ and no conductors present.
Specified sources. Any ρ or $\mathbf{J}$ inserted in Maxwell's equations in the form specified in the problem. $\mathbf{J}$ may be a neutral current ($\rho=0$) or a charged current ($\mathbf{J}=\rho\mathbf{v}$).
Sources to be found. Are any conductors present in which conduction charges and currents are induced? If so, ρ and $\mathbf{J}$ are to be found as part of the problem.

Media *In vacuo.* $\mathbf{D}=\varepsilon_0\mathbf{E}$ and $\mathbf{H}=\mathbf{B}/\mu_0$, so Maxwell can be expressed in terms of **E** and **B** only.
With media. What properties are specified? Are they LIH, in which case one or more of (13.17) applies. Note that with a ferromagnetic **M** may be *given* and be a specified source of **B**- and **H**-fields.

Boundaries and symmetry What shapes have any boundaries? What symmetry has any specified distribution of charge or current? These two factors will decide which coordinate system is most convenient.

Fields and potentials Are any of these specified, for instance as an applied uniform field or as a fixed potential of a conductor?

Note that one cannot specify just *any distribution whatever* of V, **E**, **B**, etc., in space and time—Maxwell's equations must be satisfied. Non-Maxwell fields may be adequate descriptions over limited regions only.

Finally, an important feature of Maxwell's equations is that they are linear and hence that **superposition** applies. Any number of solutions may be superposed to give another solution.

13.4 Steady solutions of Maxwell: Poisson, Laplace, and uniqueness

With no time variations at all, the electric and magnetic equations become completely uncoupled. We assume in this section that *all media are LIH*, so that with $\partial\mathbf{B}/\partial t$ and $\partial\mathbf{D}/\partial t$ both zero, we may write Maxwell's equations as

steady fields:

$$\text{Electric:}\quad \nabla\cdot\mathbf{E}=\rho/\varepsilon_r\varepsilon_0 \quad (a) \qquad \nabla\times\mathbf{E}=0 \quad (b)$$
$$\text{Magnetic:}\quad \nabla\cdot\mathbf{B}=0 \quad (c) \qquad \nabla\times\mathbf{B}=\mu_r\mu_0\mathbf{J} \quad (d) \qquad (13.19)$$

This means that, as in the earlier chapters of this book, electricity and magnetism can be treated as independent subjects called electrostatics and magnetostatics. We have already dealt extensively with simple problems in this area in earlier chapters and we now only point the way towards more complicated ones. To do this we survey the properties of potentials.

Because the curl of **E** and the divergence of **B** are zero, potentials V and **A** can be defined as follows:

$$\nabla\times\mathbf{E}=0 \text{ implies } \mathbf{E}=-\nabla V \text{ and } V \text{ satisfies } \nabla^2 V=-\rho/\varepsilon_r\varepsilon_0 \qquad (13.20)$$

$$\nabla\cdot\mathbf{B}=0 \text{ implies } \mathbf{B}=\nabla\times\mathbf{A} \text{ and } \mathbf{A} \text{ satisfies } \nabla^2\mathbf{A}=-\mu_r\mu_0\mathbf{J} \qquad (13.21)$$

(see Secs 4.7 and 8.5). Equation (13.20) is Poisson's equation and (13.21) means that all the Cartesian components of **A** also obey that equation.

Now we know that very general solutions of (13.20) and (13.21) are given by (3.17) and (8.20) with the addition of ε_r and μ_r:

steady fields:

$$V=\int_\tau \frac{\rho\,\mathrm{d}\tau}{4\pi\varepsilon_r\varepsilon_0 r},\ \text{etc.} \qquad \mathbf{A}=\int_\tau \frac{\mu_r\mu_0\mathbf{J}\,\mathrm{d}\tau}{4\pi r},\ \text{etc.} \qquad (13.22)$$

but these are only of use if the ρ and **J** are specified and even then are not always easy to apply. In many cases it may be necessary to resort to solving Poisson's or Laplace's equation by analytical or numerical methods that are beyond the scope of this volume. However, there are a number of useful methods of solution that depend heavily on the existence of a uniqueness theorem and this we now discuss.

Uniqueness theorem This theorem states that if a potential obeys Poisson's equation throughout a region and at the same time has certain specified properties over the boundaries of that region, there is only one solution that can satisfy all the conditions. The potential, and therefore the field, is unique. This result at first sight seems trivial, but is in fact of great power in the search for solutions to problems.

We shall indicate the proof of the theorem for the special case of an electrostatic potential V due to a number of charged conductors *in vacuo*. In Fig. 13.3, we have one conductor whose total charge is given as Q_0 and whose surface is called S_Q. There is another conductor whose potential is specified at V_0 over its surface S_V. The region is bounded in the figure by S_0 but we wish to specify that as S_0 recedes to infinity, V over it tends to zero, as would be the case in practice.

Consider the shaded volume τ which is bounded by S_Q, S_V and S_0, and within which V satisfies Poisson's equation if there exists any distribution of conduction charge we have not drawn, Laplace's equation if not. Hence:

$$\nabla^2 V = -\rho/\varepsilon_0; \qquad V = V_0 \text{ over } S_V; \qquad V=0 \text{ over } S_0 \text{ at infinity;}$$

$$Q_0 = \oint \sigma \, \mathrm{d}S \quad \text{ or } \quad \oint \varepsilon_0 (\partial V/\partial n) \, \mathrm{d}S \text{ over } S_Q$$

where we have used (4.12): that E_n at the surface of a conductor is σ/ε_0.

Now we suppose that there are two solutions V_1 and V_2 which satisfy all the conditions and we let $\phi = V_1 - V_2$. This new function satisfies:

$$\nabla^2 \phi = 0 \text{ over } \tau; \qquad \phi = 0 \text{ over } S_V \text{ and } S_0;$$

$$\phi \text{ is constant and } \oint \varepsilon_0 (\partial \phi/\partial n) \, \mathrm{d}S = 0 \text{ over } S_Q$$

If Gauss's divergence theorem [(B.17) of Appendix B.10] is applied to the vector $\phi\nabla\phi$, we obtain the following relation when the result is integrated over τ:

$$\int_\tau \nabla \cdot (\phi \nabla \phi) \, \mathrm{d}\tau = \oint_{S_0, S_V, S_Q} \phi (\partial \phi/\partial n) \, \mathrm{d}S$$

The right-hand side is zero because (a) over S_0 and S_V, ϕ is zero and (b) over S_Q, ϕ is constant, comes out of the integral and leaves $\oint (\partial \phi/\partial n) \, \mathrm{d}S$, which is also zero.

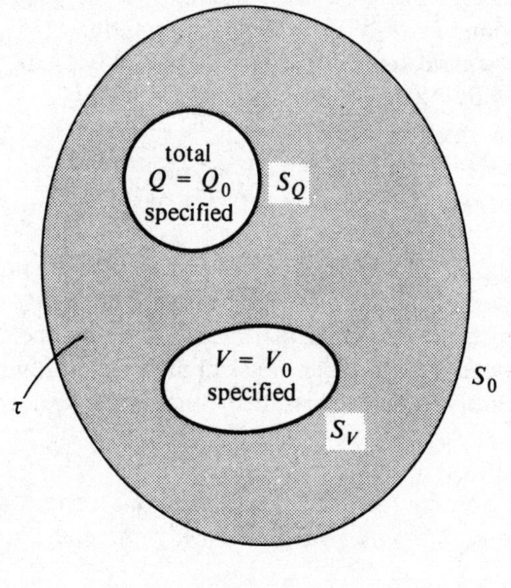

Figure 13.3 Situation in which the uniqueness theorem applies: V obeys Poisson's equation throughout the shaded volume τ. S_Q is the surface of a conductor whose total charge is specified, S_V is the surface of a conductor whose potential is specified.

The left-hand side transforms to $\phi\nabla^2\phi+(\nabla\phi)^2$ [using (B.20) of Appendix B.11], and since $\nabla^2\phi=0$ we are left with

$$\int_\tau (\nabla\phi)^2\,\mathrm{d}\tau=0$$

The integrand is essentially positive everywhere and by this equation can therefore only be zero everywhere. So $\boldsymbol{\nabla}\phi=0$. However, $\phi=0$ over S_0 and S_V, so that ϕ itself must be zero everywhere and hence $V_1=V_2$.

It is possible to take more general cases than we have or to specify fewer types of boundary conditions, but the proof broadly speaking still stands: V, and therefore **E** and the surface density of charge on any conductors, can only have one set of values. Similar theorems exist also for the magnetic scalar and vector potentials.

Importance of uniqueness and superposition In some problems, it is possible to find a solution fitting the given boundary conditions by a simple inspection or by a trial of plausible possibilities. If we can, then the uniqueness theorem tells us that it is *the* solution. For example, suppose that we wish to find the potential inside a hollow conductor maintained at V_0, a problem we looked at in Sec. 4.1. Clearly, $V=V_0$ throughout the hole not only satisfies Laplace's equation but also fits the boundary conditions as well—it is therefore the solution, and it follows that $\mathbf{E}=0$.

Since we know that uniform fields, fields of single charges, dipoles, line charges, etc., all satisfy Maxwell and yield potentials satisfying Laplace, any combination of such fields and potentials will also do so. It is often possible to fit the boundary conditions by superposing such a combination. The method of images in Sec. 4.8 is just such a method and it is ultimately validated by the uniqueness theorem. We give a few other examples.

Worked example *conducting sphere in uniform* **E**-*field* Consider a conducting sphere of radius a placed in a specified uniform electric field $\mathbf{E}_0$ (Fig. 13.4a). We wish to find how the induced charge on the surface is distributed and what effect it has on the field. Because the system has axial symmetry about a line parallel to $\mathbf{E}_0$ through the centre of the sphere, we choose polar coordinates so that a point on the surface is given by (a, θ).

There is no **E**-field in the interior of the conductor, so we concentrate on finding the additional **E**-field outside the surface which is produced by the induced charges. The positive and negative charges are equal in magnitude and suggest that their field might be matched by that of an electric dipole **p** placed at the centre as shown in Fig. 13.4b. Since a dipole potential satisfies Laplace's equation we have only to match the boundary conditions at the surface. There is, however, only one condition: that the resultant **E**-field shall be perpendicular to the surface (because E_t is continuous and E_t is zero inside). Referring to Fig. 13.4b, we see that this can be achieved if the tangential components of the dipole field and of E_0 cancel, i.e. if

$$E_0\sin\theta=p\sin\theta/(4\pi\varepsilon_0 a^3)$$

or

$$p=4\pi\varepsilon_0 a^3 E_0 \qquad (13.23)$$

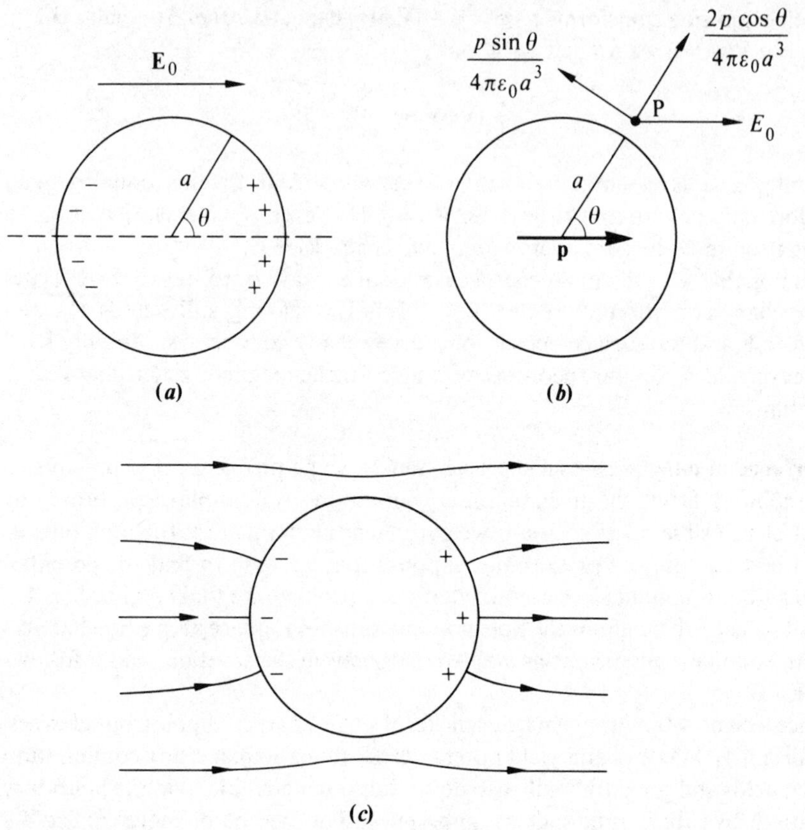

Figure 13.4 (a) Conducting sphere in a uniform **E**-field has induced charges on the surface; (b) the surface charges produce an additional **E**-field equivalent to that of a dipole at the centre: the fields at P are shown; (c) the resultant field.

Thus the polarizability (Sec. 11.9) of such a sphere is proportional to its volume. The resultant field is shown in Fig. 13.4c. The surface density of charge at any point is obtained by finding the value of $\varepsilon_0 E$ at the point [i.e. using Eq. (4.12)]. The uniqueness theorem assures us that this solution is the one we require.

Worked example *dielectric sphere in uniform* **E**-*field* The same method can be used to find the effect of a dielectric sphere with relative permittivity. We suggest the method here and leave the complete solution as a problem.

For points outside the sphere we assume that polarization produces a similar effect to electrostatic induction in the conductor, i.e. an extra field is produced equivalent to that of a dipole of moment **p** at the centre. However, inside the sphere, the field is no longer zero. In that region the simplest assumption is that the field is uniform with a constant value E' parallel to $\mathbf{E}_0$. If this can be made to work at the boundary then uniqueness applies. It is then only necessary to equate E_t and D_n inside the surface to E_t and D_n outside. Two equations are obtained from which the values of p and E' emerge. The method also yields the field in a spherical hole in a dielectric carrying a uniform field (Problem 13.6).

Worked example *floating magnet above superconductor* As a magnetic example, we consider a well-known demonstration of superconductivity. A bar magnet rests on the surface of mercury at a temperature just above critical. On lowering the temperature so that the mercury becomes superconducting, the magnet is seen to rise and float in the vapour above the surface. Why is this evidence of superconductivity?

When a metal becomes superconducting it is known that magnetic flux is completely expelled from its interior, i.e. $\mathbf{B}=0$ everywhere inside it. Since Maxwell's equations still apply, so does the boundary condition on B_n, and this means that just outside the surface, $B_n=0$. Thus the magnetic field must always be tangential to the surface.

Such a modification to the **B**-field is in fact produced by currents flowing in the surface, but the *effect* of these currents must be to produce an additional **B**-field *as if* there were an image magnet (Fig. 13.5). Such an image produces a field whose potential V_m satisfies Laplace's equation at the same time as satisfying the boundary condition that the resultant **B** is parallel to the surface. Unlike the electrostatic images in Sec. 4.8, this one has like poles opposing each other so that repulsion takes place.

Steady current flow Within a conductor of conductivity σ, we have that the current density **J** at any point is $\mathbf{J}=\sigma\mathbf{E}$. Because $\nabla\cdot\mathbf{J}=0$, the divergence of **E** is also zero. Under steady conditions, $\nabla\times\mathbf{E}=0$ so that a scalar potential V exists which satisfies Laplace's equation. This means that, provided the boundary conditions are the same, solutions of electrostatic problems and of current-flow problems will be the same.

This identity is exploited in analogies between capacitor plates *in vacuo* and metallic electrodes immersed in an electrolyte. The electrodes have a much higher conductivity than the electrolyte, and thus behave as equipotential volumes having surfaces over which V is constant. The **E**-field between the capacitor plates is identical with that between the electrodes if the geometry and voltages are the same in the two cases. Since $V=Q/C$ for the capacitance and $V=RI$ for the electrolyte, $CR=Q/I$ and it is not difficult to show that this is ε_0/σ. In two dimensions one can exploit the same analogy by using conducting paper in place of the electrolyte, thus enabling **E**-fields to be plotted for any shape of electrode.

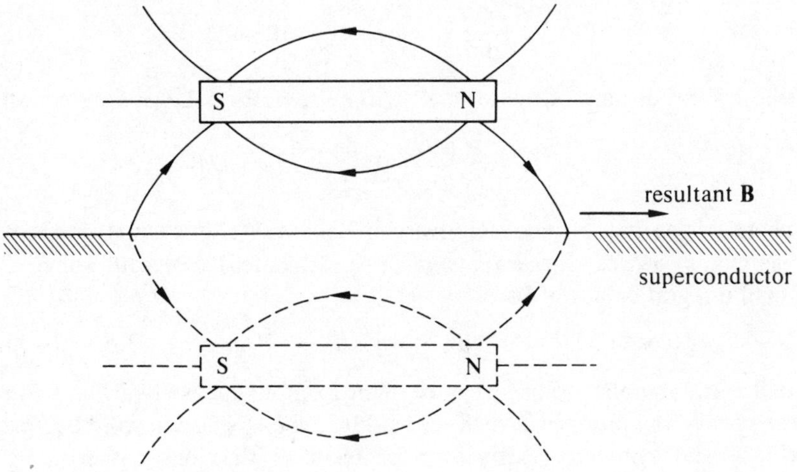

Figure 13.5 A bar magnet floating above the surface of a superconductor.

Analytical solutions of Laplace's equation We have encountered several potential functions which exist under steady conditions: the magnetic scalar potential V_m, the vector potential **A** and the electrostatic potential V which includes that occurring in current flow (see preceding paragraphs). Provided no sources are present (i.e. charge density for V, current density for V_m and **A**), they all obey Laplace's equation whose solutions are therefore clearly of great importance. When the boundaries involved in a problem are complicated, numerical methods using computer programs can be used to obtain solutions to any desired degree of accuracy, and it is beyond the scope of this book to describe these. However, if the boundaries can be conveniently and simply expressed in one of the main coordinate systems, then exact solutions can be obtained analytically. We discuss very briefly one approach to such solutions in terms of V, although it applies equally well to V_m or to any component of **A**.

When the solution is dependent on only one variable (such as x in Cartesians or r in polars), Laplace's equation degenerates into a total differential equation which can be quickly solved (e.g. see Problem 4.20). When two or three variables are involved, then a common method of solution is that of the *separation of variables*. This assumes that V can be written as the product of independent functions, e.g. as $X(x)Y(y)Z(z)$ or $R(r)\Theta(\theta)\Phi(\phi)$ say. Substitution in $\nabla^2 V=0$ then gives a separate total differential equation for each function and the solutions of these are well known. Each one of these equations has an infinite number of possible solutions and it is usually a matter of finding a combination to satisfy the given boundary conditions. The combination will often be an infinite series. We discuss briefly the types of solution where only two variables are involved.

In *Cartesian* coordinates, X is found to obey $(1/X)\,\mathrm{d}^2X/\mathrm{d}x^2+k_x^2=0$, where k_x is a constant. Y is similar, and $k_x^2+k_y^2=0$. Thus the general solution for X is of the form $X=A\exp(ik_x\mathrm{x})+B\exp(-ik_x x)$, which is sinusoidal if k_x^2 is positive (if $k_x^2=0$ we have the special case $X=Ax+B$: compare the solutions in Sec. 10.2). Boundary conditions on V frequently restrict values of k_x and k_y to integral multiples of a constant of the problem, so that the solution could be a Fourier series. The coefficient would depend on the values of V at the boundaries.

In *spherical polar* coordinates, we have $V=R(r)\Theta(\theta)$. R is found to obey the equation

$$\frac{\mathrm{d}}{\mathrm{d}r}\left(r^2\frac{\mathrm{d}R}{\mathrm{d}r}\right)-l(l+1)R=0$$

whose solutions are of the form Ar^l+B/r^{l+1}. Θ obeys Legendre's equation

$$(1-x^2)\frac{\mathrm{d}^2\Theta}{\mathrm{d}x^2}-2x\frac{\mathrm{d}\Theta}{\mathrm{d}x}+l(l+1)\Theta/0$$

where x is written for $\cos\theta$. Convergent solutions of this equation are polynomials in x known as *Legendre polynomials* and denoted by $P_l(\cos\theta)$. Some values for small integral values of l are

$$P_0(\cos\theta)=1;\qquad P_1(\cos\theta)=\cos\theta;\qquad P_2(\cos\theta=\tfrac{1}{2}(3\cos^2\theta-1)$$

so that the complete solution can be of the form $Ar^lP_l(\cos\theta)+BP_l(\cos\theta)/r^{l+1}$, or the sum of any number of such terms. This type of solution will be appropriate when axial symmetry exists in a problem so that only spherical r and θ

coordinates are needed. Among the problems of this type already solved in this book are the fields of dipoles, linear quadrupoles, spheres in uniform fields, and the reader will recognize that in each case the solution we have found does have the form given. Other problems in this category would be the fields due to discs, plane circular loops and cones.

13.5 Quasi-steady solutions of Maxwell

When time variations occur, there are two extra terms ($\partial\mathbf{B}/\partial t$ and $\partial\mathbf{D}/\partial t$) in Maxwell's equations over and above those describing the steady conditions of the last section. There is also, as a consequence, a need to use the full equation of continuity $\nabla\cdot\mathbf{J}=-\partial\rho/\partial t$ instead of $\nabla\cdot\mathbf{J}=0$. Yet there are many cases in which, even though there *are* time variations, we still seem to be able to use successfully the laws for steady conditions. In this section we examine under what circumstances we can ignore the time-varying terms.

The displacement current term appears in the same equation as the conduction current. It is thus possible to compare their magnitudes in a material possessing both a conductivity σ and a relative permittivity ε_r. Suppose that the electric E-field in the material is varying according to $\mathbf{E}=\mathbf{E}_0 \sin \omega t$. Then $\mathbf{J}_c=\sigma\mathbf{E}_0 \sin \omega t$ while $\mathbf{J}_d=\partial\mathbf{D}/\partial t=\varepsilon_r\varepsilon_0\omega\mathbf{E}_0 \cos \omega t$. The ratio of the *magnitudes* of $\mathbf{J}_c$ and $\mathbf{J}_d$ is thus $\sigma/(\varepsilon_r\varepsilon_0\omega)$. The quantity $\varepsilon_r\varepsilon_0/\sigma$ is a time τ which has a value of around 10^{-6}–10^{-12} s for poor conductors and up to 10^{-18} s for metals. Thus, until frequencies of around 10^{18} Hz are reached, the displacement current in a metallic conductor is negligible in comparison with conduction current.

For exactly the same reason, any charge density ρ that can accumulate in a good conductor disperses so rapidly that $\nabla\cdot\mathbf{J}=0$ is a good approximation. To see this, we use the full equation $\nabla\cdot\mathbf{J}=-\partial\rho/\partial t$ and put $\mathbf{J}=\sigma\mathbf{E}$ with $\nabla\cdot\mathbf{E}=-\rho/\varepsilon_r\varepsilon_0$. Then ρ satisfies the equation

$$\partial\rho/\partial t=-(\sigma/\varepsilon_r\varepsilon_0)\rho$$

which has a solution $\rho=\rho_0 \exp(-t/\tau)$, where $\tau=\varepsilon_r\varepsilon_0/\sigma$, a relaxation time which we saw was about 10^{-18} s in a good conductor.

With the information of the last two paragraphs, let us examine what we do in lumped-network theory where we seem to be able to use quasi-steady conditions most successfully. Our attention there is always concentrated on the currents and voltages in the connecting leads to the terminals of the elements while completely ignoring what is going on within the elements themselves. Thus in the case of a capacitor, we exclude the region between the plates where the displacement current predominates and deal only with the wires to the terminals in which $\partial D/\partial t$ and ρ are negligible. It follows that $\nabla\cdot\mathbf{J}=0$ can be successfully applied at any point and Kirchhoff's first law is an immediate deduction.

Kirchhoff's second law stems from the path-independence of V and thus relies on $\nabla\times\mathbf{E}$ being zero. As before, we ignore the regions where the electromotive field $\mathbf{E}_M$ might be significant and concentrate on the leads where it is negligible. If we therefore take a closed path in the network which excludes the interior of inductances, the contribution from $\partial\mathbf{B}/\partial t$ is negligible, $\nabla\times\mathbf{E}$ is effectively zero and V is path-independent.

In Chapter 9, we introduced the $\partial\mathbf{B}/\partial t$ term without the displacement current.

As a result we had to abandon the use of a scalar potential with electromotive **E**-fields. Nevertheless, Maxwell's equations can still be treated as two separate pairs: *in vacuo*, both the divergence and curl of **B** are zero and are in no way connected with the **E**-field. The situation is very different when displacement current is introduced, for now there is an association between curl **E** and **B**, on the one hand, and between curl **B** and **E**, on the other. This inevitably means that **E**- and **B**-fields are so intimately connected that they can no longer be treated independently. It is to this situation that we now turn.

13.6 Electromagnetic waves *in vacuo*

The wave equation for E and B We now wish to look for solutions to Maxwell's equations when all time variations are included. In order to bring home the effect that the displacement current term has, we restrict ourselves in this chapter to *vacua*. We can therefore put $\mathbf{D}=\varepsilon_0\mathbf{E}$ and $\mathbf{H}=\mathbf{B}/\mu_0$ and take the conductivity σ to be zero. Moreover, in this section we consider a region with no sources, so the specified ρ and $\mathbf{J}$ are both zero. Maxwell's equations then become:

fields *in vacuo*, no sources:

$$\nabla\cdot\mathbf{E}=0 \qquad \nabla\times\mathbf{E}=-\dot{\mathbf{B}} \qquad \text{a} \quad \text{b}$$
$$\nabla\cdot\mathbf{B}=0 \qquad \nabla\times\mathbf{B}=\varepsilon_0\mu_0\dot{\mathbf{E}} \qquad \text{c} \quad \text{d} \qquad (13.24)$$

An equation governing the behaviour of **E** can be obtained by eliminating **B**. To do this, take the curl of both sides of (13.24b) and use (13.24d):

$$\nabla\times(\nabla\times\mathbf{E})=-\nabla\times\dot{\mathbf{B}}=-\partial/\partial t(\nabla\times\mathbf{B})=-\varepsilon_0\mu_0\,\partial^2\mathbf{E}/\partial t^2 \qquad (13.25)$$

To see more clearly what this equation means, we may use the vector identity [(B.26) of Appendix B11]: $\nabla\times\nabla\times\mathbf{E}=\nabla(\nabla\cdot\mathbf{E})-\nabla^2\mathbf{E}$, which is useful if Cartesian coordinates are to be used. Since $\nabla\cdot\mathbf{E}=0$, (13.25) becomes

$$\nabla^2\mathbf{E}-\varepsilon_0\mu_0\,\partial^2\mathbf{E}/\partial t^2=0 \qquad (13.26)$$

Elimination of **E** from (13.24) yields the same equation for **B**:

$$\nabla^2\mathbf{B}-\varepsilon_0\mu_0\,\partial^2\mathbf{B}/\partial t^2=0 \qquad (13.27)$$

These equations mean that each Cartesian component of **E** and **B** separately satisfy them. Now if E_y, say, depended only on one space coordinate, say x, (13.26) would give

$$\partial^2E_y/\partial x^2-\varepsilon_0\mu_0\,\partial^2E_y/\partial t^2 \qquad (13.28)$$

which the reader should recognize as a one-dimensional wave equation for waves whose phase velocity is $(\varepsilon_0\mu_0)^{-1/2}$. In a similar way, (13.26) and (13.27) are three-dimensional wave equations for *electromagnetic waves in vacuo whose velocity is denoted by c* where

waves *in vacuo*:

$$c^2=1/\varepsilon_0\mu_0 \qquad (13.29)$$

The general solution of (13.28) is easily shown to be any function of $t\pm x/c$, written as $E_{y0}f(t\pm x/c)$, the sign depending on the direction of travel. With the three-dimensional equation, the general solution depends on x, y, z, and t and the type of dependence on x, y and z varies according to the shape of the wavefronts (see below for plane wavefronts). But, whatever the details, (13.26) and (13.27) mean that **all changes in E and B field are propagated *in vacuo* at a speed equal to c**.

The existence of electromagnetic waves In SI units, μ_0 is assigned the value $4\pi \times 10^{-7}$ and ε_0 has a value which can be determined by purely electrical experiments (Appendix C), so that the velocity of any electromagnetic waves may be determined from electrical measurements alone. The value of c by these methods is almost exactly 3×10^8 m s^{-1}. It was the great achievement of Maxwell not only to predict the existence of such waves but to suggest that light, which it was known travelled with this same velocity, was a form of electromagnetic wave. Heinrich Hertz was among the first to demonstrate the existence of waves generated by the oscillatory spark discharge and to show that they possessed many of the familiar properties of light such as reflection, refraction, interference and polarization. It is now known that a complete spectrum of such waves exists with wavelengths ranging from the several metres of the long radio waves, through the shorter radio waves and microwaves, the infrared, visible and ultraviolet, to the X-rays with wavelengths of only a few angstrom units (and comparable with interatomic distances) and beyond these to γ-rays (Fig. 13.6).

In physical terms this section has shown that any time variations in electric and magnetic fields are propagated *in vacuo* with the same velocity $c = 1/(\varepsilon_0\mu_0)^{1/2}$. This is immediately relevant to the discussion of transmission lines in Sec. 10.10. At that stage we talked entirely in terms of changes in current and voltage which we also found propagated with a velocity c. In such lines, however, the currents are accompanied by **B**-fields and the voltages by **E**-fields in the space between the conductors. Changes in the fields will be brought about by the changes in I and V—what we see now is that both sets of changes will travel down the line together. Transmission lines can therefore be thought of as guides either for the electromagnetic waves or for the currents and voltages.

Unbounded plane waves The wave equation is very general and the fact that **E** and **B** fields satisfy it does not tell us very much in itself. However, the fields must still also satisfy Maxwell's equations and through these it is possible to determine the properties possessed by specified types of wave.

Here we consider an important case—that of an unbounded plane wave. By a **plane** wave is meant one in which the values of all field quantities are constant over planes (the wavefronts) perpendicular to the direction of propagation. The term **unbounded** implies that the wavefronts are effectively infinite. If the x axis is taken as the direction of propagation, the plane wavefronts would be parallel to

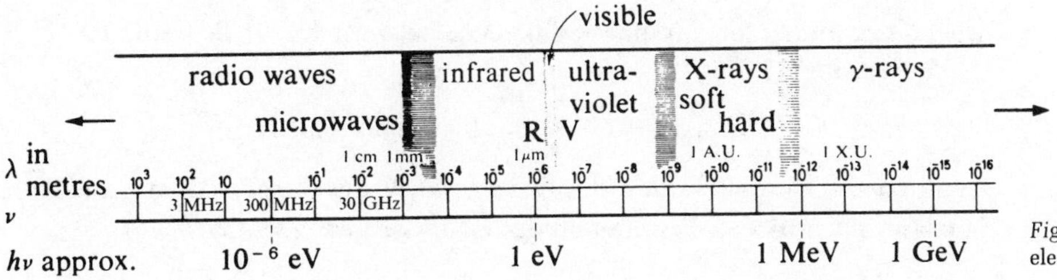

Figure 13.6 The electromagnetic spectrum.

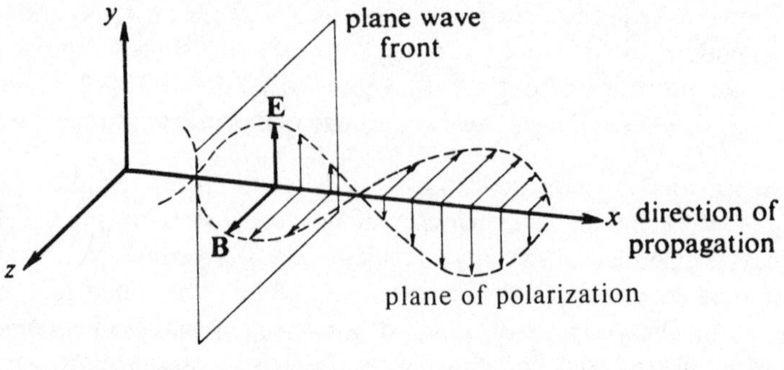

Figure 13.7 A monochromatic linearly polarized plane electromagnetic wave.

the yz plane so that $\partial/\partial y$ and $\partial/\partial z$ of any quantity would be zero. Maxwell's equations then reduce to

$$\partial E_x/\partial x=0; \qquad \partial B_x/\partial x=0; \qquad \partial B_x/\partial t=0; \qquad \partial E_x/\partial t=0 \qquad (13.30)$$

$$\left.\begin{aligned} \partial E_z/\partial x &= \partial B_y/\partial t & \partial B_z/\partial x &= -\mu_0\varepsilon_0\,\partial E_y/\partial t \\ \partial E_y/\partial x &= -\partial B_z/\partial t & \partial B_y/\partial x &= \mu_0\varepsilon_0\,\partial E_z/\partial t \end{aligned}\right\} \qquad (13.31)$$

Equations (13.30) mean that E_x and B_x vary neither with time nor in space and that at most the x components can only be steady uniform fields and cannot therefore be part of the wave. Hence **a plane wave is wholly transverse.**

Let us now further choose the y and z axes so that **E** lies along y and hence $E_z=0$. It is then clear from (13.31) that $\partial B_y/\partial x$ and $\partial B_y/\partial t$ are both zero and that B_y also does not vary in space and time. The **B** wave thus lies along z and we have the additional result that **E, B and the direction of propagation are mutually perpendicular** such that the vector $\mathbf{E}\times\mathbf{B}$ points along the direction of propagation (Fig. 13.7).

A wave in which **E** and **B** are both transverse is called a TEM wave (transverse electric and magnetic) so that **an unbounded plane wave is TEM.**

Monochromatic waves For the moment we shall keep the direction of propagation along the positive x axis so that the general forms of E_y and B_z will be functions of $t-x/c$. We are always especially interested in variations that are periodic with a definite frequency ν and wavelength λ, forming a *monochromatic* wave, because any other type of variation can be formed by superposing monochromatic waves with appropriate amplitudes, frequencies and phases (Fourier's theorem). We work in terms of the angular frequency $\omega=2\pi\nu$ and the wave number $k=2\pi/\lambda$, in which case $c=\omega/k$. The monochromatic function of $t-x/c$ can then be written as $\sin[-\omega(t-x/c)]$ or simply $\sin(kx-\omega t)$. We may therefore write

$$E_y=E_{y0}\sin(kx-\omega t); \qquad B_z=B_{z0}\sin(kx-\omega t+\alpha)$$

where α accounts for any phase difference between **E** and **B**. Using $\partial E_y/\partial x = -\partial B_z/\partial t$ from (13.31):

$$kE_{y0}\cos(kx-\omega t)=\omega B_{z0}\cos(kx-\omega t+\alpha)$$

for all x and t. The only value of α which can satisfy this is zero, so that **E** and **B** are **in phase**. The ratio of their amplitudes is ω/k or just c, so that

$$E_y=cB_z \qquad (13.32)$$

Thus **an unbounded plane monochromatic wave always has E and B in phase and with an amplitude ratio of** c.

Polarization The properties of plane waves we have deduced are those *at any point at any instant*. There is nothing to prevent *both vectors* changing direction within the yz plane at a particular point as the wave progresses (or at any instant as one progresses along the wave). If such changes in direction are quite random, the wave is said to be **unpolarized**. It is, however, possible to produce a wave whose **E** and **B** vector remain in the same planes and such a wave is **linearly polarized** or plane-polarized (we prefer the former term to avoid confusion with the plane of a wavefront). The **plane of polarization** is that containing the **E**-vector and the direction of propagation. Figure 13.7 is an attempt to show the magnitudes of **E** and **B** at any instant along the line of a linearly polarized TEM wave which is also monochromatic.

Two linearly polarized waves travelling along the x direction could be superposed. If they had the same frequency but a phase difference α, the **E**- vectors would combine to form a Lissajous figure—in general the tip of the **E**-vector would trace out an ellipse as the wave moved through a point and it would be **elliptically polarized**. A **circularly polarized** wave is a special case of elliptical polarization produced by the superposition of linearly polarized components with planes of polarization at right angles and with equal amplitudes and a phase difference of $\pi/2$. All such waves are solutions of Maxwell's equations.

13.7 Maxwell's equations with sources: retarded potentials

In the first eight chapters, the calculation of **E**- and **B**-fields due to specified steady sources of charge and current were carried out by various methods. One method, of very general application, involved the use of potentials as set out in Sec. 13.4. Suppose we now wish to find solutions to Maxwell's equations as in the last section but with specified time-varying sources added? The equations, still *in vacuo*, then become

fields *in vacuo*, with sources:

$$\nabla\cdot\mathbf{E}=\rho/\varepsilon_0 \quad \text{(a)} \qquad \nabla\times\mathbf{E}=-\dot{\mathbf{B}} \quad \text{(b)}$$
$$\nabla\cdot\mathbf{B}=0 \quad \text{(c)} \qquad \nabla\times\mathbf{B}=\dot{\mathbf{E}}/c^2+\mu_0\mathbf{J} \quad \text{(d)} \tag{13.33}$$

The differences between these equations and those of Sec. 13.4 lie in the addition of the time-varying terms and in the fact that ρ and **J** are functions of time as well as of x, y, and z. However, the zero divergence of **B** means that it is still possible to define a vector potential **A** such that

$$\mathbf{B}=\nabla\times\mathbf{A} \tag{13.34}$$

If this is substituted into (13.33b), then $\nabla\times\mathbf{E}=\nabla\times(-\dot{\mathbf{A}})$. Uncurling this, we find that **E** can still be written as $-\partial\mathbf{A}/\partial t$ together with a gradient, i.e.,

$$\mathbf{E}=-\partial\mathbf{A}/\partial t-\nabla V \tag{13.35}$$

where V must become the electrostatic potential under steady conditions. Equations (13.34) and (13.35) do not fix **A** and V even if **E** and **B** are given (cf. Chapter 9) and we still have the divergence of **A** at our disposal. In Chapter 9 we found that $\nabla\cdot\mathbf{A}=0$ was the result of choosing $\int \mu_0 \mathbf{J}\,d\tau/4\pi r$ as a solution of $\nabla^2\mathbf{A}=\mu_0\mathbf{J}$. Let us see what equations are satisfied by **A** and V here before making our choice.

Substituting (13.34) and (13.35) into the source Maxwell equations (13.33a,d), we find

$$\frac{\partial(\nabla\cdot\mathbf{A})}{\partial t}+\nabla^2 V=-\frac{\rho}{\varepsilon_0}$$

$$\nabla(\nabla\cdot\mathbf{A})-\nabla^2\mathbf{A}=-(1/c^2)\frac{\partial^2\mathbf{A}}{\partial t^2}-\left(\frac{1}{c^2}\right)\nabla(\partial V/\partial t)+\mu_0\mathbf{J}$$

in the second of which the identity (B26) of Appendix B.11 has been used for $\nabla\times\nabla\times\mathbf{A}$. Both equations involve mixtures of **A** and V, but if we choose

$$\nabla\cdot\mathbf{A}=-\left(\frac{1}{c^2}\right)\partial V/\partial t \tag{13.36}$$

known as the **Lorentz gauge**, the equations uncouple and become simply:

$$\nabla^2 V-\frac{1}{c^2}\frac{\partial^2 V}{\partial t^2}=-\frac{\rho}{\varepsilon_0} \tag{13.37}$$

$$\nabla^2\mathbf{A}-\frac{1}{c^2}\frac{\partial^2\mathbf{A}}{\partial t^2}=-\mu_0\mathbf{J} \tag{13.38}$$

These two equations can be looked at *either* as Poisson's equations with the addition of terms in $\partial^2/\partial t^2$ *or* as wave equations with sources. A general solution of Poisson itself could be written as $V=\int \rho\,d\tau/(4\pi\varepsilon_0 r)$ when ρ was steady. The corresponding solution of (13.37) is not simply the replacement of ρ_{steady} by $\rho(t)$ but, as might be expected from Sec. 13.6, by $\rho(t-r/c)$. This means that to evaluate V at a *field point* at time t we must integrate over the *sources* using the value of ρ at the source point at a time $(t-r/c)$. It is as if the effect of the change in ρ travelled with a velocity c, entirely in line with the results of the previous section. Functions of $t-r/c$ are known as **retarded values** and are written with square brackets. The same applies to **A**.

Thus the general solutions of (13.37) and (13.38) are the **retarded potentials**:

$$V=\int_\tau \frac{[\rho]\,d\tau}{4\pi\varepsilon_0 r} \tag{13.39}$$

retarded potentials:

$$\mathbf{A}=\int_\tau \frac{\mu_0[\mathbf{J}]\,d\tau}{4\pi r} \tag{13.40}$$

in which the integrations are carried out over the volumes occupied by the sources. We shall use these solutions in the next chapter.

13.8 Electromagnetic energy and the Poynting vector

General discussion We have seen in previous chapters (Secs 5.6 and 9.10) how the energy stored in electric and magnetic systems could be expressed in terms of densities $\frac{1}{2}\varepsilon_0 E^2$ and $B^2/2\mu_0$ *in vacuo*. These expressions were derived for steady fields and we were careful to point out that it does not help if we try to locate the energy since it belongs to the system as a whole. In the case of electromagnetic wave fields, however, a different situation arises for we know that energy is needed to generate the waves at the transmitting aerial (the energy input to the transmitter less the wastage as heat) and we know that energy is received at the receiving aerial. If we are to preserve the principle of the conservation of energy, we must postulate that energy is carried outwards from the aerial with the velocity c by the waves, for if we do not, there is a loss of energy from the universe in the interval between transmission and reception.

What we seek is an expression that will give us the energy flow per unit area per unit time carried by electromagnetic fields. To see how to proceed, we first consider a special case, that of the unbounded plane wave of Sec. 13.6 travelling along the x axis as in Fig. 13.8. The fields E_y and B_z satisfy what is left of Maxwell's equations:

$$\partial E_y/\partial x = -\partial B_z/\partial t; \qquad \partial B_z/\partial x = -(1/c^2)\partial E_y/\partial t \tag{13.41}$$

If we assume that the energy density is still given by $u_{\text{EM}} = \frac{1}{2}\varepsilon_0 E^2 + \frac{1}{2}B^2/\mu_0$ then this is transported by the wave. The total energy in the rectangular parallelepiped shown in the figure is $U_{\text{EM}} = \int u_{\text{EM}}\,\mathrm{d}x\,\mathrm{d}y\,\mathrm{d}z$. The rate at which energy is disappearing from the volume is $-\partial U_{\text{EM}}/\partial t$, and we seek to relate that to the outflow over the end faces at x_1 and x_2. We have

$$\begin{aligned} -\partial U_{\text{EM}}/\partial t &= \int_{\text{vol}} -\frac{\partial}{\partial t}(B_z^2/2\mu_0 + \varepsilon_0 E_y^2/2)\,\mathrm{d}x\,\mathrm{d}y\,\mathrm{d}z \\ &= \int_{\text{vol}} [-(B_z/\mu_0)\partial B_z/\partial t - (\varepsilon_0 E_y)\partial E_y/\partial t]\,\mathrm{d}x\,\mathrm{d}y\,\mathrm{d}z \\ &= \int_{\text{vol}} [(B_z/\mu_0)\partial E_y/\partial x + (E_y/\mu_0)\partial B_z/\partial x]\,\mathrm{d}x\,\mathrm{d}y\,\mathrm{d}z \end{aligned}$$

using (13.41). Since $H_z = B_z/\mu_0$, this last line can be written as

$$\int_{\text{vol}} \frac{\partial}{\partial x}(E_y H_z)\,\mathrm{d}x\,\mathrm{d}y\,\mathrm{d}z, \text{ which is } \int_{\text{surf}} [(E_y H_z)_{x_1} - (E_y H_z)_{x_2}]\,\mathrm{d}y\,\mathrm{d}z$$

This result suggests that a vector

$$\boxed{\mathbf{N} = \mathbf{E} \times \mathbf{H} \qquad \text{(definition of } \mathbf{N}\text{)}} \tag{13.42}$$

would correctly give the energy flow in magnitude and direction, because in the plane wave there is only an x component of $\mathbf{N}$ equal to $E_y H_z$. $\mathbf{N}$ is known as the **Poynting vector**.

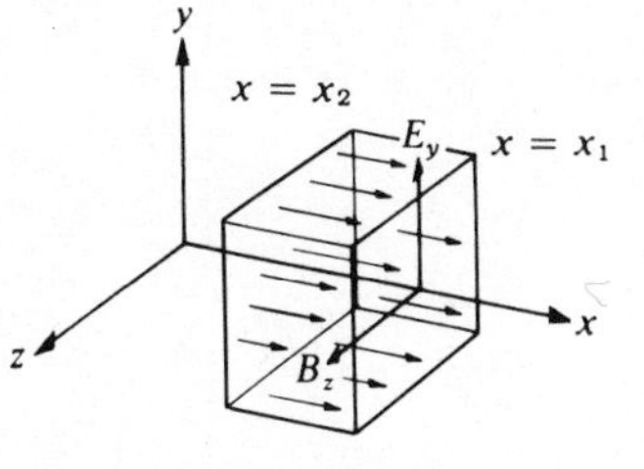

Figure 13.8 Transport of energy in a plane electromagnetic wave.

Poynting's theorem Guided by the special case in the above treatment, we now seek to generalize the result. Take a volume τ bounded by a closed surface S in which media may occur so that the full Maxwell equations must be used and not those for *vacua*. Suppose that the current density $\mathbf{J}$ at any point is produced by a field $\mathbf{E}$ so that $\mathbf{J}=\rho\mathbf{v}$, where $\mathbf{v}$ is the velocity of the charges moved by $\mathbf{E}$. The rate of working of $\mathbf{E}$ (i.e. force times velocity) is $\rho\mathbf{E}\cdot\mathbf{v}$ per unit volume, or $\mathbf{J}\cdot\mathbf{E}$. This represents the dissipation of electromagnetic energy within the volume, so that energy is being **created** at a rate $-\mathbf{J}\cdot\mathbf{E}$ per unit volume.

Now $-\mathbf{J}\cdot\mathbf{E}$ can be transformed by Maxwell since $\mathbf{J}=\nabla\times\mathbf{H}-\dot{\mathbf{D}}$, and there is a vector identity [(B.22) of Appendix B.11] that enables $\mathbf{E}\cdot(\nabla\times\mathbf{H})$ to be replaced by $\mathbf{H}\cdot(\nabla\times\mathbf{E})-\nabla\cdot(\mathbf{E}\times\mathbf{H})$. Hence

$$-\mathbf{J}\cdot\mathbf{E}=\mathbf{H}\cdot\frac{\partial\mathbf{B}}{\partial t}+\mathbf{E}\cdot\frac{\partial\mathbf{D}}{\partial t}+\nabla\cdot(\mathbf{E}\cdot\mathbf{H})$$

If we integrate over the volume τ and transform the volume integral of the last term into a surface integral of $(\mathbf{E}\times\mathbf{H})$ over S by Gauss's theorem [(B.17) of Appendix B.10], we have

$$\int_\tau -\mathbf{J}\cdot\mathbf{E}\,\mathrm{d}\tau=\int_\tau\left(\mathbf{H}\cdot\frac{\partial\mathbf{B}}{\partial t}+\mathbf{E}\cdot\frac{\partial\mathbf{D}}{\partial t}\right)\mathrm{d}\tau+\oint_S(\mathbf{E}\times\mathbf{H})\cdot\mathrm{d}\mathbf{S} \tag{13.43}$$

The left-hand side we have seen is the rate of creation of electromagnetic energy within the volume by external sources. On the right we can interpret the volume integral as the rate of increase of electromagnetic energy within the volume, the energy density being given by the sum of the expressions in Chapters 11 and 12. If energy is to be conserved, then the final term can only give the rate of outflow over the bounding surface so that the definition of (13.42) is vindicated. This interpretation of (13.43) is known as **Poynting's theorem**. Accordingly, the energy flow per unit time over any closed surface is given by the power

energy flux due to electromagnetic wave:

$$P_S=\oint_S\mathbf{N}\cdot\mathrm{d}\mathbf{S} \tag{13.44}$$

Caution is necessary if $\mathbf{N}$ is used to give energy flow over non-closed surfaces. In purely static fields, for instance, such as a uniform electrostatic field in a parallel plate capacitor crossed by a uniform magnetic field from the poles of a permanent magnet, $\mathbf{N}$ may have a flux over certain non-closed surfaces but no energy transport occurs because the flux over any *closed* surface is zero.

13.9 Electromagnetic momentum and radiation pressure

Radiation pressure Suppose the plane wave of Sec. 13.6 falls normally on the surface of a conductor with conductivity σ as in Fig. 13.9a. The **E**-field of the wave produces a current in the conductor given by $J_y=\sigma E_y$. The **B**-field of the wave then exerts a force on an element of the current given by $\mathrm{d}F_x=J_yB_z\,\mathrm{d}x\,\mathrm{d}y\,\mathrm{d}z$. Thus there is a normal pressure due to the incident radiation given by

$$\mathrm{d}p_x=J_yB_z\,\mathrm{d}x \tag{13.45}$$

To obtain this radiation pressure in a form containing only the fields, we replace J_y by using the appropriate Maxwell equation $\nabla\times\mathbf{H}=\mathbf{J}+\dot{\mathbf{D}}$. The y component of this is simply $-\partial H_z/\partial x=J_y+\varepsilon_0\partial E_y/\partial t$, and so

$$\mathrm{d}p_x=-(B_z\partial H_z/\partial x+\varepsilon_0 B_z\partial E_y/\partial t)\mathrm{d}x$$

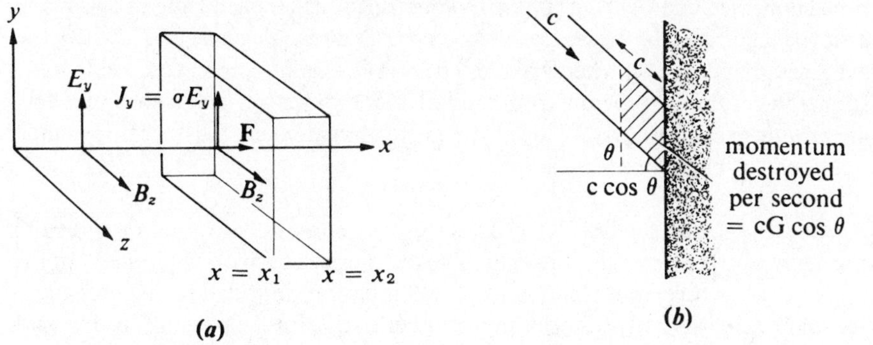

Figure 13.9 Radiation pressure. (a) An electromagnetic wave incident normally on an absorbent surface: **E** produces **J**, and this experiences a force **J** × **B** due to **B**; (b) oblique incidence of a plane wave on to an absorbent surface. G is the momentum density (momentum per unit volume).

The second term can be transformed into two terms using $\partial(B_z E_y)/\partial t = B_z \partial E_y/\partial t + E_y \partial B_z/\partial t$. However, $\partial(B_z E_y)/\partial t$ has a mean value of zero over any long period of time since in a monochromatic wave with **B** and **E** in phase it would be proportional to sin ωt cos ωt. So, if we take mean values:

$$\langle \mathrm{d}p_x \rangle = -\frac{\partial}{\partial x}(\tfrac{1}{2}B^2/\mu_0 + \tfrac{1}{2}\varepsilon_0 E^2)\,\mathrm{d}x = -\frac{\partial u_{\mathrm{EM}}}{\partial x}\mathrm{d}x$$

Hence for a block of finite thickness,

$$\langle p_x \rangle = \int_{x_2}^{x_1} -\frac{\partial u_{\mathrm{EM}}}{\partial x}\mathrm{d}x = \langle u_1 \rangle - \langle u_2 \rangle \tag{13.46}$$

where u_1 is the incident energy density and u_2 the emergent. For a material which completely absorbs radiation, the pressure is thus equal to the incident energy density for normal incidence. The existence of radiation pressure was first demonstrated by Lebedew (1901), and Nicholls and Hull (1903) showed that it was approximately equal to the incident u for a complete absorber.

Electromagnetic momentum and relationship with quantum theory The same situation arises here as with energy. If the conservation of momentum is to be obeyed then we must attribute to the electromagnetic wave momentum which is transported with velocity c, for otherwise the radiation pressure creates momentum in the universe. If the amount of momentum per unit volume possessed by the wave is G then, because a completely absorbed wave means a rate of change of momentum Gc per unit area of wavefront, the force per unit area is also Gc and hence, by (13.46),

momentum density in electromagnetic wave:

$$\boxed{G = u_{\mathrm{EM}}/c} \tag{13.47}$$

This is a relation carried over to the quantum theory of radiation in which the energy in electromagnetic radiation is carried by **photons** of energy $h\nu$ where ν is the frequency. The momentum of a photon by (13.47) is thus $h\nu/c$ or h/λ where λ is the wavelength.

For a plane wave incident at an angle θ to the normal and completely absorbed (Fig. 13.9b), the momentum destroyed per second per unit area is $Gc \cos\theta$ or $u_{\mathrm{EM}} \cos\theta$ along the direction of incidence. Thus there should be a normal pressure $u_{\mathrm{EM}} \cos^2\theta$ and a tangential pressure $u_{\mathrm{EM}} \sin\theta \cos\theta$. The existence of the latter has been confirmed by Poynting (1905). If radiation is incident at all angles on a plane absorbent surface, the pressures are $u_{EM}\langle\cos^2\theta\rangle$ and $u_{EM}\langle\sin\theta\cos\theta\rangle$, where the averages are taken over a hemisphere. The first term is $\frac{1}{3}u_{EM}$ and the second is zero so that the radiation pressure from random plane waves incident on a completely absorbing surface is $\frac{1}{3}u_{EM}$. This is an expression used in the thermodynamic treatment of black-body radiation.

Applications in astrophysics Under terrestrial conditions, radiation pressure normally exerts forces that are negligible in comparison with any others. In the universe as a whole, however, there are two situations in which such a pressure is comparable with that produced by the other major force: gravitation. The first occurs in the interior of stars, where the energy density of the radiation is so great that its pressure is a significant factor in determining stellar equilibrium. The second occurs with interstellar particles whose masses are small enough to produce a sufficient reduction in the gravitational forces on them.

In the latter case, an estimate can be made of the approximate size of a particle at which radiation pressure and gravity produce comparable forces. Assume that the particle is moving in the solar system (e.g. it may be part of the tail of a comet). If the particle has a radius r, a density ρ and is at a distance d from the sun's centre, let the mass of the sun be M and its luminosity L. The last-named quantity is the total energy radiated by it per second, estimated from terrestrial measurements to be 3.85×10^{23} kW. The gravitational force on the particle is $(GM/d^2)(4\pi r^3\rho/3)$. Since the energy from the sun has spread out over a spherical surface of area $4\pi d^2$ by the time it has reached the particle, the energy flux is $L/4\pi d^2$ and the radiation pressure is therefore $L/(4\pi d^2 c)$. The particle presents an area of roughly πr^2 to this pressure, so that the resultant force is $L(r^2/4cd^2)$.

The radiation force will therefore exceed the gravitational force if

$$r < 3L/(16\pi c r G M\rho)$$

If the particle is of similar material to that in meteorites, the relative density is about 3.5, and with the sun's mass as 2×10^{30} kg, this gives

$$r < 15\,\mu\text{m}$$

In fact, the figure is likely to be lower than this because the particles do not necessarily absorb all the radiation and because, as the size becomes comparable with the wavelength of the radiation, the process is more complex. However, it is thought that cosmic dust particles less than a few micrometres across are swept out of the solar system by the combined effect of radiation pressure and the solar wind (see McDonnell, 1978).

13.10 Relationship between electromagnetism and special relativity

The development of special relativity early in this century caused a drastic reformation of the foundations of mechanics. An apparently simple principle—that the laws of physics should take the same form in all inertial frames of reference—led to results that contradicted those of Newtonian kinematics and dynamics. Among these results we pick out the following as being particularly relevant to our discussion. In them, we use two inertial frames S and S′ having a relative uniform velocity v along their common x axis and having origins which coincide at times $t=t'=0$. We also put $\gamma=(1-v^2/c^2)^{-1/2}$ so that $\gamma \geq 1$ because $v \leq c$. Then, if y, y' and z, z' are parallel:

R1. The relation between the coordinates (x,y,z,t) of any event in S and those (x',y',z',t') of the same event in S′ is given by the Lorentz transformation $x'=\gamma(x-vt), y'=y, z'=z, t'=\gamma(t-vx/c^2)$, and *not* by the classical relation $x'=x-vt,\ t'=t$.

R2. If L is the x length of an object *stationary* in S, its length in S′ is L/γ.

R3. If T is the time interval between two events *at the same place* in S, the interval between the same events in S′ is γT.

R4. If F_x, F_y, F_z are the components of the force on a particle *stationary* in S, the force components on the same particle in S′ are $F'_x=F_x, F'_y=F_y/\gamma, F'_z=F_z/\gamma$.

The constancy of the velocity of light *in vacuo* in all frames of reference is sometimes taken as an independent postulate and sometimes as a consequence of the more general principle of the invariant form of physical laws. In either case, it bears directly on electromagnetism which predicts, as we have seen, that $c=(\varepsilon_0\mu_0)^{-1/2}$ without calling on the idea of a frame of reference. We shall see that this is because Maxwell's equations are already inherently relativistic. It is true that, unlike the laws of mechanics, those of electromagnetism have remained unscathed by special relativity. What has been changed is a viewpoint, and this we now discuss.

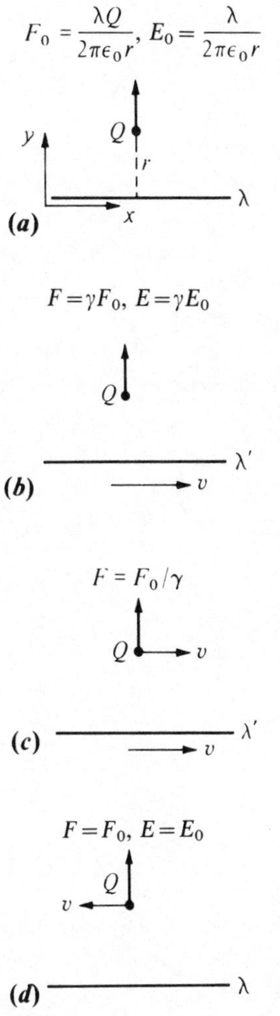

Figure 13.10 (a) Point charge Q and line charge λ both stationary; (b) the line charge in (a) moving with a velocity v, and **F** is still Q**E**; (c) both Q and the line charge moving with v: this is the same as (a) viewed from a frame moving to the left with v or the same as (b) with Q now moving; (d) only Q is moving: this is the same as (b) viewed from a frame moving to the right with v, and **F** is still Q**E**. Only in (c) is **F** = Q**E** + additional force.

Magnetism as a relativistic effect As mentioned in Sec. 7.3, it is possible to introduce the magnetic field as a consequence of Coulomb's electrostatic law together with the Lorentz transformation. The disadvantages of this approach as a method of *introducing* magnetic fields are twofold: a thorough knowledge and understanding of relativistic kinematics must first be attained; and **E**- and **B**-fields are inextricably linked from the start, which can be confusing. We have been able to keep the two types separate until time variations were introduced, and not until this chapter have we finally seen that **E** and **B** *are* intimately related so that a complete description generally involves the whole *electromagnetic* field. However, at the stage we have now reached, it is quite illuminating to see the relationship between **E** and **B**, as illustrated by a special case.

Consider a charge Q situated a distance r from a stationary infinite line charge of density λ along x as in Fig. 13.10a. The **E**-field at Q is $\lambda/(2\pi\varepsilon_0 r)=E_0$, say, and the force on Q is $\lambda Q/(2\pi\varepsilon_0 r)=F_0$, say, both E_0 and F_0 being along y. What we shall do is to look at three different modifications of this arrangement: case (b) with just the line charge moving, case (d) with just Q moving and case (c) with both moving. We notice particularly that case (c) is that of two currents.

First take case (b) in which the line charge of (a) is now moving with v along the x-axis. We know from Comment C3.3 that charge itself is invariant when it moves relative to an observer, but the length occupied by unit charge is decreased by a factor γ from R2 above. The new charge density is thus $\lambda'=\gamma\lambda$. The **E**-field is γ times its value in (a), i.e., is γE_0, and the force on Q is γF_0. Thus, we learn that the movement of the *sources* of an electric field may well alter the value of the field, but that nevertheless the expression $F=QE$ is still correct with the new value of E.

Suppose, however, that (b) was being observed in a frame S that is stationary with respect to Q. Now observe it instead from a frame S′ moving to the *right* with velocity v relative to S: we arrive at case (d) in which only Q is moving. According to R4, the y-force in S′ is $1/\gamma$ of that in S, i.e., it becomes simply F_0 again. The E-field is consequently E_0 so that $F=QE$ still holds. This removes one of our worries in Sec. 3.3: we may calculate the path of *moving* charges in an E-field still using $\mathbf{F}=Q\mathbf{E}$ and the electrostatic field. **E** is unaffected by the motion of test charges.

Finally, we turn to case (c) where both Q and the line charge are moving so that, as in (b), the linear charge density becomes $\lambda'=\gamma\lambda$. This case could be regarded as case (a) viewed from a frame S′ moving to the *left* with velocity v relative to Q. By R4, the y force becomes F_0/γ. This is a highly significant result as we can see by comparing (b) and (c). In both of those cases we have a *current* $I=\lambda'v$. The force on Q when it is *stationary* in (b) is γF_0. The force on Q when it is *moving with* v in (c) is F_0/γ. There is an extra force $F_0/\gamma-\gamma F_0$ due to the motion of Q. This extra force is attractive, perpendicular to the velocity v and we call it the **magnetic force**. If we define a magnetic field **B** by equating the additional force to $Q\mathbf{v}\times\mathbf{B}$, then in our example the magnitude of **B** would be given by

$$B=\frac{F_0}{Qv}\left(\frac{1}{\gamma}-\gamma\right)=\frac{\gamma F_0}{Qv}\left(\frac{1}{\gamma^2}-1\right)$$

$$=\frac{\gamma\lambda Q}{2\pi\varepsilon_0 rQv}\left(-\frac{v^2}{c^2}\right)$$

$$=-\lambda'v/(2\pi\varepsilon_0 rc^2)$$

$$=-\mu_0 I/(2\pi r)$$

as given by (7.19) for the **B**-field of an infinite straight current.

To summarize, we have started with a direct consequence of Coulomb's law ($E=\lambda/2\pi\varepsilon_0 r$) and have shown that the application of special relativity leads to a direct consequence of the Biot–Savart law ($B=\mu_0 I/2\pi r$). Among other things, this means that the magnetic laws for steady currents may be taken as established with as much accuracy as Coulomb's law.

Invariance of Maxwell's equations Note particularly in the above discussion that the fields observed in any situation depend on the frame of reference of the observer: in case (a) there was only an **E**-field at Q, but when viewed from S′ to give case (c), a **B**-field was observed as well.

In general, it can be shown that the electric and magnetic field components transform from a frame of reference S to S′ (as defined above) through the relations

$$\boxed{\begin{array}{lll} E'_x=E_x; & E'_y=\gamma(E_y-vB_z); & E'_z=\gamma(E_z+vB_y) \\ B'_x=B_x; & B'_y=\gamma(B_y+vE_z/c^2); & B'_z=\gamma(B_z-vE_y/c^2) \end{array}} \qquad (13.48)$$

where the force law $\mathbf{F}=Q(\mathbf{E}+\mathbf{v}\times\mathbf{B})$ defines **E** and **B** in any frame. If these are substituted into Maxwell's equations and the Lorentz transformation used, it can be shown that the equations take the same form in S and S′. Thus the principle of relativity is satisfied and in particular the velocity of light *in vacuo* $c=1(/\varepsilon_0\mu_0)^{1/2}$ is the same in all inertial frames and is thus the same for all observers.

For a detailed treatment of the relativistic approach to electromagnetism see Rosser (1968).

13.11 Revision summary

Maxwell's equations are completed by the addition of a displacement current I_d to the conduction current I_c, where I_d and the corresponding current density $\mathbf{J}_d$ are given by

$$\mathbf{J}_d = \partial\mathbf{D}/\partial t; \qquad I_d = \int_S \mathbf{J}_d \cdot d\mathbf{S} \qquad \text{from (13.9)}$$

over any surface S. The complete equations then become, in integral form,

$$\oint_S \mathbf{D}\cdot d\mathbf{S} = \sum_\tau Q_c \qquad \oint_L \mathbf{E}\cdot d\mathbf{L} = -\partial\Phi_S/\partial t \qquad \text{a} \quad \text{b}$$

$$(13.14)$$

$$\oint_S \mathbf{B}\cdot d\mathbf{S} = 0 \qquad \oint_L \mathbf{H}\cdot d\mathbf{L} = I_c + I_d \qquad \text{c} \quad \text{d}$$

and in differential form,

$$\nabla\cdot\mathbf{D} = \rho_c \qquad \nabla\times\mathbf{E} = -\partial\mathbf{B}/\partial t \qquad \text{a} \quad \text{b}$$

$$(13.15)$$

$$\nabla\cdot\mathbf{B} = 0 \qquad \nabla\times\mathbf{H} = \mathbf{J}_c + \partial\mathbf{D}/\partial t \qquad \text{c} \quad \text{d}$$

In these equations **D** and **E** on the one hand and **B** and **H** on the other may be related through the constitutive relations of Chapters 11 and 12. For example, $\mathbf{D} = \varepsilon_r\varepsilon_0\mathbf{E}$ and $\mathbf{H} = \mathbf{B}/\mu_r\mu_0$ in LIH media. The current $\mathbf{J}_c$ may be specified or may be related to the E-field in a conductor through $\mathbf{J} = \sigma\mathbf{E}$.

The following **boundary conditions** result from Maxwell's equations

$E_{\text{tangential}}$, B_{normal} are continuous   b c

(13.18)

D_{normal} is discontinuous by σ_c a

$H_{\text{tangential}}$ is discontinuous by J_{sc} d

where σ_c is the surface density of conduction charge and J_{sc} the surface density of conduction current.

The equations need to be supplemented by the Lorentz force law

$$\mathbf{F} = Q(\mathbf{E} + \mathbf{v}\times\mathbf{B}) = \int_\tau (\rho\mathbf{E} + \mathbf{J}\times\mathbf{B})\,d\tau \qquad (13.1)$$

and by the definitions of **D** and **H** if the media are not LIH:

$$\mathbf{D} = \varepsilon_0\mathbf{E} + \mathbf{P}; \qquad \mathbf{H} = \mathbf{B}/\mu_0 - \mathbf{M} \qquad (13.16)$$

Conservation of charge is still embodied in the equation of continuity which is now a consequence of Maxwell's equations.

Solutions to Maxwell's equations have two important properties:

- they **superpose** because of the linearity of the equations,
- they are **unique** if certain boundary conditions are satisfied.

***In vacuo* without sources** the new solutions of Maxwell's equations include electromagnetic waves: i.e. the propagation of changes in **E**- and **B**-fields with a velocity $c = 1/(\varepsilon_0\mu_0)^{1/2}$, the velocity of light *in vacuo*. Unbounded plane waves are TEM with $\mathbf{E}\times\mathbf{B}$ in the direction of propagation and **E** and **B** at right angles. If monochromatic, **E** and **B** are in phase and have amplitudes related by $E = cB$.

***In vacuo* with sources**, the vector potential **A** and scalar potential V can be used to find the fields, where

$$\mathbf{B}=\nabla\times\mathbf{A} \quad \text{and} \quad \mathbf{E}=-\partial\mathbf{A}/\partial t-\nabla V \qquad (13.34)\ (13.35)$$

If the Lorentz condition $\nabla\cdot\mathbf{A}=-(1/c^2)\partial V/\partial t$ is chosen, then **A** and V satisfy the equations

$$\nabla^2 V-(1/c^2)\partial^2 V/\partial t^2=-\rho/\varepsilon_0; \qquad \nabla^2\mathbf{A}-(1/c^2)\partial^2\mathbf{A}/\partial t^2=-\mu_0\mathbf{J} \qquad (13.37)\ (13.38)$$

whose solutions are the retarded potentials

$$V=\int_\tau \frac{[\rho]\,\mathrm{d}\tau}{4\pi\varepsilon_0 r} \qquad \mathbf{A}=\int_\tau \frac{\mu_0[\mathbf{J}]\,\mathrm{d}\tau}{4\pi r} \qquad (13.39)(13.40)$$

Electromagnetic energy and momentum The flux of electromagnetic energy is found to be given generally by the Poynting vector

$$\mathbf{N}=\mathbf{E}\times\mathbf{H} \qquad (13.42)$$

so that the outward energy flow per unit time over any closed surface is given by

$$P=\oint_S \mathbf{N}\cdot\mathrm{d}\mathbf{S} \qquad (13.44)$$

This is consistent with a total electromagnetic energy density associated with electromagnetic fields given by

$$u_{\mathrm{EM}}=\tfrac{1}{2}\mathbf{D}\cdot\mathbf{E}+\tfrac{1}{2}\mathbf{B}\cdot\mathbf{H}$$

At the same time, momentum must also be associated with electromagnetic waves if the conservation laws are to be retained. The momentum per unit volume G is found to be given by

$$G=u_{\mathrm{EM}}/c \qquad (13.47)$$

which is consistent with the quantum theory of radiation in which the momentum of a photon is $h\nu/c$.

Maxwell's equations are **relativistically invariant** under a Lorentz transformation. The **E**- and **B**-fields resulting from given charges depend on the motion of the observer relative to those charges. In general, fields transform between frames of reference in standard configuration according to

$$E_x'=E_x; \quad E_y'=\gamma(E_y-vB_z); \quad E_z'=\gamma(E_z+vB_y)$$
$$B_x'=B_x; \quad B_y=\gamma(B_y+vE_z/c^2); \quad B_z'=\gamma(B_z-vE_y/c^2)$$

Problems

Section 13.1

13.1 Derive the expression (7.27) for the magnetic **B**-field due to a moving charge by using the displacement current.

13.2 A straight wire of negligible cross-section is effectively infinite in length but has a break of length $2l$ between points A and B on it. All parts of the wire carry a steady conduction current I so that charge accumulates at the points A and B. Find the **B**-field at a point a distance R along the perpendicular bisector of AB. Use both the current element formula (Biot–Savart law) and the circuital law and check that the same result is obtained.

13.3 An imperfect insulator of conductivity σ and relative permittivity ε_r fills the space between the plates of a parallel plate capacitor. An alternating voltage $V_0 \sin \omega t$ is applied across the plates. Compare the magnitudes of the displacement and conduction current densities. Show that for many actual materials the displacement current is negligible for frequencies in the radio wave region.

Section 13.3

13.4 Show that a magnetic field specified by $B_x = B_0 x \sin \omega t$, $B_y = 0$, $B_z = 0$ cannot exist.

Section 13.4

13.5 In the conducting sphere of Fig. 13.4, show that the maximum surface density of charge is $3\varepsilon_0 E_0$.

13.6 A spherical cavity of radius a is found in an LIH dielectric of relative permittivity ε_r. If the electric field far from the cavity is uniform and of magnitude E, what is the field in the cavity itself?

13.7 A spherical capacitor has the space between the plates filled with a material of conductivity σ and a relative permittivity ε_r, and has its outer plate earthed. Show that the resistance R between the plates and the capacitance C are related by $CR = \varepsilon_r \varepsilon_0 / \sigma$.

Section 13.6

13.8 Show that the *intrinsic impedance of free space*, defined as E/H for a plane wave *in vacuo*, has the value 376.6 Ω.

Section 13.8

13.9 Show that the mean electric and magnetic energies in a plane electromagnetic wave *in vacuo* are equal.

13.10 Show that the Poynting vector in a plane electromagnetic wave *in vacuo* has a magnitude $E^2/\mu_0 c$.

13.11 Estimate the E-field at 1 m from a 30 W lamp, assuming that the energy is radiated equally in all directions and that no losses occur by conduction or convection of heat.

13.12 A long straight uniform wire carries a steady current I. If the potential difference across a length l is V, find the value of the Poynting vector at points a distance r from the axis of the wire. Show from that result that the energy flowing into the wire is VI per unit time. Comment on this as a view of the energy exchanges taking place in a steady current circuit.

Section 13.9

13.13 Estimate the force on 1 cm² of a perfect absorber of electromagnetic waves due to radiation pressure from the lamp of Problem 13.11. The surface of the absorbing material is 1 m from the lamp and is receiving the radiation normally.

13.14 A plane electromagnetic wave with energy density u_{EM} is incident on the surface of a material at an angle θ to the normal. A fraction f of the radiation is reflected and the rest is absorbed. What is the radiation pressure?

Section 13.10

13.15 By using Eqs (13.48) show that $E - cB$ and $\mathbf{E} \cdot \mathbf{B}$ are invariant. What implication does this have for a plane wave *in vacuo*?

14

Electromagnetic waves

This chapter is devoted entirely to an investigation of electromagnetic waves whose propagation *in vacuo* was discussed in Sec. 13.6. We first look at the propagation of such waves in insulating and conducting media together with the effects occurring at boundaries (Secs 14.1–14.3). Since light waves are electromagnetic, we expect to be able to deduce optical properties of various media from Maxwell's equations and we find that this is indeed the case. When propagated over short distances, transmission is more effective if there is lateral confinement and in Sec. 14.4 the characteristics of metallic waveguides and cavities are examined.

Finally, the next three sections deal with the origin of the wave fields and their generation from time-varying sources. Here at last we are able to place in perspective all the fields encountered in previous chapters.

14.1 Propagation in LIH non-conducting media

Refractive index In Sec. 13.6 we showed how Maxwell's equations *in vacuo* without sources led to a wave equation for **E** or **B**, predicting waves with a velocity $c = 1/(\varepsilon_0\mu_0)^{1/2}$. Some properties of unbounded plane waves were also deduced, for instance that they are TEM and, if monochromatic, have **E** and **B** in phase and with magnitudes related by $E = cB$.

In this section we consider similar waves in a non-conducting LIH medium with relative permittivity ε_r and relative permeability μ_r. The only modifications we need make to the Maxwell equations used in Sec. 13.6 are the replacement of ε_0

and μ_0 by $\varepsilon_r\varepsilon_0$ and $\mu_r\mu_0$, respectively. It follows that all the general conclusions about plane waves (TEM, **E** in phase with **B**, **E** perpendicular to **B**) still hold and only the values of certain quantities will be affected.

The first effect to be noticed is that the velocity of the waves in the medium, which we shall denote by c_m, now becomes

$$c_m = 1/(\varepsilon_r\varepsilon_0\mu_r\mu_0)^{1/2} = c/(\varepsilon_r\mu_r)^{1/2} \tag{14.1}$$

Since we are considering only LIH media, ferromagnetics are excluded and μ_r is very nearly 1 so that c_m becomes almost equal to $c/\varepsilon_r^{1/2}$. Now in general the refractive index n of a medium is given by

$$n = c/c_m \qquad \text{(definition of } n\text{)} \tag{14.2}$$

so that we ought to find that

$$n^2 \sim \varepsilon_r \tag{14.3}$$

This is a relationship that can be checked by independent measurements of n and ε_r, and which is found generally to be obeyed. Apparent violations of it are largely due to the measurement of n and ε_r at different frequencies, for just as ε_r varies with frequency (Fig. 11.11) so does n. For non-polar substances in particular, where variation with frequency is small, the agreement is good (e.g. diamond has $\varepsilon_r = 5.68$, $n^2 = 5.66$; air has $\varepsilon_r = 1.000\,586$, $n^2 = 1.000\,588$). For polar substances, ε_r rises markedly at low frequencies and agreement with n^2 measured at optical wavelengths would not be expected (e.g. water has $\varepsilon_r = 81$ at low frequencies, $n^2 = 1.78$ in the optical region rising to about 80 at very low frequencies).

Note that, because ε_r is always greater than unity, c_m is always less than c. Although special relativity shows that particles cannot attain the velocity c unless they have zero rest mass, there is nothing in principle to prevent velocities in excess of c_m. So high-energy particles passing through an insulator may have velocities greater than the velocity of light *in the medium*. If they do, a bluish light known as *Čerenkov radiation* is emitted which can be used as a measure of the energy of the particles. This type of radiation is analogous to the shock wave or bow wave occurring when an object travels through a fluid at a speed greater than that of sound waves in the fluid.

Relation between *E* and *B* in plane waves The same steps that led to Eq. (13.32)—that $E = cB$ in a plane wave *in vacuo*—now lead to

$$E = c_m = cB/n \tag{14.4}$$

We shall, in the next section, find that it is useful to have a relationship between H and E. Therefore, putting $H = B/\mu_r\mu_0 \sim B/\mu_0$, we have

plane wave in refracting medium: $$H = nE/c\mu_0 \tag{14.5}$$

and the intrinsic impedance of the medium (Problem 13.8) becomes $377/n\ \Omega$.

14.2 Reflection and refraction at dielectric boundaries

General discussion We turn now to the behaviour of electromagnetic waves when they encounter a plane boundary between two non-conducting media, i.e. the way in which the waves are reflected and refracted. We should realize that, because Maxwell's equations (13.15) are macroscopic, the model of the medium is also macroscopic: it is treated as a continuum defined only by the properties ε_r and μ_r. In that case we should not expect the treatment to yield an explanation of reflection and refraction in atomic terms.

In fact, such an explanation is possible though quite intricate. When a beam of radiation meets any non-conducting substance, the atoms absorb some of the energy and re-radiate the rest as scattered waves. For homogeneous substances the re-radiation from a large number of atoms may recombine to give a reflected beam and a transmitted beam. The latter may itself combine with the remainder of the incident radiation to give refraction. The analysis in this form is quite complicated and we prefer at this stage to ignore the processes occurring on an atomic scale and obtain very general relations using the theory developed so far.

Mathematical forms of a plane wave For the first time we shall encounter plane waves whose direction of propagation is not along a principal coordinate axis. They cannot therefore be expressed in quite such a simple form as $E = E_0 \sin(kx - \omega t)$. However, that form can be used as a guide in finding the one we need. We note that $E_0 \sin(kx - \omega t)$ is appropriate for a plane wave travelling along the x axis because kx = constant is the equation of planes perpendicular to x, and E_0 does not vary over these planes. Thus E has the same phase and magnitude at any instant over any wavefront.

For other directions of propagation, kx would be replaced by a more general expression of the form $k(lx + my + nz)$ because the general equation of a plane in Cartesian coordinates is $lx + my + nz$ = constant. Here, l, m and n are the direction cosines of the *normal* to the plane, i.e. the cosines of the angles between the normal and the x, y and z axes, respectively. The replacement for kx could also be written in the form $\mathbf{k} \cdot \mathbf{r}$, where $\mathbf{k}$ is a wave vector and $\mathbf{r}$ the position vector of any point on the plane. The wave vector $\mathbf{k}$ has a direction along that of the normal to the wavefront (the ray) and a magnitude $k = 2\pi/\lambda$.

To illustrate all this, consider the incident wave coming in from the bottom left of Fig. 14.1. The ray direction makes an angle θ_i with the y axis so that the plane wavefronts have equations $lx + my$ = constant, with $l = \sin\theta_i$ and $m = \cos\theta_i$ (no z term because the planes always contain z, i.e. the ray lies in the xy plane). It follows that the incident plane monochromatic wave would have an $\mathbf{E}$-field of the form

$$\mathbf{E} = \mathbf{E}_{i0} \sin[k_1(x \sin\theta_i + y \cos\theta_i) - \omega t] \qquad (14.6)$$

where $k_1 = 2\pi/\lambda_1$ for medium 1. The cosine form would be equally appropriate, and the complex form could also be used because $e^{j\alpha} = \cos\alpha + j\sin\alpha$ for any α:

$$\mathbf{E} = \mathbf{E}_{i0} e^{j[k_1(x \sin\theta_i + y\cos\theta_i) - \omega t]} \qquad (14.7)$$

Here we must take the real or imaginary part should we need instantaneous values (compare the use of complex forms in Chapter 10).

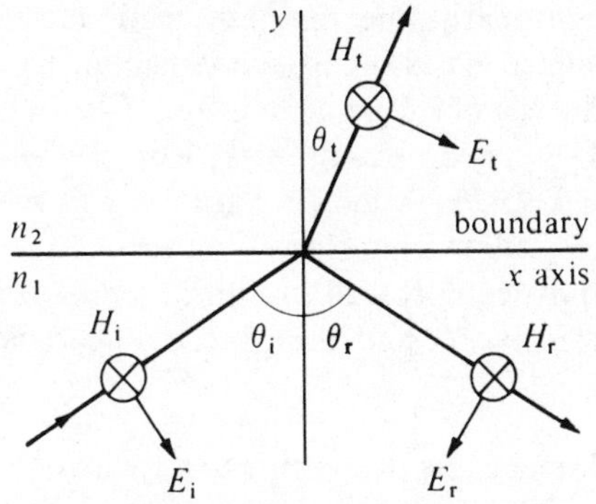

Figure 14.1 Reflection and refraction of an electromagnetic wave at a boundary between two dielectric media. The wave illustrated is polarized with its **E** vector parallel to the plane of incidence.

Finally, we note the general relationship between k and the refractive index stemming from their definitions. Since $k = 2\pi/\lambda = \omega/c_m$, we have

non-conductor: $$k = n\omega/c \qquad (14.8)$$

and k can be replaced by this expression with the appropriate n in any of the above expressions. Should k or n become complex, the imaginary part would, according to (14.7), yield an exponential factor with a real negative index. This would mean that the amplitude decays exponentially with distance, i.e. the wave is **absorbed**.

Reflection and refraction: frequency and direction Consider a plane wave incident at an angle θ_i on a boundary between two dielectric media of refractive indices n_1 and n_2. The incident **E**- and **H**-fields are of amplitudes E_i and H_i and are directed as shown in Fig. 14.1: a linearly polarized wave with **E parallel** to the plane of incidence. In general there will be a reflected beam at an angle θ_r and a transmitted beam at θ_t with appropriate **E**- and **H**-fields. In all cases $H = nE/\mu_0 c$ by (14.5). Although we ought not to assume that the latter two beams are also in the plane of incidence, it will become immediately apparent that they are.

We first apply the boundary condition that the tangential component of the total **E**-field is continuous at the boundary $y = 0$. In medium 1 the total E_{tan} is the sum of the incident and reflected contributions:

$$E_{i0} \cos\theta_i \sin(k_1 x \sin\theta_i - \omega_1 t) - E_{r0} \cos\theta_r \sin(k_1 x \sin\theta_r - \omega t) \qquad (14.9)$$

and in medium 2 is

$$E_{t0} \cos\theta_t \sin(k_2 x \sin\theta_t - \omega_2 t) \qquad (14.10)$$

The total fields (14.9) and (14.10) must be equal for *all* values of x and t and this can only be true if the coefficients of x and t in the sines are all equal. (This also means that if no term in z occurs in the incident wave, there can be none in the others, i.e. the three rays drawn in Fig. 14.1 are coplanar, so that we are justified in drawing them so.) Thus

$$\omega_1 = \omega_2; \qquad k_1 \sin\theta_i = k_1 \sin\theta_r = k_2 \sin\theta_t \qquad (14.11)$$

The first relation would be expected from the linearity of the medium and shows that the frequency of light is not changed in passing from one medium to another. The second relation gives $\theta_i = \theta_r$, the **law of reflection** in geometrical optics. The third relation gives $\sin\theta_i/\sin\theta_t = k_2/k_1$. From (14.8), $k_2/k_1 = n_2/n_1$, the ratio of refractive indices. Hence, $\sin\theta_i/\sin\theta_t = n_2/n_1$, which is **Snell's law of refraction**. The laws of geometrical optics thus follow from the wave theory.

All of the above would clearly apply equally well to a linearly polarized wave with **E perpendicular** to the plane of incidence and, therefore, by superposition to any incident wave.

Reflection and refraction: amplitude and phase Returning to the case of Fig. 14.1, in which **E** is **parallel** to the plane of incidence, all that remains of the equality between (14.9) and (14.10) is

$$E_i \cos\theta_i - E_r \cos\theta_i = E_t \cos\theta_t \qquad (14.12)$$

where we have dropped the zero subscripts on the understanding that we are dealing with amplitudes. We now obtain a second equation by using the continuity of H_{tan} across the boundary to give $H_i + H_r = H_t$. From (14.5), $H = nE/\mu_0 c$ so that

$$n_1 E_i + n_1 E_r = n_2 E_t \qquad (14.13)$$

Equations (14.12) and (14.13) can be used to find what proportions of the incident amplitude are reflected or transmitted. The ratio E_r/E_i, known as the reflection coefficient $r_\parallel$ for **E** parallel to the plane of incidence, is found by eliminating E_t:

$$\boxed{r_\parallel = \frac{E_r}{E_i} = \frac{(n_2/n_1)\cos\theta_i - \cos\theta_t}{(n_2/n_1)\cos\theta_i + \cos\theta_t} = \frac{\tan(\theta_i - \theta_t)}{\tan(\theta_i + \theta_t)}} \qquad (14.14)$$

where the second form has been obtained by using Snell's law. Similarly, by eliminating E_r, we have

$$\boxed{t_\parallel = E_t/E_i = \frac{2\cos\theta_i}{(n_2/n_1)\cos\theta_i + \cos\theta_t}} \qquad (14.15)$$

A similar calculation with the E-vector **perpendicular** to the plane of incidence gives reflection and transmission coefficients:

$$\boxed{r_\perp = \frac{\cos\theta_i - (n_2/n_1)\cos\theta_t}{\cos\theta_i + (n_2/n_1)\cos\theta_t} = -\frac{\sin(\theta_i - \theta_t)}{\sin(\theta_i + \theta_t)}} \qquad (14.16)$$

$$\boxed{t_\perp = \frac{2\cos\theta_i}{\cos\theta_i + (n_2/n_1)\cos\theta_t}} \qquad (14.17)$$

Equations (14.14)–(14.17) are known as **Fresnel's relations**. As long as r and t are real quantities they also give the phases of the beams. If positive, there is no change of phase; if negative, the phase changes by π.

Vacuum-dielectric interface: reflecting power To bring out the physical significance of the Fresnel relations, we slightly specialize the situation to that of a vacuum and a dielectric of refraction index n, and concentrate on the reflection coefficients. We also define a **reflecting power** R as the ratio of reflected and incident *intensities*, so that $R = r^2$.

For **incidence from the vacuum**, we have that $n_2/n_1 = n$ and hence

$$r_{\parallel} = \frac{n\cos\theta_i - \cos\theta_t}{n\cos\theta_i + \cos\theta_t}; \qquad r_{\perp} = \frac{\cos\theta_i - n\cos\theta_t}{\cos\theta_i + n\cos\theta_t} \tag{14.18}$$

These are plotted against θ_i in Fig. 14.2a and we pick out three main features.

First, the reflecting power. At normal incidence this is given for both states of polarization by

$$R_n = \left(\frac{n-1}{n+1}\right)^2 \tag{14.19}$$

which predicts, for instance, that glass with $n = 1.5$ would have R_n about 4 per cent. As the angle of incidence increases towards grazing angles, R_n tends to unity for both states of polarization. This is a phenomenon that can be observed quite easily with most surfaces.

Secondly, there is a particular angle of incidence, the **Brewster angle** θ_B, at which $r_{\parallel}$ becomes zero while $r_{\perp}$ does not. At this angle, any incident light can only be reflected with **E** perpendicular to the plane of incidence, giving a method of producing linearly polarized light from an unpolarized beam. At other angles, the difference between the values of $r_{\parallel}$ and $r_{\perp}$ means that reflected light will be partially polarized. The value of θ_B is obtained by setting $r_{\parallel}$ to zero and using $n = \sin\theta_i/\sin\theta_t$ so that at the Brewster angle, $\sin 2\theta_i = \sin 2\theta_t$. This means that $2\theta_i = \pi - 2\theta_t$ or that $\theta_i = \frac{1}{2}\pi - \theta_t$. Hence n becomes $\sin\theta_i/\cos\theta_i$ for $\theta_i = \theta_B$, i.e.

$$\tan\theta_B = n \tag{14.20}$$

Finally, there is a phase change of π on reflection except for $r_{\parallel}$ when θ_i is less than θ_B.

For **incidence within the dielectric**, the expression for $r_{\parallel}$ and $r_{\perp}$ of (14.14) and

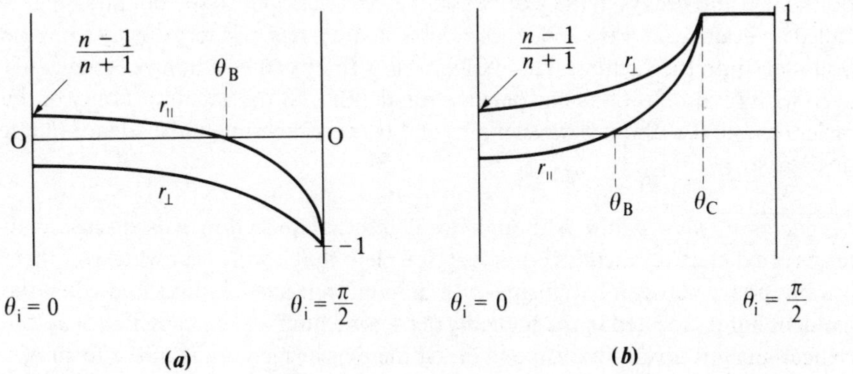

Figure 14.2 Variation of reflected wave amplitude with angle of incidence for (a) waves incident from a vacuum on to a dielectric with refractive index n and (b) waves incident from inside a dielectric on to a dielectric–vacuum boundary.

(14.16) need only be interchanged, but with the significant difference that n is now $n_1/n_2 = \sin\theta_t/\sin\theta_i$. This means that the tangent and sine forms in (14.14) and (14.16) still apply correctly but with $\theta_i < \theta_t$ and with the Brewster angle still in $r_{\parallel}$. More importantly, as can be seen from the plot of Fig. 14.2b, there comes an angle of incidence at which both reflection coefficients become unity because θ_t becomes $\frac{1}{2}\pi$. This is the **critical angle** θ_c given by

$$\boxed{\sin\theta_c = \frac{1}{n}} \tag{14.21}$$

At $\theta_i = \theta_c$, the transmission coefficients do not become zero (see below) but the 100 per cent reflection means that we have **total internal reflection**.

What happens when θ_i becomes greater than θ_c? In that case θ_t does not exist as a real angle given by Snell's law $n\sin\theta_i = \sin\theta_t$. We can, however, retain at least the algebraic *forms* of $r_{\parallel}$ and $r_{\perp}$ by writing

$$\cos\theta_t = (1-\sin^2\theta_t)^{1/2} = (1-n^2\sin^2\theta_i)^{1/2} = \mathrm{j}b, \text{ say} \tag{14.22}$$

where j is the imaginary operator ($\mathrm{j}^2 = -1$) and b is real when $\theta_i > \theta_c$. With this substitution, both $r_{\parallel}$ and $r_{\perp}$ become expressions of the form $(a - \mathrm{j}b)/(a + \mathrm{j}b)$, i.e. complex numbers with amplitude unity and phase angle ϕ between 0 and π. Such a complex number can be written as $\mathrm{e}^{\mathrm{j}\phi}$.

What does this mean physically? First, the reflected beam has an amplitude given by rE_i and so, using the complex form of E_i:

$$E_r = E_i\mathrm{e}^{\mathrm{j}\phi} = E_{i0}\mathrm{e}^{\mathrm{j}(\mathbf{k}\cdot\mathbf{r} - \omega t + \phi)}$$

So the reflected beam has the same amplitude as the incident beam but a phase difference ϕ lying between 0 and π. What of the transmitted beam? Something odd must occur because the transmission coefficients do *not* go to zero at and beyond the critical angle, so that there is evidently some sort of wave on the vacuum side of the boundary. We can write the transmitted wave formally as

$$E_t = E_{t0}\mathrm{e}^{\mathrm{j}(k_0 x\sin\theta_t + k_0 y\cos\theta_t - \omega t)}$$

where $k_0 = \omega/c$. Because $\sin\theta_t = n\sin\theta_i$ and $\cos\theta_t = \mathrm{j}b$ from (14.22), it follows that

$$E_t = E_{t0}\mathrm{e}^{-\omega b y/c}\mathrm{e}^{\mathrm{j}(k_m x\sin\theta_i - \omega t)} \tag{14.23}$$

where $k_m = n\omega/c$. This is a wave that *travels* in the x direction (i.e. *along* the boundary) and **decays** in the y direction (i.e. *perpendicular* to the boundary). It is called an **evanescent wave** and, once established, carries no energy away from the boundary into the vacuum. This explains how there can be a non-zero value of t even with total reflection. The penetration depth into the vacuum is only of the order of $c/\omega b$ (i.e. about a wavelength) and there is experimental evidence that it does exist.

Application *fibre optics* Although total internal reflection was discussed in terms of a dielectric–vacuum boundary, it is clear that it will occur whenever there is a boundary between two media with different refractive indices and when the incident light is located in the optically denser medium. In this case, if $n_1 > n_2$, the critical angle is given by $\sin\theta_c = n_2/n_1$. If the denser medium is in the form of a

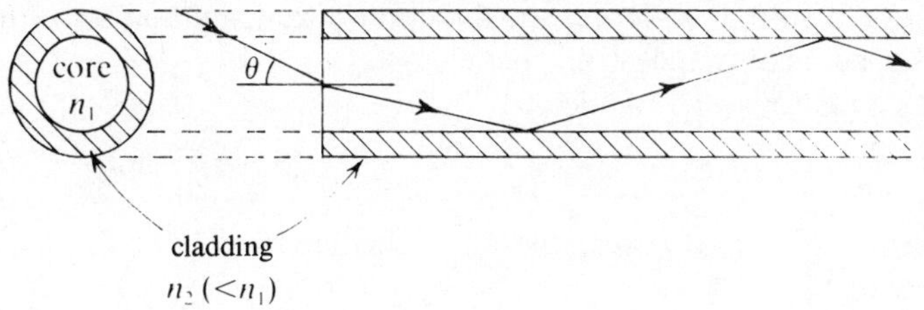

Figure 14.3 Principle of an optical fibre. Waves incident on the end of the fibre at angles less than the acceptance angle are totally internally reflected at the boundary of the core and are propagated to the far end of the fibre. The maximum θ is given by $\sin\theta = (n_1^2 - n_2^2)^{1/2}$.

long cylinder or fibre and is coated with the less dense medium, light entering one end at a suitable angle is trapped and propagated to the far end (Fig. 14.3). This effect has been known for a long time but, until the last few decades, was not of much practical use: although no losses occur during total internal reflection, the absorption of light propagated in typical glasses is of the order of 1000 dB km^{-1} and very little emerges from the end of long fibres.

However, the development of ultra-pure glasses and plastics with absorption as low as 1 dB km^{-1} has enabled optical fibres to be used not only for imaging (particularly in medical applications) but also for telecommunications. Provided the curvature does not lead to angles of incidence less than the critical angle, a fibre can follow any path, however convoluted, without loss.

The fibres themselves consist of a core with a refractive index typically between 1.5 and 1.6 and with a diameter that can vary from 2 μm to 0.5 cm depending on the application. The core is coated with a material of lower refractive index between 1.4 and 1.5. The core can be selected to transmit over the ultraviolet, visible or infrared ranges, but dispersion can be a problem.

Fibre bundles are used for optical imaging while the narrowest fibres of a few micrometres diameter are used for telecommunications. In the latter case, because the dimensions are now comparable with the wavelength of the radiation, the propagation must be treated as if the fibre were a waveguide (Sec. 14.4). The very narrow fibres are preferred in this case because they propagate in a single mode.

14.3 Propagation and surface reflection in metals

Plane waves in metals We first introduce a conductivity σ as an extra property of the LIH medium of Sec. 14.1 and later consider the effect of having a metallic medium in which σ is very high. If we still work in terms of **E** and **B**, and with $\mu_r \sim 1$, Maxwell's equations are now

fields in a conductor:

$$\nabla\cdot\mathbf{E}=0 \qquad \nabla\times\mathbf{E}=-\partial\mathbf{B}/\partial t \qquad \text{a} \quad \text{b}$$
$$\nabla\cdot\mathbf{B}=0 \qquad \nabla\times\mathbf{B}=(1/c_m)^2\partial\mathbf{E}/\partial t+\mu_0\sigma\mathbf{E} \qquad \text{c} \quad \text{d} \qquad (14.24)$$

We can derive a type of wave equation by the same techniques as in Sec. 13.6, but to see more clearly what is happening let us use the same geometry as we did in that section to find the properties of plane waves propagated along x. By exactly the same arguments as before, we find that E_x and B_x cannot have a wavelike variation, so that the waves are again TEM. If we then choose **E** along y we find that **B** is along z, so that **E**, **B** and the direction of propagation form the same orthogonal triad as in a dielectric.

Maxwell's equations then reduce to

$$\partial E_y \partial x = -\partial B_z/\partial t; \qquad \partial B_z/\partial x = -(1/c_m{}^2)\partial E_y/\partial t - \mu_0 \sigma E_y \tag{14.25}$$

E_y or B_z can be eliminated between these to yield the same wave equation in each case, for example

$$\frac{\partial^2 E_y}{\partial x^2} = (1/c_m^2)\frac{\partial^2 E_y}{\partial t^2} + \mu_0 \sigma \partial E_y/\partial t \tag{14.26}$$

To go further we investigate monochromatic variations of the form

$$E_y = E_{y_0} e^{j(kx-\omega t)}; \qquad B_z = B_{z_0} e^{j(kx-\omega t)} \tag{14.27}$$

and when E_y is substituted into (14.26) we obtain

conductor:
$$\boxed{\mathbf{k}^2 = \omega^2/c_m^2 + j\omega\mu_0\sigma} \tag{14.28}$$

instead of just $k^2 = \omega^2/c_m^2$ as in a non-conductor. The wave number has become complex. Its imaginary part in the exponent of (14.27) produces a factor with a negative real exponent indicating absorption.

In a metallic conductor the conduction current is normally much greater than the displacement current so that the second term on the right of (14.28) would be the only significant one. We then have $\mathbf{k}^2 = j\omega\mu_0\sigma$. Using the identity $j^{1/2} = (1+j)/2^{1/2}$, it follows that

metal:
$$\boxed{\mathbf{k} = (1+j)/\delta \quad \text{where} \quad \delta = (2/\mu_0\sigma\omega)^{1/2}} \tag{14.29}$$

When this is used in (14.27), the variation of **E** takes the form

$$E_y = E_{y_0} e^{-x/\delta} e^{j(x/\delta - \omega t)} \tag{14.30}$$

This is a wave that attenuates as it progresses in the x direction with a **penetration depth** δ. For a metal like copper and incident radiation with microwave frequencies, the value of δ is only about 100 μm. The wave and the induced currents are only appreciable in the surface layers and this is the same skin effect that was discussed in Sec. 9.6 from a different viewpoint.

Apart from the great attenuation, there are other properties of the wave in the metal which differ considerably from that in a dielectric. The effective wavelength is obtained from the imaginary exponent in (14.30). The effective wavelength is obtained from the imaginary exponent in (14.30) and is $\delta/2\pi$, which is considerably shorter than the corresponding free space wavelength of the incident beam. Moreover, if we use (14.25) to obtain a relation between **E** and **B**, we obtain

$$B_z = \mathbf{k} E_y/\omega \tag{14.31}$$

This means that **B** and **E** are $\pi/4$ out of phase, using the value of **k** in (14.29). The relation between H and E can be expressed using a complex refractive index from (14.8), $\mathbf{n} = \mathbf{k}c/\omega$, and $H = B/\mu_0$:

plane wave in a metal: $$H = \mathsf{n}E/\mu_0 c \tag{14.32}$$

Reflection at a metal surface We consider in detail only the case of normal incidence so that the plane of polarization of the **E** vector in Fig. 14.1 becomes immaterial. Moreover, because **E** and **H** are transverse, they are already tangential to the boundary, so that the application of boundary conditions yields the simple equations $E_i + E_r = E_t$, $H_i - H_r = H_t$. Using (14.32), a reflection coefficient can be obtained:

metallic reflection: $$\mathsf{r} = E_r/E_i = (1-\mathsf{n})/(1+\mathsf{n}) \tag{14.33}$$

where n is $\mathsf{k}c/\omega$ or $(1+j)c/\omega\delta$. The real part of n, that is $c/\omega\delta$, is very large compared with 1 so that the reflection coefficient given by (14.33) is very nearly -1. The reflecting power of metals at normal incidence is thus very high, in accordance with observation. Even the slight variation of reflecting power with frequency predicted by (14.33) has been confirmed at infrared wavelengths, assuming that the conductivity retains its static value. At shorter wavelengths it appears that σ deviates from this value, and (14.33) is then used to obtain σ by measurements of reflecting power.

We note that for very good conductors the fact that $\mathsf{r} \sim -1$ and thus that $E_i \sim -E_r$ means that at the boundary we have an E-field which is nearly zero, a condition we use in the next section.

14.4 Metallic waveguides and cavities

Laterally bounded waves Hitherto, we have only discussed wave solutions of Maxwell's equations with plane wavefronts unlimited in extent, a situation commonly found in practice. However, the transmission of electromagnetic power sometimes involves the use of waves confined to the interior of hollow metallic tubes known as waveguides and in this section we examine this type of propagation.

Although a guide may have any cross-section, we shall fix our attention mainly on a rectangular section as in Fig. 14.4 and consider propagation in the x direction. Since the guide is hollow, the appropriate form of Maxwell's equations is that used in Sec. 13.6:

fields *in vacuo*, no sources:

$$\begin{array}{llr} \nabla\cdot\mathbf{E}=0 & \nabla\times\mathbf{E}=-\partial\mathbf{B}/\partial t & \text{a} \quad \text{b} \\ & & (14.34) \\ \nabla\cdot\mathbf{B}=0 & \nabla\times\mathbf{B}=(1/c^2)\partial\mathbf{E}/\partial t & \text{c} \quad \text{d} \end{array}$$

and the equation obeyed by both **E** and **B** is the wave equation

$$\nabla^2\mathbf{E} - (1/c^2)\partial^2\mathbf{E}/\partial t^2 = 0; \qquad \nabla^2\mathbf{B} - (1/c^2)\partial^2 B/\partial t^2 = 0 \tag{14.35}$$

However, we now have an entirely different situation from that of the unlimited wavefronts, for we shall take the conductivity of the walls of the guide to be high enough for the condition $E_{\tan} = 0$ to apply at the surface. If we now tried to find the properties of plane waves propagated down the guide in the x direction as in

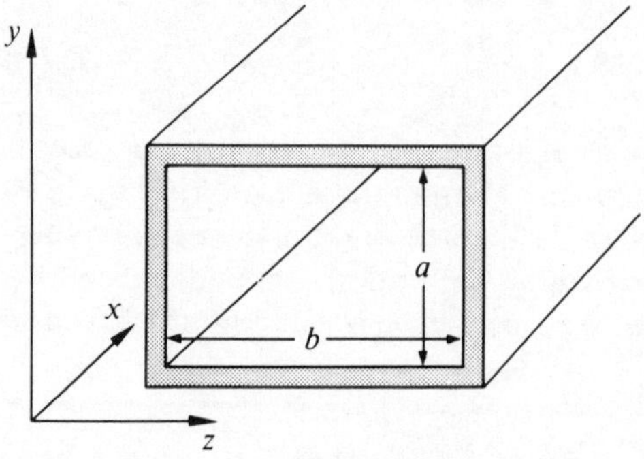

Figure 14.4 Cross-section of a rectangular waveguide.

Sec. 13.6, we should be led to the following conclusions. Firstly, such a wave would have E_x and B_x zero as before so that the wave would be TEM. But secondly, since $\partial/\partial y$ and $\partial/\partial z$ of all quantities are zero, and since E_y and E_z must be zero at *some* point round the boundary ($E_{\tan}=0$), then E_y and E_z must be zero everywhere. So all the components of **E**, and therefore of **B**, are zero: *a plane wave with wavefronts perpendicular to the axis of a guide cannot be propagated.*

We then ask what type of wave *can* be propagated. Can we, for instance, have TEM waves ($E_x=0$, $B_x=0$) that do *not* have plane wavefronts (i.e., $\partial/\partial y$ and $\partial/\partial z \neq 0$)? It turns out that this is only possible when the **E**-field has the same configuration as an electrostatic field within the guide. A hollow conductor can have no such field inside it and *it cannot therefore propagate TEM waves.* However, the introduction of a second conductor, as in a coaxial cable, does permit an electrostatic field to exist and therefore can be used to propagate TEM waves. As we mentioned in Sec. 10.13, the transmission of electromagnetic energy can then be thought of as being carried either by the current and voltage variations or by the **E**- and **B**-wave fields. Note particularly, however, that the waves, while TEM, are not plane: in a coaxial cable both **E** and **B** vary as $1/r$ over any cross-section, r being the distance from the axis.

TM waves We are now forced to look for solutions of Maxwell's equations and the wave equation that do not have plane wavefronts in the yz-plane and are not TEM. We first consider the possibility of monochromatic TM waves ($B_x=0$) so that the **B**-field is transverse but the **E**-field has an x-component

$$E_x = E_{x0} e^{j(k_g x - \omega t)} \tag{14.36}$$

All the transverse components will have similar forms but we should be clear that E_{x0}, B_{y0}, etc., are not constant as they were in a plane wave, but vary across the yz-plane and are thus functions of y and z. On the other hand, the variations with x and t are taken care of by the exponential factor. We cannot assume that the wavelength λ_g in the guide will be the same as the free space wavelength λ_0 corresponding to ω, so we use k_g, where

$$k_g = 2\pi/\lambda_g \quad \text{and} \quad k_0 = 2\pi/\lambda_0 = \omega/c \tag{14.37}$$

The first task is to find E_x in the rectangular guide of Fig. 14.4. Substitution of (14.36) into the wave equation for E_x from (14.35) shows that

$$\partial^2 E_{x0}/\partial y^2 + \partial^2 E_{x0}/\partial z^2 = (k_g^2 - k_0^2)E_{x0} \tag{14.38}$$

The boundary condition $E_{\tan}=0$ gives $E_{x_0}=0$ at $y=0$, a and at $z=0$, b. It is well known that such conditions imply standing-wave solutions to (14.38) of the form

$$E_{x_0}=E_0 \sin(l\pi y/a)\sin(m\pi z/b) \tag{14.39}$$

in which E_0 is a constant and l and m are integers satisfying

$$l^2/a^2+m^2/b^2=(k_0^2-k_g^2)/\pi^2 \tag{14.40}$$

Each pair of integers (l, m) specifies a possible *mode* denoted by TM_{lm}. It is clear that TM_{00}, TM_{01} and TM_{10} modes do not exist but that TM_{11} and all others do. The y and z components of **E** and **B** can be found from E_x using Maxwell's equations. For instance B_z is given by

$$B_z=\frac{jk_0/c}{k_0^2-k_g^2}\,\partial E_x/\partial y \tag{14.41}$$

and thus has a variation in the yz plane like $\cos(l\pi y/a)\sin(m\pi z/b)$. *E and B are always found to be orthogonal.*

Cut-off wavelength It seems at first sight as if an infinite number of TM modes can be propagated down a given guide, but this is not so. The relation (14.40) can be put into the form

$$\frac{1}{\lambda_g^2}=\frac{1}{\lambda_0^2}-\frac{1}{4}\left(\frac{l^2}{a^2}+\frac{m^2}{b^2}\right) \tag{14.42}$$

If we define a **critical wavelength** λ_c given by

$$\boxed{1/\lambda_c^2=(l/2a)^2+(m/2b)^2} \tag{14.43}$$

we obtain

$$\boxed{1/\lambda_g^2=1/\lambda_0^2-1/\lambda_c^2} \tag{14.44}$$

and this shows that when $\lambda_0=\lambda_c$, λ_g becomes infinite. Moreover, when λ_0 becomes greater than λ_c, λ_g does not exist and propagation cannot occur (in fact a rapidly attenuated field is established). Thus a given guide will only propagate modes for which $\lambda_c>\lambda_0$ and the guide dimensions can always be chosen so that only one mode is propagated at a given frequency.

TE waves Here we have $E_x=0$ and $B_{x0}e^{j(k_gx-\omega t)}$, so that it is now B_x that satisfies the wave equation corresponding to (14.38):

$$\partial^2 B_{x0}/\partial y^2+\partial^2 B_{x0}/\partial z^2=(k_g^2-k_0^2)B_{x0} \tag{14.45}$$

The boundary conditions on B_x are not quite so simple as those on E_x and have to be obtained through Maxwell's equations and $E_{\tan}=0$. Remembering that all components vary as $\exp(jk_gx-j\omega t)$, we take the y component of $\nabla\times\mathbf{E}=-\dot{\mathbf{B}}$ and the z component of $\nabla\times\mathbf{B}=(1/c^2)\dot{\mathbf{E}}$ to give

$$B_y=-k_gE_z/\omega; \qquad \partial B_x/\partial y=j(k_gB_y+\omega E_z/c^2)$$

Substitution of the first into the second gives $\partial B_x/\partial y=(jE_z/\omega)(k_0^2-k_g^2)$. So $\partial B^x/\partial y$

is proportional to E_z and is thus zero at $y=0, a$. Similarly, $\partial B_x/\partial z$ is zero at $z=0, b$. This means that solutions to (14.45) are again standing-wave solutions but this time of the form

$$B_{x_0}=B_0 \cos(l\pi y/a)\cos(m\pi z/b) \tag{14.46}$$

where l and m are again integers satisfying (14.40) and specifying the TE mode. Although TE_{00} does not exist, TE_{01} and TE_{10} do. There is a critical wavelength given by the same formulae as for TM modes. For any given guide, either the TE_{01} or TE_{10} mode will have the largest λ_c of any of the TE or TM modes so that it is possible to choose λ_0 so that only one is propagated, known as the *dominant mode*.

Guides in practice If we take as an example a guide with $a=3$ cm, $b=4$ cm, then the values of λ_c for the various modes are

TE_{01}, 8 cm; TE_{10}, 6 cm; TE_{11}, TM_{11}, 4.8 cm; TE_{02}, 4 cm, etc.

so that a free space wavelength of 7 cm would only propagate in the TE_{01} mode. In practice, this is the mode invariably used and its field patterns are shown in Figs 14.5a and 14.5b. The method of injecting power into, or extracting power from, a guide is shown in Figs 14.5c and 14.5d, using a probe to interact with the **E**-field or a loop that is linked by the **B**-field. The figure also shows the method of feeding two guides from one by using a junction.

Electromagnetic cavities If a short length of waveguide were blocked at both ends by plane conducting walls perpendicular to the axis, then a cavity is formed in which the further boundary conditions produce stationary wave patterns in all three directions. Although practical cavities are most often circular in section, the principles are exhibited just as well by considering a rectangular section as with waveguides. Suppose a length d of the guide of Fig. 14.4 is used as a cavity, then a solution for a TE_{lmn} mode would be

$$B_x=B_0 \cos(l\pi y/a)\cos(m\pi z/b)\sin(n\pi x/d)e^{-j\omega t} \tag{14.47}$$

where

$$\boxed{\nu^2=\frac{c^2}{4}\left(\frac{l^2}{a^2}+\frac{m^2}{b^2}+\frac{n^2}{d^2}\right)} \tag{14.48}$$

which is a relation giving the frequencies at which solutions exist. The cavity thus responds or resonates at certain specific frequencies and can play the part of a resonant network element at microwave frequencies just as *LC* circuits do with low frequency alternating currents.

14.5 Generation of electromagnetic waves from an accelerated charge

Until now, we have been examining the *propagation* of electromagnetic waves without enquiring into their *generation* from sources. We have, in fact, so far in this chapter put any specified sources, ρ (or Q) and **J** (or I), equal to zero in

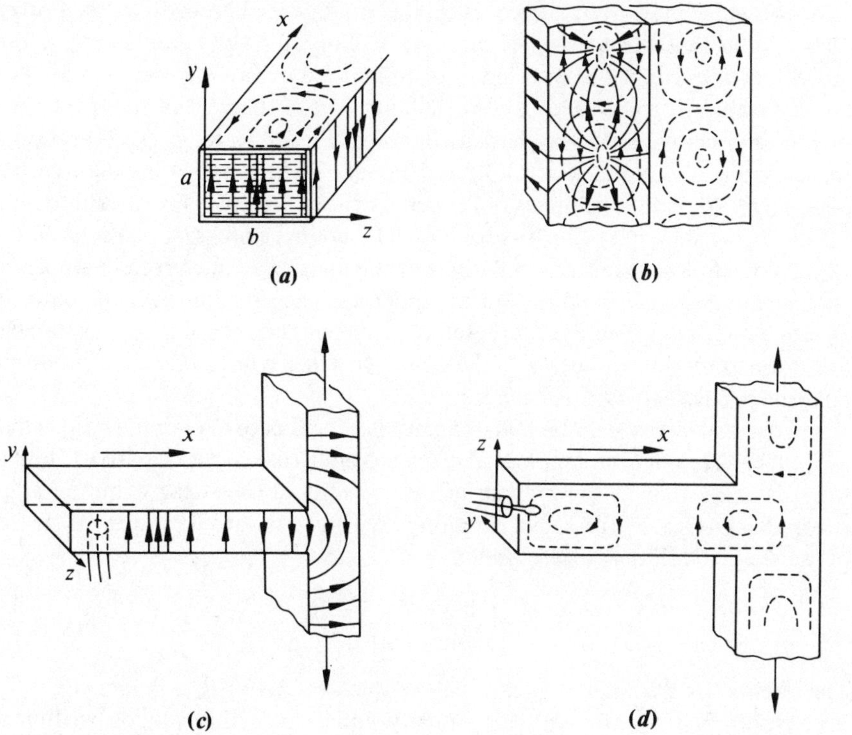

Figure 14.5 (a) The TE_{01} mode in a rectangular waveguide. The **E**-field is shown by the full lines, the **B**-field by broken lines; (b) lines of current flow in the walls of the waveguide for the TE_{01} mode. The **B**-field is shown by broken lines. The **B**-field pattern for a TE_{02} mode is shown by the double width guide; (c) a waveguide junction and injection using the **E**-field; (d) a waveguide junction and injection using the **B**-field.

Maxwell's equations. However, some of the particular solutions of the equations with specified ρ and **J** are already known to us. Firstly, if ρ is constant in time and **J** is therefore zero, only a *steady* **E**-field is generated (Chapter 3). Secondly, if ρ is allowed to move uniformly so that steady currents **J** arise, then a *steady* **B**-field is produced as well as the **E**-field (Chapter 7). It seems clear that the *time-varying* **E**- and **B**-fields in a wave can only be generated if currents vary in time and hence if charges *accelerate*. A rigorous derivation of the fields due to an accelerated charge would take us beyond the standard of this volume. Instead, we take two particular examples which prove more tractable: in the next section, we look at an oscillating electric dipole, while here we take the case of an accelerated charge, but one whose velocity is at all times much less than c.

Fields due to accelerating charge Consider a charge Q at rest at O (Fig. 14.6) with an **E**-field like that of Fig. 3.1a diverging from O and varying as $1/r^2$. At time $t=0$, an acceleration a is given the charge for a short time Δt so that a velocity $v=a\Delta t$ is reached. Thereafter Q moves with a constant velocity v and after a further time t is at O′ where $OO'=vt$. The **E**-field in its vicinity then diverges from O′ as shown and still varies as $1/r^2$.

Previous work has shown us that changes in fields are propagated with a velocity c, so that outside a sphere of radius $c(t+\Delta t)$ the **E**-field will still be as if the charge were at O. Inside a sphere of radius $r=ct$, however, the field will be as if the charge were at O′: the fact that $v \ll c$ means that the field even at a distance r is already diverging from O′. We wish to investigate what is happening in the region of thickness $c\Delta t$ between the spheres.

At this stage many derivations using this treatment (Thomson, 1904; Purcell, 1984) argue that the lines of **E** are kinked, but we have taken the view that individual lines cannot be labelled or distinguished in any way that would allow any assumptions to be made about the direction of the field in the intersphere region. Instead, consider the shaded volume in Fig. 14.6a formed by rotation of the diagram about the axis OO′, and use its surface as a Gaussian surface enclosing no charge. The fluxes of **D** over the caps of the two spheres subtending θ_1 and θ_2 as shown in Fig. 14.6b are $\frac{1}{2}Q(1-\cos\theta_1)$ and $\frac{1}{2}Q(1-\cos\theta_2)$. These expressions follow from the solid angle at the apex of a cone given in Appendix A.1. Because $\theta_1 > \theta_2$, the flux over the inner cap exceeds that over the outer by $\frac{1}{2}Q(\cos\theta_2 - \cos\theta_1)$. There must therefore be an equal outward flux over the strip at P formed by the revolution of the figure. In other words, *there is a transverse field component between the spheres.*

Let this transverse field be E_t in magnitude. The flux of **D** over the strip is then $\varepsilon_0 E_t$ times the area $(2\pi r \sin\theta)c\Delta t$ (here we assume that θ_1 and θ_2 are so close to their mean, θ, that in the limit of $v/c \to 0$ we may use θ as the common angle subtended at OO′ by the caps). Equating the total outward flux over the surface of the shaded volume to zero yields

$$E_t = \frac{Q(\cos\theta_2 - \cos\theta_1)}{4\pi\varepsilon_0 rc \sin\theta\, \Delta t}$$

The factor $(\cos\theta_2 - \cos\theta_1)$ can be written as $2\sin\frac{1}{2}(\theta_1-\theta_2)\sin\frac{1}{2}(\theta_1+\theta_2)$. Because θ_1 and θ_2 are both very nearly equal to θ, this can be written as $(\theta_1-\theta_2)\sin\theta$. From Fig. 14.5b, $(\theta_1-\theta_2) = (v \sin\theta_1)/c$ and in the limit this is $(v\sin\theta)/c$. Finally, therefore, $(\cos\theta_2 - \cos\theta_1)$ is $(v\sin^2\theta)/c$. Using, in addition, $\Delta t = v/a$ where a was the acceleration, we have for E_t the expression $(Qa\sin\theta)/(4\pi\varepsilon_0 rc^2)$. This is the field propagated outwards with velocity c and it is therefore a wave field. At large distances the wavefront becomes almost plane and we then expect the magnetic field to be given in magnitude by $B = E/c$ from (13.32). Its direction should be related to that of **E** and to the direction of propagation as in the plane wave of Fig. 13.7.

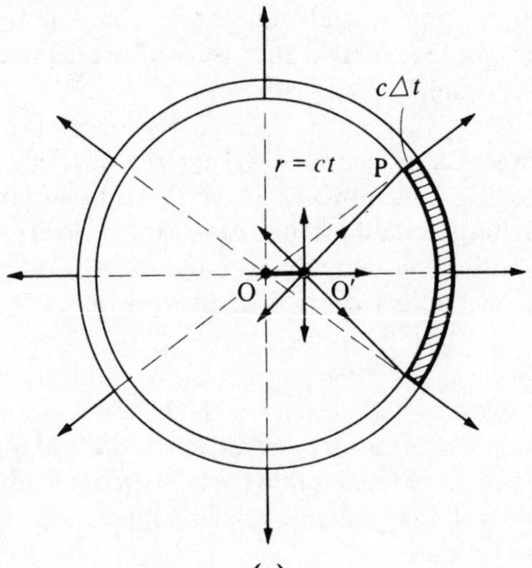

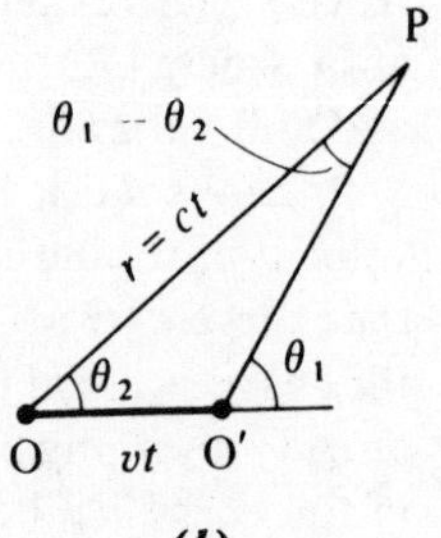

Figure 14.6 (a) Fields of an accelerated charge: the distances OO′ (= vt) and $c\Delta t$ are exaggerated compared with r for clarity; (b) detail of (a).

We thus have for the wave field from an accelerated charge:

$$\text{accelerating charge:} \qquad \mathbf{E}_{acc}=\frac{Q[a]\sin\theta}{4\pi\varepsilon_0 rc^2}\hat{\boldsymbol{\theta}}; \qquad B_{acc}=\frac{Q[a]\sin\theta}{4\pi\varepsilon_0 rc^3}\hat{\boldsymbol{\phi}} \qquad (14.49)$$

where $\hat{\boldsymbol{\theta}}$ and $\hat{\boldsymbol{\phi}}$ are unit vectors in the directions shown in Fig. 14.7a. The square brackets round a indicate that the value to be inserted, should the acceleration vary, is the *retarded value* as in Sec. 13.7 and Eqs (13.39) and (13.40), i.e. the value at a time r/c before that at which the fields at P are to be calculated. The *polar diagram* of Fig. 14.7b shows the relative magnitudes of the fields in different directions and it should be noted that no wave is propagated along the line of the acceleration.

Equations (14.49) show that the wave fields vary with distance as $1/r$, whereas we know that the static fields vary as $1/r^2$. At sufficiently large distances, therefore, the wave field will predominate. Comparing (14.49) with $E=Q/(4\pi\varepsilon_0 r^2)$ shows that this will occur when $r \gg c^2/a$.

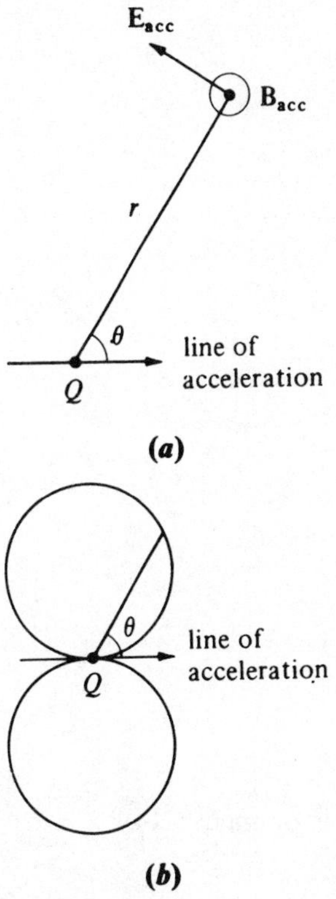

Figure 14.7 (a) Directions of wave fields due to an accelerated charge with $v \ll c$; (b) polar diagram showing the variation with θ of the magnitude of the wave fields. The distance from Q to the curve in any direction θ gives the relative magnitude of **E** or **B** in that direction ($\propto \sin\theta$).

Fields due to oscillating charges Suppose the accelerating charge above is in fact oscillating with a displacement $x=x_0 \sin\omega t$ about a mean position as it might be in a transmitting aerial. The acceleration at any instant is then $\ddot{x}=-\omega^2 x_0 \sin\omega t$, so that in magnitude the fields at a distance r are

$$E_{acc}=E_0 \sin\omega(t-r/c); \qquad B_{acc}=B_0 \sin\omega(t-r/c) \qquad (14.50)$$

where

$$E_0=\frac{Q\omega^2 x_0 \sin\theta}{4\pi\varepsilon_0 rc^2}; \qquad B_0=\frac{Q\omega^2 x_0 \sin\theta}{4\pi\varepsilon_0 rc^3} \qquad (14.51)$$

We must be clear about the conditions under which these apply. First, the amplitude x_0 must be much less than the wavelength λ corresponding to ω. If this is not so, the retarded times at various stages of the oscillation are different. If our oscillating charge is an electron in an atom, x_0 is not likely to exceed the atomic dimensions of about 0.1 nm (10^{-10} m), so that $x_0 \ll \lambda$ until the X-ray region is reached. If the oscillating charge, on the other hand, is one of the electrons in a transmitting aerial, then the length of the aerial must be $\ll \lambda$ for (14.51) to apply. Most aerials are in fact comparable in size with λ so that calculations of the resultant fields must take into account the different phases of the oscillations arriving at a point from the various parts of the aerial.

The second condition concerns the distance at which the wave field will predominate. In the cases both of the atom and of the aerial there are also positive charges Q which are almost stationary. The displacement of the negative charges thus produces an electric dipole moment of amplitude $p_0=Qx_0$. The magnitude of the electric wave field from (14.51) is $(p_0\omega^2 \sin\theta)/(4\pi\varepsilon_0 rc^2)$. The corresponding static dipole field is of the order of $p_0/(4\pi\varepsilon_0 r^3)$ by (3.39). The wave field thus predominates greatly if $r^2 \gg c^2/\omega^2$, and this will be true if $r \gg \lambda$. This condition is always fulfilled in regions of interest to us when the radiating systems are either atoms or aerials.

Radiated power from accelerated charges The direction of the Poynting vector in the wave fields of accelerating or oscillating charges is always radially away from the source, as can be seen from Fig. 14.7a. For an oscillating charge with amplitude x_0, the magnitude of **N** is obtained at a distance r from (14.50) and (14.51):

$$N = \frac{Q^2 \omega^4 x_0^2 \sin^2 \theta}{16\pi^2 \varepsilon_0 r^2 c^3} \sin^2 \omega(t - r/c)$$

The mean value of $\sin^2 \omega t$ over a period is $\frac{1}{2}$. By integrating over a sphere of radius r surrounding the charge, the mean power passing out over the surface is

$$P = \frac{Q^2 x_0^2 \omega^4}{12\pi\varepsilon_0 c^3} = \frac{p_0^2 \omega^4}{12\pi\varepsilon_0 c^3} \tag{14.52}$$

where p_0 is the magnitude of the dipole moment. This expression shows that the wave fields do lead to a loss of energy from the source and justifies the use of the word *radiation* for electromagnetic waves.

A source will be a more effective radiator at high frequencies as is clear from (14.52). It will also be more effective if the fields are not confined to a small region as they are between the plates of a laboratory capacitor. Consequently, efficient radiators are those which are open systems, such as an aerial and the earth beneath it which behave like the two plates of a capacitor with a large distance separating them.

Comment C14.1

Accelerating charges in free space are also open systems and radiate freely. In the betatron the radiation constitutes a limit to the energy attainable by the electrons, while the retardation of fast electrons by a metallic target produces continuous spectra of X-rays (bremsstrahlung). Orbiting electrons in the Bohr–Rutherford nuclear atom would also be expected to radiate continuously and thus to lose energy: this was one of the difficulties of the classical atomic theory.

Reception of waves by a system is most efficient when it is tuned to the same frequency as the incoming waves. Thus an aerial regarded as a capacitance should have connected between it and earth an inductance so that the two form a resonant circuit.

Finally, it should be noted that even an a.c. circuit operating at 50 or 60 Hz will radiate energy, but at such low frequencies the loss is small enough to be neglected and no account need be taken of it in network theory.

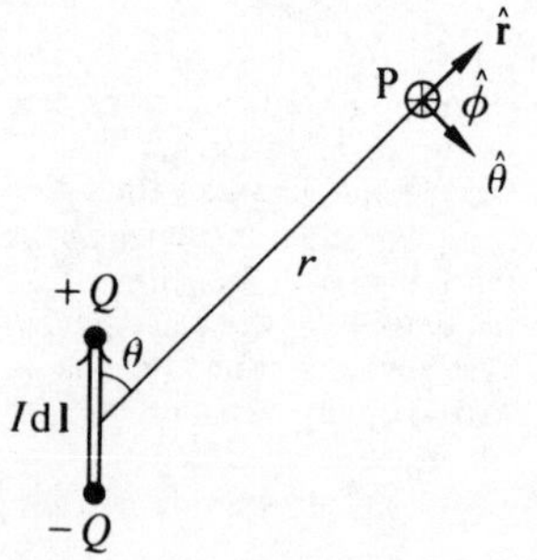

(*a*)

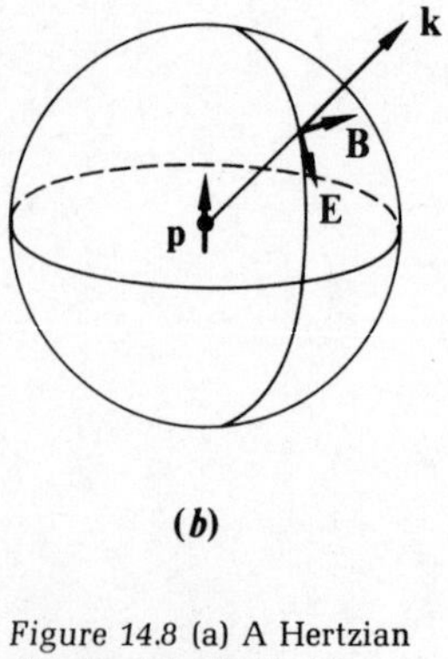

(*b*)

Figure 14.8 (a) A Hertzian dipole with $dl \ll \lambda$; (b) radiation fields from a Hertzian dipole.

14.6 Generation of fields from an oscillating dipole

The Hertzian dipole The treatment in the last section contains many provisos and approximations and does not give a full picture of the relationship between various types of **E**- and **B**-field. We therefore look at a system whose complete fields can be obtained without too much difficulty through solutions of Maxwell's equations via the retarded potentials of Sec. 13.7. This system is the Hertzian dipole, consisting of an electric dipole with charges $\pm Q$ separated by a distance dl much less than the distance of the field point r (Fig. 14.8a). The charges vary because a current I flows along a wire or spark, etc., which connects $+Q$ and $-Q$, and clearly $I = dQ/dt$. The dipole moment **p** is $Q\,\mathbf{dl}$ and therefore

$$\dot{\mathbf{p}} = \dot{Q}\,\mathbf{dl} = I\,\mathbf{dl} \tag{14.53}$$

the dot denoting differentiation with respect to time.

The centre of the dipole is situated at the origin of polar coordinates, but the axial symmetry means that we can express the field at P in terms of (r, θ) only. If λ is the wavelength corresponding to any periodic time variations, we also stipulate that $dl \ll \lambda$ so that all source points have the same regarded time and no phase differences occur at P between waves from different parts of the dipole.

By (13.40), using $I\,\mathbf{dl}$ instead of $\mathbf{J}\,d\tau$,

$$\mathbf{A}=\frac{\mu_0[I]\,\mathbf{dl}}{4\pi r}=\frac{\mu_0[\dot{\mathbf{p}}]}{4\pi r} \tag{14.54}$$

Since the direction of **p** is constant, **A** lies along the axis of symmetry and has components

$$A_r=\frac{\mu_0[\dot{p}]\cos\theta}{4\pi r}; \qquad A_\theta=-\frac{\mu_0[\dot{p}]\sin\theta}{4\pi r} \tag{14.55}$$

Thus, using Eq. (B.35) from Appendix B.12, and remembering that $\partial[\dot{p}]/\partial r=-[\ddot{p}]/c$ because $[\dot{p}]$ is a function of $(t-r/c)$ and not just of t,

$$\nabla\cdot\mathbf{A}\equiv\operatorname{div}\mathbf{A}=-\frac{\mu_0\cos\theta}{4\pi}\left(\frac{[\dot{p}]}{r^2}+\frac{[\ddot{p}]}{rc}\right)$$

By the Lorentz gauge (13.36), this is equal to $-(1/c^2)\partial V/\partial t$ and so

$$V=\frac{\cos\theta}{4\pi\varepsilon_0}\left(\frac{[p]}{r^2}+\frac{[\dot{p}]}{rc}\right) \tag{14.56}$$

The fields are now obtained from (13.34) and (13.35) using the values of **A** and V from (14.55) and (14.56). The working is not difficult using (B.34) and (B.36) from Appendix B.12, and the results are:

$$\mathbf{B}=\left(\frac{\mu_0[\dot{p}]\sin\theta}{4\pi r^2}\right)\hat{\boldsymbol{\phi}}+\left(\frac{\mu_0[\ddot{p}]\sin\theta}{4\pi rc}\right)\hat{\boldsymbol{\phi}} \tag{14.57}$$

$$\mathbf{E}=\left(\frac{2[p]\cos\theta}{4\pi\varepsilon_0 r^3}\hat{\mathbf{r}}+\frac{[p]\sin\theta}{4\pi\varepsilon_0 r^3}\hat{\boldsymbol{\theta}}\right)$$
$$+\left(\frac{2[\dot{p}]\cos\theta}{4\pi\varepsilon_0 r^2 c}\hat{\mathbf{r}}+\frac{[\dot{p}]\sin\theta}{4\pi\varepsilon_0 r^2 c}\hat{\boldsymbol{\theta}}\right)+\left(\frac{[\ddot{p}]\sin\theta}{4\pi\varepsilon_0 rc^2}\right)\hat{\boldsymbol{\theta}} \tag{14.58}$$

The first bracketed term in **E** is the only **static field**: it is the field of an electric dipole as would be given by (3.39) and because it falls off as $1/r^3$ it is the dominant field only at small distances. The first term in **B** and the second in **E** depend on $\dot{p}$ or the current I and are fields which would arise from steady currents: they fall off as $1/r^2$ and are called **induction fields**. The important new terms are the last in (14.57) and (14.58): they fall off only as $1/r$ and predominate at large distances. They are known as **wave** or **radiation fields**, or as the **far field**:

oscillating dipoles: $$\mathbf{E}_{\text{rad}}=\frac{[\ddot{p}]\sin\theta}{4\pi\varepsilon_0 rc^2}\hat{\boldsymbol{\theta}}; \qquad \mathbf{B}_{\text{rad}}=\frac{[\ddot{p}]\sin\theta}{4\pi\varepsilon_0 rc^3}\hat{\boldsymbol{\phi}} \tag{14.59}$$

These expressions should be compared with (14.49), and it should be noted that these **E**- and **B**-fields are related to each other in magnitude and direction exactly as they are in a plane wave (Fig. 14.8b).

The term *radiation field* is appropriate because it implies a resultant loss of energy from the source, unlike the other fields. This energy is obtained by integrating the Poynting vector over a sphere round the dipole. With the static and induction fields, the resultant energy decreases to zero as the radius of the sphere tends to infinity, whereas with the radiation fields the value remains constant. The Poynting vector for the fields of (14.59) is

$$\mathbf{N}=\mathbf{E}\times\mathbf{H}=\frac{[\ddot{p}]^2\sin^2\theta}{16\pi^2\varepsilon_0 r^2c^3}\hat{\mathbf{r}} \tag{14.60}$$

and when integrated over a sphere, this gives the total energy radiated from the dipole per unit time:

$$P=\left(\frac{\mu_0}{4\pi}\right)\frac{2[\ddot{p}]^2}{3c}=\frac{2[\ddot{p}]^2}{3(4\pi\varepsilon_0)c^3} \tag{14.61}$$

If the time variations in the source dipole are sinusoidal, then the retardation in $\ddot{p}$ is immaterial. When $p=p_0\sin\omega t$, the mean value of $(\ddot{p})^2$ is $\omega^4p_0^2/2$, and so the mean radiated power is

$$\boxed{\langle P\rangle=\frac{\omega^4p_0^2}{12\pi\varepsilon_0c^3}} \tag{14.62}$$

as in (14.52). In terms of the current 1, $\ddot{p}=\dot{I}\,\mathrm{d}l$ and if $I=I_0\sin\omega t$, then (14.62) gives

Hertzian dipole: $$\langle P\rangle=\frac{2\pi}{3\varepsilon_0c}\left(\frac{\mathrm{d}l}{\lambda}\right)^2I_{\mathrm{rms}}^2 \tag{14.63}$$

14.7 Aerials (antennae)

Radiation resistance The mean power radiated from a source such as an oscillating dipole has the same form as the power RI^2 consumed in a resistor. By analogy with this, the coefficient of I^2 in the expression for the radiated power is known as the **radiation resistance** R_{rad} of the source:

$$\boxed{\langle P\rangle=R_{\mathrm{rad}}I_{\mathrm{rms}}^2 \qquad \text{(definition of } R_{\mathrm{rad}})} \tag{14.64}$$

When used for the transmission of electromagnetic waves through free space, such a source forms a transmitting **aerial** (US **antenna**) and a similar receiving aerial is used for the detection and reception of the signals. The radiation resistance gives an indication of the power radiated by an aerial in contrast to its ohmic resistance, which is proportional to the power dissipated in it.

The radiation resistance of a Hertzian dipole is, from (14.63):

Hertzian dipole: $$R_{\mathrm{rad}}=\frac{2\pi}{3\varepsilon_0c}\left(\frac{\mathrm{d}l}{\lambda}\right)^2\sim 800\left(\frac{\mathrm{d}l}{\lambda}\right)^2 \quad \Omega \tag{14.65}$$

so that such a dipole, with $\mathrm{d}l/\lambda=0.01$, say, has $R_{\mathrm{rad}}\approx 0.08\,\Omega$. This may be considerably smaller than its ohmic resistance, particularly at long wavelengths, in which case it forms a relatively inefficient radiator. Moreover, the cable feeding such a dipole will have a characteristic impedance of about 75 Ω so that bad mismatching will occur, with a consequent loss of power transferred to the aerial from the transmitting circuits, unless some form of impedance matching is incorporated.

Types of aerial A more satisfactory aerial design is achieved by increasing the length of the dipole to $\lambda/2$ to form a **half-wave dipole**. If the transmitter circuits feed such an aerial at its centrepoint, there will be a distribution of current along the dipole such that there is a maximum flow at the midpoint and zero at the ends. The radiated power in any direction can be obtained by treating the whole dipole as a set of Hertzian dipoles of different amplitudes producing signals at any point in space which differ in phase due to the different distances to the elements of the aerial. Such a calculation is complex, but leads to a radiation resistance:

half-wave dipole: $$R_{rad} \approx 73\,\Omega$$

A half-wave dipole is therefore both a more efficient radiator than a Hertzian dipole and provides a better match to connecting cables, which usually have a characteristic impedance of the order of 75 Ω. An exact $\lambda/2$ dipole can be shown to present a slightly inductive load to the feeding line, but the reactance can be made zero by using a length slightly shorter than $\lambda/2$ and this is invariably done.

Other forms of aerial, consisting of loops, rhombi, apertures in waveguides, dielectric rods and metallic horns and dishes are used in different systems, the choice being dictated by the wavelength involved and the directivity required.

Directivity of aerials The concentration of the transmitted energy into a given direction is important in many applications and enables the power to be used most effectively. When communications satellites are being used, for example, extreme directivity is clearly essential. In some cases, however, such a concentration is undesirable, e.g. when all-round broadcasting over a wide area is required. The variation of the field strengths or energy with direction is displayed in a **polar diagram** or **radiation pattern** of the aerial (there is a reciprocity theorem which states that an aerial has the same directive properties for reception as transmission, so that a receiving aerial has the same polar diagram as the corresponding transmitting aerial.

The polar diagrams in a vertical plane for vertical Hertzian and half-wave dipoles are shown in Fig. 14.9a. In practice, as with all polar diagrams, there is a complication due to the presence of the earth's surface which acts as a reflector of the waves. In broad terms this can be taken into account by means of an image dipole below the surface: e.g. a $\lambda/4$ dipole mounted vertically just above the earth's surface would create the effect of a $\lambda/2$ dipole.

However, the horizontal polar diagram is also important, and for vertical Hertzian and half-wave dipoles this is simply a circle: there is no directivity at all. To achieve a high degree of directivity at long wavelengths with a single element is not easy: a **loop aerial** with its plane vertical has moderate directional properties because it behaves like a radiating or receiving magnetic dipole and will have a horizontal polar diagram somewhat similar to the vertical diagram of the electric dipoles in Fig. 14.9a. At microwave frequencies, a dipole located at the focus of a

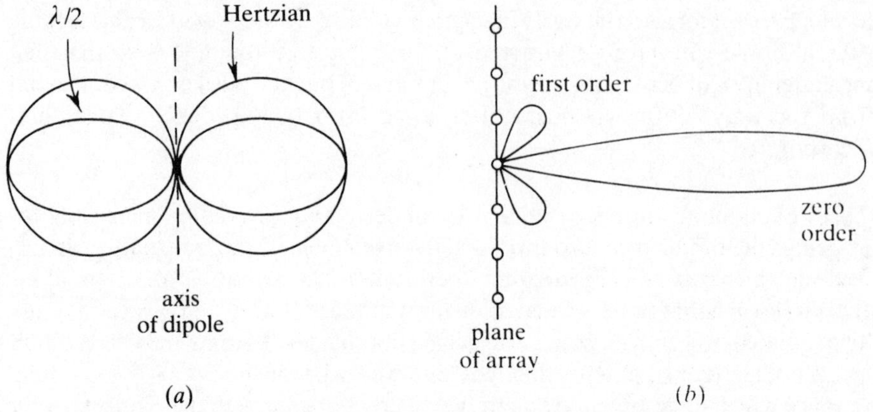

Figure 14.9 (a) Vertical polar diagrams for a Hertzian dipole and a half-wave dipole; (b) horizontal polar diagram for a line of equally spaced vertical dipoles fed in phase.

paraboloidal reflecting dish can transmit or receive over a narrow range of angles about the axis of the paraboloid. This is because reflection of the waves takes place at the paraboloidal surface just as it does in producing a searchlight beam.

Aerial arrays The greatest directivity is achieved by using a linear or planar array of dipoles fed in certain phase relationships so that reinforcement occurs in some directions and cancellation in others. The problem of calculating a polar diagram in these cases is very similar to that in calculating the Fraunhofer diffraction patterns produced by one- or two-dimensional gratings in optics. For instance, an array consisting of a horizontal line of parallel vertical dipoles fed in phase behaves like a linear diffraction grating and produces an intense zero-order beam at right angles to the plane of the array (Fig. 14.9b). There are also other beams of much smaller intensity at angles corresponding to the first, second, etc., orders, but these are usually unwanted and are eliminated by adjusting the spacing of the dipoles.

Comment C14.2

The direction of transmission or reception can be varied in two different ways: one is by mechanically rotating the whole system about a vertical axis so as to sweep the beam through all the required azimuth angles, a method often used in radar systems of many different types. The other method is to feed the various elements, not in phase, but with a phase lag or lead which is the same between any pair of adjacent elements. This can be achieved electronically and shifts the direction of the zero-order beam away from its direction perpendicular to the array. Electronic scanning is clearly capable of being carried out much more quickly than mechanical scanning.

Radio wave propagation The electromagnetic spectrum in the radio wave region is conventionally divided into ranges described by the terms in Table 14.1.

The path taken between transmitter and receiver varies according to the frequency range. The lower frequencies can pass directly through the atmosphere with only moderate absorption, helped round the curvature of the earth by atmospheric refraction and diffraction, and are known as ground waves: they have several components, a surface wave, a direct wave and a ground-reflected wave. 'Long-wave' and 'medium-wave' radio broadcasting in the LF and MF ranges takes place in this way. Television broadcasting uses ground waves in the

Table 14.1 Radio frequency bands

Band	Frequency range	Wavelength range
Extremely low frequency (ELF)	<3 kHz	>100 km
Very low frequency (VLF)	3–30 kHz	10–100 km
Low frequency (LF)	30–300 kHz	1–10 km
Medium frequency (MF)	300 kHz–3 MHz	100 m–1 km
High frequency (HF)	3–30 MHz	10–100 m
Very high frequency (VHF)	30–300 MHz	1–10 m
Ultra high frequency (UHF)	300 MHz–3 GHz	10 cm–1 m
Superhigh frequency (SHF)	3–30 GHz	1–10 cm
Extremely high frequency (EHF)	30–300 GHz	1 mm–1 cm

VHF and UHF bands, but the shorter wavelength means that there is more need for a direct line of sight between transmitting and receiving aerials.

The ionosphere has a different effect on the various ranges. At MF and below, it absorbs the radiation, while waves in the HF region are reflected and can be used for long distance 'short-wave' radio broadcasting. The frequency range roughly between 10 MHz and 50 GHz (wavelengths 6 mm to 30 m) passes through the ionosphere and forms a **radio window** of frequencies that can be used for communication satellites and radio astronomy (Fig. 14.10).

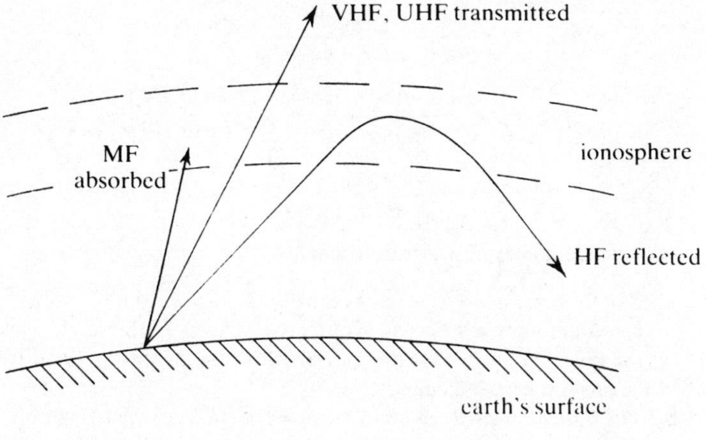

Figure 14.10 Effect of the ionosphere on electromagnetic waves in the various frequency bands.

14.8 Revision summary

Plane electromagnetic waves may be represented generally by **E**- and **B**-fields of the form

$$\mathbf{E}=\mathbf{E}_0\exp(\mathbf{k}\cdot\mathbf{r}-\omega t); \qquad \mathbf{B}=\mathbf{B}_0\exp(\mathbf{k}\cdot\mathbf{r}-\omega t)$$

where $\mathbf{E}_0$ and $\mathbf{B}_0$ are constants. Here **k** is the wave vector, normal to the plane wavefronts $\mathbf{k}\cdot\mathbf{r}=$constant, with a magnitude $k=2\pi/\lambda$. The equation of the wavefronts can also be written as $k(lx+my+nz)=$constant, where l, m, n are the direction cosines of the normal to the planes. If instantaneous values of **E** and **B** are required, or if a non-linear function like power is involved, then the real or imaginary parts of the exponential form must be taken.

Unbounded monochromatic plane waves of angular frequency ω propagated in **dielectric media** having relative permittivity ε_r possess the following properties:

- A phase velocity $c_m = c/(\varepsilon_r \mu_r)^{1/2} \approx c/\varepsilon_r^{1/2}$, and hence
- a refractive index $n(=c/c_m) \sim \varepsilon_r^{1/2}$ (14.3).
- A wave vector of magnitude $k = n\omega/c$ (14.8)
- Transverse **E**- and **B**-fields (i.e. it is a TEM wave) with **k**, **E**, **B** forming a right-handed orthogonal triad.
- Magnitudes of **E** and **B** related by $B = E/c_m = nE/c = kE/\omega$ (from 14.4)
- **E** and **B** in phase.

At a boundary between two dielectrics of refractive indices n_1 on the incident side and n_2 on the other, the boundary conditions applied to the field vectors yield the optical laws of reflection and refraction:

$$\theta_i = \theta_r; \qquad n_1 \sin\theta_i = n_2 \sin\theta_t; \qquad \omega_1 = \omega_2 \qquad \text{(from 14.11)}$$

and also yield the Fresnel relations giving the amplitudes of reflected and transmitted beams as proportions of the incident amplitude. We have reflection and transmission coefficients which differ according to the polarization, i.e. whether the **E**-vector is parallel or perpendicular to the plane of incidence:

$$r_\parallel = \frac{(n_2/n_1)\cos\theta_i - \cos\theta_t}{(n_2/n_1)\cos\theta_i + \cos\theta_t} = \frac{\tan(\theta_i - \theta_t)}{\tan(\theta_i + \theta_t)} \tag{14.14}$$

$$t_\parallel = \frac{2\cos\theta_i}{(n_2/n_1)\cos\theta_i + \cos\theta_t} \tag{14.15}$$

$$r_\perp = \frac{\cos\theta_i - (n_2/n_1)\cos\theta_t}{\cos\theta_i + (n_2/n_1)\cos\theta_t} = -\frac{\sin(\theta_i - \theta_t)}{\sin(\theta_i + \theta_t)} \tag{14.16}$$

$$t_\perp = \frac{2\cos\theta_i}{\cos\theta_i + (n_2/n_1)\cos\theta_t} \tag{14.17}$$

Two special angles of incidence occur:

- *The Brewster angle* θ_B given by $\tan\theta_B = n_2/n_1 = n$ (14.20), for incidence from a vacuum on to a dielectric of refractive index n. At this angle, $r_\perp$ is zero and the reflected light is linearly polarized, while the incident and transmitted beams are perpendicular.
- *The critical angle* θ_c given by $\sin\theta_c = 1/n$ (14.21), which can only occur for incidence within a medium with the greater refractive index. At θ_c, $\theta_t = \pi/2$ and for $\theta_i \geq \theta_c$, total internal reflection occurs. There is then an evanescent wave on the other side of the boundary that carries away no energy, that travels along the boundary and is strongly attenuated perpendicular to the boundary.

In **metallic media** with conductivity σ and at such a frequency that displacement current is negligible compared with conduction current, we denote the quantity $(2/\mu_0\sigma\omega)^{1/2}$ by δ, and find that δ is very much smaller than λ_0, the free-space wavelength $2\pi c/\omega$. Unbounded monochromatic waves in such a medium have the following properties:

- A wave vector $\mathbf{k} = (1+j)\delta$ and hence
- a refractive index $\mathbf{n} = \mathbf{k}c/\omega$, so that the wave is strongly absorbed with a penetration depth δ (14.29).

- A phase velocity $c_m \approx 2\pi(\delta/\lambda_0)c$, i.e. very much less than c.
- Transverse **E**- and **B**-fields (TEM) with **k**, **E**, **B** forming a right-handed orthogonal triad.
- Magnitudes of **E** and **B** related by $B = \mathsf{k}E/\omega$ so that **E** and **B** are $\pi/4$ out of phase (14.31).
- At a boundary, the reflection coefficient at normal incidence is $(1-\mathsf{n})/(1+\mathsf{n})$ which is very nearly of magnitude unity.

In **metallic waveguides** of constant cross-section:

- A plane wave with wavefronts perpendicular to the axis of the guide cannot be propagated.
- TEM waves cannot be propagated unless there is at least one inner conductor down the guide

but both TE and TM, or a combination of the two, can be propagated provided the free-space wavelength λ_0 is less than a critical value λ_c, the cut-off wavelength. For rectangular guides of sides a and b, λ_c is given by

$$1/\lambda_c^2 = (l/2a)^2 + (m/2b)^2 \tag{14.43}$$

where l and m are integers specifying the particular *mode* TE_{lm} or TM_{lm}. The patterns of **E**- and **B**-field lines in a guide are those of standing waves in planes perpendicular to the axis and of travelling waves along the axis. In a rectangular guide, l and m indicate the number of antinodes in the field patterns across the guide.

In a **rectangular cavity**, the patterns of **E**- and **B**-field lines are those of standing waves in all three directions. Solutions of Maxwell's equations for a box of sides a, b and d occur only at frequencies given by

$$\nu^2 = \frac{c^2}{4}\left(\frac{l^2}{a^2} + \frac{m^2}{b^2} + \frac{n^2}{d^2}\right) \tag{14.48}$$

where l, m and n are integers.

Generation of E- and B-fields Static charges generate only steady **E**-fields studied in Chapters 3 and 4. Charges moving with constant velocity generate *in addition* **B**-fields studied in Chapters 7 and 8. Both of these vary with distance from their source as $1/r^2$. When charges accelerate, or currents vary in time, we find that additional fields are produced called *radiation* or *wave* fields. If the direction of acceleration is along a spherical polar axis $\theta = 0$, then the radiation fields are

$$\mathbf{E}_{\text{rad}} = \frac{Q[a]\sin\theta}{4\pi\varepsilon_0 rc^2}\,\hat{\boldsymbol{\theta}}; \qquad \mathbf{B}_{\text{rad}} = \frac{Q[a]\sin\theta}{4\pi\varepsilon_0 rc^3}\,\hat{\boldsymbol{\phi}} \tag{14.49}$$

where $[a]$ is the retarded acceleration and r the distance of the field point from the charge Q.

If Q oscillates with an amplitude small compared both with r and with the wavelength λ of the wave fields generated, we have a Hertzian dipole of instantaneous electric dipole moment **p**. Its radiation fields are the same as those for the accelerated charge with [a] replaced by $[\ddot{p}]$. For a sinusoidally varying Hertzian dipole of amplitude p_0, the total power radiated is $p_0^2\omega^4/12\pi\varepsilon_0 c^3$. The maximum power is radiated perpendicular to the dipole axis and there is zero power radiated along the direction of the axis.

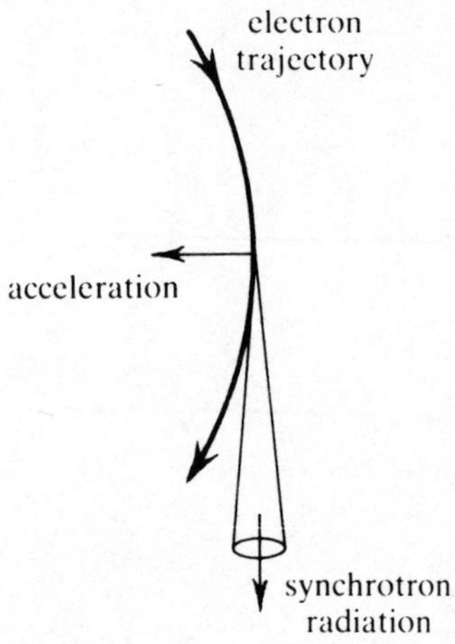

Figure 14.11 Synchrotron radiation.

Commentary

C14.1 On Synchrotron Radiation Electrons moving in a circular orbit with $v \ll c$, as in a fixed-frequency cyclotron, are being accelerated and therefore radiate electromagnetic waves. This is sometimes called **cyclotron radiation**: it is relatively weak and not highly directional, as can be seen from Fig. 14.7b.

However, electrons in a synchrotron are moving with relativistic velocities and the radiation emitted has an entirely different character. Not only is it much more intense, but it becomes concentrated into a narrow conical beam emitted along its direction of motion and it is almost 100 per cent linearly polarized with its **E**-vector parallel to the direction of the acceleration (Fig. 14.11). This is known as **synchrotron radiation**.

Synchrotron radiation from particle accelerators has become a powerful research tool because of the intensity and excellent collimation of the beam and because it covers a broad spectrum of wavelengths from about 0.01 nm to 10 μm. Any required wavelength can be selected by using monochromators.

Synchrotron radiation is also thought to be the origin of the radio emissions from extra-galactic sources such as pulsars and supernova remnants and of the galactic radio background radiation. The latter arises from cosmic ray particles spiralling about the magnetic lines of the interstellar **B**-field. Pulsars are thought to be rotating neutron stars whose synchrotron radiation is emitted by electrons moving in the enormous **B**-fields of these stars ($\approx 10^8$ T).

See Rowe and Weaver (1977) for a general description of the uses of synchrotron radiation.

C14.2 On Radar Systems for detecting the range and direction of conducting objects, radar systems, use many of the principles developed in this chapter. They depend basically on the transmission of a pulse of radio waves towards an object and the reception of a reflected or triggered pulse from the object back to the radar system, which now acts as a receiver and displays the received signal on a cathode ray tube. The range of the object is given by the time taken for the pulse to traverse its outward and return journey, while the direction is easily obtained if the aerials used have a good directivity (Sec. 14.7), but otherwise its determination requires the use of several stations.

Objects with high electrical conductivity reflect radio waves well, as we saw in Sec. 14.3, and are relatively easy to detect, whereas non-conducting objects are relatively 'invisible' to a radar system. In order to enhance the received signal from an object such as an aircraft in situations where cooperation is required (e.g. in landing control systems), the transmitted pulse can be used to trigger a pulsed signal which is much stronger than any reflection is likely to be and thus provides a much greater contrast with unwanted reflections such as those from hills, clouds, etc. Differences in the conductivity of land and sea make it possible to use radar for navigation in fog or at night.

One of the problems frequently encountered is related to the vertical polar diagram of many aerial systems (e.g. Fig. 14.9a). The power transmitted at low angles of elevation is greatly reduced and this makes it more difficult to detect objects at low altitudes, particularly at sea (ships and buoys). The use of microwaves and even millimetre waves for short ranges enables much higher directivity to be achieved and improves the accuracy and resolution of the received signals.

Electronic scanning as described in Sec. 14.7 under 'Aerial arrays', known as **phased-array radar**, is described in Brookner (1985).

Problems

Section 14.1

14.1 Compare the electric and magnetic energy densities in an electromagnetic wave travelling in a dielectric medium of refractive index 1.5.

Section 14.2

14.2 An electromagnetic wave of unit amplitude is incident on one side of a boundary between two media, and the amplitudes of the reflected and transmitted waves are r and t, respectively. If the rays are all reversed in direction, then r and t should combine to give just the original unit amplitude. Use this argument to show that $r' = -r$ and $r^2 + tt' = 1$, where r' and t' are the values of r and t for incidence on the other side of the boundary. Show that the Fresnel coefficients obey these relations.

14.3 A plane electromagnetic wave is incident on the boundary between two dielectric media at an oblique angle. Show that the sum of the reflected and transmitted power per unit area of the interface is equal to the incident power per unit area, using the Fresnel relations.

14.4 Derive Eqs (14.16) and (14.17) for the reflection and transmission coefficients of waves with the **E**-field perpendicular to the plane of incidence. What will the general state of polarization of a reflected beam be when the incident beam is linearly polarized but with the **E**-vector neither parallel nor perpendicular to the plane of incidence?

Section 14.3

14.5 Compare the electric and magnetic energy densities for an electromagnetic wave travelling in a metallic conductor.

14.6 Find expressions for the phase velocity and wavelength of an electromagnetic wave of angular frequency ω propagating in a metal of conductivity σ. How do the values compare with the corresponding values in free space at the same frequency?

Section 14.4

14.7 If v_p is the phase velocity of a guided wave and v_g its group velocity, show that $v_p v_g = c^2$.

14.8 Find expressions for the transverse components of the **E**- and **B**-fields in the guide of Fig. 14.3 transmitting in a TE_{01} mode.

Section 14.5

14.9 Show that the mean power radiated by an oscillating dipole of amplitude p_0 is given by (14.52).

Appendix A

Sundry mathematical ideas

A.1 Plane and solid angles (Chapters 2, 4, 7 and 8)

A **plane** angle θ in radians (Fig. A.1) is given by the ratio s/r, where s is the length of the arc of a circle centred at O cut off by the arms of the angle and r is the radius of this circle. For a small angle $d\theta$, the chord and arc differ in length by a second order of smallness and $d\theta$ is given equally well by ds/r or dl/r.

A **solid** angle Ω is an extension to three dimensions of the above idea. To define a solid angle (Fig. A.2), a sphere of radius r is constructed with centre at the apex of the angle O. Suppose the area cut off on the surface of this sphere by the generators of the angle is S. The solid angle is given in *steradians* by

$$\Omega = S/r^2 \qquad \text{(definition of } \Omega\text{)} \tag{A.1}$$

It follows that the complete solid angle about a point is 4π because the surface area of a sphere is $4\pi r^2$. Another useful result which also follows from the definition is that the solid angle at the apex of a cone of semi-vertical angle θ is

$$\Omega_{\text{cone}} = 2\pi(1 - \cos\theta) \tag{A.2}$$

For a small solid angle $d\Omega$, the plane area dA differs from dS by a second order of smallness so that $d\Omega = dS/r^2$ of dA/r^2. If an area dS is not normal to the radius r, then it must be projected on to a plane which *is* normal to r as an area $dS\cos\theta$. The solid angle subtended by it at O is then given by

$$d\Omega = \frac{dS\cos\theta}{r^2} \tag{A.3}$$

A.2 Coordinate systems, particularly polar coordinates (Chapter 3 onwards)

Two dimensions Instead of specifying the position of a point in a plane by its Cartesian coordinates x and y, it is sometimes convenient to use plane polar coordinates (r, θ). Figure

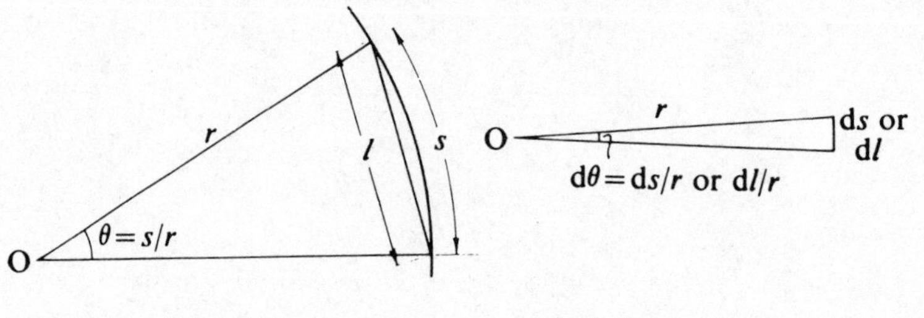

Figure A.1 Plane angles: θ and $d\theta$ are in radians.

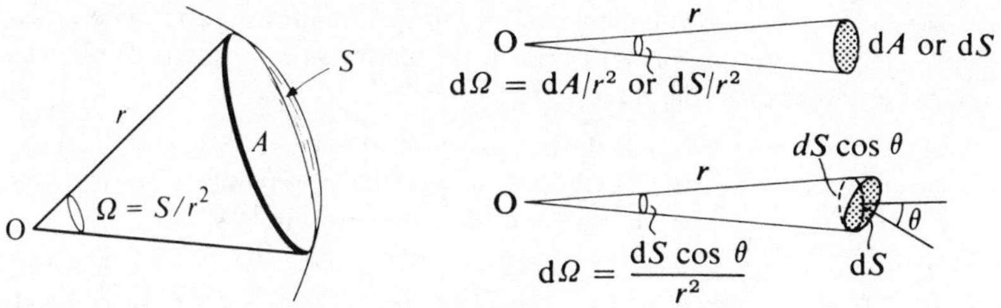

Figure A.2 Solid angles: Ω and $d\Omega$ are in steradians.

A.3 shows that the relationship between the two sets will be

$$x = r\cos\theta; \qquad y = r\sin\theta \tag{A.4}$$

or

$$r = (x^2 + y^2)^{1/2}; \qquad \theta = \tan^{-1}(y/x) \tag{A.5}$$

Three dimensions To obtain the **Cartesian coordinates** (x, y, z) of a point in three-dimensional space, a z axis is added at right angles to the x and y axes. The positive direction of the z axis can be chosen in one of two senses (up or down in Fig. A.4, in which the x axis points out of the page). It is conventional to choose the *right-handed* system illustrated in Fig. A.4, so called because the rotation of a right-handed screw from the x direction to the y direction about z as axis would cause the screw to move along z. A left-handed system would have the z axis pointing downwards.

Polar coordinates are sometimes simpler to use than Cartesians. The **cylindrical polar coordinates** (r, θ, z) of a point are obtained by using a z axis as in Cartesians but adding it to plane polar (r, θ) coordinates in the xy plane as shown in Fig. A.4a. The relationships between (r, θ, z) and (x, y, z) are as in Eqs (A.4) and (A.5) above, the z's being the same.

Alternatively, the position of a point can be expressed in **spherical polar coordinates** (r, θ, ϕ) defined in relation to Cartesians as shown in Fig. A.4b. The relationships between (r, θ, ϕ) and (x, y, z) are

$$x = r\sin\theta\cos\phi; \qquad y = r\sin\theta\sin\phi; \qquad z = r\cos\theta \tag{A.6}$$

or

$$r = (x^2 + y^2 + z^2)^{1/2}; \qquad \theta = \cos^{-1}(z/r); \qquad \phi = \tan^{-1}(y/x) \tag{A.7}$$

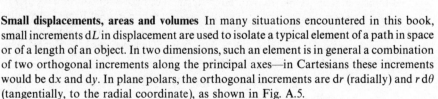

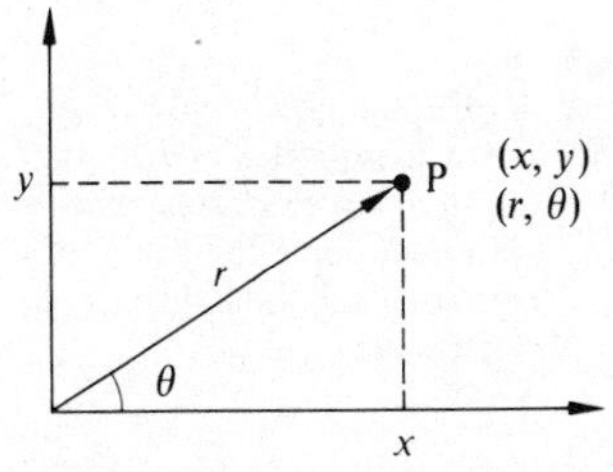

Figure A.3 Plane Cartesian and plane polar coordinates.

Note that the r in spherical polars is the distance from the *origin*, while in cylindrical polars it is the *perpendicular distance from the z axis.*

Which of the coordinate systems is best often depends on the symmetry of the situation. For instance, the potential due to a point charge at the origin depends only on the distance from the charge so that it can be written most simply in spherical polars as $V = Q/(4\pi\varepsilon_0 r)$, there being no dependence on θ or ϕ. In Cartesians, this would become $V = Q/[4\pi\varepsilon_0(x^2 + y^2 + z^2)^{1/2}]$, a more cumbersome form that would only be used if a subsequent problem demanded the use of x, y, z.

Small displacements, areas and volumes In many situations encountered in this book, small increments dL in displacement are used to isolate a typical element of a path in space or of a length of an object. In two dimensions, such an element is in general a combination of two orthogonal increments along the principal axes—in Cartesians these increments would be dx and dy. In plane polars, the orthogonal increments are dr (radially) and $r\,d\theta$ (tangentially, to the radial coordinate), as shown in Fig. A.5.

In three dimensions, the orthogonal increments are

$$\begin{aligned} &dx,\ dy,\ dz\text{—Cartesians} \\ &dr,\ r\,d\theta,\ dz\text{—cylindrical polars} \\ &dr,\ r\,d\theta,\ r\sin\theta\,d\phi\text{—spherical polars} \end{aligned} \tag{A.8}$$

[These are in fact the components of an elementary vector displacement—see Sec. B.8 of Appendix B and the gradient in polars given in Eqs (B.30) and (B.34).]

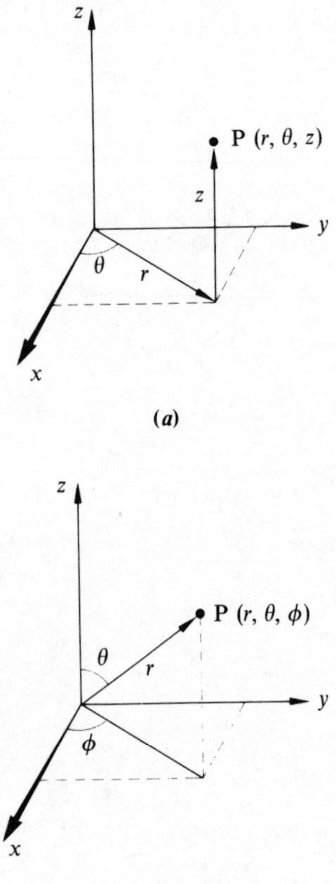

Figure A.4 (a) Cylindrical polar coordinates and their relationship to Cartesians; (b) spherical polar coordinates and their relationship to Cartesians.

Elementary areas and volumes are also used and are constructed from the increments in (A.8). Thus a typical elementary area in the Cartesian xy plane is $\mathrm{d}x\,\mathrm{d}y$, while typical volume elements $\mathrm{d}\tau$ in the three systems are

$$\begin{aligned} \mathrm{d}\tau &= \mathrm{d}x\,\mathrm{d}y\,\mathrm{d}z && \text{—Cartesians} \\ \mathrm{d}\tau &= r\,\mathrm{d}r\,\mathrm{d}\theta\,\mathrm{d}z && \text{—cylindrical polars} \\ \mathrm{d}\tau &= r^2 \sin\theta\,\mathrm{d}r\,\mathrm{d}\theta\,\mathrm{d}\phi && \text{—spherical polars} \end{aligned} \tag{A.9}$$

A.3 Partial differentiation, partial derivatives (Chapter 3 onwards)

When a quantity Z is a well-behaved† function of one other quantity x, it can be represented by a curve $Z=f(x)$ as in Fig. A.6a: $\mathrm{d}Z/\mathrm{d}x$ can be evaluated at any point such as P, and is equal to the slope of the tangent to the curve at that point.

If, however, Z is a function of two quantities, x and y, then $Z=f(x, y)$ is a *surface* in x, y, Z coordinates, part of which is drawn in Fig. A.6b. In this case, $\mathrm{d}Z/\mathrm{d}L$ can have an infinite number of values at a point P depending on the direction of L in the xy plane. An example is the rate of change of height of a hill with distance: the result depends entirely on the direction of travel.

When we have $Z=f(x, y)$ we commonly use two rates of change—that of Z with x when y is held constant, denoted by $(\partial Z/\partial x)_y$, and of Z with y when x is held constant, $(\partial Z/\partial y)_x$. In Fig. A.6b these correspond to the slopes at P and Q of sections of the surface cut by xZ and yZ planes respectively. They are known as *partial differential coefficients* or *partial derivatives*.

We often wish to know the increment in Z when both x and y vary by $\mathrm{d}x$ and $\mathrm{d}y$, respectively. Since the increment due to $\mathrm{d}x$ alone is $(\partial Z/\partial x)_y\,\mathrm{d}x$ and that due to $\mathrm{d}y$ alone is $(\partial Z/\partial y)_x\,\mathrm{d}y$, the total differential $\mathrm{d}Z$ is

$$\mathrm{d}Z = \left(\frac{\partial Z}{\partial x}\right)_y \mathrm{d}x + \left(\frac{\partial Z}{\partial y}\right)_x \mathrm{d}y \tag{A.10}$$

For functions of more than two variables, say $Z=f(x, y, z)$, the use of $(\partial Z/\partial x)$ implies that all the other variables are to be held constant. For instance, if $Z=x^2y+2y^2+3z^2$, then $\partial Z/\partial x = 2xy$; similarly, if $Z=xy+yz$ then $\partial Z/\partial x = y$. The function Z cannot now be drawn but the total increment in Z due to increments in x, y and z is, by an extension of (A.10),

$$\mathrm{d}Z = \left(\frac{\partial Z}{\partial x}\right)\mathrm{d}x + \left(\frac{\partial Z}{\partial y}\right)\mathrm{d}y + \left(\frac{\partial Z}{\partial z}\right)\mathrm{d}z \tag{A.11}$$

As an example, the potential V of Chapter 3 is *in general* a function of three variables because its value depends on position in space. In Cartesians it would be a function of x, y and z, and its three principal gradients would be denoted by $\partial V/\partial x$, $\partial V/\partial y$, $\partial V/\partial z$.

Occasionally a quantity may be expressed in terms of one set of coordinates and the variation with respect to another is wanted. For instance, the potential V due to a dipole is usually in plane polar coordinates and $\partial V/\partial x$ may be required. In carrying out such a calculation, expressions like $\partial r/\partial x$, $\partial \theta/\partial x$, etc., arise. They are easily evaluated when needed from Eqs (A.4) and (A.5). For example,

$$\begin{aligned} &\partial x/\partial r = \cos\theta = x/r; && \partial x/\partial \theta = -r\sin\theta = -y \\ &\partial r/\partial x = x/r = \cos\theta; && \partial \theta/\partial x = -y/r^2 = -(\sin\theta)/r \end{aligned} \tag{A.12}$$

with corresponding expressions for y. Note particularly that $\partial x/\partial r \neq 1/(\partial r/\partial x)$.

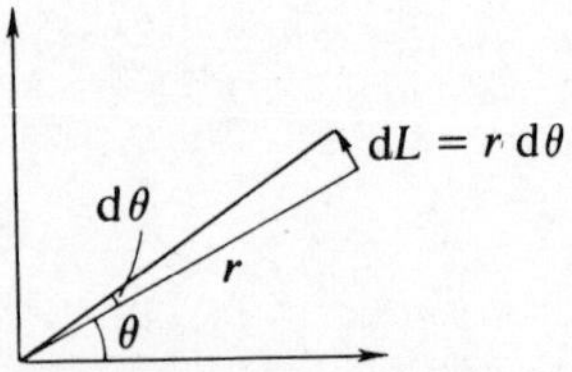

Figure A.5 An element of displacement perpendicular to the radial coordinate.

† We usually assume *at least* that the function and its first derivative are single-valued continuous functions of x.

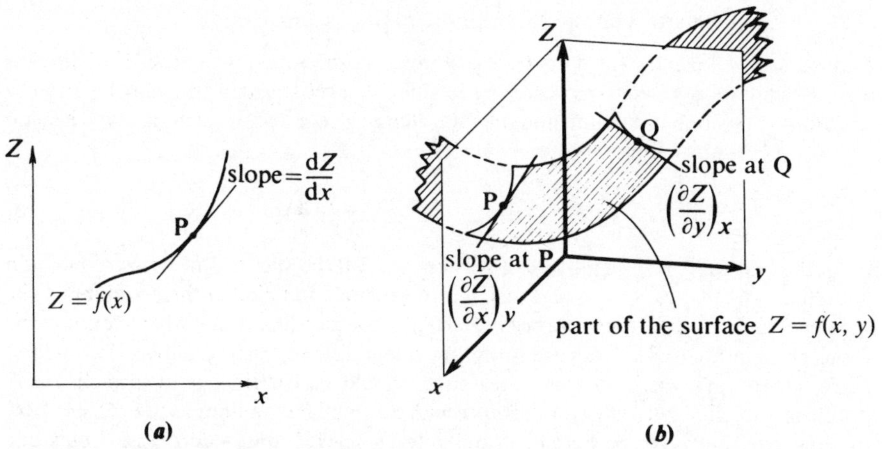

Figure A.6 (a) The line $Z = f(x)$; (b) the surface $Z = f(x, y)$.

A.4 Line integrals (Chapter 2 onwards)

I find the best way to understand line integrals is to see them as generalizations of ordinary definite integrals like $\int_1^2 x^2\,dx$. Instead of interpreting this as an area under a $y = x^2$ curve, think of it as a summation along the x axis: each element of the x axis, dx, has a value x^2 associated with it. Form the product $x^2\,dx$ for that element and sum all such products from $x = 1$ to $x = 2$. This is in fact a line integral taken along the x axis.

To generalize the idea to any path, suppose that a scalar quantity F has a value at every point in a region of space. It will therefore have a value at every point along a line L in that region between the points A and B (Fig. A.7a). For a small element of the path dL form the product of dL with the value of F at the element: $F\,dL$. The *line integral* of F between A and B along L is the sum of all the $F\,dL$'s from A to B and is written $\int_A^B F\,dL$.

Line integrals occur most frequently when the quantity varying along the line is a vector **F**, for instance a force or electric field. In that case, the line integral is formed by taking the product of dL with the resolved part of **F** along dL, $F \cos\theta$, and summing as before. Thus the work done by a force F along the path L from A to B is $\int_A^B F\cos\theta\,dL$. (This could be written as $\int \mathbf{F}\cdot d\mathbf{L}$ using the notation for a scalar product as in Appendix B.2.)

The evaluation of line integrals is a matter best left to mathematical texts, but it is evident that the result will in general depend both on the positions of A and B and on the path L. In certain cases involving vector quantities, however, the line integral is independent of the path. For instance, we saw in Sec. 2.7 that, if **F** is a central force, then

$$\int_A^B F\cos\theta\,dL = \int_{r_A}^{r_B} F(r)\,dr = f(r_B) - f(r_A) \tag{A.13}$$

where r is the radial distance from the centre of force and $f(r)$ is the indefinite integral of $F(r)$, and this depends only on A and B. The physical reason for the path-independence has been discussed in Sec. 2.7. *Where path-independence occurs, we choose a path for integration which makes the integration as simple as possible.*

Sometimes a line integral round a closed path (as in Fig. A.7b) is required and it is conventional to indicate this by the symbol $\oint$. For a vector quantity **F**, $\oint \mathbf{F}\cdot d\mathbf{L}$ is called the *circulation* of **F** round the path. When a function of position F has a *path-independent* line integral, $\oint F\,dL$ is zero since A and B in Eq. (A.13) coincide.

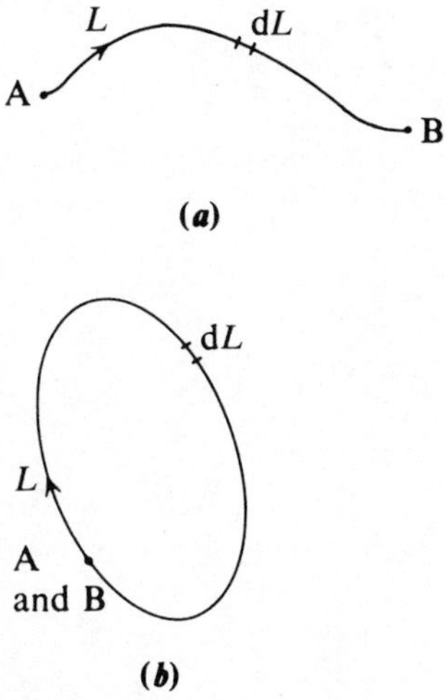

Figure A.7 (a) An open path along which the line integral of any function F of position can be evaluated; (b) a closed path: A and B now coincide.

A.5 Surface and volume integrals (Chapter 1 onwards)

Summations frequently have to be made over surfaces and volumes. The simplest examples are those of the type occurring in Chapter 1: charge densities may be given as functions of position in a region and the total charge is required. In such cases the general form of the result is either

$$Q=\int \rho \,\mathrm{d}\tau \qquad \text{or} \qquad Q=\int \sigma \,\mathrm{d}S$$

Take the first case. The quantity ρ, if it is known, will be specified in terms of position coordinates like (x, y, z) or (r, θ, ϕ), i.e. ρ is in general a function of three variables. The volume element $\mathrm{d}\tau$ must therefore be given in the same coordinates and will have one of the forms given in Eq. (A.9). This reveals that the integral sign is really an abbreviation for a triple integral $\iiint$. Each separate integration should be carried out in turn. In many problems with high symmetry (for instance with no angular variations as in Problem 1.14) the volume elements can be simplified, and only a single integration is necessary over one remaining variable. In the same way, integration over a surface would involve a *double integral.*

In this book we shall use the single integration sign as a recognized abbreviation for both volume and surface integrals written in terms of $\mathrm{d}\tau$ or $\mathrm{d}S$. Where the volume element or surface element is specifically written out as $\mathrm{d}x\,\mathrm{d}y\,\mathrm{d}z$, for example, then the full triple or double integral sign may be used. Surface integrals over a *closed* surface will be denoted by $\oint_S$.

See Appendices B.3 and B.9 for further discussion in connection with vector quantties.

A.6 Use of potential diagrams (Chapters 3, 4 and 6)

In Fig. A.8, the curve represents the potential energy of a particle at various distances from a centre of force. If the total mechanical energy is conserved, it can be represented by a horizontal line, W. Because the curve gives the potential energy, the kinetic energy is given by $W-U$ as at P in the figure.

A particle projected from infinity (where $U=0$) with a kinetic energy equal to W cannot approach nearer than R, at which point the kinetic energy, and hence the velocity, becomes zero. Another particle projected with an energy W_1 will suffer the same fate if it starts from a great distance: it will approach no nearer than r_1, but if captured by the centre of force it may oscillate between r_2 and r_3.

The particles with energies W and W_1 are said to encounter *potential barriers* at R and

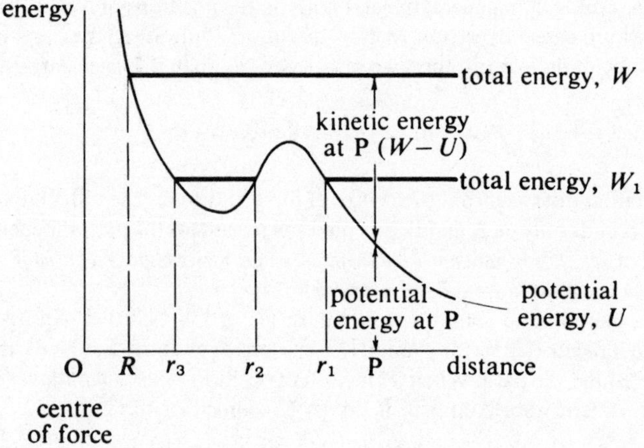

Figure A.8 Illustrating the use of a potential diagram.

r_1, respectively, while the particle with energy W_1 is said to be within a *potential well* between r_2 and r_3. Thus, Fig. 4.9a illustrates the type of potential barrier that a positive particle, such as an α-particle, would meet when approaching the positive charge of an atom. On the other hand, the diagram would have to be inverted for an electron near a nucleus, forming a potential well—the electron is then bound to the atom.

Quantum mechanics modifies the classical picture in two ways. Firstly, there is a finite probability that a particle will penetrate a potential barrier of finite height (for instance, the particle with W_1 in Fig. A.8 could *tunnel* through from r_1 to r_2). Secondly, particles are allowed to possess only certain discrete values of W unless they are completely free. This gives rise to energy levels or stationary states of systems such as atoms, represented by lines at the appropriate heights.

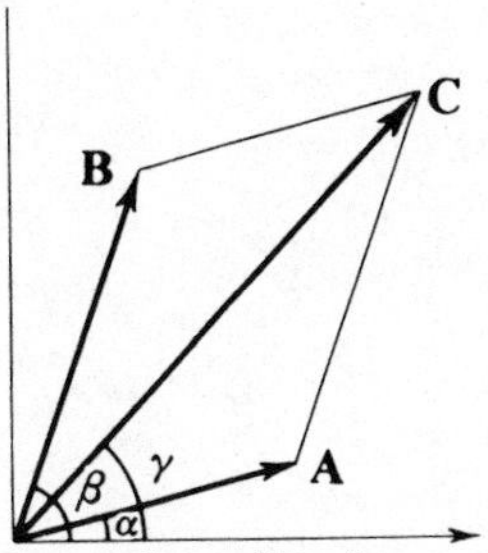

Figure A.9 Phase–amplitude diagram for the addition of sinusoidally varying quantities (phasors).

A.7 Phasor diagrams and complex numbers (Chapter 10 onwards)

A common problem encountered in physics is that of adding a number of scalar quantities each varying sinusoidally with time at the same frequency but having various amplitudes and phases. Such quantities are known as *phasors*. The methods used will be illustrated by adding two such quantities $A\sin(\omega t+\alpha)$ and $B\sin(\omega t+\beta)$: the extension to more than two is trivial.

Trigonometrical method In order to justify the geometrical and complex number methods which follow, we first establish a suitable expression for the sum by standard trigonometrical formulae. If the two quantities are expanded and the resultant terms collected appropriately, it is straightforward to show that $A\sin(\omega t+\alpha)+B\sin(\omega t+\beta)$ can be expressed as $C\sin(\omega t+\gamma)$ provided that

$$C=[A^2+B^2+2AB\cos(\beta-\alpha)]^{1/2}$$

and

$$\tan\gamma=\frac{A\sin\alpha+B\sin\beta}{A\cos\alpha+B\cos\beta} \tag{A.14}$$

The resultant is thus also a quantity with angular frequency ω but with an amplitude and phase given by (A.14). Rather than use these complicated expressions, one of the following two methods is employed in a.c. and vibration theory.

Geometrical addition We represent each of the quantities by a line whose length is equal to the amplitude A or B and whose angle made with a base line is equal to the initial phase angle α or β (Fig. A.9). If we then complete the parallelogram as shown, the diagonal from the origin has a length C and makes an angle γ with the base line where C and γ are given by (A.14), as can be obtained by trigonometrical methods. This, of course, is precisely the method used to find the resultant of two vectors and for this reason such a diagram is often referred to as a 'vector diagram', although we shall call it a *phasor diagram.* Note that for a series of additions, the polygon method of addition as illustrated in Fig. 10.9c is often more convenient.

As it stands, this method gives only the *relative* phases and amplitudes (hence the name 'phase–amplitude diagram' encountered in optics). Should the values *at any instant* be required, they are obtained by imagining the whole set of phasors as rotating counterclockwise about the origin with angular velocity ω. The projections on the vertical stationary base line then give $A\sin(\omega t+\alpha)$, etc., provided the initial positions are as shown. Equally well, of course, do the projections on the horizontal axis give acceptable values $A\cos(\omega t+\alpha)$, etc., if we prefer the cosine form. Mostly, however, we are only interested in the *relative* amplitudes and phases so that the rotation can be completely ignored.

Addition by complex numbers The geometrical method can be translated into algebraic terms by treating the phasor diagram as an Argand diagram with the imaginary axis vertical and the real axis horizontal. The reader is assumed to be familiar with the representation of a complex number $\mathbf{z}$, say, in two ways: firstly, in terms of its real and imaginary parts (x, y) so that, with $j^2 = -1$:

$$\mathbf{z} = x + jy$$

secondly, in terms of its modulus $|\mathbf{z}|$ and argument θ, giving a polar form

$$\mathbf{z} = |\mathbf{z}|e^{j\theta}$$

which is possible because $e^{j\theta} = \cos\theta + j\sin\theta$, so that $|\mathbf{z}|^2 = x^2 + y^2$ and $\theta = \tan^{-1} y/x$. Figure A.3 can be used as an Argand diagram to illustrate the relation between the two representations, using $r = |\mathbf{z}|$.

Now in the geometrical representation, a phasor like $A\sin(\omega t + \alpha)$ was represented by the projection on one of the axes of a rotating line with length A, angular velocity ω and initial phase angle α. On the Argand diagram, the rotating line would be the complex number $\mathbf{A} = Ae^{j(\omega t + \alpha)}$, and if we want the instantaneous value then we take the imaginary part (corresponding to a projection on to the real axis). Similarly, $B\sin(\omega t + \beta)$ would be represented by $\mathbf{B} = Be^{j(\omega t + \beta)}$. Because complex numbers are added by adding separately their real and imaginary parts, the sum of $\mathbf{A}$ and $\mathbf{B}$ will be $[A\cos(\omega t + \alpha) + B\cos(\omega t + \beta)] + j[A\sin(\omega t + \alpha) + B\sin(\omega t + \beta)]$. However, this is just the complex number $\mathbf{C} = Ce^{j(\omega t + \gamma)}$, where C and γ are again given by (A.14). This means that the complex number given by

$$\mathbf{C} = \mathbf{A} + \mathbf{B} \tag{A.15}$$

is the correct sum of the phasors. The use of bold sans-serif type indicates that the complex representation is meant rather than the magnitude.

Complex numbers as operators The above allows us to add quantities of a *similar* kind, such as several currents at a node or several voltages round a mesh, and justifies Kirchhoff's laws for a.c. There is, however, a further advantage in using complex numbers when the relation between phasors of *different* kinds is needed. Suppose one number $\mathbf{z} = ze^{j\theta}$ multiplies another, $\mathbf{A}$. Then the modulus of $\mathbf{A}$ is increased z times and its argument is increased by an amount θ: on the Argand diagram, $\mathbf{A}$ is increased in magnitude z times and turned counterclockwise through an angle θ. $\mathbf{z}$ is said to operate on $\mathbf{A}$. If we now have a voltage $\mathbf{V} = Ve^{j(\omega t + \beta)}$ and a current $\mathbf{I} = Ie^{j(\omega t + \alpha)}$, the ratio $\mathbf{V}/\mathbf{I}$ is a complex number $\mathbf{Z} = (V/I)e^{j(\beta - \alpha)}$. Hence in the relationship $\mathbf{V} = \mathbf{ZI}$ we can regard $\mathbf{Z}$ as operating on $\mathbf{I}$ to produce $\mathbf{V}$. Note that although $\mathbf{Z}$ is complex and can appear in an Argand diagram of its own, it is *not* a phasor because the $e^{j\omega t}$ factors always cancel from the ratio of $\mathbf{V}$ to $\mathbf{I}$ and it should not therefore appear on a phasor diagram.

A useful relation to remember is that if $\mathbf{z} = (a + jb)/(c + jd)$ then

$$|\mathbf{Z}| = \left(\frac{a^2 + b^2}{c^2 + d^2}\right)^{1/2} \quad \text{and} \quad \arg\mathbf{Z} = \tan^{-1}\frac{bc - ad}{ac + bd} \tag{A.16}$$

Appendix B

Scalars, vectors and fields

B.1 Vector notation and addition (Chapter 2 onwards)

Notation Physical quantities specified in terms of a unit by one number only are known as *scalars*: examples are mass and electric charge. Many familiar quantities, however, need three numbers to specify them: examples are velocity, acceleration and forces of many kinds. These are *vectors*, usually defined as quantities with magnitude and direction and represented diagrammatically by a line whose length is proportional to its magnitude and whose direction and sense (indicated by an arrowhead) represent its direction. Figure B.1 shows that the direction could be unambiguously given with reference to a three-dimensional coordinate system by two angles θ and ϕ. The three numbers needed to specify the vector may thus be its magnitude A together with θ and ϕ. Alternatively, and much more commonly, the projections of the vector A on to the x, y and z axes will also specify it completely. These projections are noted by A_x, A_y, A_z and are called the *Cartesian components*.

We shall distinguish vectors by using heavy type thus: **A**. In handwritten work it is usual to indicate them by some form of underlining. The magnitude of **A** will be denoted by $|\mathbf{A}|$ or, if no ambiguity arises, simply by A.

Equations involving vectors are more concisely written using such a notation—see Eqs (2.3)–(2.5). In any case, however, the habit of indicating vectors consistently in equations is advisable so that the directions of quantities are not forgotten.

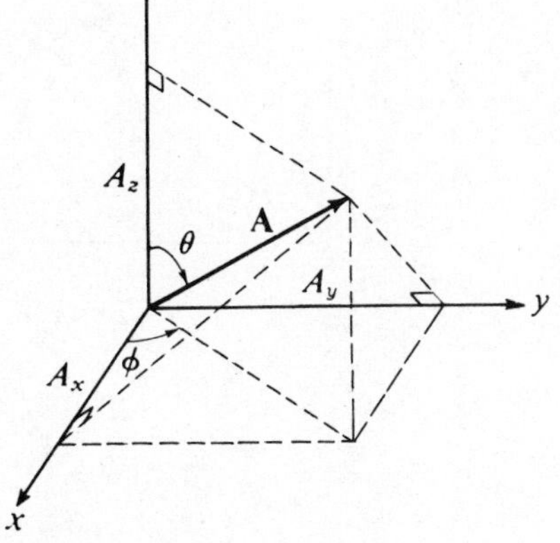

Figure B.1 A vector **A** from the origin of a reference coordinate system.

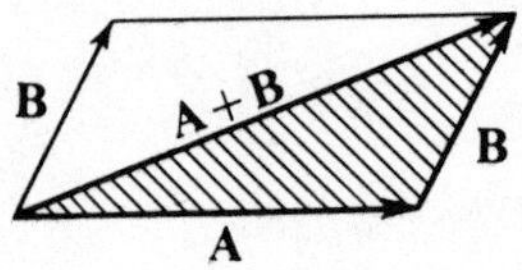

(*a*)

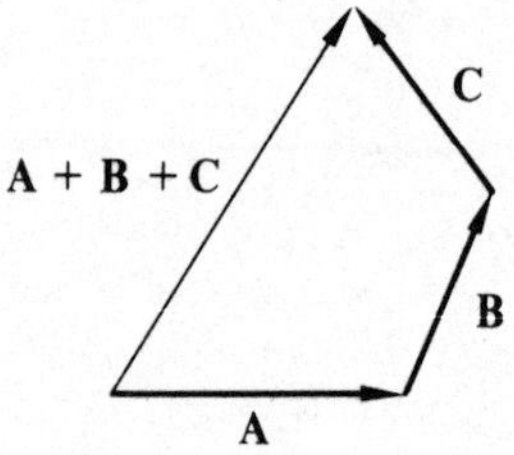

(*b*)

Figure B.2 (a) Addition of two vectors by parallelogram or triangle; (b) addition of many vectors by polygon.

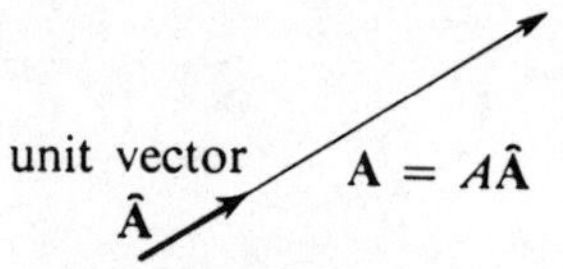

Figure B.3 Expression of a vector in terms of a unit vector.

Vector addition The sum of two vectors, $\mathbf{A}+\mathbf{B}$, is found by the parallelogram rule familiar in the case of forces and velocities. An equivalent method of addition is illustrated by the shaded triangle of Fig. B.2a in which **A** and **B** are drawn head to tail to give the resultant. The latter method is easily extended to any number of vectors as shown in Fig. B.2b.

By a vector $-\mathbf{B}$ we mean one whose sense is opposite to that of **B** but is otherwise equal in all respects to **B**. The subtraction of vectors, $\mathbf{A}-\mathbf{B}$, is defined as the addition of **A** to $-\mathbf{B}$. The associative rule for addition, $\mathbf{A}+(\mathbf{B}+\mathbf{C})=(\mathbf{A}+\mathbf{B})+\mathbf{C}$, is easily proved, and the commutative rule, $\mathbf{A}+\mathbf{B}=\mathbf{B}+\mathbf{A}$, must also apply.

Unit vectors, vector components A vector of unit length in any direction is denoted by adding a circumflex or 'hat' to any vector in the same direction. Thus $\hat{\mathbf{A}}$ is a unit vector along the same direction as **A**. The vector **A** itself could thus be written as $A\hat{\mathbf{A}}$ (Fig. B.3), a device used in Eq. (2.3) where **F** is written as $Q_1Q_2/(4\pi\varepsilon_0 r^2)$ times a unit vector $\hat{\mathbf{r}}$.

In *Cartesian coordinates*, unit vectors parallel to the x, y and z axes are denoted conventionally by **i**, **j** and **k**, although some authors use the equivalent $\hat{\mathbf{x}}$, $\hat{\mathbf{y}}$ and $\hat{\mathbf{z}}$. If the Cartesian components of **A** are A_x, A_y and A_z, then the vector can be written as

$$\mathbf{A}=\mathbf{i}A_x+\mathbf{j}A_y+\mathbf{k}A_z \tag{B.1}$$

Any number of vectors can be added by adding the corresponding components, an algebraic method equivalent to the parallelogram or polygon methods of Fig. B.2.

A vector can also be expressed in terms of components in polar coordinates. In two dimensions, plane polar coordinates (r, θ) can be used as in Appendix A.2. The orthogonal components of a vector **E** are then the radial E_r and tangential E_θ as shown in Fig. B.4a. In three-dimensional spherical polar coordinates, the directions of orthogonal components are as shown in Fig. B.4b. In this case the vector **E** could be written as

$$\mathbf{E}=\hat{\mathbf{r}}E_r+\hat{\boldsymbol{\theta}}E_\theta+\hat{\boldsymbol{\phi}}E_\phi \tag{B.2}$$

in which $\hat{\mathbf{r}}$, $\hat{\boldsymbol{\theta}}$ and $\hat{\boldsymbol{\phi}}$ are unit vectors in the directions of E_r, E_θ and E_ϕ in the figure

B.2 Scalar product of two vectors (Chapter 3 onwards)

Although we have attached a meaning to the addition and subtraction of vectors and to the multiplication of a vector by a number, there is no obvious meaning to be given to the *product* of two vectors. We find in physical applications, however, that there are two functions of pairs of vectors which are common and useful: one is itself a vector and this we leave till later (Appendix B.5); the other is a scalar known as the *scalar product*.

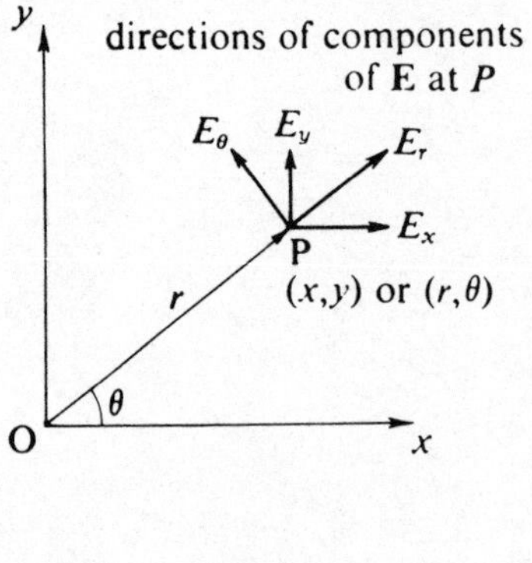

(*a*)

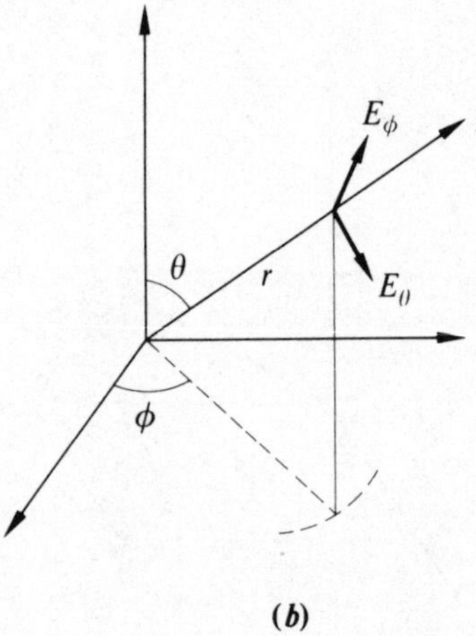

(*b*)

Figure B.4 Components of a vector **E**. (a) Two dimensions: Cartesian components E_x and E_y or plane polar components E_r and E_θ; (b) three dimensions: spherical polar components E_r, E_θ and E_ϕ.

Two vectors **A** and **B** (Fig. B.5) with an angle θ between them are said to have a scalar product $\mathbf{A}\cdot\mathbf{B}$ (read as '*A* dot *B*') defined by

$$\mathbf{A}\cdot\mathbf{B}=AB\cos\theta \tag{B.3}$$

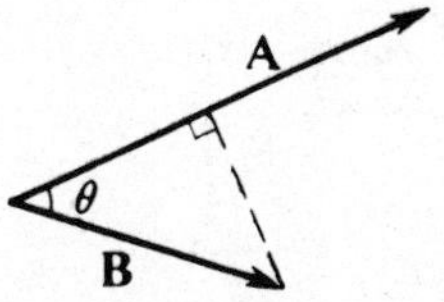

Figure B.5 Scalar product $\mathbf{A}\cdot\mathbf{B}$ is defined as $AB\cos\theta$.

so that it is the product of the magnitude of either vector and the projection on it of the other.

Scalar products occur whenever one vector has to be resolved in the direction of another: for example, the work done by a force **F** when its point of application moves a distance **L** is $F_L\cos\theta$ or $\mathbf{F}\cdot\mathbf{L}$ and if $\hat{\mathbf{L}}$ is a unit vector in a direction **L** then any vector **A** has a component $A_L=\mathbf{A}\cdot\hat{\mathbf{L}}$.

The following properties are easily derived from the definition: (a) $\mathbf{A}\cdot\mathbf{B}=\mathbf{B}\cdot\mathbf{A}$, (b) $\mathbf{A}\cdot\mathbf{B}=0$ if **A** is perpendicular to **B**, (c) $\mathbf{A}\cdot\mathbf{B}=AB$ if **A** is parallel to **B**, (d) $\mathbf{A}\cdot(\mathbf{B}+\mathbf{C})=\mathbf{A}\cdot\mathbf{B}+\mathbf{A}\cdot\mathbf{C}$, (e) $(k\mathbf{A})\cdot\mathbf{B}=k(\mathbf{A}\cdot\mathbf{B})$, (f) $\mathrm{d}(\mathbf{A}\cdot\mathbf{B})/\mathrm{d}t=\mathbf{A}\cdot(\mathrm{d}\mathbf{B}/\mathrm{d}t)+\mathbf{B}\cdot(\mathrm{d}\mathbf{A}/\mathrm{d}t)$.

Finally, if **A** and **B** are expressed in terms of Cartesian components by $\mathbf{A}=\mathbf{i}A_x+\mathbf{j}A_y+\mathbf{k}A_z$, etc., then since $\mathbf{i}\cdot\mathbf{j}=0$, $\mathbf{i}\cdot\mathbf{i}=1$, etc.,

$$\mathbf{A}\cdot\mathbf{B}=A_xB_x+A_yB_y+A_zB_z \tag{B.4}$$

This can be useful in evaluating line integrals of vector quantities (Appendix A.4). For instance, in Cartesian coordinates $\int\mathbf{F}\cdot\mathrm{d}\mathbf{L}$ can be written as $\int(F_x\,\mathrm{d}x+F_y\,\mathrm{d}y+F_z\,\mathrm{d}z)$ and each of the three integrals evaluated separately.

B.3 Flux of a vector over a surface (Chapter 4 onwards)

A small surface area $\mathrm{d}S$ can be treated as a vector whose direction is that of a line normal to it. If $\mathrm{d}S$ is part of a larger area so that its two sides can be distinguished as *inside* and *outside*, then let $\hat{\mathbf{n}}$ be a unit vector along the outward normal. The *vector area* $\mathrm{d}\mathbf{S}$ is defined as $\hat{\mathbf{n}}\,\mathrm{d}S$ (Fig. B.6a).

Now let **A** be any vector quantity which has a value at all points on the surface of which $\mathrm{d}\mathbf{S}$ is a part, and suppose $\mathrm{d}\mathbf{S}$ to be small enough for **A** to be constant over it. *The flux of* **A** *over* $\mathrm{d}\mathbf{S}$ is defined as the product of the normal component of **A** and the area, that is $A\,\mathrm{d}S\cos\theta$ or $\mathbf{A}\cdot\mathrm{d}\mathbf{S}$ (Fig. B.6b). It is thus a scalar quantity and will be negative if **A** crosses $\mathrm{d}\mathbf{S}$ from outside to inside.

If we require the flux of **A** over a large area *S*, we must add the elementary fluxes and obtain $\int\mathbf{A}\cdot\mathrm{d}\mathbf{S}$, where the integral is a double or surface integral because an element of *S* is the product of two elements of distance, say $\mathrm{d}x$ and $\mathrm{d}y$. The evaluation of surface integrals, like line integrals, should be looked up in mathematical texts if needed. We usually try to choose the surface in such a way that **A** is constant over some parts of it and zero over others, as in Sec. 4.2.

In Chapter 1, we could have regarded current as the flux of current density over a surface because, if **J** is the current density in magnitude and direction, the total current *I* over a surface *S* is

$$I=\int_S\mathbf{J}\cdot\mathrm{d}\mathbf{S} \tag{B.5}$$

so that, like any flux, current *I* may have a *sign* but is a scalar; current density is the vector.

It should be made clear that flux does not necessarily mean the flow of anything even though the term derives from this idea.

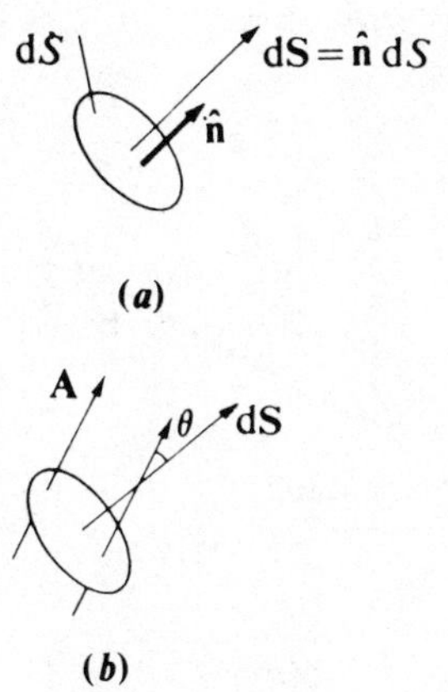

Figure B.6 (a) Surface area as a vector; (b) the flux of any vector **A** over $\mathrm{d}\mathbf{S}$ is $A\,\mathrm{d}S\cos\theta$ or $\mathbf{A}\cdot\mathrm{d}\mathbf{S}$.

B.4 Axial and polar vectors (Chapter 7 onwards)

Many vector quantities are associated not with a direction in space but with rotation about an axis and if we wish to add or resolve them we find that the axis itself can be used as the direction of the vector (Fig. B.7). These are known as *axial vectors* as opposed to the

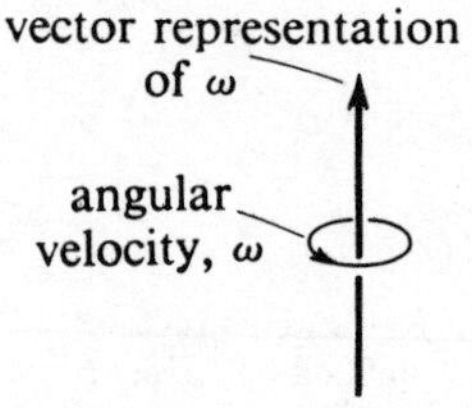

(a)

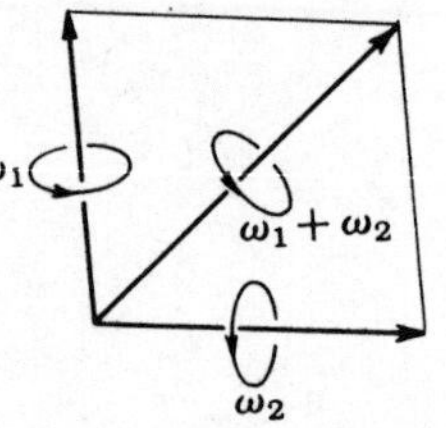

(b)

Figure B.7 (a) Representation of an axial vector, angular velocity; (b) addition of angular velocities.

polar vectors like force and velocity hitherto encountered. Angular velocity, angular momentum, and couple or torque are examples of axial vectors.

A convention is adopted that the positive direction of the vector representation is given by the following rule: rotate a right-handed screw in the same sense as that of the quantity and it will proceed in the positive direction along the axis (Fig. B.7a). An area associated with a rotation as in the current-loop dipole is also an axial vector whose direction is given by the same convention; other areas are also axial vectors whose directions are assigned as in Appendix B.3.

B.5 Vector product of two vectors (Chapter 7 onwards)

Two vectors **A** and **B** as in Fig.B.8a define a plane, that of the parallelogram between them. The *vector product* of **A** and **B**, denoted by **A × B** or **A ∧ B** (read 'A cross B'), is defined as a vector whose *magnitude* is $AB \sin\theta$ and whose *direction* is normal to the plane containing **A** and **B** and with a positive sense such that **A**, **B** and **A × B** form a *right-handed system*. The term 'right-handed system' has the same meaning as in Appendix A.2: the rotation of a right-handed screw from the first-named quantity (**A**) to the second (**B**) through the smaller angle between their positive directions (θ) produces movement in the direction of the third quantity (**A × B**) as in Fig. B.8.

While this appears complicated at first sight, it is well worth grasping for the simplification of notation that it brings. We use it frequently in Chapter 7 and the reader unfamiliar with the idea should check that the directions in formulae such as (7.2) are correctly given.

A number of results follow from the definition. It is most important to realize that commutation does not occur in the product, since $\mathbf{A}\times\mathbf{B} = -(\mathbf{B}\times\mathbf{A})$. Other results are (a) $\mathbf{A}\times\mathbf{B}=0$ if **A** and **B** are parallel, (b) $\mathbf{A}\times(\mathbf{B}+\mathbf{C})=(\mathbf{A}\times\mathbf{C})+(\mathbf{A}\times\mathbf{C})$, (c) $d(\mathbf{A}\times\mathbf{B})/dt = (\mathbf{A}\times d\mathbf{B}/dt)+(d\mathbf{A}/dt\times\mathbf{B})$, keeping strictly to the correct order. Note also that $AB\sin\theta$ is the area of the parallelogram contained by **A** and **B**, so that **A × B** could be said to represent the area as a vector quantity.

Finally, using the Cartesian component forms of **A** and **B** as in Appendix B.1, we have

$$\mathbf{A}\times\mathbf{B}=\mathbf{i}(A_yB_z-A_zB_y)+\mathbf{j}(A_zB_x-A_xB_z)+\mathbf{k}(A_xB_y-A_yB_z) \qquad \text{(B.6)}$$

since $\mathbf{i}\times\mathbf{i}=0$, $\mathbf{i}\times\mathbf{j}=1$, etc. This is easy to remember provided the cyclic order in the first term of each component is noted, e.g., the **x**-component starts with A_yB_z.

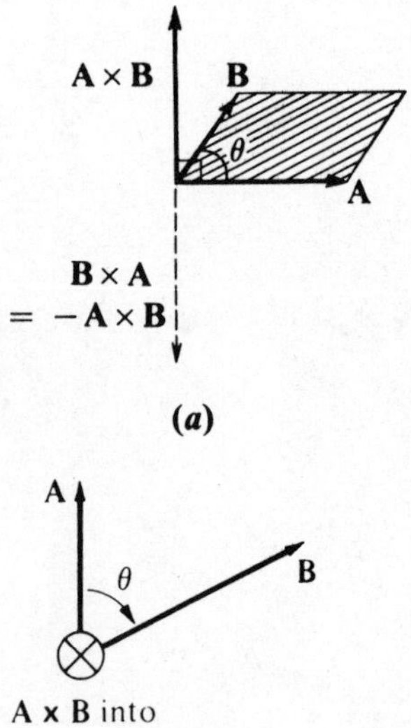

(a)

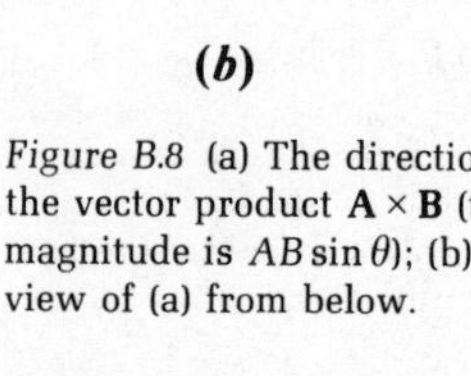

(b)

Figure B.8 (a) The direction of the vector product **A × B** (the magnitude is $AB\sin\theta$); (b) view of (a) from below.

B.6 Triple products (Chapter 7 onwards)

(**A × B**) can form either a scalar product or a vector product with another vector **C**. The *scalar triple product*, (**A × B**)·**C**, is equal in magnitude to the volume of a parallelepiped with sides **A**, **B**, and **C**. This is because the volume is (base area × perpendicular height): the vector area of the base is (**A × B**) as in Fig. B.8a, and the scalar product of this with **C** multiplies that area by the resolved part of **C** perpendicular to the base. All other scalar triple products between **A**, **B**, and **C** give the volume of the same parallelepiped and so are equal in magnitude. At most there is a difference of sign between them and it is not difficult to show that:

$$\begin{aligned}\mathbf{A}\cdot(\mathbf{B}\times\mathbf{C})&=\mathbf{B}\cdot(\mathbf{C}\times\mathbf{A})=\mathbf{C}\cdot(\mathbf{A}\times\mathbf{B})=-\mathbf{A}\cdot(\mathbf{C}\times\mathbf{B})\\&=-\mathbf{B}\cdot(\mathbf{A}\times\mathbf{C})=-\mathbf{C}\cdot(\mathbf{B}\times\mathbf{A})\end{aligned} \qquad \text{(B.7)}$$

so that the sign is unchanged as long as the cyclic order of the three vectors is also unchanged.

The *vector triple product*, (**A × B**) × **C**, must have a direction normal to the vector (**A × B**) and so must be in the plane containing **A** and **B**. In fact, as can be checked using cartesian component forms like (B.1):

$$(\mathbf{A}\times\mathbf{B})\times\mathbf{C}=\mathbf{B}(\mathbf{C}\cdot\mathbf{A})-\mathbf{A}(\mathbf{B}\cdot\mathbf{C}) \tag{B.8}$$

with corresponding expressions for $(\mathbf{B}\times\mathbf{C})\times\mathbf{A}$, etc.

B.7 Scalar and vector fields (Chapter 3 onwards)

When a *scalar* quantity has a unique magnitude at every point in a region of space it is said to form a *scalar field* over that region. Examples are the temperature in a room, the pressure in the atmosphere or the potential around a charged body. A two-dimensional example is the height of the land above sea level. The only way to picture such a field is to imagine a series of surfaces over which the scalar has the same value, spacing the surfaces at intervals such that the change in magnitude of the scalar is the same between any two adjacent ones. These would form, for instance, the isothermal surfaces in a temperature field, the isobaric surfaces in a pressure field and the equipotential surfaces in a potential field. Because these are three-dimensional they are still difficult to draw and it is common practice to use cross-sections of the surfaces producing lines—the isotherms, isobars and equipotentials of many diagrams. In the two-dimensional example of the height above sea level, the curves through points at the same height are of course the contours. In general, the scalar is a function of position, say $V(x,y,z)$, and the equations of the isotherms, equipotentials, etc., are given by $V(x,y,z)=\text{constant}$.

When a *vector* quantity has a unique magnitude and direction at every point in a region it is said to form a *vector field* over that region. Examples are the velocity of fluid in a stream, the gravitational force per unit mass, and the electric force per unit charge. Here the pictorial representation is conventionally by curved lines whose tangent at any point is in the *direction* of the vector at that point. It is difficult, except in special circumstances, to incorporate into such a picture any indication of the *magnitudes* of the quantity. Although the density of lines can sometimes be used as an indication of magnitude (e.g. the **E**-field of charges) it is not a valid method for every vector field.

B.8 The gradient, the ∇ operator, the Laplacian (Chapters 3, 4 and 8 onwards)

An important property of any scalar field is the variation of its magnitude with position. If we denote a general scalar field quantity by V, the rate of change of V in any direction $\hat{\mathbf{L}}$ is given by $\partial V/\partial L$. The value of $\partial V/\partial L$ at any point will be different for different directions of $\hat{\mathbf{L}}$, but it will have a maximum value $\partial V/\partial L_m$ in some direction $\hat{\mathbf{L}}_m$. This maximum value defines a vector called the *gradient* and is denoted by **grad**, so that

$$\mathbf{grad}\ V=\frac{\partial V}{\partial L_m}\hat{\mathbf{L}}_m \quad \text{(definition of grad)} \tag{B.9}$$

An argument similar to that in Sec. 3.6 shows that the direction of the gradient is always normal to the surfaces giving constant values of V, e.g. it is normal to the equipotential surfaces in a potential field and at right angles to the contours on land.

Since **grad** is a vector with a value at every point it forms a vector field. By virtue of (B.9) the Cartesian components of **grad** V are $\partial V/\partial x$. $\partial V/\partial y$, $\partial V/\partial z$. We can therefore write

$$\mathbf{grad}\ V=\mathbf{i}\frac{\partial V}{\partial x}+\mathbf{j}\frac{\partial V}{\partial y}+\mathbf{k}\frac{\partial V}{\partial z}=\left(\mathbf{i}\frac{\partial}{\partial x}+\mathbf{j}\frac{\partial}{\partial y}+\mathbf{k}\frac{\partial}{\partial z}\right)V=\nabla V \tag{B.10}$$

The bracketed expression is a differential operator like the ordinary d/dx and in (B.10) is denoted by the symbol ∇ (pronounced 'del'). In Cartesians, therefore, ∇ stands for

$$\nabla=\mathbf{i}\frac{\partial}{\partial x}+\mathbf{j}\frac{\partial}{\partial y}+\mathbf{k}\frac{\partial}{\partial z} \tag{B.11}$$

Although the gradient is given in Cartesians in (B.10) it does not depend on any particular coordinate system and will have other forms, for instance in polars (see Appendix B.12). For use in Chapter 3 it should be noted that if V is given in plane polar coordinates (r, θ), the components of **grad** V are $\partial V/\partial r$ (radially) and $(1/r)\partial V/\partial\theta$ (tangentially). Note the latter expression particularly—it follows from the fact that the element of displacement $\mathrm{d}L$ in a tangential direction is $r\,\mathrm{d}\boldsymbol{\theta}$ as in Fig. A.5, so that $\partial V/L$ becomes $(1/r)\partial V/\partial\theta$.

In Chapter 3 the gradient of the potential field was identified as a vector field equal to $-\mathbf{E}$. Vector fields that can be expressed as the gradient of a scalar field have special properties (see Appendix B.11).

The operator formed by the scalar product of $\boldsymbol{\nabla}$ with itself, $(\boldsymbol{\nabla}\cdot\boldsymbol{\nabla})$, is written $\boldsymbol{\nabla}^2$ and called the *Laplacian.* Equations (B.11) and (B.4) show that in Cartesians

$$\boldsymbol{\nabla}^2 = \frac{\partial^2}{\partial x^2} + \frac{\partial^2}{\partial y^2} + \frac{\partial^2}{\partial z^2} \qquad \text{(B.12)}$$

In other coordinate systems the Laplacian takes different forms (see Appendix B.12).

B.9 Divergence and curl (Chapter 4 and Chapter 8 onwards)

A vector field has two important properties known as its divergence and curl. The divergence at a point tells us what flux of the field originates there while the curl at a point tells us what circulation there is about it.

Divergence is intimately connected with the *flux* of the vector as defined in Appendix B.3. Take a small volume $\Delta\tau$ bounded by a closed surface S in a vector field $\mathbf{A}$. The outward flux of $\mathbf{A}$ over S is denoted by $\oint \mathbf{A}\cdot\mathrm{d}\mathbf{S}$ and the flux per unit volume is this divided by $\Delta\tau$. If $\Delta\tau$ is shrunk to an infinitesimal volume about a point, then the divergence is defined as the limiting value of the outward flux per unit volume:

$$\operatorname{div}\mathbf{A} = \lim_{\Delta\tau\to 0} \frac{\oint \mathbf{A}\cdot\mathrm{d}\mathbf{S}}{\Delta\tau} \qquad \text{(definition of div)} \qquad \text{(B.13)}$$

If Cartesian coordinates are used, a method similar to that used for $\mathbf{E}$ in Sec. 4.5 can be used to show that

$$\operatorname{div}\mathbf{A} = \frac{\partial A_x}{\partial x} + \frac{\partial A_y}{\partial y} + \frac{\partial A_z}{\partial z} \qquad \text{(B.14)}$$

By using the $\boldsymbol{\nabla}$ operator defined in (B.11) together with the scalar product of (B.4), we see that div $\mathbf{A}$ can be written as $\boldsymbol{\nabla}\cdot\mathbf{A}$. Note that div $\mathbf{A}$ forms a scalar field, e.g. the divergence of the E-field is shown in Chapter 4 to be proportional to the charge density.

Curl is intimately connected with the *line integral* of a vector field as defined in Appendix A.4. Take a small plane area ΔS bounded by a closed path L in a vector field $\mathbf{A}$. The *circulation* of $\mathbf{A}$ round the path L is given by $\oint \mathbf{A}\cdot\mathrm{d}\mathbf{L}$, and the circulation per unit area is this divided by ΔS. Here, however, the circulation depends on the orientation of ΔS and before proceeding to the limit we find that area $\Delta\mathrm{S_m}$ giving the maximum circulation. The curl is then defined as the limit

$$\mathbf{curl}\,\mathbf{A} = \lim_{\Delta\mathbf{S}_\mathrm{m}\to 0} \frac{\oint \mathbf{A}\cdot\mathrm{d}\mathbf{L}}{\Delta\mathbf{S}_\mathrm{m}} \qquad \text{(definition of curl)} \qquad \text{(B.15)}$$

If Cartesian coordinates are used, a method similar to that for $\mathbf{E}$ in Sec. 4.6 shows that **curl A** is a vector field of the form

$$\mathbf{curl}\,\mathbf{A} = \mathbf{i}\left(\frac{\partial A_z}{\partial y} - \frac{\partial A_y}{\partial z}\right) + \mathbf{j}\left(\frac{\partial A_x}{\partial z} - \frac{\partial A_z}{\partial x}\right) + \mathbf{k}\left(\frac{\partial A_y}{\partial x} - \frac{\partial A_x}{\partial y}\right) \qquad \text{(B.16)}$$

The $\boldsymbol{\nabla}$ operator in conjunction with (B.6) shows that **curl A** can be written as $\boldsymbol{\nabla}\times\mathbf{A}$.

B.10 Stokes's and Gauss's vector field theorems (Chapters 13 and 14)

The definitions of divergence and curl can be used to provide useful theorems applying to large volumes and surfaces instead of points. We make no attempt to prove them here and refer the reader to mathematical texts for details.

Gauss's divergence theorem (not to be confused with Gauss's law in electrostatics) states that if a closed surface S bounds a volume τ, then the volume integral of the divergence of **A** over τ is equal to the surface integral of **A** over S, or

$$\int_\tau \nabla\cdot\mathbf{A}\,d\tau = \oint_S \mathbf{A}\cdot d\mathbf{S} \tag{B.17}$$

Stokes's theorem states that if a closed path L bounds a surface S, then the surface integral of **curl A** over S is equal to the line integral of **A** round L, or

$$\int_S (\nabla\times\mathbf{A})\cdot d\mathbf{S} = \oint_L \mathbf{A}\cdot d\mathbf{L} \tag{B.18}$$

B.11 Vector field relationships (Chapters 8, 13 and 14)

Grad, div and curl of products Since **grad**, div and **curl** are differentiating operators, care has to be taken if they should operate on products. We quote here some identities involving such products, concentrating on those most useful in the later chapters of this book. U and V represent general scalar field quantities, and **A** and **B** general vector field quantities:

$$\mathbf{grad}\, UV = U\,\mathbf{grad}\, V + V\,\mathbf{grad}\, U \qquad \nabla(UV) = U\nabla V + V\nabla U \tag{B.19}$$

$$\text{div}\, V\mathbf{A} = V\,\text{div}\,\mathbf{A} + \mathbf{A}\cdot\mathbf{grad}\, V \qquad \nabla\cdot(V\mathbf{A}) = V\nabla\cdot\mathbf{A} + \mathbf{A}\cdot\nabla V \tag{B.20}$$

$$\mathbf{curl}\, V\mathbf{A} = V\,\mathbf{curl}\,\mathbf{A} - \mathbf{A}\times\mathbf{grad}\, V \qquad \nabla\times(V\mathbf{A}) = V\nabla\times\mathbf{A} - \mathbf{A}\times\nabla V \tag{B.21}$$

$$\text{div}(\mathbf{A}\times\mathbf{B}) = \mathbf{B}\cdot\mathbf{curl}\,\mathbf{A} - \mathbf{A}\cdot\mathbf{curl}\,\mathbf{B} \qquad \nabla\cdot(\mathbf{A}\times\mathbf{B}) = \mathbf{B}\cdot(\nabla\times\mathbf{A}) - \mathbf{A}\cdot(\nabla\times\mathbf{B}) \tag{B.22}$$

Combinations of div, grad and curl in pairs The differential operators may operate on each other but only in certain combinations. Possible ones are **grad** div, div **grad**, **curl grad**, **curl curl**. The following are true for any scalar field V and vector field **A**:

$$\mathbf{curl}\,\mathbf{grad}\, V = 0 \qquad \nabla\times\nabla V = 0 \tag{B.23}$$

$$\text{div}\,\mathbf{curl}\,\mathbf{A} = 0 \qquad \nabla\cdot\nabla\times\mathbf{A} = 0 \tag{B.24}$$

$$\text{div}\,\mathbf{grad}\, V = \nabla^2 V \qquad \nabla\cdot\nabla V = \nabla^2 V \tag{B.25}$$

$$\mathbf{curl}\,\mathbf{curl}\,\mathbf{A} = \mathbf{grad}\,\text{div}\, A - \nabla^2\mathbf{A} \qquad \nabla\times\nabla\times\mathbf{A} = \nabla\nabla\cdot\mathbf{A} - \nabla^2\mathbf{A} \tag{B.26}$$

Identity (B.26) is useful only in Cartesian coordinates where

$$\nabla^2\mathbf{A} = \mathbf{i}\nabla^2 A_x + \mathbf{j}\nabla^2 A_y + \mathbf{k}\nabla^2 A_z \tag{B.27}$$

In other systems of coordinates, (B.26) merely defines $\nabla^2\mathbf{A}$.

Special fields: (a) irrotational or lamellar We have already mentioned in Appendix B.8 that a vector field which is the gradient of a scalar field has special properties. The **E**-field of static charges in Chapter 3 is an example. Before discussing such fields, however, we must distinguish between regions of space that are *simply connected* and those that are not. By a simply connected region we mean one in which any closed curve can be shrunk to a point without at any stage passing outside the region. Thus the interior of a sphere would be simply connected but the space between two concentric spheres would not be.

If a vector field $\mathbf{A}_L$ is expressible as the gradient of a scalar field V over the whole of a simply connected region, then (B.23) ensures that $\mathbf{curl}\,\mathbf{A}_L$ is zero everywhere in the region. Stokes's theorem (B.18) then shows that $\oint \mathbf{A}_L\cdot d\mathbf{L}$ is zero for every closed path and hence that the line integral of $\mathbf{A}_L$ between any two points is independent of the path. In fact, any one of the properties cited implies all the others so that

$$\mathbf{A}_\mathrm{L}=\nabla V \Leftrightarrow \nabla\times\mathbf{A}=0 \Leftrightarrow \oint \mathbf{A}_\mathrm{L}\cdot d\mathbf{L}=0 \Leftrightarrow \int_A^B \mathbf{A}_\mathrm{L}\cdot d\mathbf{L} \text{ is path-independent} \tag{B.28}$$

in a region that is simply connected. A vector field for which the curl is zero is said to have no vortices and to be *irrotational*, *lamellar* or *conservative*, and V is called its *scalar potential.*

When the region is not simply connected, the implications of (B.28) *may* still operate in both directions. However, the argument that $\nabla\times\mathbf{A}=0\Leftrightarrow\oint\mathbf{A}\cdot d\mathbf{L}=0$ *could* now be false for paths that cannot be shrunk to a point without passing outside the region. The **B**-field of currents is just such a field (see Chapter 8 for a discussion of this point).

Special fields: (b) solenoidal A vector field expressible as the curl of another also has special properties. Suppose $\mathbf{B}_S$ is the curl of a field **A** over the whole of a region. Then (B.24) ensures that the divergence of $\mathbf{B}_S$ is zero in the region and therefore, by Gauss's theorem (B.17), that the flux of $\mathbf{B}_S$ over any closed surface in the region is zero as well. For an open surface S, bounded by a closed path L, the flux of $\mathbf{B}_S$ will depend only on the boundary L and will be the same for all surfaces bounded by L. In fact, any one of the properties cited implies all the others so that

$$\mathbf{B}_S=\nabla\times\mathbf{A}\Leftrightarrow\nabla\cdot\mathbf{B}_S=0\Leftrightarrow\oint\mathbf{B}_S\cdot d\mathbf{S}=0$$

$$\Leftrightarrow\int_S \mathbf{B}_S\cdot d\mathbf{S} \text{ is independent of } S \text{ for a given boundary} \tag{B.29}$$

A field for which $\nabla\cdot\mathbf{B}=0$ is said to have no sources and to be *solenoidal.* For such a field, **A** is said to be its *vector potential.*

Helmholtz's theorem The two types of vector field just discussed have very special properties, one having sources and no vortices and the other having vortices but no sources. It can be shown, however, that *any* vector field can be expressed as the sum of an irrotational and a solenoidal field, a result known as Helmholtz's theorem. It follows that when the divergence and curl of a vector field are given (so that both its sources and vortices are fixed) it is completely determined, apart from an additive constant.

B.12 Polar coordinate forms (Chapters 4, 8, 13 and 14)

We list here for reference the cylindrical and spherical polar coordinate forms of the gradient, the divergence, the curl and the Laplacian.

Cylindrical polar forms:

$$\nabla V\equiv\mathbf{grad}\,V=\hat{\mathbf{r}}\frac{\partial V}{\partial r}+\hat{\boldsymbol{\theta}}\frac{1}{r}\frac{\partial V}{\partial\theta}+\hat{\mathbf{z}}\frac{\partial V}{\partial z} \tag{B.30}$$

$$\nabla\cdot\mathbf{A}\equiv\mathrm{div}\,\mathbf{A}=\frac{1}{r}\frac{\partial}{\partial r}(rA_r)+\frac{1}{r}\frac{\partial A_\theta}{\partial\theta}+\frac{\partial A_z}{\partial z} \tag{B.31}$$

$$\nabla\times\mathbf{A}\equiv\mathbf{curl}\,\mathbf{A}=\hat{\mathbf{r}}\left(\frac{1}{r}\frac{\partial A_z}{\partial\theta}-\frac{\partial A_\theta}{\partial z}\right)+\hat{\boldsymbol{\theta}}\left(\frac{\partial A_r}{\partial z}-\frac{\partial A_z}{\partial r}\right)+\hat{\mathbf{z}}\frac{1}{r}\left[\frac{\partial}{\partial r}(rA_\theta)-\frac{\partial A_r}{\partial\theta}\right] \tag{B.32}$$

$$\nabla^2 V=\frac{1}{r}\frac{\partial}{\partial r}\left(r\frac{\partial V}{\partial r}\right)+\frac{1}{r^2}\frac{\partial^2 V}{\partial\theta^2}+\frac{\partial^2 V}{\partial z^2} \tag{B.33}$$

Spherical polar forms:

$$\nabla V \equiv \mathbf{grad}\, V = \hat{\mathbf{r}}\frac{\partial V}{\partial r} + \hat{\boldsymbol{\theta}}\frac{1}{r}\frac{\partial V}{\partial \theta} + \hat{\boldsymbol{\phi}}\frac{1}{r\sin\theta}\frac{\partial V}{\partial \phi} \tag{B.34}$$

$$\nabla \cdot \mathbf{A} \equiv \operatorname{div} \mathbf{A} = \frac{1}{r^2}\frac{\partial}{\partial r}(r^2 A_r) + \frac{1}{r\sin\theta}\frac{\partial}{\partial\theta}(A_\theta \sin\theta) + \frac{1}{r\sin\theta}\frac{\partial A_\phi}{\partial\phi} \tag{B.35}$$

$$\begin{aligned}\nabla \times \mathbf{A} \equiv \mathbf{curl}\, \mathbf{A} = {} & \hat{\mathbf{r}}\frac{1}{r\sin\theta}\left[\frac{\partial}{\partial\theta}(A_\phi \sin\theta) - \frac{\partial A_\theta}{\partial\phi}\right] \\ & + \hat{\boldsymbol{\theta}}\frac{1}{r}\left[\frac{1}{\sin\theta}\frac{\partial A_r}{\partial\phi} - \frac{\partial}{\partial r}(rA_\phi)\right] + \hat{\boldsymbol{\phi}}\frac{1}{r}\left[\frac{\partial}{\partial r}(rA_\theta) - \frac{\partial A_r}{\partial\theta}\right]\end{aligned} \tag{B.36}$$

$$\nabla^2 V = \frac{1}{r^2}\frac{\partial}{\partial r}\left(r^2\frac{\partial V}{\partial r}\right) + \frac{1}{r^2\sin\theta}\frac{\partial}{\partial\theta}\left(\sin\theta\frac{\partial V}{\partial\theta}\right) + \frac{1}{r^2\sin^2\theta}\frac{\partial^2 V}{\partial\phi^2} \tag{B.37}$$

Appendix C

Electromagnetic standards and units

Introduction It is evident from Chapter 1 that the establishment of agreed laws needs only an arbitrary set of units for the various quantities involved. The only proviso is that the meaning of the quantities should be clear, either by definition (like $R = V/I$) or by a measurable phenomenon (like Q = deflection of an electroscope). It is important, however, to be able to communicate, not just laws, but the results of measurements and the values of constants for technical, commercial and legal reasons as well as scientific ones. To do this, an internationally agreed system of units must be set up in terms of which all measurements can be expressed. These units must be embodied in *standards*, adequately maintained, which can be widely used to calibrate apparatus.

In this appendix we trace briefly the steps by which everyday electrical measurements are related to the basic SI units of mass, length and time through the activities of national standardizing laboratories such as the National Physical Laboratory (NPL) in the UK, the National Bureau of Standards (NBS) in the USA or the National Measurement Laboratory (NML) in Australia. Since the base units for mechanical measurements are by agreement the kilogramme, metre and second as defined in SI, these laboratories all maintain copies of the prototype kilogramme kept at Sèvres and all maintain interferometers and clocks to realize the metre and second. High-quality oscillators allow frequency to be realized to a few parts in 10^{13}, a greater precision than that attainable with any other physical quantity, and this is an important factor in electrical measurements. The acceleration due to gravity, g, at each standardizing laboratory is measurable to about 1 part in 10^5 and this enables a standard to be provided for force or weight, while the velocity of light *in vacuo* can be measured to a few parts in 10^7. Details of these matters can be found in the references at the end of the appendix.

Turning now to the measurement of electrical quantities, we might reasonably expect that a method for realizing the ampere would be sufficient to relate all electrical quantities to mechanical ones, since energy (or power) are common to both electromagnetism and mechanics. However, although the **joule** (or watt) and the **ampere** are sufficient to *define* the rest of the electrical units, the energy or power involved in any standardizing experiments must be electrical in nature. In practice, any two of the four quantities, ohm, amp, volt, watt, are enough and most standardizing laboratories concentrate their efforts in this way. The scheme adopted in such laboratories is illustrated in Table C.1, although the reader should be aware that advances in techniques mean that changes are continually taking place in this area. We amplify some of these steps.

Transfer standards, mechanical to electrical: the current balance The definition of the ampere is equivalent to choosing a value of $4\pi \times 10^{-7}$ H m^{-1} for the constant μ_0 so that, if the laws of magnetic interaction of Chapters 7, 8 and 9 are correct, we can use the force between any pair of conductors to determine a current flowing through both as in the

Table C.1 Measurement of electromagnetic quantities in terms of the standards of mass, length and time

I	*Mechanical standards*	length	time	mass
		g, c	frequency	
II	*Transfer, mechanical to electrical*	current balance (I)	calculable inductor (M) calculable capacitor (C)	
III	*Primary d.c. working standards* (international comparison and checks for drifting)	standard cells (V)	standard resistors (R)	
IV	*Transfer, d.c. to a.c.*	electrostatic voltmeter (V) at NPL	dynamometer (I) at NBS	calculable inductors (M) and capacitors (C) at NPL and NSL
V	*Secondary working standards*	calibrated inductors, resistors, capacitors, cells, meters, etc.		

current balance of Chapter 1. Two forms of balance using very different coil designs have been described in some detail in the literature: the Rayleigh type used at the NBS and the Ayrton-Jones type used at the NPL (Fig. C.1). In both, the principle of the method is the balancing of the magnetic force $I^2\,\partial M/\partial x$ [Eq. (9.46)] against a weight mg, where M is the mutual inductance between fixed and movable coils and x is the direction of displacement of the movable coil. Thus

$$I = \sqrt{\left(\frac{mg}{\partial M/\partial x}\right)}$$

and the right-hand side can be measured entirely in mechanical units.

The Rayleigh balance uses one movable flat coil of many turns situated coaxially between two fixed coils and at such a distance that the force is a maximum. The directions of current are such that the initial force is a downward one from both fixed coils and a weight is added to the opposite arm to restore balance. The current in the fixed coils is then reversed and a standard mass m added to the weight already present to restore balance. The weight mg is thus twice the magnetic force. If the current through the two fixed coils is in the same direction the force on the movable coil should be zero when at the midpoint and this is used in alignment. The important geometrical factor is the ratio of the mean diameters of the fixed and movable coils and this is determined electrically. Allowance is made for temperature effects and the density of the air.

The Ayrton-Jones balance uses two movable solenoids on opposite balance arms and two pairs of fixed coaxial solenoids. The symmetrical arrangement reduces temperature effects but allowance now has to be made for cross-forces between the fixed circuits on one side and the movable one on the other. All the solenoids are single-layer with the winding set in grooves cut into marble cylinders, thus locating them very precisely and permitting accurate measurement of dimensions.

The Rayleigh balance is capable of measuring a current in amperes to 3 or 4 parts in 10^5 and the Ayrton-Jones to a few parts in 10^6. The amperes measured by the two types of balance have been shown to agree to a few parts per million by interchange between the

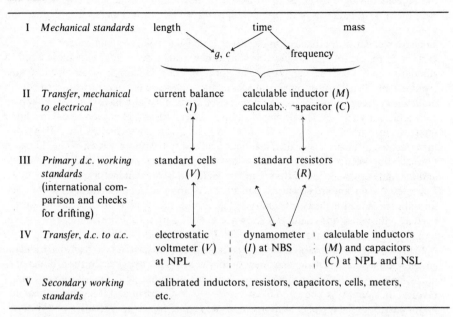

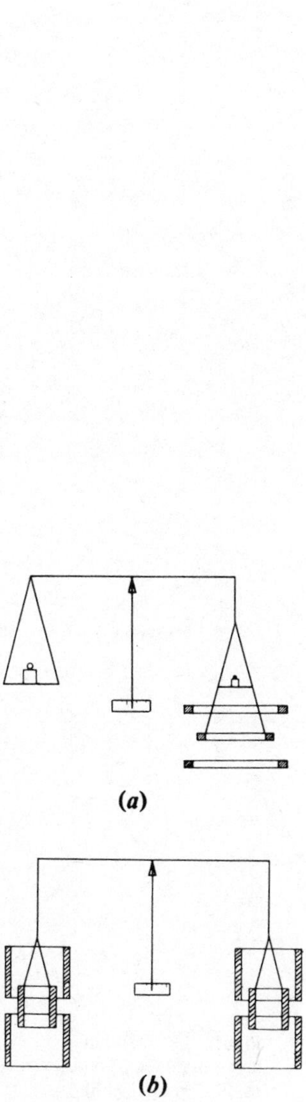

Figure C.1 (a) The Rayleigh current balance; (b) the Ayrton–Jones current balance.

national laboratories of the primary working standards of voltage and resistance (below): this agreement can be taken as experimental verification of the magnetic laws. (Note that were the current balances of the same design this would not be an allowable deduction.)

Transfer standards, mechanical to electrical: calculable inductors and capacitors Of the various forms of transfer **inductance** standard we shall deal only with the Campbell standard maintained at the NPL. The form of the inductor is determined by the requirements that both primary and secondary shall have enough turns to provide an inductance large enough for use in normal circuits yet must also have dimensions known with sufficient accuracy. The primary is a single-layer coil wound in grooves cut in a fused quartz former and is in two parts as shown in Fig. C.2(a). The secondary is a coil of many turns located in a region where the magnetic field due to the primary is zero. The change in mutual inductance resulting from deviations from the neutral point is then at its smallest and the finite cross-section of the secondary is not an important factor in the value of M. The latest NPL standard constructed in 1960 is of about 10 mH. Corrections to the ideal formula are made for the permeability of the quartz and for non-uniform distribution of current in the wires. As a result M is known to about 1 part in 10^5 from formulae giving it in terms only of μ_0 and linear dimensions [cf. Eqs (9.30) and (9.31)].

In principle, it should be possible to use a **capacitor** in the same way as a mutual inductor since the capacitance of an arrangement of conductors is always given by the product of ε_0 and a function of the linear dimensions. It is true, of course, that the velocity of light is involved in ε_0 but it is known to a few parts in 10^7 at present and it has never been the major barrier to the use of capacitors as primary standards. Until recently, it has just not been possible to design and construct any arrangement whose capacitance could be calculated from dimensions with sufficient accuracy. However, a remarkable theorem of Thomson and Lampard (1956) has altered the situation, and calculable capacitors of a new pattern have since been constructed at most standardizing laboratories.

The theorem shows that if a hollow cylindrical conductor of any cross-section is divided into four parts separated by small gaps as in Fig. C.2b, the cross-capacitances per unit length C_1 and C_2 are related by

$$e^{-\pi C_1/\varepsilon_0} + e^{-\pi C_2/\varepsilon_0} = 1$$

for a system of infinite length *in vacuo*. If, therefore, a symmetrical shape is chosen so that $C_1 = C_2 = C$, then $e^{-\pi C/\varepsilon_0} = \frac{1}{2}$, and hence the capacitance per unit length C is simply $(\varepsilon_0 \log_e 2)/\pi$ whatever the symmetrical shape chosen. A practical capacitor (Fig. C.2c) consists of four insulated metal cylinders of circular section nearly touching each other, the relevant capacitance being that between one of the heavily marked quadrants and its opposite. The length is made definite by two earthed metal tubes, one pushed in at each end of the space as shown by the dotted circle in the figure. End effects are eliminated by

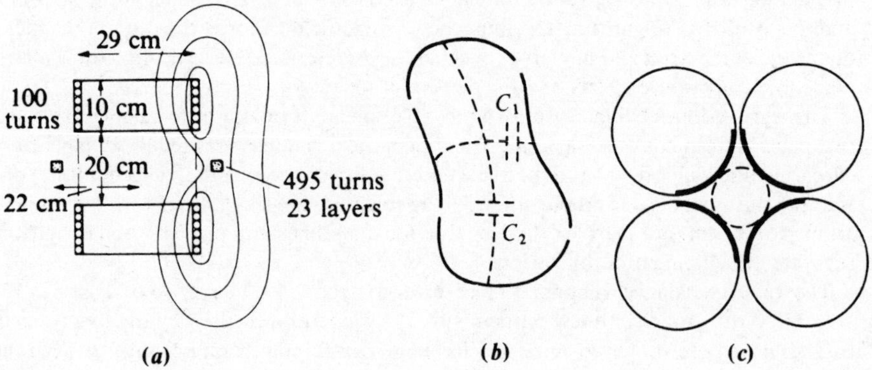

Figure C.2 Calculable components. (a) The Campbell mutual inductor; (b) illustrating the theorem of Thomson and Lampard; (c) cross-section of a calculable capacitor.

moving one tube in by a measured distance and using the change in capacitance. The distance moved is the only dimension involved and it can be measured very precisely by an interferometric method. The capacitances involved are very small, that at the NML for instance being about 0.25 pF.

Transfer standards, mechanical to electrical: electrometers and wattmeters More recently, both the NPL and the NML have developed other methods for relating electrical to mechanical units. The NPL method is described in Sec. 9.2 and involves a special form of wattmeter in which the same coil is subjected to a force when carrying a steady current in a **B**-field and is then moved with a known velocity in the same **B**-field. The product of the induced electromotance and the original current is obtained in terms of mechanical quantities (Kibble, Smith and Robinson, 1983). At the NML, a special form of attracted disc electrometer has been designed with a mercury surface as the lower plate (Sec. 5.7; see Sloggett, Clothier and Ricketts, 1986).

Primary and secondary working standards Although the methods and components described in the last section could in principle be used directly to standardize other components for everyday use, the experiments involved are laborious to perform with the necessary accuracy and do not lend themselves to international comparison, although the calculable capacitor is the least limited in these respects. The national laboratories therefore maintain primary working standards which are suitable for use in purely electrical methods of the highest precision and convenience. These standards are invariably sets of standard cells and resistors at the moment, although the pace of technical progress is such that this situation may not persist for much longer.

For reference back to the mechanical standards, a resistor can be compared with the calculable M or C, a process requiring several stages because the inductance and capacitance are so small. A set of resistors is made to the pattern shown in Fig. 6.16a and each in turn compared with the standardized one. The mean value for the set determines the standard ohm for a particular national laboratory in terms of basic SI units. A resistor is replaced when unacceptable variations relative to the mean occur, usually because of age.

The standard cells are all of the saturated Weston cadmium type (Fig. C.3a) and a set is constructed, maintained and compared as in the case of the resistors. The mean of the set determines the standard volt. For reference back to the SI units, a cell is used in conjunction with a standard resistor in the circuit of Fig. C.3b. The current measured by the current balance passes through R and the known potential difference RI is balanced by the electromotance of the cell.

Every three years or so, each national laboratory sends a few resistors and cells to the International Bureau of Weights and Measures (BIPM) at Sèvres for intercomparison so

Figure C.3 (a) The saturated Weston cadmium cell; (b) circuit for calibration of a standard cell.

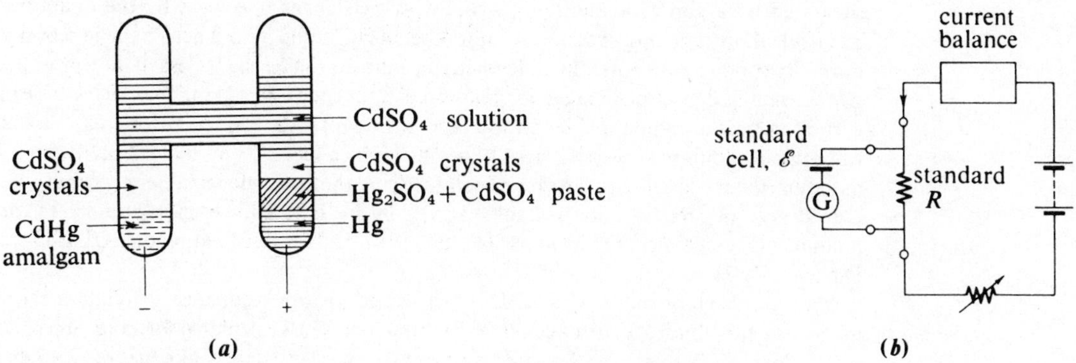

that each country is aware of discrepancies between its own set of standards and others. Such intercomparison is also valuable as a check on the general drift which might occur in a whole batch of standards. The need to make measurements with greater and greater precision means that drifts of a few parts per million which might have been acceptable in the past are not small enough today. A more stable voltage standard than the Weston cell, for instance, is a particularly pressing requirement.

For a.c. work, resistors form a suitable primary standard and thus provide a link with the d.c. standards. The calculable M and C can also be used for this link. However, no precisely reproducible source of alternating electromotance akin to the standard cell exists, so that a transfer standard from d.c. to a.c. has to be used for the volt or ampere to supplement the ohm. At the NPL an electrostatic voltmeter has been used as an a.c. voltage standard after calibration in terms of the d.c. standard, but thermal converters using thermoelectric electromotances are now increasingly found to be more accurate. At the NBS an electrodynamometer provides a standard of a.c. current through the torque exerted between two coils carrying the same current (as in Fig. 7.23).

We now pass to the final step in the chain of measurement: the connection between the standardizing laboratories and the numerous instrument manufacturers and other laboratories all of whom need their own set of standards calibrated in terms of the national ones. For this purpose, the standardizing laboratories maintain a whole set of secondary standards which can easily be compared with the primary standards on the one hand and with components submitted for calibration on the other. The methods used are principally based on the bridge and potentiometric circuits outlined in Secs 6.9 and 10.11. The secondary standards themselves consist of resistors, capacitors, etc., covering such a range of values that intercomparison with both large and small magnitudes is simplified.

Recent advances The methods of standardizing measurements described above involve a long chain of tedious processes and are subject to uncertainties which seem small but which may not be adequate in the future. One problem has always been the variability of any standards which are *constructed* rather than naturally occurring. The abandonment of constructed standards is exemplified by the obsolescent metal bar as the standard metre and the mercury column as the ohm. We tend all the time to move towards natural phenomena, particularly properties of atoms, partly because we feel inherently that these are likely to be 'unchanging' (whatever that word may mean when applied to a standard) but also because they are properties available to all workers without continual intercomparison. At the moment some of these properties are being used to check on the drift of material standards rather than as standards themselves. As an example, the NBS has kept its standard ampere constantly checked for 10 years by using it to provide a magnetic field in which the proton precession frequency is monitored (Sec. 8.7).

A recent development is in the use of a device known as the Josephson junction to provide a standard for voltage. The junction consists of a pair of superconductors separated by a gap of about 1 nm across which electrons may pass by the quantum-mechanical process known as tunnelling. When a current is passed across the junction it normally produces no potential difference since the flow is resistanceless. If, however, the gap is irradiated by electromagnetic radiation of frequency ν, it is found that an increase in current is now accompanied by an increase in potential difference. The increase is not smooth but jumps in successive steps of height V given by $2eV = h\nu$, a result predicted by quantum theory applied to superconductors. Since the frequency can be measured to a high degree of precision, the voltage V given by $h\nu/2e$ is known as accurately as the fundamental constants are known (see Petley, 1987, and for information on SQUIDS see Dickens, 1987).

Another development at the NPL is the use of proton resonance to relate a large magnetic field (about 1 T) to a weaker field (about 1 mT) in a calculable solenoid. Since the strong field can be related to the ampere by a much simpler balance (like that of Fig. 7.4b) there is a possibility that the need for the cumbersome current balances currently used can be eliminated. Indeed, the use of quantum standards (see Hartland, 1988) coupled with a method such as that outlined in Sec. 9.2 have already made it unlikely that conventional current balances will ever be used again for establishing electrical units.

Systems of units The only systems of units in electromagnetism used extensively by modern writers are the versions of SI called after Sommerfeld and Kennelly and the Gaussian system arising from CGS e.m.u. and e.s.u. The equations and formulae throughout this book have been expressed in a general form by using the constants ε_0 and μ_0 which take different values according to the units adopted. It should be clear, however, that the values of ε_0 and μ_0 cannot be independent of each other in a consistent system. This is because ε_0 fixes the unit of Q, μ_0 fixes that unit of I and it is agreed that $I = \mathrm{d}Q/\mathrm{d}t$. It is easy to show that the dimensions of the *constant quantity* $1/(\varepsilon_0\mu_0)^{1/2}$ must be those of a velocity. Hence in any cm–g–s system it must have the same value, say c_0 cm s^{-1}, where $c_0 \approx 3 \times 10^{10}$, while in any m–kg–s system it must also have the same value, say c_p m s^{-1}, where $c_p \approx 3 \times 10^8$.

SI units These are favoured throughout this book and are defined by the choice

$$\mu_{0\mathrm{SI}} = 4\pi \times 10^{-7}\ \mathrm{H\,m^{-1}}$$

and hence

$$\varepsilon_{0\mathrm{SI}} = 10^7/4\pi c_p^2 = \frac{1}{36\pi \times 10^9}\ \mathrm{F\,m^{-1}} \tag{C.1}$$

The CGS e.m.u. system This is defined by the choice

$$\mu_{0\mathrm{emu}} = 4\pi \qquad \text{and hence} \qquad \varepsilon_{0\mathrm{emu}} = 1/4\pi c_0^2 \tag{C.2}$$

Coulomb's law in e.m.u. thus reads $F = c_0^2 Q_1 Q_2 / r^2$. Let there be n coulombs in 1 e.m.u. of charge. Then the force between 1 e.m.u. and another 1 cm away is c_0^2 dyn or $c_0^2/10^5$ N. Using (2.2) and (C.1), the *same* force is also $n^2 c_p^2/10^3$ N. Thus $n = 10$ exactly and a coulomb is 1/10 of an e.m.u. of charge. This was the original definition of the coulomb.

In addition to the substitutions (C.2) we must remember that the definitions of **D**, **H** (and V_m), χ_e and χ_m differ in CGS systems from those in the MKS. Comments C11.1 and C12.3 showed that

$$\mathbf{H} \to \mathbf{H}/4\pi \ (\text{and } V_m \to V_m/4\pi), \qquad \mathbf{D} \to \mathbf{D}/4\pi, \qquad \chi_e \to 4\pi\chi_e, \qquad \chi_m \to 4\pi\chi_m \tag{C.3}$$

must also be used to convert our formulae to CGS e.m.u. If the Kennelly form is preferred (Comment C.12.4) then $\mathbf{I} \to 4\pi\mathbf{I}$ as well.

The CGS e.s.u. system This is defined by the choice

$$\varepsilon_{0\mathrm{esu}} = 1/4\pi \qquad \text{and hence} \qquad \mu_{0\mathrm{esu}} = 4\pi/c_0^2 \tag{C.4}$$

and the ratio of the e.m.u. of charge to the e.s.u. of charge is easily shown to be c_0 or $1/(\varepsilon_0\mu_0)^{1/2}$. This is why the determination of ε_0 by electrical methods was often known as the determination of the ratio of the e.m.u. to the e.s.u. Substitutions (C.3) must also be made.

Determination of ε_0 In describing the calculable capacitor above as a means of standardizing the ohm, we noted that a value for ε_0 is needed and that this is obtained from c, the velocity of light *in vacuo*. Of course, it would be quite permissible for us to give up the capacitor as a standard for measurement and regard it as a means of obtaining an experimental value for ε_0 and thus for c. We should use a bridge method to obtain the capacitance in farads in terms of standard resistors in ohms, and a measurement of length

Table C.2. Conversion to Gaussian units

Replace	By	Replace	By
ε_0	$1/4\pi$	μ_0	$4\pi/c^2$
$\mathbf{D}$	$\mathbf{D}/4\pi$	$\mathbf{H}$	$c\mathbf{H}/4\pi$
χ_e	$4\pi\chi_e$	χ_m	$4\pi\chi_m$
$\mathbf{B}$	$\mathbf{B}/c$	V_m	$cV_m/4\pi$
Φ	Φ/c	$\mathbf{m}$	$c\mathbf{m}$
$\mathbf{M}$	$c\mathbf{M}$	P (pole)	cP
$\mathbf{I}$	$4\pi\mathbf{I}/c$		

would enable us to calculate ε_0. We should then have the value of c from purely electrical and mechanical measurements. This method was used in the early days of the century: Rosa and Dorsey (1907) obtained $c = 2.9978 \times 10^8\ \mathrm{m\,s^{-1}}$ as compared with the recent value of $2.99793 \times 10^8\ \mathrm{m\,s^{-1}}$.

The Gaussian system The CGS e.m.u. and e.s.u. systems are rarely, if ever, used throughout the whole of theoretical electricity and magnetism. Texts and original papers not using an MKS system usually adopt the CGS Gaussian system. The clearest way to define this is to divide quantities into two classes designated, for this purpose only, as 'electric' and 'magnetic'. Electric quantities are Q, I, V, $\mathscr{E}$, C, R, L, M, $\mathbf{p}$, $\mathbf{E}$, $\mathbf{D}$, $\mathbf{P}$ and χ_e; magnetic quantities are $\mathbf{H}$, $\mathbf{B}$, Φ, V_m, $\mathbf{M}$, $\mathbf{m}$, χ_m, $\mathscr{H}$ and $\mathscr{R}$. The Gaussian unit of an electric quantity is equal in all cases to the CGS e.s.u. and that of a magnetic quantity to the CGS e.m.u. The advantage of this is that any relation between quantities belonging to the first group only is the same as if they were all in e.s.u. with (C.3) and (C.4) while any relation between quantities belonging only to the second group is the same as if they were all in e.m.u. with (C.2) and (C.3).

Any relation involving members from both groups must also involve the numerical factor c_0 expressing the relation between e.m.u. and e.s.u. There are two basic relations involving these cross-connections: (a) the $\mathbf{B}$ and $\mathbf{H}$ of conduction and displacement currents and (b) the induced electromotance accompanying a changing magnetic flux.

Maxwell's equations in Gaussian units then become (vector operator form):

$$\operatorname{div}\mathbf{D} = 4\pi\rho; \qquad \operatorname{div}\mathbf{B} = 0; \qquad \operatorname{curl}\mathbf{E} = -\frac{1}{c_0}\frac{\partial \mathbf{B}}{\partial t};$$

$$\operatorname{curl}\mathbf{H} = \frac{4\pi\mathbf{J}}{c_0} + \frac{1}{c_0}\frac{\partial \mathbf{D}}{\partial t}$$

with $\mathbf{D} = \mathbf{E} + 4\pi\mathbf{P}$ and $\mathbf{B} = \mathbf{H} + 4\pi\mathbf{M}$.

Conversion of formulae from SI to Gaussian units The main sections of this book have all formulae and equations written in SI units (Sommerfeld version). To convert consistently to Gaussian units, the substitutions given in Table C.2 should be made, other quantities being left unchanged. The constant c has the value 3×10^{10}.

References For general information on standards in electromagnetism see Vigoureux (1971). The authoritative source for up-to-date details of the realization and comparison of standards is the journal *Metrologia*, together with the publications of the national standardizing laboratories. Articles giving more details of developments in the field of standards are Petley (1971), Dix (1975) and Petley (1987). For electrical measurements and instrumentation, see Lion (1975) and B.E. Jones (1977).

References

Aitken, M.J., *Contemp. Phys.*, **3**, 344, 1962.
American Association of Physics Teachers, 'Report of the Coulomb's Law Committee', *Amer. J. Phys.*, **18**, 1 and 69, 1950.
Atkins, P.W., *Physical Chemistry*, 3d edn, Oxford University Press, 1986.
Bartlett, D.F., Goldhagen, P.E. and Phillips, E.A., *Phys. Rev. D*, **2**, 483, 1970.
Becker, R., *Introduction to Theoretical Mechanics*, McGraw-Hill Book Co., New York, 1954.
Bennett, G.A.G., *Electricity and Modern Physics*, 2d edn, Edward Arnold, London, 1974.
Boyd, R.N. *et al.*, *Phys. Rev. Letters*, **43**, 1288, 1979.
Bright, A.W. and Makin, B., *Contemp. Phys.*, **10**, 331, 1969.
Brookner, E., *Sci. Amer.*, **252**, 2, 76, 1985.
Brown, S. and Barnett, S.J., *Phys. Rev.*, **87**, 601, 1952.
Bucherer, A.H., *Ann. d. Phys.*, 28, 513, 1909.
Burge, E.J., *Phys. Educ.*, **22**, 375, 1987.
Cabrera, B., *Phys. Rev. Letters*, **48**, 1378, 1982.
Campbell, N.R., *Physics, The Elements*, Cambridge University Press, 1920. (Republished as *Foundations of Science*, Dover Publications, New York, 1957.)
Campbell, N.R., *An Account of the Principles of Measurement and Calculation*, Longmans, Green and Co., London, 1928.
Carden, P.O., *Rep. Prog. Phys.*, **39**, 1017, 1976.
Carrigan, R.A. and Trower, W.P., *Sci. Amer.*, **246**, 4, 91, 1982.
Catt, I., Davidson, M.F. and Walton, D.S., *Wireless World*, p. 51, Dec. 1978.
Chambers, R.G., *Phys. Educ.*, **12**, 374, 1977.
Chrystal, G., *British Association Report*, p. 36, 1876.
Close, F., *The Cosmic Onion*, Heinemann, London, 1983.
Cohen, E.R. and Taylor, B.N., CODATA Bulletin 63, 1986.
Cook, A.H., *Contemp. Phys.*, **18**, 393, 1977.
Crangle, J., *The Magnetic Properties of Solids*, Edward Arnold, London, 1977.
Cross, J.A., *Electrostatics: Principles, Problems and Applications*, Adam Hilger, Bristol, 1987.
Davies, P.C.W., *The Forces of Nature*, 2d edn, Cambridge University Press, 1986.
Dickens, M., *Phys. Bull.*, **38**, 296, 1987.
Dix, C.H., *Proc. IEE*, **122**, 1018, 1975.
Ehrlichson, H., *Amer. J. Phys.*, **38**, 1252, 1970.
Eichenwald, A., *Ann. d. Physik*, **xi**, 1, 1903.
Ellis, B., *Basic Concepts of Measurement*, Cambridge University Press, 1966.
Fairs, R.A., *Wireless World*, p. 510, Dec., 1974.

Fairs, R.A., *Wireless World*, p. 187, Oct. 1975.

Faraday, M., *Experimental Researches in Electricity*, 3 vols, Richard John Edward Taylor, London, 1839.

Faraday, M., *Experimental Researches in Electricity* (Ed. F. Sherwood Taylor), J.M. Dent and Sons, London; E.P. Dutton and Co., New York, 1951.

Gass, I.G., Smith, P.J. and Wilson, R.C.L., *Understanding the Earth*, Artemis Press for the Open University, 1971.

Goldhaber, A.S. and Nieto, M.M., *Rev. Mod. Phys.*, **43**, 277, 1971.

Goldhaber, A.S. and Smith, J., *Rep. Prog. Phys.*, **38**, 731, 1975.

Gough, C.E. *et al.*, *Nature*, **326**, 855, 1987.

Gribbin, J., *New Scientist*, **117**, 1601, 35, 25 Feb. 1988.

Gubbins, D., *Contemp. Phys.*, **25**, 269, 1984.

Guinier, A. and Jullien, R., *The Solid State from Superconductors to Superalloys*, Oxford University Press, 1989.

Hall, S.H., *Contemp. Phys.*, **8**, 447, 1967.

Hartland, A., *Contemp. Phys.*, **29**, 477, 1988.

Hazen, R., *The Breakthrough*, Unwin Hyman, London, 1988.

Henry, P.S.H., Preface to Institute of Physics below, 1967.

Herlach, F., *Rep. Prog. Phys.*, **31**, 341, 1968.

Herlach, F. (Ed.), *Strong and Ultrastrong Magnetic Fields*, Topics in Applied Physics, vol. 57, Springer-Verlag, Berlin, 1985.

Hillas, A.M. and Cranshaw, T.E., *Nature*, **184**, 892, 1957.

Hoffman, K.A., *Sci. Amer.*, **258**, 5, 50, 1988.

Holden, A., *Bonds between Atoms*, Clarendon Press, Oxford, 1971.

Holt, C., *Phys. Bull.*, **39**, 3, 99, 1988.

Hones, E.W., *Sci. Amer.*, **254**, 3, 32, 1986.

Hopper, V.D. and Laby, T.H., *Proc. Roy. Soc. A*, **178**, 243, 1941.

Institute of Physics, London, Conference Series on *Static Electrification*: no. 4, 1967; no. 11, 1971; no. 27, 1975; no. 48, 1979; no. 66, 1983.

Jefimenko, O., *Amer. J. Phys.*, **30**, 19, 1962.

Jones, B.E., *Instrumentation, Measurement and Feedback*, McGraw-Hill Book Co., London, 1977.

Jones, L.W., 'A review of quark search experiments', *Rev. Mod. Phys.*, **49**, 717, 1977.

Kalvius, G.M. and Tebble, R.S. (Eds), *Experimental Magnetism*, John Wiley and Sons, New York, 1979.

Kettering, C.F. and Scott, C.G., *Phys. Rev.*, **66**, 257, 1944.

Kibble, B.P., Smith, R.C. and Robinson, I.A., *IEEE Trans. Inst. Meas.*, **IM-32**, 141, 1983.

King, J.G., *Phys. Rev. Letters*, **5**, 562, 1960.

Kip, A.F., *Contemp. Phys.*, **1**, 355, 1960.

Kite, L.V., *An Introduction to Linear Electric Circuits*, Longman Group, London, 1974.

Kittel, C., *Introduction to Solid State Physics*, 6th edn, John Wiley and Sons, New York, 1986.

Laithwaite, E.R., *Phys. Educ.*, **4**, 96, 1969.

LaRue, G.S., Philips, J.D. and Fairbank, W.M., *Phys. Rev. Letters*, **46**, 967, 1981.

Latham, J., *Weather*, **21**, 120, 1966.

Lebedew, P., *Ann. Phys. Lpz.*, **6**, 433, 1901.

Lenz, J., *Phys. Educ.*, **14**, 45, 1979.

Lion, K.S., *Elements of Electrical and Electronic Instrumentation*, McGraw-Hill Kogakusha, Tokyo, 1975.

McCaig, M. and Clegg, A.G., *Permanent Magnets in Theory and Practice*, 2d edn, John Wiley and Sons, 1987.

McDonald, K.L., *Amer. J. Phys.*, **22**, 586, 1954.

McDonnell, J.A.M. (Ed.), *Cosmic Dust*, John Wiley and Sons, Chichester, 1978.

McGraw-Hill Encyclopedia of Science and Technology, 6th edn, vol. 9, pp. 360 and 363, McGraw-Hill, New York, 1987.
Magie, W.F., *A Source Book in Physics*, Oxford University Press, 1964.
Malan, D.J., *Physics of Lightning*, English Universities Press, London, 1963.
Martin, T., *Faraday's Discovery of Electromagnetic Induction*, Edward Arnold and Co., London, 1949.
Mason, M. and Weaver, W., *The Electromagnetic Field*, Dover Publications, New York, 1929.
Maxwell, J.C., *A Treatise on Electricity and Magnetism*, 3d edn, Dover Publications, New York, 1904.
Mayo, J.S., *Sci. Amer.*, **255**, 4, 58, 1986.
Mendelssohn, K., *The Quest for Absolute Zero*, Weidenfield and Nicolson, London, 1966.
Moorcroft, D.R., *Amer. J. Phys.*, **37**, 221, 1969; **38**, 376, 1970.
Moore, A.D., 'Electrostatics', *Sci. Amer.*, **226**, 3, 47, 1972.
Moore, A.D., *Electrostatics and Its Applications*, John Wiley and Sons, New York, 1973.
Nelkon, M. and Parker, P., *Advanced Level Physics*, 6th edn, Heinemann, London, 1987.
Nicholls, E.F. and Hull, G.F., *Phys. Rev.*, **17**, 26 and 91, 1903.
Nicola, M., *Amer. J. Phys.*, **40**, 189, 1972.
Parker, E.N., *Sci. Amer.*, **249**, 2, 36, 1983.
Parkinson, D.H. and Mulhall, B.E., *The Generation of High Magnetic Fields*, Heywood Books, London; Plenum Press, New York, 1967.
Petley, B.W., *Contemp. Phys.*, **12**, 453, 1971.
Petley, B.W., *Phys. Bull.*, **38**, 375, 1987.
Petley, B.W. and Morris, K., *Nature*, **213**, 586, 1967.
Plimpton, S.J. and Lawton, W.E., *Phys. Rev.*, **50**, 1066, 1936.
Poynting, J.H., *Phil. Mag.*, **9**, 169, 1905; **9**, 393, 1905.
Purcell, E.M., *Electricity and Magnetism*, Berkeley Physics Series, 2d edn, McGraw-Hill, New York, 1984.
Quigg, C., *Sci. Amer.*, **252**, 4, 84, 1985.
Rosa, E.B. and Dorsey, B.E., *Bull. Bur. Stand.*, **3**, 433, 1907.
Rose, P.H. and Wittkower, A.B., *Sci. Amer.*, **223**, 2, 24, 1970.
Rosser, W.G.V., *Classical Electromagnetism via Relativity*, Butterworths, London, 1968.
Rosser, W.G.V., *Amer. J. Phys.*, **38**, 265, 1970.
Rowe, E.M. and Weaver, J.H., *Sci. Amer.*, **236**, 6, 32, 1977.
Rutherford, E., *Phil. Mag.*, **21**, 669, 1911.
Sanders, J.H., Tittel, K.F. and Ward, J.F., *Proc. Roy. Soc. A*, **272**, 103, 1963.
Schwarzchild, B.M., *Physics Today*, **37**, 4, 17, April 1984.
Scott, C.G., *Phys. Rev.*, **83**, 656, 1951.
Scott, W.T., *The Physics of Electricity and Magnetism*, 2nd edn, John Wiley and Sons, New York, 1966.
Shaw, R., *Amer. J. Phys.*, **33**, 300, 1965.
Shoop, S., *Chemistry in Britain*, **24** (1), 31, 1988.
Shull, C.G., Billman, K.W. and Wedgwood, F.A., *Phys. Rev.*, **153**, 1415, 1967.
Slepian, J., *Amer. J. Phys.*, **19**, 87, 1951.
Sloggett, G.J., Clothier, W.K. and Ricketts, B.W., *Phys. Rev. Letters*, **57**, 3237, 1986.
Smith, P.F., Homer, G.J., Lewin, J.D. and Walford, H.E., *Phys. Lett. B.*, **197**, 3, 447, 1987.
Stacey, F. and Tuck, C., *Physics World*, **1**, 29, Dec. 1988.
Stevenson, D.J., *Rep. Prog. Phys.*, **46**, 555, 1983.
Stover, R.W., Moran, T.I. and Trischka, J.W., *Phys. Rev.*, **164**, 1599, 1967.
Stow, C.D., *Rep. Prog. Phys.*, **32**, 1, 1969.
Strnad, J., *Contemp. Phys.*, **12**, 187, 1971.
Thomas, H.A., Driscoll, R.L. and Hipple, J.A., *Phys. Rev.*, **78**, 787, 1950.
Thomson, A.M. and Lampard, D.G., *Nature*, **177**, 888, 1956.

Thomson, J.J., *Phil. Mag.*, **44**, 293, 1897.

Thomson, J.J., *Phil. Mag.*, **48**, 547, 1899.

Thomson, J.J., *Electricity and Matter*, Archibald Constable and Co., London, 1904.

Tolman, R.C., Osgerby, E.W. and Stewart, T.D., *J. Amer. Chem. Soc.*, **36**, 466, 1914.

Tolman, R.C. and Stewart, T.D., *Phys. Rev.*, **9**, 164, 1917.

Trotter, D.M., *Sci. Amer.*, **259**, 1, 58, 1988.

Vigoureux, P., *Units and Standards of Electromagnetism*, Wykeham Publications, London, 1971.

Walker, J.D., *The Flying Circus of Physics*, with answers, John Wiley and Sons, New York, 1978.

Whittaker, E.T., *History of the Theories of the Aether and Electricity*, vol. I: *The Classical Theories*, Thomas Nelson and Sons, London, 1951; reprint published by *American Journal of Physics*, and Adam Hilger, Bristol, 1987.

Williams, E.R., *Sci. Amer.*, **259**, 5, 48, 1988.

Williams, E.R., Faller, J.E. and Hill, H.A., *Phys. Rev. Letters*, **26**, 721, 1971.

Williams, L.P., *Contemp. Phys.*, **5**, 28, 1963.

Witteborn, F.C. and Fairbank, W.M., *Phys. Rev. Letters*, **19**, 1049, 1967.

Zemansky, M.H. and Dittman, R.H., *Heat and Thermodynamics*, 6th edn, McGraw-Hill, New York, 1981.

Answers to problems

Chapter 1

1.1 A method less practicable than the electroscope or electrometer. A balance similar to that of Fig. 1.7a needs a fixed charge on one conducting body, so that a force W defines unit charge on the other. One similar to Fig. 1.7b needs two identical bodies so that any charge on one is shared equally on contact, assuming conservation. In that case, a force W defines unit charge on both and n^2W a charge of n units on both. Unfortunately geometrical factors intrude (as they do not in an electroscope) rather as in the current balance, but less tractably.
1.2 They are magnets: one turned through 180° should attract the other.
1.3 Practically: measurement of heat relatively inaccurate; only the magnetic effect is based on a universal law and not on a material standard (see Appendix C). Theoretically: heating effect does not allow allocation of sign to I; having $I = \mathrm{d}Q/\mathrm{d}t$ and Kirchhoff's first law (Chapter 6) is preferable to $I_H = (\mathrm{d}Q/\mathrm{d}t)^2$ without the law.
1.4 $I = (Q_0/T)\mathrm{e}^{-t/T}$ into conductor. In the second case $I = -(Q_0/T)\mathrm{e}^{-t/T}$, i.e. out of the conductor.
1.5 $Q = I_0(1 - \mathrm{e}^{-\alpha_0 t})/\alpha$.
1.6 5.7×10^{-9} g. A good chemical balance weighing to 10 g could not detect much better than 10^{-5} g.
1.7 $N_A = (\text{mass of 1 u in g})^{-1} = 6.022 \times 10^{23}\,\mathrm{mol}^{-1}$.
1.8 $F = 9.65 \times 10^4\,\mathrm{C\,mol}^{-1}$.
1.9 Just less than $0.01\,\mathrm{cm\,s}^{-1}$.
1.10 1 year is about 3×10^7 s. Current is about $3 \times 10^{-18}\,\mathrm{A\,mm}^{-2}$ and contains about 20 electrons per second per square mm.
1.12 Element $\mathrm{d}l$ has a charge $\lambda\,\mathrm{d}l$ or $\lambda_0 l^2\,\mathrm{d}l/L^2$. Integrating from $l=0$ to $l=L$ gives $Q = \lambda_0 L/3$. So $\langle\lambda\rangle = \lambda_0/3$.
1.13 In each case divide disc into annular rings, typically of inner radius r, outer radius $r + \mathrm{d}r$. Area of ring is $2\pi r\,\mathrm{d}r$ and charge on it is $\sigma 2\pi r\,\mathrm{d}r$. This expression is then integrated from 0 to a with the appropriate σ. (a) $2\pi\sigma_0 a^2/3$; (b) $\pi\sigma_0 a^2/2$; (c) $2\pi\sigma_0 a^2(1 - 2/\mathrm{e})$.
1.14 In each case divide sphere into concentric shells radius r, thickness $\mathrm{d}r$, each with volume $4\pi r^2\,\mathrm{d}r$, charge $4\pi r^2\rho\,\mathrm{d}r$. Then integrate from 0 to a with appropriate ρ. (a) $\pi\rho_0 a^3$, ρ_0 is density at outside; (b) $8\pi\rho_0 a^3/15$, ρ_0 is density at centre; (c) $4\pi\rho_0 a^3(2 - 5/\mathrm{e})$, ρ_0 is density at centre. Last case requires integration by parts twice.
1.15 $\pi a^2 J_0/3; \langle J\rangle/J_0/3$.
1.16 For the neutral current, $J = n_p e_p(v_p - v_n) = \rho_p \Delta v$, say. Both ρ_p and Δv are independent of the observer. If the resultant charge density in a charged current $n_p e_p + n_n e_n = \rho$, say, then $J_0 = J_P - \rho v$. Only if $\rho = 0$ does $J_0 = J_P$ for all v.
1.17 Time for one revolution $= 2\pi/\omega$, so number of revolutions per unit time $= \omega/2\pi$. Hence charge e passes every point in orbit $\omega/2\pi$ times per second, equivalent to a current $e\omega/2\pi$.

Chapter 2

2.1 Forces must be equal and opposite by Newton's third law and thus form a couple. For the equilibrium of the complete system, the charges would have to exert torques on each other whose total moment would be equal and opposite to that of the forces (compare that with the action between dipoles in Fig. 3.23d). Non-central forces introduce directions other than the joining line so that charges would have vector properties (like current elements or dipoles). Thus the central nature of the force shows the scalar nature of charge.

2.2 The torque on the rod and charge is proportional to $(\sin\theta)/d^2$, where θ is the deflection of the rod. For small oscillations, $\ddot{\theta}=k\theta/d^2$ and hence the period is proportional to d.

2.3 Between charges, about 9×10^4 tonnes; between thundercloud charges, about 600 tonnes.

2.4 Say weight of paper is about 10^{-3} g or 10^{-5} N. For a distance apart of 1 cm, Q_1 and Q_2 in Coulombs law would need to be about 3×10^{-4} μC each. Of course, the paper has equal and opposite induced charges which will reduce the force, but allowing a factor of 10 for this means that Q will be $10^{1/2}$ smaller or about 10^{-4} μC. Even this charge is sufficient to raise the potential of an electroscope to about 100 V because of its small capacitance (Sec. 5.7).

2.5 F_{elec}/F_{grav} is about 10^{43} for electrons, 10^{36} for protons. Comment C2.1 answers the last part.

2.6 $(\sqrt{2}+\frac{1}{2})Q^2/(4\pi\varepsilon_0 a^2)$ outward along diagonal direction.

2.7 Yes. $-Q(1+2\sqrt{2})/4$. No.

2.8 One position, a from $-Q$, $2a$ from $+4Q$. Unstable.

2.9 $Q\lambda L/[4\pi\varepsilon_0 h(h+L)]$. As $L\to0$, $F\to Q\lambda L/(4\pi\varepsilon_0 h^2)$, as if rod were a point charge λL at h from Q. As $L\to\infty$, $F\to Q\lambda h/(4\pi\varepsilon_0 h^2)$, as if rod were a point charge λh at h from Q.

2.10 If y is vertical and x horizontal, force on Q has $F_x=[Q\lambda/(4\pi\varepsilon_0 h)][1-h/(h^2+L^2)^{1/2}]$, $F_y=Q\lambda L/[(4\pi\varepsilon_0 h)(h^2+L^2)^{1/2}]$. This gives resultant force of magnitude $[2Q\lambda/(4\pi\varepsilon_0 h)]\times[1-h/(h^2+L^2)^{1/2}]^{1/2}$, and making an angle ϕ with h such that $\tan\phi=[(h^2+L^2)^{1/2}-h]/L$. Force on rod is equal and opposite.

2.11 $(Q\sigma/2\varepsilon_0)[1-a/(a^2+b^2)^{1/2}]$. As $b\to0$, $F\to Q(\sigma\pi b^2)/(4\pi\varepsilon_0 a^2)$ as if sheet were a point charge $\pi b^2\sigma$; as $b\to\infty$, $F\to Q\sigma/2\varepsilon_0$ independent of a.

2.12 Potential energy $=[-Q^2/(4\pi\varepsilon_0 a)](a^2+x^2)/(a^2-x^2)$.

2.14 $(2e^2\log_e 2)/(4\pi\varepsilon_0 a)$ in middle; half that value at the end. So U lies between N times these values. Needs $\log_e(1+x)=x-x^2/2+x^3/3-\cdots$ for $-1<x\le1$.

2.15 0.82 aJ.

Chapter 3

3.1 $a(1+\sqrt{2})$ from $-Q$.

3.2 $(2-\sqrt{2})a$ from $2Q$ towards Q.

3.3 $Q(\frac{1}{2}+\sqrt{2})/(4\pi\varepsilon_0 a^2)$ along diagonal.

3.4 Nearly 6×10^9 V m^{-1}; nearly 6×10^{11} V m^{-1}. These are enormous fields by man-made standards. Even at 0.1 mm from an electron, the E-field is 0.1 V m^{-1}.

3.5 (a) $4rQl/[4\pi\varepsilon_0(r^2+l^2)^2]$; (b) $2Ql/[4\pi\varepsilon_0(r^2+l^2)^{3/2}]$.

3.6 $\lambda l/[2\pi\varepsilon_0(a^2-l^2)]$.

3.8 $Q/(8\pi\varepsilon_0 r^2)$ along axis.

3.9 By superposing the field of an infinite sheet and that of a negative disc (to subtract its contribution) we obtain an E-field of $\sigma a/[2\varepsilon_0(a^2+b^2)^{1/2}]$.

3.10 In Cartesians, $y=Cx$; in polars, $\theta=C'$, where C and C' are constants determining the particular line of force.

3.11 E lines are orthogonal to the equipotentials of Problem 3.18.

3.12 See derivation of Eq. (5.19) in Sec. 5.5.

3.13 $(2+1/\sqrt{2})Q/(4\pi\varepsilon_0 a)$; $(2+1/\sqrt{2})Q^2/(4\pi\varepsilon_0 a)$; four times the last result.

3.14 $(\lambda/4\pi\varepsilon_0)\log_e[(a+l)/(a-l)]$. Use (3.17) rather than (3.13).

3.17 (a) $E_x=-3y$, $E_y=-3x$; (b) $E_r=-2r\cos\theta$, $E_\theta=+r\sin\theta$.

3.18 Equipotentials are $xy=\pm1$, $xy=\pm2$, i.e. in the xy plane they are rectangular hyperbolae.

3.20 100 eV or 1.6×10^{-17} J (or 16 aJ); 5.9×10^6 m s^{-1}; 1.06×10^7 cm^{-3}.

3.21 $v=5.93\times10^5 V^{1/2}$ for electrons; $v=1.38\times10^4 V^{1/2}$ for protons, where V is in volts and v in m s^{-1}; 2.6 kV; 4.7 MV; i.e. electrons approach relativistic velocities at much lower energies than protons.

3.22 $[QV_0/(md\omega^2)]/(\omega t-\sin\omega t)$. No. No.

3.23 3.4×10^{-9} s or 3.4 ns.

3.24 168 mm, 8 mm between the plates, 160 mm beyond them.

3.25 Trace becomes a parabola $y^2=A^2-2Ax$.

3.27 The resultant torque on the whole system, including the effect of the F_θ's, is zero.

Chapter 4

4.1 By symmetry, total flux Q/ε_0 gives $Q/6\varepsilon_0$ over each face. With Q in corner, flux is zero over three faces adjacent to corner, $Q/24\varepsilon_0$ over the other three.
4.2 About 4.4×10^5 C.
4.4 E has only an x component equal to $3kx^2$. Hence outward flux over surface of cube is that over only one face at $x=a$. From this, enclosed charge is $3\varepsilon_0 ka^4$.
4.5 0.09 μC; $10^5\,\mathrm{V\,m^{-1}}$; $2.5\times10^4\,\mathrm{V\,m^{-1}}$; 4.5 kV.
4.6 2.7 μC.
4.7 3 cm.
4.8 4.55×10^{-12} cm.
4.9 (a) $\rho_0 r^2/4\pi\varepsilon_0 a$; (b) $(\rho_0/\varepsilon_0)(r/3-r^3/5a^2)$; (c) $(\rho_0/\varepsilon_0)[2a^3/r^2-ae^{-r/a}(1+2a/r+2a^2/r^2)]$.
4.10 Outside, $E=\lambda/(2\pi\varepsilon_0 r)$; inside, $E=\lambda r/(2\pi\varepsilon_0 a^2)$.
4.11 If σ is shared between both sides, each has a density of $\sigma/2$ and (4.12) then correctly gives $E=\sigma/2\varepsilon_0$ on both sides. In a parallel plate capacitor, the whole charge is attracted to one side with the result shown in Fig. 3.5b.
4.12 Use $E=\sigma(1-a/r)/2\varepsilon_0$ for the field a distance a from a finite circular sheet whose rim is at r from the field point, P. Hence if P is 1/100 mm from a surface, all but 1 per cent of $\sigma/2\varepsilon_0$ is produced by charge less than 1 mm from P.
4.13 The charges near to a point P on the surface produce fields $\sigma/2\varepsilon_0$ directed both inwards and outwards. Since the field inside the conductor is zero, the rest of the charges which are more distant must produce a field at P of $\sigma/2\varepsilon_0$ directed outwards. This will make the total field inside zero, and that outside σ/ε_0.
4.15 Because $E_y=E_z=0$ everywhere, $\partial E_y/\partial y=\partial E_z/\partial z=0$, and hence $\partial E_x/\partial x=0$ from (4.19).
4.16 $\rho=6\varepsilon_0 kx$.
4.17 (a) Yes, $V=kxy$; (b) Yes, $V=\frac{1}{2}k(x^2+y^2+z^2)$; (c) No, $\partial E_y/\partial z$ is not equal to $\partial E_z/\partial y$.
4.18 (a) $\rho=0$; (b) $\rho=-3\varepsilon_0 k$.
4.19 Solve Poisson's equation in one dimension using the substitution $p=\mathrm{d}V/\mathrm{d}x$ so that $\mathrm{d}^2V/\mathrm{d}x^2=p\,\mathrm{d}p/\mathrm{d}V$. Solution gives $\rho=-a^2x^2/12\varepsilon_0$.
4.20 Write the first two terms of Poisson's equation as $(1/r)\,\partial(r\,\partial V/\partial r)/\partial r$ and equate to constant ρ. General solution is $V=-\rho r^2/2\varepsilon_0+k_1\log_e r+k_2$.
4.21 $(2\sqrt{2}-1)Q^2/(32\pi\varepsilon_0 a^2)$ towards the intersection of the planes.
4.23 $2\pi\{l/[g+Q^2/(16\pi\varepsilon_0 a^2 m)]\}^{1/2}$, where g is the acceleration due to gravity.
4.24 $[(a+d)/(a-d)]^3$.
4.25 $3p^2/(32\pi\varepsilon_0 r^4)$.
4.26 $R=\sqrt{(2)}H$ for zero field at A.

Chapter 5

5.1 (a) About 1 pF; (b) 760 pF assuming the second formula of Problem 5.3; (c) 0.8 pF; (d) 800 μF.
5.2 $r_1=r_2/e$.
5.3 Calculation of V at the midpoint of the axis of the cylinder yields (a) assuming uniform distribution, $C=4\pi\varepsilon_0 l/\log_e\{[l+(a^2l^2)^{1/2}]/a\}$; (b) assuming charge only on rim of ends. $C=4\pi\varepsilon_0(a^2+l^2)^{1/2}$. As $l\to\infty$, any end effects produce negligible changes at the centre and (a) becomes more accurate. For large l, this tends to $4\pi\varepsilon_0/\log_e(2l/a)$.
5.4 Uniform distribution gives $C_u=\varepsilon_0 2a/x$ per unit length; edge distribution gives $C_e=2\pi\varepsilon_0/\log_e(1+x^2/a^2)$ per unit length.
5.5 Ideal two-dimensional capacitor has capacitance given by C_u of Problem 5.4. $C_u/C_e=[\log_e(1+p^2)]/\pi p$, where $p=x/a$ from Problem 5.4 and this is less than 1 for all p. Since the real C lies between C_u and C_e, edge effects increase C.
5.6 Use the electrical image. $C=\pi\varepsilon_0/\log_e(2h/a)$ per unit length.
5.7 In series, $V_A=40$ V, $V_B=100$ V, $Q=20$ μC on both; in parallel, $Q_A=70$ μC, $Q_B=28$ μC, $V=140$ V on both.
5.8 0.5 V; $Q_1=0.05$ μF, $Q_2=0.15$ μF.
5.9 2 μF.
5.10 $V=Qx(a-x)/(\varepsilon_0 Aa)$; $Q_1=Q(1-x/a)$, $Q_2=Qx/a$.
5.13 Treat the capacitor as two in parallel (common V) of capacitances $4\pi\varepsilon_0 ab/(b-a)$ and $4\pi\varepsilon_0 a$. Hence charges are in the ratio $a/(b-a)$.
5.14 (a) 159 pF; (b) 0.21 μF; (c) 0.0053 μF.

5.15 0.0106 μF.
5.16 No. E is only about $6\,\mathrm{kV\,m^{-1}}$ just outside such a drop.
5.17 In series: stored 1.4×10^{-3} J, supplied 2.4×10^{-3} J; in parallel: stored 6.9×10^{-3} J, supplied 1.4×10^{-2} J.
5.18 48 V; increased four times.
5.22 $\varepsilon_0 AV^2/4x$.
5.23 $F=\frac{1}{2}V^2\,\mathrm{d}C/\mathrm{d}x=\pi\varepsilon_0 V^2/\log_e(b/a)$.
5.24 Voltage measurer needs $C_i \ll C_e$, charge measurer needs $C_i \gg C_e$.

Chapter 6
6.1 (a) 162 000 C; (b) 1.94×10^6 J; (c) $22\frac{1}{2}$ h; (d) 20 min.
6.2 The fact that the double layer is finite in area is critical. In that case there will always be a leakage field outside: if there were not, it could only be because the double layer was infinite in extent and no *external* circuit could exist.
6.3 $G=\mathrm{d}I/\mathrm{d}V=(e/kT)(I+I_0)$.
6.4 $(40-0.71n)\Omega$; 360 W delivered, of which 234 W are dissipated in resistances. The potential difference across the cells is 42.6 V and *not* 42.0 V.
6.5 Power consumed in R is $R\,\mathscr{E}^2/(R+r)^2$ in both cases. The Thevenin or constant voltage source consumes $r\,\mathscr{E}^2/(R+r)^2$, the Norton or constant current source consumes $R^2\,\mathscr{E}^2/r(R+r)^2$. Not a surprising result since the equivalent sources are not concerned with actual happenings inside the sources but only with the relationship between V and I at the terminals.
6.6 Both supply 1/3 A.
6.7 $\mathscr{E}/2R$ in all but two opposite arms: zero in these. Use superposition.
6.8 19/24.
6.9 1/4880 A.
6.10 This is the *minimum heat theorem.* It can be deduced generally from Kirchhoff's laws.
6.11 40 Ω.
6.13 $5R/6$; $7R/12$.
6.14 $3R$, $2.75R$, etc. Since an infinite ladder is still infinite when one section at the beginning is omitted, the resistance looking in at AA is the same as that looking in at the original terminals. This yields a simple equation giving the input resistance as $(\sqrt{3}+1)R$.
6.15 (a) No; (b) 1/3 Ω. Use the symmetry as follows: if 1 A is injected at any node and spreads out through the network to infinity, what are the currents in the six branches radiating from the node? Now, quite separately, remove 1 A from an adjacent node without injecting current anywhere and repeat the argument. Now superpose the two and consider the current (and hence the voltage) in the branch connecting the two nodes; (c) see figure. I am indebted to Professor Maurice Stewart for pointing out the inadmissibility of the solution to this problem given in previous editions. I also wish to thank Dr Roy Dennett for proving that the problem is insoluble with eight resistors and for providing me with the solution for nine resistors given in the figure.

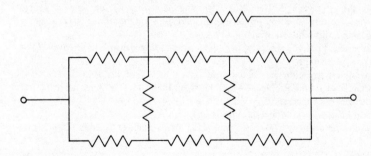

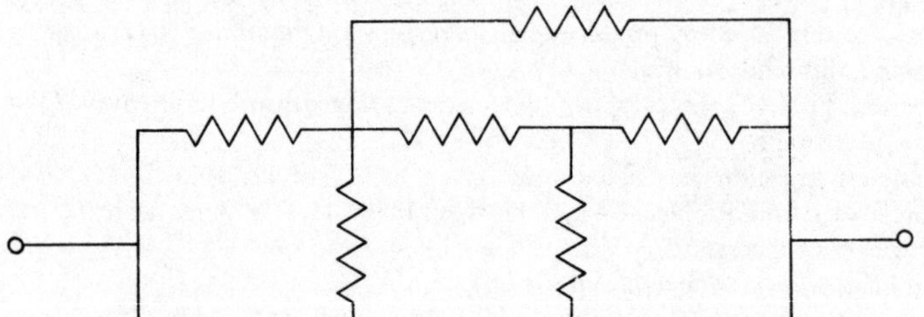

Answer to Problem 6.15(c).

6.17 6/7 A.
6.20 35/12 A.
6.23 $RC_1C_2/(C_1+C_2)$.
6.24 Sensitivity $=\mathscr{E}/(n+1)[(m+1)r+R_G(1+1/n)]$. Large r makes sensitivity low, small r makes it tend to a limiting value. Very small or very large n makes sensitivity low, so it should be around unity.

m should not be large but can be small. $m=n=1$ is a good all-round compromise and is easy to remember.

6.25 Resistance $1/[\pi\sigma(b^2-a^2)]$; conductance $2\pi\sigma/\log_e(b/a)$. For radial currents the resistance decreases with increasing length and to quote a value of resistance *per* unit length would be incorrect.

6.26 $\varepsilon_r\varepsilon_0/\sigma$. For water about 3 μs. Note that for good conductors it could be as low as 10^{-18} s.

6.27 1.25 miles from A.

6.28 $R-R_1^2/(R_1+R_L)$. This input resistance is variable for small loads.

6.30 $4.34\times10^{-3}\,\mathrm{m^2\,V^{-1}\,s^{-1}}$.

6.31 $\tan\theta_1/\tan\theta_2=\sigma_1/\sigma_2$. Current always enters electrolyte at nearly 90° because its conductivity is much less than that of the metal.

6.33 λ is about 10 nm at room temperature.

Chapter 7

7.1 Suspend disc with plane vertical so that it can oscillate freely; it will come to rest in earth's horizontal **B**-field. Then suspend similarly with plane horizontal: it will come to rest with the diameter required along the earth's **B**-field.

7.2 $\sin^{-1}(IB_0/mg)$.

7.4 IBA.

7.5 (a) $2\sqrt{(2)}\mu_0 I/\pi a$; (b) $4\mu_0 Ia^2/[\pi(a^2+4x^2)(2a^2+4x^2)^{1/2}]$; (c) $\mu_0 nI\tan(\pi/n)/2\pi a$, all perpendicular to the plane of the coil.

7.6 (a) $(\mu_0 J_s/2\pi)\log_e[(R+L)/(R-L)]$ perpendicular to the plane of the strip; (b) $(\mu_0 J_s/\pi)\tan^{-1}(L/R)$ parallel to the plane of the strip but perpendicular to the direction of current. For infinite width, (b) becomes $\mu_0 J_s/2$.

7.7 $\pi/50$ N of attraction.

7.8 $\mu_0 Q\omega/4\pi a$.

7.9 $\frac{1}{2}\mu_0\sigma\omega a$.

7.10 2.05 m.

7.11 5600 MHz: in the microwave region with λ about 5.4 cm.

7.12 $4\times10^5\,\mathrm{m\,s^{-1}}$; about 840 eV.

7.15 $(\mu_0 Ia/2\pi)\log_e(1+b/d)$.

7.18 $10^{-3}\pi/4\,\mathrm{A\,m^2}$; $3\times10^{-9}\pi^2/8$ N.

7.19 4.3×10^{-5} N of attraction.

7.21 $(\mu_0 m\sin^2\theta)/2r$.

7.22 $p^2/2mB$; p^2/eB.

7.23 A field is *rate of change* of potential with position.

7.24 For $r<a$: $-2\mu_0 M/3$. For $r>a$: as if all dipoles were at the centre, i.e. a total moment $4\pi a^3 M/3$ at the centre.

Chapter 8

8.1 If x is the distance from the axis: $\mu_0 Ix/2\pi a^2$ for $x\le a$; $\mu_0 I/2\pi x$ *for* $a\le x\le b$; $\mu_0 I(c^2/x-x)/2\pi(c^2-b^2)$ for $b\le x\le c$; 0 for $x\ge c$.

8.2 $\mu_0 J_s/2$.

8.3 Force on electron $=\frac{1}{2}\mu_0 ne^2v^2r$.

8.4 $B_\theta=\frac{1}{2}\mu_0 nevr$; $E_r=ner/2\varepsilon_0$. Resultant force on electron is $(ne^2r/2\varepsilon_0)(1-\varepsilon_0\mu_0 v^2)$ away from the axis. As we shall see in Chapter 13, $\varepsilon_0\mu_0=1/c^2$, where c is the velocity of light *in vacuo*, so $v<c$ means that the resultant force is always outwards from the axis. For neutral currents, no electric force exists and the magnetic force tends to contract the beam: the **pinch effect**.

8.5 In any xy plane, **B**-lines are straight lines originating from the origin. Such a field would give a non-zero flux of **B** over a closed surface including any part of the z axis and so violate Gauss's law for **B**.

8.6 (a) If B_θ were not zero, $\oint \mathbf{B}\cdot d\mathbf{L}\neq0$ for a closed circular path centred on the z axis and lying in the xy plane. As no *resultant* current links this path, $\oint \mathbf{B}\cdot d\mathbf{L}=0$. Hence $B_\theta=0$. (b) If B_r were not zero, there would by symmetry be a field away from the axis as in Problem 8.5. Such a field violates Gauss's law for **B**. (c) Take a rectangular closed path entirely outside the solenoid with one pair of sides along the B_z direction and the other along any B_r direction. We know $B_r=0$ so the circuital law tells us that B_z is the same for both the other two sides, i.e. it has a constant value all the way to infinity. This constant value can only be zero. Finally, take a rectangular path like that of (c) but with the inner side in the interior of the solenoid. Apply the circuital law.

8.7 The arguments are similar to those in Problem 8.6.
8.8 Apply $\nabla \times \mathbf{B} = \mu_0 \mathbf{J}$ in Cartesian form to give $J_x = 0$, $J_y = 0$, $J_z = -2k/\mu_0$.
8.9 (a) No; (b) yes; (c) yes.
8.12 The difference between the two **A**-fields is a vector field whose x component is $\frac{1}{2}B_0 y$ and whose y component is $\frac{1}{2}B_0 x$, and it is thus expressible as the gradient of $\frac{1}{2}B_0 xy$. Since **B** is given by the curl of **A**, and since curl grad is always zero, the two **A**-fields will yield the same **B**-field.
8.14 Force is a maximum for $x = \frac{1}{2}a$ as in the Helmholtz coils.
8.15 11.2 GHz; about 5.6 GHz.

Chapter 9
9.2 1.6 mV.
9.3 $I = (\mathscr{E} - \Phi\omega/2\pi)R$. From this, $\mathscr{E}I = \Phi\omega I/2\pi - RI^2$ and the first term on the right is the mechanical power output of the motor.
9.4 $v = v_0 e^{-t/\tau}$, where $\tau = mR/B^2a^2$.
9.6 $I = (\mathscr{E} - \Phi\omega \sin \omega t)/R$; power $= \Phi I \sin \omega t$.
9.8 $\pi a^2 \omega B_0 (\cos \omega t)/R$.
9.9 Current is $\sigma\omega B_0 b a^2 (\cos \omega t)/4$. Mean power loss is $\pi\sigma\omega^2 B_0^2 b a^4/8$. The **B**-field would in fact be considerably changed by the induced currents, but the expression shows the importance of reducing σ, ω and a to cut down eddy current losses.
9.10 $\pi a^2 B(\sin \delta)/R$; about 113 μC.
9.11 $V_R = RI_0(1 - e^{-Rt/L})$, $V_L = RI_0 e^{-Rt/L}$.
9.13 320 μH.
9.14 Mutual inductance = 4 mH; electromotance = 2 mV.
9.15 $L_1/4L_2$. Coefficient of coupling $= (L_1/L_2)^{1/2}/4$.
9.16 $\mu_0 a \log_e[(b/d) + 1]/2\pi$.
9.17 $I_1 I_2 \partial M/\partial t$.
9.18 Magnetic energy stored is $\frac{1}{2}\mu_0 n^2 l A I^2$.
9.19 $\mu_0 a b I^2/[2\pi d(b+d)]$.
9.20 Use Problem 7.21 and equate $2\pi a E$ to $d\Phi/dt$ with Φ expressed in terms of the variable x only and $v = dx/dt$. $E = -3\mu_0 m a x v/r^5$, where $r = (a^2 + x^2)^{1/2}$.

Chapter 10
10.1 For the example suggested, the equation is $R_2 LC\ddot{I} + (R_1 R_2 C + L)\dot{I} + (R_1 + R_2)I = V_0 \sin \omega t + \omega C R_2 V_0 \cos \omega t$.
10.2 10^{-4} s. Circuit is critically damped: maximize current in (10.14).
10.3 $10^{4.5}$ Hz.
10.4 $\omega_N = \omega_0(1 - \frac{1}{2}\delta^2)$ while $\Lambda = 2\pi\delta(1 + \delta^2)$ so that ω_N only differs from ω_0 by a very small quantity and Λ is nearly proportional to δ.
10.5 ω_0 and $\sqrt{(3)}\omega_0$, where $\omega_0 = 1/(LC_1)^{1/2}$.
10.6 At $\omega = 1/CR$. Phase difference is $\pi/4$ or 45°.
10.7 4.6 μF.
10.8 (a) $1/R$ and $1/\omega L$; (b) $R/(R^2 + \omega^2 L^2)$ and $-\omega L(R^2 + \omega^2 L^2)$.
10.9 1000 Ω; 7.3 H.
10.10 For purely resistive Z, $\arg Z = 0$. Hence $(\omega^2 LC - 1)(R^2 - L/C) = 0$. Conditions are (a) $R = (L/C)^{1/2}$; (b) $\omega = 1/(LC)^{1/2}$.
10.14 (a) $(I_2^2 + \frac{1}{2}I_1^2)^{1/2}$; (b) $(\frac{1}{2}I_1^2 + \frac{1}{2}I_2^2 + I_1 I_2 \cos \alpha)^{1/2}$; (c) $(\frac{1}{2}I_1^2 + \frac{1}{2}I_2^2)^{1/2}$.
10.15 $2I_0/\pi$.
10.16 Resistance $= 33\frac{1}{3}\,\Omega$; reactance $= 100/3\sqrt{(2)}\,\Omega$.
10.17 $60/\pi$ μF or 19.1 μF.
10.18 1 V; 206 kHz; 53.1°.
10.21 Resonant angular frequency ω given by $\omega^2 = [\omega_0^2(\omega_0^2 + 2R^2/L^2)]^{1/2} - R^2/L^2$.
10.23 A is $(1 - j/(\omega CR))/[1 + 1/(\omega^2 C^2 R^2)]$ and in the $re^{j\theta}$ form, we have that $r^2 = 1/[1 + 1/(\omega^2 C^2 R^2)]$ and $\tan \theta = -1/(\omega CR)$. It follows that $r = \cos \theta$, the equation of a circle, radius $\frac{1}{2}$, centre at $(r, \theta) = (\frac{1}{2}, 0)$.
10.25 $M^4/L^2C^2 - 4M^2R^2/LC > R^4$.
10.31 Note that $(\sqrt{(2)} \pm 1)^2 = 3 \pm 2\sqrt{2}$.
10.32 $e^{\alpha} = 38$ approximately. Hence attenuation is $-20 \log_e 38$ or about 116 dB.

10.33 $60 \log_e 3$ or about $66\,\Omega$. $Z_L = 165\,\Omega$.
10.34 The jet behaves as a transmission line and the *change* in voltage will travel along it with the speed of light.
10.35 $L \approx$ a few $\mu H\,m^{-1}$ in series, using Eq. (9.22); $C \approx$ a few $pF\,m^{-1}$ in parallel from Problem 5.1c. Both ωL and $1/\omega C$ are about $1\,k\Omega$ at 100 MHz. Even at 1 MHz, strays are 1 per cent and thus significant.
10.36 $2.34\,\mu F$; connect in series.
10.37 (a) $L = CR_1R_2$, $r = R_1R_2/R_3$; (b) $L = R_1R_2C_1$, $r = R_1C_1/C_2$; (c) $\omega^2 = 1/(R_1R_2C_1C_2)$, $C_1/C_2 = R_4/R_3 - R_2/R_1$.
10.38 $Z(s) = sL + R + 1/sC$.
10.40 19.3 cm; capacitance of 20.4 pF.

Chapter 11
11.2 $P \cos\theta$ at a point whose radius makes θ with **P**. For the second part, see Sec. 11.8.
11.4 $0.015\,\mu C$; $0.005\,\mu C$ and $16\frac{2}{3}$ V.
11.5 $0.009\,\mu C$ and $0.006\,\mu C$; 30 V.
11.6 2:3.
11.7 $4\pi\varepsilon_0/3 \log_e 2$.
11.10 $\tan\theta_1/\tan\theta_2 = \sigma_1/\sigma_2$.
11.11 (a) $-\varepsilon_0(\varepsilon_r - 1)AtV^2/2\varepsilon_r x^2$; (b) $+\varepsilon_0(\varepsilon_r - 1)AtV^2/2\varepsilon_r x^2$.
11.12 $\pi\varepsilon_r\varepsilon_0 a^2 K^2 \log_e(b/a)$.
11.13 $CV^2(\varepsilon_r - 1)/2a$.
11.16 $\varepsilon_r\varepsilon_0/\sigma$. For water, about $3\,\mu s$.

Chapter 12
12.1 (a) Increases 10^4 times; (b) decreases 100 times; (c) increases 100 times.
12.2 $J_{sc} = 150\,A\,m^{-1}$; $J_{sM,\,iron} =$ nearly $150\,000\,A\,m^{-1}$; $J_{sM,\,Cu} = 0.0015\,A\,m^{-1}$.
12.3 1.2 H; 0.30 T; $749\,500/\pi\,A\,m^{-1}$.
12.5 $500/\pi\,A\,m^{-1}$; halved when 500 turns are reverse wound.
12.6 $\mu_0(\mu_r + 1)(\log_e 2)/2\pi$.
12.9 0.0433 T.
12.10 $2\pi \times 10^{-4}$ Wb.
12.14 $10^{-11}\,kg\,m^2\,s^{-1}$.

Chapter 13
13.2 $(\mu_0 I/2\pi R)[1 - l/(R^2 + l^2)^{1/2}]$.
13.3 $J_d/J_c = \varepsilon_r\varepsilon_0\omega/\sigma$.
13.6 $3\varepsilon_r E_0/(2\varepsilon_r + 1)$.
13.11 $30\,V\,m^{-1}$.
13.12 Poynting vector has a magnitude $VI/2\pi rl$.
13.13 $10^{-11}/4\pi$ N.
13.14 Normal pressure $u_{EM}(1+f)\cos^2\theta$; tangential pressure $u_{EM}(1-f)\sin\theta\cos\theta$.
13.15 The important point is that the relation $E = cB$ between the *magnitudes* of **E** and **B**, and the orthogonality of their *directions* as indicated by $\mathbf{E} \cdot \mathbf{B} = 0$, are not dependent on the frame of reference any more than c is. So a plane wave *in vacuo* looks the same in all respects to any observer.

Chapter 14
14.1 They are equal.
14.5 u_E/u_M is approximately $2\pi^2\varepsilon_r\mu_r(\delta/\lambda_0)^2$, where δ is the penetration depth and λ_0 the free space wavelength. Since δ/λ_0 is very small, this indicates that the energy is predominantly in the **B**-field.
14.6 $c_{metal} = (2\omega/\mu_0)^{1/2}$ and $c_{metal}/c = 2\pi(\delta/\lambda_0)$, i.e. very small, $\lambda_{metal} = 2\pi\delta$, much less than λ_0.
14.7 Use phase velocity $= \omega/k$, group velocity $= d\omega/dk$.
14.8 $E_y = -j(k_0 cbB_0/\pi)\sin(\pi z/b)\exp[j(k_g x - \omega t)]$; $E_z = 0$; $B_y = 0$; $B_z = -j(k_g bB_0/\pi) \times \sin(\pi z/b)\exp[j(k_g x - \omega t)]$. $k_g^2 - k_0^2 = \pi^2/b^2$.

Index

Physical constants

Elementary charge	$e = 1.602\,18 \times 10^{-19}\,\mathrm{C}$
Electron rest mass	$m_e = 9.109\,39 \times 10^{-31}\,\mathrm{kg}$ ($=.511\,\mathrm{MeV}/c^2$)
Proton rest mass	$m_p = 1.672\,62 \times 10^{-27}\,\mathrm{kg}$ ($=938\,\mathrm{MeV}/c^2$
Unified atomic mass constant	$m_u = 1.660\,54 \times 10^{-27}\,\mathrm{kg}$
	$e/m_e = 1.758\,82 \times 10^{11}\,\mathrm{C\,kg^{-1}}$
	$m_p/m_e = 1.836\,15 \times 10^3$
Speed of light *in vacuo*	$c = 2.997\,92 \times 10^8\,\mathrm{m\,s^{-1}}$
Avogadro constant	$N_A = 6.022\,14 \times 10^{23}\,\mathrm{mol^{-1}}$ ($N'_A = 10^3 N_A$)
Planck constant	$h = 6.626\,07 \times 10^{-34}\,\mathrm{J\,s}$
Boltzmann constant	$k = 1.380\,66 \times 10^{-23}\,\mathrm{J\,K^{-1}}$
Gravitational constant	$G = 6.672\,60 \times 10^{-11}\,\mathrm{N\,m^2\,kg^{-2}}$
Electric constant	$\varepsilon_0 = 8.854\,19 \times 10^{-12}\,\mathrm{F\,m^{-1}}$
Magnetic flux quantum, $h/2e$	$\Phi_0 = 2.067\,83 \times 10^{-15}\,\mathrm{Wb}$

Table of units in mechanics

The last column gives the number of CGS units in one SI unit.

Quantity	*SI Unit and Symbol*	*CGS Unit and Symbol*
Mass, m	kilogramme, kg	10^3 gramme, g
Length, l	metre, m	10^2 centimetre, cm
Time, t	second, s	1 second, s
Frequency, ν	hertz, Hz	1 Hz
Velocity, $\mathbf{v}$	$\mathrm{m\,s^{-1}}$	$10^2\ \mathrm{cm\,s^{-1}}$
Linear momentum, $\mathbf{p}$	$\mathrm{kg\,m\,s^{-1}}$	$10^5\,\mathrm{g\,cm\,s^{-1}}$
Density, ρ_m	$\mathrm{kg\,m^{-3}}$	$10^{-3}\,\mathrm{g\,cm^{-3}}$
Force, $\mathbf{F}$	newton, N	10^5 dyne, dyn
Work, W; energy (potential, U, kinetic, K)	joule, J	10^7 erg
Power, P	watt, W	$10^7\,\mathrm{erg\,s^{-1}}$
Couple, torque, $\mathbf{T}$	N m	10^7 dyn cm
Angular momentum, $\mathbf{L}$	$\mathrm{kg\,m^2\,s^{-1}}$	$10^7\,\mathrm{g\,cm^2\,s^{-1}}$
Moment of inertia, I_m	$\mathrm{kg\,m^2}$	$10^7\,\mathrm{g\,cm^2}$
Plane angle, θ	radian, rad	1 rad
Solid angle, Ω	steradian, sr	1 sr
Angular velocity, ω	$\mathrm{rad\,s^{-1}}$	$1\,\mathrm{rad\,s^{-1}}$
Pressure, p	pascal, Pa ($1\,\mathrm{N\,m^{-2}}$)	$10\,\mathrm{dyn\,cm^{-2}}$